Research in Mathematics Education

Series editors
Jinfa Cai
James Middleton

More information about this series at http://www.springer.com/series/13030

James A. Middleton • Jinfa Cai • Stephen Hwang
Editors

Large-Scale Studies in Mathematics Education

 Springer

Editors
James A. Middleton
Arizona State University
Tempe, AZ, USA

Jinfa Cai
Department of Mathematical Sciences
University of Delaware
Newark, DE, USA

Stephen Hwang
Department of Mathematical Sciences
University of Delaware
Newark, DE, USA

Research in Mathematics Education
ISBN 978-3-319-07715-4   ISBN 978-3-319-07716-1   (eBook)
DOI 10.1007/978-3-319-07716-1

Library of Congress Control Number: 2015932484

Springer Cham Heidelberg New York Dordrecht London

Printed on acid-free paper

Springer International Publishing AG Switzerland is part of Springer Science+Business Media
(www.springer.com)

Foreword

Although this is the third book in the *Research in Mathematics Education* series, this is the volume, as series editors, we first thought about editing ourselves. There are three reasons why we chose this topic for our first coedited volume. First, we both have experience conducting large-scale studies. While Cai has used mixed methods in all of his three related lines of research (cross-national comparative studies, mathematical problem solving and posing, and curriculum studies), his recent funded research has been longitudinally investigating the effect of curriculum on algebra learning (LieCal). Middleton began his career studying teacher collaboration in systemic initiatives and has continued to utilize large-scale methods as both the backdrop and focus of inquiry. Second, we both see the need for the field of mathematics education to encourage more large-scale studies. Third, while in recent years a significantly greater number of books have been published in mathematics education [See Hwang, S. & Cai, J. (2014). Book series in mathematics education: An overview of a growing resource. *Journal for Research in Mathematics Education*, *45*(1), 139–148], none, to date, has focused on large-scale studies in mathematics education. Our knowledge of the field as well as our knowledge of the scholars who grapple with issues inherent to large-scale studies led us to believe that it is the right time to put such a volume together.

The goals we have set for the series are to publish the latest research in the field in a timely fashion. This design is particularly geared towards highlighting the work of promising graduate students and junior faculty working in conjunction with senior scholars. The audience for this monograph series consists of those in the intersection between researchers and mathematics education leaders—people who need the highest quality research, methodological rigor, and potentially transformative implications ready at hand to help them make decisions regarding the improvement of teaching, learning, policy, and practice. In addition to meeting the goals we set for the series, this volume has at least the following three features. The first feature is its broad coverage of topics, illustrating the use of large-scale methods in a variety of lines of inquiry. The studies reported in this volume involve various aspects of mathematics education, such as curriculum, teaching, learning, teacher learning, assessment, equity, and policy. The second feature is the variety of empirical bases used. Some studies in this book have used existing large-scale data sets; in others, researchers collected original data. The third feature of this volume is its discussion of methodological issues that cut across particular studies, thus highlighting key factors anyone interested in large-scale work must ponder. Readers of this volume should gain methodological insights for conducting large-scale studies. Because of these features, this book should be useful for graduate seminars related to mathematics education research.

We thank all the authors' for the generous contribution of their time and intellectual effort to this volume. In particular, we would like to extend our gratitude to Stephen Hwang for joining us in the final stage of editing this volume.

Finally, we are grateful for the support of Melissa James and her assistants from Springer in developing and publishing this book series, as well as their support in publishing this volume.

Newark, DE, USA   Jinfa Cai
Tempe, AZ, USA   James Middleton

Contents

Why Mathematics Education Needs Large-Scale Research 1
James A. Middleton, Jinfa Cai, and Stephen Hwang
What Is Meant by "Large Scale?" 3
Sample Size 3
    Purpose of the Study 5
    Generalizability and Transportability of Results 7
    Type and Complexity of Data Analysis 8
Summary 12
References 13

Part I   Curriculum

**A Lesson for the Common Core Standards Era
from the NCTM Standards Era: The Importance
of Considering School-Level Buy-in When Implementing
and Evaluating Standards-Based Instructional Materials** 17
Steven Kramer, Jinfa Cai, and F. Joseph Merlino
Background 18
Method 23
    Achievement Measures 23
    Will-to-Reform Scale 24
    Comparison Schools 27
    Statistical Model 30
Results 33
    Overall Treatment Effects 33
    Buy-in Effects 34
    Effects of Will-to-Reform Subcomponents 37
Discussion 39
References 41

**Longitudinally Investigating the Impact of Curricula
and Classroom Emphases on the Algebra Learning
of Students of Different Ethnicities** 45
*Stephen Hwang, Jinfa Cai, Jeffrey C. Shih, John C. Moyer, Ning Wang,
and Bikai Nie*
Background 46
    The LieCal Project 46
    Algebra Readiness 47
    Conceptual and Procedural Emphases 47
Method 49
    Sample 49
    Assessing Students' Learning 49
    Conceptual and Procedural Emphases as Classroom-Level Variables 51
    Quantitative Data Analysis 51
Results 52
    State Standardized Tests 53
    Student-Level and Curriculum Cross-Sectional HLM Models 53
    Student-Level, Classroom-Level, and Curriculum HLM Models 55
Discussion 57
References 58

**Exploring the Impact of Knowledge of Multiple Strategies
on Students' Learning About Proportions** 61
*Rozy Vig, Jon R. Star, Danielle N. Dupuis, Amy E. Lein,
and Asha K. Jitendra*
Theoretical Background 62
Method 64
    Participants 64
    Intervention 65
    Measures 65
    Strategy Coding 67
Results 68
    Strategy Use at Pretest 68
    Relationship Between Strategy Profile and Posttest Performance 70
Discussion 70
References 72

**Challenges in Conducting Large-Scale Studies of Curricular
Effectiveness: Data Collection and Analyses in the COSMIC Project** 75
James E. Tarr and Victor Soria
The COSMIC Project 75
Results of the COSMIC Project 76
Issues Related to Collection and Analysis of Student Data 78
    Lack of a Common Measure of Prior Achievement 78
    Composition of Student Sample 81

Issues Related to Collection and Analysis of Teacher Data............................ 86
Issues Related to Modeling Student Outcomes 87
Conclusion 90
References.. 91

Part II Teaching

**Turning to Online Courses to Expand Access: A Rigorous
Study of the Impact of Online Algebra I for Eighth Graders**................... 95
Jessica B. Heppen, Margaret Clements, and Kirk Walters
Overview.. 95
Background and Rationale.. 98
  Significance of Algebra I 98
  Use of Online Courses to Expand Offerings.. 100
  Prior Research on Online Course Effectiveness 101
Study Design and Methodological Considerations.. 102
  Goals and Research Questions.. 102
  Study Design.. 103
  Analytic Methods.. 106
Study Sample 106
  School Recruitment.. 106
  Description of Participating Schools 108
  Description of Students in Participating Schools.. 109
Measures 112
  Implementation Measures.. 112
  Outcome Measures.. 113
The Online Algebra I Course: Course Content,
Online Teachers, and On-Site Proctors 115
  Course Content.. 116
  Online Teachers 117
  On-Site Proctors 118
Implementation Findings 118
  Online Course Activity 118
  On-Site Proctors.. 120
  Online Course Completion Rates... 121
  Course Content in Control Schools: Treatment Contrast............................ 121
  Summary of the Implementation Findings 122
Impact Findings 123
  Impacts on Algebra-Ready Students' Algebra Scores
  and High-School Coursetaking.. 123
  Impacts on Algebra-Ready Students' General Math
  Achievement and Non-Algebra-Ready Students' Outcomes....................... 125
Conclusions and Future Directions.. 128
  Limitations of the Study and Future Research Directions 129
References.. 130

A Randomized Trial of Lesson Study with Mathematical
Resource Kits: Analysis of Impact on Teachers' Beliefs
and Learning Community 133
Catherine C. Lewis and Rebecca Reed Perry
Background on Lesson Study 134
Method 136
    The Study Design and Conditions 136
    Measurement of Teachers' Beliefs and Teacher
    Learning Community 139
    Measurement of Teachers' and Students' Fractions Knowledge 140
    Data Collection 141
    Data Analysis 143
Results 144
Discussion 151
Conclusions 153
Appendix: Scales to Measure Teachers' Beliefs and Teacher
Learning Community 154
References 155

Conceptualizing Teachers' Capacity for Learning
Trajectory-Oriented Formative Assessment in Mathematics 159
Caroline B. Ebby and Philip M. Sirinides
Conceptual Framework: Learning Trajectory-Oriented
Formative Assessment 160
The TASK Instrument 161
    Scoring Rubric 166
    Ongoing TASK Development 167
Large-Scale Field Trial Results 168
    Descriptive Statistics 168
    Instrument Properties 170
Pathways Analyses 171
Building Capacity for Effective Mathematics Instruction 174
References 174

Part III   Learning

Using NAEP to Analyze Eighth-Grade Students'
Ability to Reason Algebraically 179
Peter Kloosterman, Crystal Walcott, Nathaniel J.S. Brown,
Doris Mohr, Arnulfo Pérez, Shenghi Dai, Michael Roach,
Linda Dager Hall, and Hsueh-Chen Huang
The NAEP Assessments 180
    NAEP Framework and Scoring 181
    Access to NAEP Data 182
What Can NAEP Tell Us About Students' Algebraic
Reasoning Skills? 182

Method 184
Results 190
NAEP as a Database of Student Understanding 197
    Themes in the Algebra Data 197
Conceptual, Logistical, and Methodological Issues
in the Use of NAEP Data 198
    Analyses Are Limited by the Data Available... 199
    Access to Secure NAEP Data 200
    Using Statistical Software with NAEP Data.. 201
    What Does It Mean to Say That a Certain Percentage
    of Students Answered an Item Correctly? 202
    Limitations on Analyses by Demographic Subgroup 202
Looking Forward.. 203
    Subscales for Specific Mathematics Skills 203
    Psychometric Issues.. 204
References.. 206

**Homework and Mathematics Learning: What Can We Learn
from the TIMSS Series Studies in the Last Two Decades?** 209
Yan Zhu
Homework Is an Important Issue Inside
and Outside of Academia.. 209
Effects of Homework Are Inclusive... 211
Changes in TIMSS Investigations About Homework
from 1995 to 2011.. 213
Is There a System-Level Homework Policy Available?.................................. 214
How Often Do Students Receive Homework
in Mathematics from Teachers? 215
How Much Time Do Students Spend
on Mathematics Homework? 217
What Types of Mathematics Homework Do Teachers Assign?...................... 220
What Mathematics Homework-Related Activities
Were Carried Out in Classes? 223
How Much Were Parents Involved in Students' Homework?.......................... 227
Summary and Conclusions 229
References.. 232

**Effect of an Intervention on Conceptual Change of Decimals
in Chinese Elementary Students: A Problem-Based
Learning Approach**.. 235
Ru-De Liu, Yi Ding, Min Zong, and Dake Zhang
A Conceptual Change Approach to Explain Children's
Difficulties with Decimals 236
Existing Interventions for Teaching Decimals.. 237
Problem-Based Learning and Self-Efficacy... 239
Method 240

Design.. 240
Participants and Setting... 240
Dependent Measures 241
Coding and Scoring... 242
Procedures... 243
Treatment Fidelity 247
Results... 247
Pretreatment Group Equivalency 247
Quantitative Measure of Students' Conceptual Change in Decimals 247
Students' Self-Efficacy and Academic Interest 249
Qualitative Measure of Students' Conceptual Change in Decimals 250
Students' Computation Errors.. 250
Analysis of Relations Between Whole Number
and Decimal Computation 252
Discussion 253
PBL and Improvement in Computation Skills............................. 253
Effects on Enhancing Students' Self-Efficacy
and Academic Interest... 253
Effects on Enhancing Students' Metacognition............................. 254
Effects on Conceptual Change in Decimals................................. 254
Limitations and Conclusions... 256
Appendix 1: Teaching Scripts for Teaching New Decimal Division 257
Appendix 2: Teaching Scripts for Reviewing Previous Contents
(Decimal Division Error Clinic) 258
Appendix 3: PBL Procedure 258
Appendix 4: Treatment Fidelity Checklists 259
Appendix 5: Sample Problems from the Curriculum 259
References.. 260

**A Longitudinal Study of the Development of Rational Number
Concepts and Strategies in the Middle Grades** 265
*James A. Middleton, Brandon Helding, Colleen Megowan-Romanowicz,
Yanyun Yang, Bahadir Yanik, Ahyoung Kim, and Cumali Oksuz*
Introduction.. 265
Longitudinal Analysis.. 266
Issues in Mapping Students' Growing Knowledge...................... 267
Method 268
Setting and Participants.. 268
Data Collection Procedures: Interviews
and Classroom Observations... 269
Assessment of Students' Rational Number Performance 271
Results.. 273
Comparison of Performance of Sample
to a National/International Sample 273
Comparison of Performance at Different Grade Levels 274

Describing Students' Mathematics Achievement over Time....................... 275
Interview Results... 277
Summary of Interview Data... 281
Discussion.. 284
Conclusions... 285
Commentary on the Issue of Scale in Intensive
Interview and Observational Methods 286
References.. 287

Part IV Methodology

**Measuring Change in Mathematics Learning with Longitudinal
Studies: Conceptualization and Methodological Issues**............................ 293
Jinfa Cai, Yujing Ni, and Stephen Hwang
Two Longitudinal Studies Examining Curricular Effect
on Student Learning... 293
Conceptualizing and Measuring Change in Student Learning 294
Analyzing and Reporting Change... 296
Analyzing and Reporting Change Quantitatively 297
Analyzing and Reporting Change Qualitatively 299
Analyzing and Reporting Change Beyond the Grade Band 301
Interpreting Change in Mathematics Achievement 302
Equivalence of Student Sample Groups... 302
Initial Conceptual Model 303
Conclusion 305
References.. 307

**A Review of Three Large-Scale Datasets Critiquing Item Design,
Data Collection, and the Usefulness of Claims**... 311
Darryl Orletsky, James A. Middleton, and Finbarr Sloane
Introduction.. 311
Education Longitudinal Study of 2002 312
National Assessment of Educational Progress.. 313
Trends in International Mathematics and Science Study............................ 314
Sampling Issues 315
Validity... 316
Statistical Validity 318
Internal Validity 320
Construct Validity 324
External Validity 329
Conclusions on Validity 330
Discussion: Usefulness 330
References.. 332

**Methodological Issues in Mathematics Education Research
When Exploring Issues Around Participation and Engagement**................ 335
Tamjid Mujtaba, Michael J. Reiss, Melissa Rodd, and Shirley Simon
Background.. 335
The Context of This Study.. 336
Introduction to Findings... 337
   Multi-level Findings: Intention to Participate in Mathematics
   Post-16 Amongst Year 8 Students 340
   The Emergence of the Importance of Teachers via
   Qualitative Work 345
   Deconstructing What Our Original Constructs Actually
   Measured: Perceptions of Mathematics Teachers,
   Mathematics and Mathematics Lessons... 351
   Multi-level Re-analysis to Explore the Importance
   of Students' Perceptions on Intended Post-16 Mathematics
   Participation (Using Items from the Survey Rather
   than Constructs) 357
Methodological Conclusions 361
References... 362

**Addressing Measurement Issues in Two Large-Scale
Mathematics Classroom Observation Protocols** 363
Jeffrey C. Shih, Marsha Ing, and James E. Tarr
Methods... 364
   Observational Protocols 364
   Analysis.. 366
Results... 366
   Lessons (CLE) 366
   Lessons and Raters (MQI) 367
Discussion 369
References... 370

**Engineering [for] Effectiveness in Mathematics Education:
Intervention at the Instructional Core in an Era of Common
Core Standards**.. 373
Jere Confrey and Alan Maloney
The Process of "Engineering [for] Effectiveness" 373
Intervening at the Instructional Core 377
Curricular Effectiveness Studies 378
   Case One: Single-Subject vs. Integrated Mathematics
   (COSMIC Study) 379
   Case Two: Comparing Effects of Four Curricula
   on First- and Second-Grade Math Learning 385
   Case Three: The Relationship Among Teacher's Capacity, Quality
   of Implementation, and the Ways of Using Curricula 390
Overall Conclusions from the Three Cases... 393

Engineering [for] Effectiveness: Summary and Recommendations 397
    Steps in a Strategic Plan to Strengthen the Instructional Core in Relation
    to Curricular Use, Implementation, and Outcomes 400
References 401

The Role of Large-Scale Studies in Mathematics Education 405
Jinfa Cai, Stephen Hwang, and James A. Middleton
Benefits of Large-Scale Studies in Mathematics Education 406
    Understanding the Status of Situations and Trends 406
    Testing Hypotheses 407
    Employing Sophisticated Analytic Methods 408
Pitfalls of Large-Scale Studies in Mathematics Education 410
    Resources and Time 410
    Complexity of Authentic Research Settings 411
    Methodology 411
Looking to the Future 412
References 413

Index 417

About the Authors

Nathaniel J.S. Brown has been a Visiting Professor in the Educational Research, Measurement, and Evaluation (ERME) department in the Lynch School of Education at Boston College since 2012. He earned a B.S. in Chemistry from Harvey Mudd College and an M.Sc. in Chemistry from Cambridge University, receiving fellowships from the National Science Foundation and the Winston Churchill Foundation. He earned a Ph.D. in Science and Mathematics Education from the University of California Berkeley, having received a dissertation fellowship from the Spencer Foundation, and joined the Learning Sciences faculty at Indiana University Bloomington in 2007. He specializes in a broad range of quantitative and qualitative methodologies, including measurement, psychometrics, statistics, research design, video analysis, interaction analysis, and cognitive ethnography. His most recent work concerns the development of meaningful and useful embedded classroom assessments for science learning, aligned with the *Next Generation Science Standards* and measuring conceptual understanding and scientific practice.

Jinfa Cai is a Professor of Mathematics and Education and the director of Secondary math education at the University of Delaware. He is interested in how students learn mathematics and solve problems, and how teachers can provide and create learning environments so that students can make sense of mathematics. He received a number of awards, including a National Academy of Education Spencer Fellowship, an American Council on Education Fellowship, an International Research Award, and a Teaching Excellence Award. He has been serving on the Editorial Boards for several international journals, such as the Journal for Research in Mathematics Education. He was a visiting professor in various institutions, including Harvard University. He has served as a Program Director at the U.S. National Science Foundation (2010–2011) and a co-chair of American Educational Research Association's Special Interest Group on Research in Mathematics Education (AERA's SIG-RME) (2010–2012). He will be chairing a plenary panel at the ICMI-12 in Germany in 2016.

Margaret (Peggy) Clements is a Senior Research Scientist at Education Development Center's (EDC) Center for Children and Technology. Drawing on her training in development psychology, community psychology, and quantitative research methods, she conducts research investigating the effective and equitable use of educational technology to support teaching and learning. Dr. Clements is also the lead researcher for the Regional Educational Laboratory Midwest's Virtual Education Research Alliance, overseeing multiple collaborative research and analytic technical support projects exploring K-12 online learning, A central focus of this work is developing a better understanding of how online teachers and school staff can best support the educational experiences of their students, and developing innovative strategies to develop and disseminate the research in close collaboration with policymakers and practitioners. With colleagues at EDC, Dr. Clements also leads a program of research guiding the development of an online assessment system targeting middle-grades students' rational number misconceptions.

Jere Confrey is the Joseph D. Moore Distinguished University Professor at North Carolina State University. She teaches in mathematics education and offers courses on learning sciences, curriculum development and evaluation, and related topics in mathematics education. Her current research interests focus on analyzing national policy, synthesizing research on rational number, designing diagnostic assessments in mathematics focused on student thinking, building innovative software linking animation and mathematics, and studying school improvement for under-served youth at the high school level in rural and urban settings.

Shenghi Dai is currently a doctoral student in the Inquiry Methodology Program in the Department of Counseling and Educational Psychology at Indiana University. He is also pursuing a master's degree in Applied Statistics at IU. His research interests lie in psychometric theories and their implementations in educational contexts. He is particularly interested in Item Response Theory (IRT), Cognitive Diagnostic Models (CDMs), multi-level modeling (MLM), and structural equation modeling (SEM). Prior to coming to IU, he received his B.A. and M.A. in Language Teaching and Applied Linguistics at the Beijing Language and Culture University.

Yi Ding is an associate professor of school psychology at Fordham University. She went to University of Minnesota Medical School for her internship in Pediatric Psychology and Pediatric Neuropsychology rotations. She received her Ph.D. in School Psychology from The University of Iowa. She is a certified school psychologist in the state of Iowa, Ohio, and New York and a Nationally Certified School Psychologist (NCSP). Her research interests include reading disabilities, mathematics learning disabilities, and special education and school psychology issues based on a multicultural perspective. She has published two book translations, more than 20 peer-reviewed journal articles, and two invited book chapters. She has presented more than 30 conference papers. She has been elected as a board reviewer for two peer-reviewed journals and invited to serve as an ad-hoc reviewer for several other journals such as *Developmental Psychology* and *Journal of Learning Disabilities*.

She has served as a standing board reviewer for the Research Grants Council under the University Grants Committee sponsored by the Hong Kong government.

Danielle N. Dupuis is currently a doctoral candidate in the Department of Educational Psychology at the University of Minnesota. Her research interests include value-added modeling for teachers, test development, and mathematics education.

Caroline B. Ebby is a senior researcher at the Consortium for Policy Research in Education (CPRE) and an adjunct associate professor at the Graduate School of Education at the University of Pennsylvania. She has extensive experience bridging research and practice in K-8 mathematics teacher education, professional development, and assessment. Her research focuses on the use of formative assessment to improve mathematics instruction, creating and sustaining professional learning communities focused on analysis of student learning, and using locally relevant contexts to enhance student learning of mathematics.

Linda Dager Hall is currently a fifth grade mathematics teacher in Washington, DC. After teaching high school mathematics, she earned a Ph.D. from the University of Wisconsin and then served on the faculty of the University of Delaware. She has worked on many different NAEP-related research and development projects since 1999, when she headed the writing team for a revision of the NAEP mathematics framework. More recently, she participated in a comparative analysis of mathematics assessments in NAEP and PISA. Her research work has also included several projects in mathematics education at Project 2061 of the American Association for the Advancement of Science.

Brandon Helding is a statistical consultant for Boulder Language Technologies in Boulder, Colorado. He graduated in 2009 from Arizona State University with a Ph.D. in mathematics education. His work consists of computer-based learning theories, models, and techniques for measurement of human knowledge, and dynamic modeling processes for social complexity.

Jessica B. Heppen is a Managing Researcher at the American Institutes for Research. Her research focuses on the use and impact of technology and mathematics innovations, particularly for at-risk students. She has designed and led multiple federally funded randomized control trials (RCTs) testing the impact of interventions on at-risk or underserved students. She is Principal Investigator of three current or recent studies of online/blended Algebra and is also directing a study of mathematics professional development for in-service teachers. In addition, she has led multiple studies on the use of data from online systems for instructional improvement, targeted intervention, and dropout prevention, and she has developed guides and tools to help states, districts, and schools establish early warning indicator systems for at-risk students. Prior to working at AIR, Dr. Heppen conducted evaluation studies of educational technology programs in secondary schools New York City. She has a Ph.D. in social psychology from Rutgers University.

Hsueh-Chen Huang is a doctoral student in the Mathematics Education Program in the Department of Curriculum and Instruction at Indiana University. After earning a B.S. in pure mathematics in Chung Yuan Christian University, Taiwan, she taught elementary and middle school mathematics and worked as an executive mathematics editor for school textbooks and supplementary learning materials. She also earned an M.S. in secondary education at IU in 2012. From 2011 to 2014, she was involved in the analysis of NAEP data and her current research investigates 15-year-olds' mathematics performance on OECD's Program for International Student Assessment (PISA). In particular, she is analyzing student performance on released PISA items for selected countries.

Stephen Hwang is currently a post-doctoral researcher working with Jinfa Cai in the Department of Mathematical Sciences at the University of Delaware. His research interests include the teaching and learning of mathematical justification and proof, the nature of practice in the discipline of mathematics, the development of mathematical habits of mind, and mathematics teacher preparation.

Marsha Ing is an Assistant Professor in the Graduate School of Education at University of California, Riverside. Her research focuses on methods for measuring and linking student performance and instructional opportunities and increasing the instructional utility of student assessments. Her methodological experiences include applications of generalizability theory, item response theory, multilevel modeling, and latent variable modeling. Her work has appeared in the *American Educational Research Journal*, *Contemporary Educational Psychology*, and *Educational Measurement: Issues and Practice*.

Asha K. Jitendra received her Ph.D. in Curriculum and Instruction from the University of Oregon. She is currently the Rodney Wallace Professor for Advancement of Teaching and Learning at the University of Minnesota. Her research interests include academic and curricular strategies in mathematics and reading for students with learning disabilities and those at risk for academic failure, assessment practices to inform instruction, instructional design, and textbook analysis.

Ahyoung Kim is a middle school teacher and faculty at Ewha Women's University in Seoul, Korea. She received her Ph.D. in Mathematics Education at Arizona State University. She has studied children's understanding of fractions and rational number.

Peter Kloosterman (http://education.indiana.edu/dotnetforms/Profile.aspx?u=klooster) is a Professor of Mathematics Education and the 2010–2015 Martha Lea and Bill Armstrong Chair for Teacher Education at Indiana University. After earning a B.S. in mathematics from Michigan State University, he taught high school mathematics and science before completing an M.S. and Ph.D. from the University of Wisconsin. In 1984, he joined the faculty at IU where he served as chair of the Department of Curriculum and Instruction from 1996 to 2001 and executive associate dean of the School of Education from 2003 to 2008. He has been the director or principal investigator for 22 externally funded projects and is the coeditor of four books and the author or coauthor of more than 60 articles and book chapters. His most recent work documents performance of 4th, 8th, and 12th grade students on the NAEP mathematics assessment from the 1970s to the present.

Steven Kramer is a Senior Research Analyst for the twenty-first Century Partnership for STEM Education. He received his Ph.D. in Curriculum and Instruction from the University of Maryland, College Park in 2002. He has been a fourth grade classroom teacher and a middle school mathematics teacher. For the past 15 years, he has been an educational research analyst with a focus on quantitative program evaluation, math and physics curriculum, and formative and summative assessment. His research has been published in Educational Policy, the Journal for Research in Mathematics Education, the Journal for Research in Science Teaching, and presented at national and international conferences.

Amy E. Lein is currently a doctoral candidate in the Department of Educational Psychology at the University of Minnesota. She has over 10 years of experience teaching high school and middle school special education and mathematics. Her research interests include educational interventions for struggling students and professional development for teachers.

Catherine C. Lewis in the School of Education at Mills College has served as principal investigator on a series of lesson study research projects funded by the NSF and federal Department of Education, including the recent randomized controlled trial reported in this volume, which demonstrates the impact of lesson study supported by mathematical resources on both teachers' and students' mathematical knowledge. Fluent in Japanese, Lewis is author of the prize-winning book *Educating Hearts and Minds* (Cambridge University Press, 1995). Lewis introduced many English speakers to lesson study through articles such as "A Lesson is Like a Swiftly Flowing River" (*American Educator*, 1998), handbooks such as *Lesson Study Step-by-Step* (Heinemann, 2011, now in its sixth printing), and video cases such as "How Many Seats?" and "Can You Lift 100 kg?" Video clips and further information about lesson study can be found at the website of the Mills College Lesson Study Group, www.lessonresearch.net.

Ru-De Liu is a professor of psychology in the School of Psychology at Beijing Normal University in China. He is interested in students' learning process of mathematics and its underlying cognitive mechanism. In addition, he explores students' learning strategies and achievement motivation. He has published more than 90 peer-reviewed journal articles and authored more than 20 books. He is a recipient of the Plan for Supporting the New Century Talents sponsored by Ministry of Education in China. He was a visiting scholar at Harvard University and a Fulbright research scholar at the University of Missouri-Columbia.

Alan Maloney is Senior Research Fellow at the Friday Institute, and Extension Associate Professor of Mathematics Education at North Carolina State University.

His interests include synthesis of research on, and the design and implementation of diagnostic assessments for, rational number reasoning, and the design of innovative technology for mathematics and science education. He is currently research coordinator for the Friday Institute's DELTA math educational group, co-PI of the North Carolina Integrated Mathematics project, and coordinator of the FI's Brown Bag and Voices of Innovation seminar series. He received his B.S. and Ph.D. in Biological Sciences from Stanford University.

A 20+ year veteran high school teacher, in 2007 **Colleen Megowan-Romanowicz** earned her Ph.D. in Physics Education Research at Arizona State University in 2007 under the guidance of David Hestenes. She joined the faculty at ASU in 2008 and in 2009 she secured funding to create an MNS degree program for middle school science and mathematics teachers. In spring of 2011, Megowan-Romanowicz accepted an appointment as Executive Officer for the American Modeling Teachers Association (AMTA). She retired from ASU in 2014.

F. Joseph Merlino is President of The twenty-first Century Partnership for STEM Education. For the past 25 years, he has served as the principal investigator to National Science Foundation, Institute of Education Science and USAID projects impacting nearly 5,000 secondary math and science teachers. These projects include directing the curriculum and assessment team for the Education Consortium for the Advancement of STEM in Egypt (2012–2016); the 21st Center for Cognition and Science Instruction (2008–2014); the Math Science Partnership of Greater Philadelphia (2003–2012); the Greater Philadelphia Secondary Mathematics Project (1998–2003); and the Philadelphia Interactive Mathematics Program (IMP) Regional Dissemination Site (1992–1998). In addition, he has taught graduate courses in curriculum design. He started his career as a high school math teacher. He received his B.A. in Psychology from the University of Rochester, his M.A. in Education from Arcadia University, and did doctoral studies in cognitive developmental psychology at Catholic University.

James A. Middleton is Professor of Mechanical and Aerospace Engineering at Arizona State University. Prior to these appointments, Dr. Middleton served as Associate Dean for Research for the Mary Lou Fulton College of Education at Arizona State University and as Director of the Division of Curriculum and Instruction. He received his Ph.D. in Educational Psychology from the University of Wisconsin-Madison in 1992, where he also served in the National Center for Research on Mathematical Sciences Education as a postdoctoral scholar for 3 years. Jim's research interests focus in the following areas where he has published extensively: Children's mathematical thinking; Teacher and Student motivation in mathematics; and Teacher Change in mathematics. He has served as Senior co-Chair of the Special Interest Group for Mathematics Education of the American Educational Research Association, and as chair of the National Council of Teachers of Mathematics *Research Committee*.

Doris Mohr is an Associate Professor in the Department of Mathematics at the University of Southern Indiana in Evansville. She received a B.S. in Mathematics from USI and an M.S. in Mathematics from Indiana University in Bloomington. After working as a math instructor at USI for 8 years, she returned to IU to complete a Ph.D. in Curriculum and Instruction, specializing in Mathematics Education. She now teaches mathematics content and methods courses for prospective elementary teachers. Her interests include student thinking, teacher preparation, and large-scale assessment. She is currently serving as coeditor of the peer-reviewed journal *Indiana Mathematics Teacher*.

John C. Moyer is a professor of mathematics specializing in mathematics education in the Department of Mathematics, Statistics, and Computer Science at Marquette University. He received his M.S. in mathematics in 1969, and his Ph.D. in mathematics education in 1974, both from Northwestern University. Since 1980, he has been an investigator or director of more than 70 private- and government-funded projects. The majority of the projects have been conducted in formal collaboration with the Milwaukee Public Schools to further the professional development of Milwaukee-area middle school teachers and the mathematics development of their students.

Tamjid Mujtaba works on a range of research activities at the Institute of Education, University of London which includes research in science and mathematics education. Her current work involves projects at the Learning for London@IOE Research Centre (www.ioe.ac.uk/LearningforLondon) and a 5-year Royal Society of Chemistry funded project to evaluate the effectiveness of science interventions. Her work in science education, which complements the findings in this chapter includes: Mujtaba, T., & Reiss, M. J. (2013). Inequality in experiences of physics education: Secondary school girls' and boys' perceptions of their physics education and intentions to continue with physics after the age of sixteen. *International Journal of Science Education*, 35(17), 1824–1845.

Yujing Ni is a professor in the Department of Educational Psychology at the Chinese University of Hong Kong. Her research areas include cognitive development, numerical cognition, classroom processes, educational assessment, and evaluation. She served as an editor for the volume "Educational Evaluation" of International Encyclopedia of Education, 3rd edition (2010, Elsevier, UK) and a guest editor for the special issue "Curricular effect on the teaching and learning of mathematics: Findings from two longitudinal studies from the USA and China" for International Journal of Educational Research. She obtained her Ph.D. degree from the University of California at Los Angeles.

Bikai Nie is a research associate and an instructor at the University of Delaware. Previously, he taught at the School of Mathematics and Statistics in Central China Normal University. He has 10 years of experience teaching mathematics in both middle school and high school. He received his Ph.D. in Mathematics Education from East China Normal University.

Cumali Oksuz is Assistant Professor in the Department of Elementary Mathematics at Adnan Menderes University. His research has focused on rational number learning, children's mathematical thinking, technology integration and curriculum development in mathematics, and pre- and in-service preparation of mathematics teachers.

Darryl Orletsky is a doctoral candidate in Mathematics Education at Arizona State University and has a keen interest in the philosophy of science and epistemology. His master's thesis examined mathematical complexity and its application to mathematics education research, whereas his doctoral research is examining how adults in the workplace reason proportionally. For the past 9 years, he has taught mathematics and physics for the Phoenix-based Great Hearts Academies.

Arnulfo Pérez is an Assistant Professor of Mathematics Education at The Ohio State University. He earned a B.S. in Applied Mathematics and taught high school mathematics for 3 years prior to completing an M.A. in Mathematics and a Ph.D. in Curriculum and Instruction at Indiana University. He has contributed to multiple projects focused on the teaching and learning of mathematics and the use of reform-based practices in urban settings. His current research focuses on data from the grade-12 National Assessment of Educational Progress (NAEP) and the relationship between performance on function items and student- and school-level factors. He is also working with NAEP data to explore the college preparedness of grade-12 students, particularly those students who are from historically underperforming backgrounds.

Rebecca Reed Perry is a Senior Program Associate with WestEd's Innovation Studies program, where she is currently studying implementation of the mathematics Common Core State Standards in ten California school districts. Prior to joining WestEd, Rebecca was co-Principal Investigator for the Mills College Lesson Study Group, where she conducted research on lesson study implementation and the use of lesson study and mathematical content materials to support learning. Her research has utilized both quantitative and qualitative analyses, including: co-design and implementation of randomized controlled experiments; conducting site visits to assess implementation of policy and instructional ideas; measuring and conducting statistical analyses of teacher and student learning, and: identifying and coding instances of learning during classroom and teacher professional community discussions. Rebecca received a B.S. in cognitive science from the Massachusetts Institute of Technology and a Ph.D. in education administration and policy analysis from Stanford University.

Michael J. Reiss is Professor of Science Education at the Institute of Education, University of London, Vice President of the British Science Association, Visiting Professor at the Universities of Leeds and York and the Royal Veterinary College, Docent at the University of Helsinki and a Fellow of the Academy of Social Sciences. Books of his include: Reiss & White (2013) *An Aims-based Curriculum*,

IOE Press; Jones, McKim & Reiss (Eds) (2010) *Ethics in the Science and Technology Classroom*: *A New Approach to Teaching and Learning*, Sense; Jones & Reiss (Eds) (2007). *Teaching about Scientific Origins*: *Taking Account of Creationism*, Peter Lang; Braund & Reiss (Eds) (2004) *Learning Science Outside the Classroom*, RoutledgeFalmer; Levinson & Reiss (Eds) (2003) *Key Issues in Bioethics*: *A Guide for Teachers*, RoutledgeFalmer; and Reiss (2000) *Understanding Science Lessons*: *Five Years of Science Teaching*, Open University Press. For further information, see www.reiss.tc.

Michael Roach is a doctoral student in mathematics education at Indiana University. He received his B.S. and M.A. at Ball State University. Michael has worked as a mathematics teacher at a small rural high school, a mathematics consultant at a state department of education, and a mathematics coach in a large urban district. His research interests include secondary analyses of assessment data and the impact of mathematics standards on schools, teachers, and students.

Melissa Rodd Melissa Rodd's areas of interest center around learning and teaching mathematics at secondary and early university level across a wide range of courses and levels of specialism. She is particularly interested in the interplay between affect (emotion) and cognition and how this affects a person's mathematics learning and their participation in mathematics (do they choose to study mathematics). A related area of interest is teaching mathematics and how a teacher's mathematical and pedagogical knowledge are intertwined and how their affect (emotion) enhances or inhibits their performance as a teacher. This has led to research on identities of teachers who are becoming mathematics specialists. Her favorite branch of mathematics is geometry (in which she specialized for her M.A.), and she is active both in teaching and researching the teaching of geometry.

Jeffrey C. Shih is an associate professor of elementary mathematics education at the University of Nevada, Las Vegas. He currently serves on the editorial boards of the *Elementary School Journal* and *Mathematics Teacher Educator*, as well as the coordinator for the Nevada Collaborative Teaching Improvement Program (NeCoTIP), the Improving Teacher Quality State Grant program. His research focuses on the effect of curriculum on student achievement. His work has been published in the *Journal for Research in Mathematics Education* and *Teaching Children Mathematics*.

Shirley Simon is Professor of Science Education at the Institute of Education, University of London. She began her career as a chemistry teacher and taught in inner-city London schools before becoming a full-time researcher and lecturer in science education. Shirley's doctoral research focused on the implementation of new assessment practices, and she has since undertaken research in scientific inquiry, cognitive acceleration, teacher learning, and professional development. Her current research focuses on argumentation in science and attitudes to science. She teaches on two masters programs and supervises doctoral students working in many aspects of science education in secondary and elementary school contexts. She is also a visiting professor at Umea University in Sweden, where she advises on research projects and supervises doctoral study.

Philip M. Sirinides is a statistician and researcher with expertise in quantitative research methods and the development and use of integrated data systems for public-sector planning and evaluation. His interests include early childhood programing, educational leadership, program evaluation, and comprehensive school reform. As a Senior Researcher at the Consortium for Policy Research in Education at the University of Pennsylvania, he plans and conducts studies in the areas of estimating intervention impacts and measuring teacher effectiveness. Formerly, as Director of Research and Evaluation for Early Childhood at the Pennsylvania Department of Education, he developed and implemented a broad research program to inform and promote the effectiveness of Pennsylvania's children and family services.

Finbarr Sloane a native of Ireland, received his Ph.D. in Measurement, Evaluation, and Statistical Analysis from the University of Chicago with specialization in Mathematics Education and Multilevel Modeling. He is currently a program director at the National Science Foundation Directorate for Education and Human Resources. There he oversees national efforts to conduct research on the scaling of educational interventions in STEM disciplines. While at the NSF he provided institutional direction to Trends in International Mathematics and Science Study (TIMSS), the Board on International Comparative Studies in Education (BICSE), and has presented to the National Research Council of National Academies of Science. His research has appeared in Educational Researcher, Reading Research Quarterly, and Theory into Practice.

Victor Soria is a Ph.D. candidate at the University of Missouri in Mathematics Education. Prior to pursuing his doctorate he taught secondary mathematics and received an M.S. of Applied Mathematics. As a graduate student at MU, he has taught undergraduate courses in mathematics and an introductory course for preservice mathematics teachers. His research includes examining the influence of curriculum on student learning and classrooms assessments.

Jon R. Star is current the Nancy Pforzheimer Aronson Associate Professor in Human Development and Education in the Graduate School of Education at Harvard University. He earned his Ph.D. in Education and Psychology from the University of Michigan and is a former middle school and high school mathematics teacher. His research focuses on mathematics teaching and learning at all levels.

James E. Tarr is Professor of Mathematics Education at the University of Missouri. He is active in the mathematics education research community, including serving one term on the NCTM Research Committee (2010–2013) that developed guidelines for the Linking Research and Practice Outstanding Publication Award. His research interests include the development of probabilistic reasoning and the impact of mathematics curriculum on student learning and the classroom-learning environment. He is currently or has been a Principal Investigator or Co-Principal Investigator on several multiyear, federally funded research grants. His work has been published in a variety of national and international research and practitioner journals, books, and edited books. At MU, he serves as Associate Division Director for the Department of Learning, Teaching, and Curriculum, teaches methods courses for preservice middle and secondary mathematics teachers, courses in an alternative certification Master's program, and research seminars in mathematics education.

Rozy Vig recently completed a postdoctoral fellowship at Harvard University. She earned her Ph.D. in mathematics education at the University of California, Berkeley. Her research interests include student cognition, mathematics models, and design-based research.

Crystal Walcott is an Associate Professor in the Division of Education at Indiana University Purdue University Columbus (IUPUC). After receiving a B.S. in mathematics from the University of North Dakota she taught mathematics in middle and high schools in North Dakota and Indiana. She then enrolled as a graduate student at Indiana University Bloomington (IUB), where she received an M.I.S. and a Ph.D. in Curriculum and Instruction. While at IUB, she served as a content specialist at the Center for Innovation in Assessment, working on early versions of Indiana's Academic Standards and Resources. She also worked for the Indiana School Mathematics Project, conducting secondary analyses of National Assessment data. She currently serves as the Elementary Education Program Coordinator and teaches content and methods courses in the elementary teacher education program at IUPUC.

Kirk Walters is a Principal Researcher at AIR. Dr. Walters leads rigorous evaluations of programs designed to improve K-12 mathematics teaching and learning, including experimental evaluations of mathematics teacher professional development programs and online and blended Algebra I programs. Walters currently directs the Promoting Student Success in Algebra I Study, funded by the U.S. Department of Education (ED), and serves as Deputy Director of the Math Professional Development Study, funded by ED's Institute of Education Sciences. He also serves as a Principal Investigator on two studies funded by the National Science foundation, the Intensified Algebra Efficacy Study and the Developing Teaching Expertise in K-5 Mathematics Study. Before entering graduate school, Walters spent 9 years as an urban middle and high school teacher, department chair, and professional development facilitator.

Ning Wang received her Ph.D. in Educational Measurement and Statistics from the University of Pittsburgh. She also received a master's degree in Research Methodology and another master's degree in mathematics education. Currently, she is a Professor at Widener University, teaching research methodology courses at the Master's and Doctoral level. Dr. Wang has extensive experience in the validation of assessment instruments in various settings, scaling using Item Research Theory (IRT), and conducting statistical data analysis using Hierarchical Linear Modeling and Structural Equation Modeling. In particular, she is interested in the applications of educational measurement, statistics, and research design techniques into the exploration of the issues in the teaching and learning of mathematics.

Yanyun Yang is an associate professor in the Department of Educational Psychology and Learning Systems at Florida State University. Her research interests include structural equation modeling, reliability estimation methods, factor analysis, and applications of advanced statistical techniques to applied research. She has published in journals such as *Behavioral Research methods, Educational and Psychological Measurement, Journal of Psychoeducational Assessment, Methodology, Psychometrika*, and *Structural Equation Modeling*. She is an associate editor for the *Journal of Psychoeducational Assessment*.

Bahadir Yanik has a Ph.D. in Mathematics Education from Arizona State University. Currently, he is working as an associate professor in the Department of Elementary Education at Anadolu University, Eskisehir, Turkey. He is interested in children's mathematical thinking and teacher education.

Dake Zhang is an assistant professor in the Department of Educational Psychology at Rutgers University. She received her Ph.D. in Special Education from Purdue University in 2011. Dr. Zhang's primary research is about the assessment and intervention for students with learning difficulties in mathematics. She has published in the *Journal of Special Education*, the *Journal of Teacher Education*, the *Journal of Learning Disabilities*, the *Journal of Educational Research*, *The Journal of Experimental Education*, *Learning Disability Quarterly*, *Learning Disabilities Research and Practice*, and other peer-reviewed journals. Dr. Zhang has been serving on the editorial board for *Journal of Disability Policy Studies* and has been guest reviewers for multiple peer-reviewed journals, such as *Remedial and Special Education*, *Educational Psychologist*, *The Journal of Teacher Education*, and *Learning and Individual Differences*. Dr. Zhang is certified to teach secondary-level mild to moderate special education students in Indiana.

Yan Zhu is an Associate Professor in the Department of Curriculum and Instruction, School of Education Science, East China Normal University. She received her B.Sc. and M.Ed. from East China Normal University and her Ph.D. from Nanyang Technological University, Singapore. Her research interest includes education equity, comparative studies, mathematics problem solving, mathematics assessment, and secondary data analysis.

Min Zong is a lecturer of psychology and supervises the counseling center at China Foreign Affairs University in China. Her current research interests include mathematics education, problem-based learning, and college students' mental health. She received a number of awards, including The Beijing Pioneer of Education Award, Excellent Teaching Award, and Excellent Young Teacher Award at China Foreign Affairs University. She was a visiting scholar in National Changhua University of Education in Taiwan. She is a program director of the Fundamental Research Funds for the Central Universities.

Why Mathematics Education Needs Large-Scale Research

James A. Middleton, Jinfa Cai, and Stephen Hwang

Over the years our community has benefitted greatly from the application of large-scale methods to the discernment of patterns in student mathematics performance, attitudes, and to some degree, policies and practices. In particular, such research has helped us discover differential patterns in socioeconomic, gender, and ethnic groups and point out that, as a system, mathematics curriculum and instruction has hardly been equitable to all students. From the National Center on Education Statistics (in the US), large scale studies such as High School and Beyond, the Longitudinal Study of American Youth, and the National Assessment of Educational Progress came important calls to focus attention on improving instruction for marginalized populations and to increase emphasis on more complex problem solving than had typically been the norm (Dossey & Wu, 2013).

But these studies have been less useful, historically, in helping us design and implement our responses to their call. Mathematics curriculum design has been, typically, an intense form of educational engineering, wherein units or modules are developed and piloted in relatively insular settings, with large-scale field tests held at or near the end of development. Arithmetic materials, for example, have been informed by a large *body* of small to medium *scale* studies of the development of children's mathematical thinking. Algebra, which has many fewer studies of learners' thinking, is even more dependent upon small-scale studies. Towards the end of the 1990s and into the early 2000s, policy devoting more research funding on efficacy studies renewed interest in experimental and quasi-experimental methods, sample size, and generalizability of results (Towne & Shavelson, 2002). The push

J.A. Middleton (✉)
Arizona State University, Tempe, AZ, USA
e-mail: jimbo@asu.edu

J. Cai • S. Hwang
University of Delaware, Ewing Hall 523, Newark, DE 19716, USA
e-mail: jcai@udel.edu; hwangste@udel.edu

© Springer International Publishing Switzerland 2015
J.A. Middleton et al. (eds.), *Large-Scale Studies in Mathematics Education*,
Research in Mathematics Education, DOI 10.1007/978-3-319-07716-1_1

has been to demonstrate the impact of different education interventions on mathematics performance and achievement statistically.

One notable study conducted in this period evaluated the impact of SimCalc, a computer-facilitated system for representing and manipulating functions and coordinating representations with simulation (animations) and real-world data (Roschelle & Shechtman 2013; Roschelle, Tatar, Hedges, & Shechtman 2010). In this set of studies, the authors examined implementation of SimCalc in over 100 schools (150 teachers, 2,500 students) throughout Texas.

This study is a good example of many of the issues facing large-scale research today. For example, the authors took care to select schools from urban as well as rural areas and admitted that urban schools were under-sampled as well as those that served African-American students. In particular, the reality of working with intact classrooms, in schools drawn non-randomly from widely different communities, forced the authors to utilize statistical controls to equilibrate experimental versus control groups across a variety of demographic and attitude variables, to insure that performance differences are meaningfully attributed to the intervention rather than to presage variables.

In addition, fidelity of implementation is an issue impacting the internal validity of a study. The authors had to implement a wide battery of assessments to determine the degree to which SimCalc-as-implemented reflected SimCalc-as-intended. What is noteworthy in this study is the use of multiple indices to understand the implementation of the program as integral to assessing its impact. The authors used a pre-post design to assess student performance and collected teacher knowledge assessments and tests of teacher mathematical knowledge for teaching, teacher attitude questionnaires, teacher logs, teacher interviews, and coordinated this data with demographic data. Such a wide geography of implementation, as well as a wide demography showed that, despite variation in implementation, the structure of the tools themselves constrained student and teacher behavior to be roughly in line with the design intent.

Hierarchical Linear Modeling (HLM) was used to preserve the levels of nested effects (students within classes). Results showed that students who utilized SimCalc in their classes outperformed a control group with effect sizes ranging from .6 to .8 or .9 for complex items (focusing on proportionality for younger students and functions for older students). Even low-complexity items showed significant effect sizes, though lower than those found for complex items (ranging from .1 to .19).

So, this study and others (see Romberg & Shafer, 2008) show that interventions can be developed, theoretically, and analyzed experimentally at a large enough scale to give us some confidence that, if employed elsewhere, there is a good probability the intervention will result in meaningful improvement of teacher practice and student learning.

But also, large-scale studies can help us theoretically, by providing a check against a set of findings drawn from a number of diverse, small-scale exploratory studies. Even a body of data as coherent and long-standing as that for proportional reasoning can be found wanting. In "Exploring the Impact of Knowledge of Multiple Strategies on Students' Learning about Proportions," Vig, Star, Depuis, Lein, and Jitendra (this volume), for example, show us that large-scale data can provide a cross-check on the continued utility of some models developed across many small-scale studies. For example, they found in a study of implementation of a unit designed to teach proportional reasoning and associated content that cross-multiplication as a wide-ranging strategy may be much less prevalent now than in previous years due to changes in curriculum and instruction. This illustrates the potential from large data to see new things that are impossible to discern on the small scale and to judge the generality of findings of small-scale research in the larger population.

The Institute for Education Sciences and NSF (U.S. Department of Education & National Science Foundation, 2013) present six classes of research in their *Common Guidelines for Education Research and Development*: (1) Foundational Research, (2) Early Stage or Exploratory Research, (3) Design and Development Research; (4) Efficacy, (5) Effectiveness, and (6) Scale-up. From exploring new phenomena in Exploratory research, or development of new theories in Foundational research, to the examination of the effectiveness of interventions across a wide range of demographic, economic, and implementation factors in Scale-Up research, scale is a critical factor to establish the believability of our conceptual models and the potential efficacy of our designed innovations in mathematics education.

What exactly is the scale that would constitute compelling evidence of intervention efficacy? What is the appropriate scale that would convince the field that inequities exist? That those same inequities have been ameliorated significantly? What scale would convince us that a long-standing research finding may no longer be as prevalent? These are unanswered questions that the chapters in this book can help us answer.

What Is Meant by "Large Scale?"

In this chapter, we introduce this book by asking the fundamental question, "What is meant by Large Scale?" In particular, the word "Large" is problematic, as it must be compared with something "small" to be meaningful. Anderson and Postlethwaite (2007), in their comparison of small versus large scale program evaluation research, provide a convenient taxonomy of factors that distinguish issues of scale: (1) Sample size, (2) purpose of the research, (3) generalizability of results, (4) type and complexity of data analysis, and (5) cost. Anderson and Postelthwaite's discussion is limited to studies of program evaluation, but the issues they raise are clearly relevant to curriculum, teaching, learning, and other more basic research foci. We will introduce chapters in this volume utilizing the first four of these issues. Cost is a factor that is determined, in part, by each of the first four and will be woven into our discussion as appropriate.

Sample Size

At first pass, we can define "Large" in terms of the sheer size of the sample(s) being examined. Chapters in this book address samples on the order of 2,500 participants, to three orders of magnitude greater for international data sets. Small, therefore, would include the undertaking just held up as an exemplar for large scale, the work of Rochelle and colleagues. The scale of such studies, in terms of sample size, therefore, must be tempered with the kinds of methods used. Complex methods that employ multiple measures, including qualitative approaches such as interviews and observation, can be considered "Large" with samples in the hundreds, as opposed to relatively "simple" studies that may employ only a single measure.

Thomas, Heck, and Bauer (2005) report that many large-scale surveys must develop a complex method for determining the sampling frame so that important subpopulations with characteristics of interest (SES, ethnicity, grade level, for example) will be insured representation. A simple random sample, in many cases, will not yield enough members of the target subpopulation to generate adequate confidence intervals. In "Longitudinally Investigating the Impact of Curricula and Classroom Emphases on the Algebra Learning of Students of Different Ethnicities," Hwang, Cai, Shih, Moyer, and Wang (this volume) illustrate this issue clearly. Their work on curriculum implementation required a sample large enough disaggregate results for important demographic groups. Their research shows that while achievement gaps tended to lessen for middle school students engaged in reform-oriented, NSF-sponsored curricula, the performance gap between White students and African-American students remained robust to the intervention. These results show demonstrably that curriculum implementation is not uniform for all ethnic groups, and that we have much work to do to create tasks and sequences that *do* address the cultural and learning needs of all students.

Likewise, in "A Randomized Trial of Lesson Study with Mathematical Resource Kits: Analysis of Impact on Teachers' Beliefs and Learning Community," Lewis and Perry (this volume) performed a randomized control trial of lesson study implementation, examining the impact of a set of support materials that provide imbedded professional development, and a structure for neophytes to implement lesson study on fractions with fidelity. The authors took great pains in their sampling frame, to establish the equivalence of control versus treatment groups. Their results show that such support improves teachers' knowledge of fractions, their fidelity of implementation of lesson study, and subsequent student performance on fractions. Their use of multiple methods, across multiple levels (students, teachers, teacher lesson study groups) also highlights how studies across diverse geography and demography can explore the efficacy of locally organized professional development (with nationally designed support) versus larger policy-level organization.

For government agencies, these sampling issues are often addressed through multi-stage cluster sampling, often including oversampling of under-represented groups. These strategies have implications for the calculation of standard errors in subsequent analyses, particularly if within-group variation is smaller than cross-group variation. "A Review of Three Large-Scale Datasets Critiquing Item Design, Data Collection, and the Usefulness of Claims," "Using NAEP to Analyze Eighth-Grade Students' Ability to Reason Algebraically," and "Homework and Mathematics Learning: What Can We Learn from the TIMSS Series Studies in the Last Two Decades?" (Orletsky, Middleton & Sloane, this volume; Kloosterman et al., this volume; and Zhu, this volume) each deal with these complex sampling issues, both practically as the authors implement studies that must take sample weighting into account and by methodological critique of secondary databases.

Researchers without the financial wherewithal of government agencies often must resort to other methods for insuring the sample of a study effectively represents some general population. Matching participants across experimental units on a variety of important covariates is a statistical method for making the case that experimental units are *functionally* equivalent prior to an intervention, and therefore, that any differences found after the study are due solely to the intervention. Such methods do not account for *all* preexisting variation in groups; some systematic variation is inevitably unaccounted for. Lakin and Lai (2012), for example, show that the generalizability of standardized tests can be much lower than for non-ELLs. Their study showed that ELL students would have had to respond to more than twice as many mathematics items and more than three times as many verbal items for the instrument to show the same precision as non-ELL students. Care must be taken, then, to not underestimate the standard error of measurements for subpopulations.

In "A Lesson for the Common Core Standards Era from the NCTM Standards Era: The Importance of Considering School-level Buy-in When Implementing and Evaluating Standards Based Instructional Materials, Kramer, Cai, & Merlino (this volume) performed a quasi-experiment examining the impact of school-level attitudes and support on the efficacy of two NSF-supported middle school curricula. Using data from a Local Systemic Change project, they assessed "Will to Reform," a survey-proxy for fidelity of implementation, roughly defined as teacher buy-in to the curriculum, and principal support for the curriculum. Carefully matching, statistically, schools implementing either *Connected Mathematics Project* or *Mathematics in Context*, they found that choice of material did not matter so much as the degree to which schools supported the curricula, teachers showed buy-in to the methods, and principals supported teachers' reform. The ability to match schools across several potential nuisance factors (such as prior mathematics and reading scores, demographics, SES) *requires* a large enough sample to provide adequate variability across all matching factors.

Regardless of the techniques used, the point is to reduce the overall systematic variation between experimental units enough to claim that the residual variation has a relatively minor effect. Careful choice of covariates is critical to make this claim, in addition to randomization or other equilibration techniques.

Purpose of the Study

Small-scale studies tend to be used towards the beginning of a research program: To explore new phenomena for which existing measures are not yet developed. Many focus on developing measures, drafting tasks for curriculum and assessment, or for exploring new teaching practices. Large-scale studies, in contrast, tend to be employed after such methods or instruments have been piloted and their use justified, and the phenomena to which they apply have been adequately defined. Anderson and Postlethwaite (2007) define the purpose of large-scale studies as describing a system as a whole and the role of parts within it. But the complexity of the system and the type of understanding to be gained from the study greatly impact how large the scale must be.

When examining a relatively simple system (say, performance on a test of proportional reasoning), the relatively low cost of administering a single measure, relative to the necessary power for detecting a particular effect, makes a "large" scale smaller, proportionally, than when examining the interaction between a set of variables. In general, the more variables one is interested in, the larger the scale one must invest in. But this is even more crucial if the *interaction* among variables is under study. Even relatively simple factorial designs to test interaction effects require a polynomially increasing sample size as the number of interactions increases. When ordered Longitudinally, concepts assessed in year 1 of a study, for example, do not ordinarily have a one-to-one relationship with concepts in subsequent year. Thus, the combinatorial complexity of human learning requires a huge sample size if the researcher is interested in mapping the potential trajectories learners may travel across a domain (Confrey & Maloney, this volume; Hwang, et al., this volume; Lewis & Perry, this volume).

In contrast, when the number of potentially interacting variables is high, the analysis is fine-grained (such as interviews of individual learning trajectories or observation of classroom interactions), and the purpose of the study is to create a new model of the phenomenon, smaller sample sizes may be needed to distinguish subtle differences in effects of tasks, questioning techniques, or other relevant factors. Middleton et al. (this volume), in "A Longitudinal Study of the Development of Rational Number Concepts and Strategies in the Middle Grades," show that, with only about 100 students, intense interview and observation techniques over several years can be considered large scale due to the purpose of the study as modeling student development of rational number understanding. The authors found that, contrary to their initial hypotheses, students' understanding grew less complex over time due to key biases in models used to teach fractions and ratios.

In "Engineering [for] Effectiveness in Mathematics Education: Intervention at the Instructional Core in an Era of Common Core Standards," Confrey and Maloney (this volume) provide a reconciliation of these extremes. They make the case that findings across *many* such studies can highlight the interplay among factors central to what they term the "instructional core"—the complex system bounded by curriculum, assessment, and instruction. The scale here is defined as the extent of the common efforts across studies. They call for the creation of collaborative efforts among large-scale development and implementation projects and the development of technologically-facilitated systems of data collection and sharing to facilitate analysis of this complex system *as* a complex system, using modern analytics.

Generalizability and Transportability of Results

Generalizability is a valued outcome of most large-scale studies. We report data not just as a description of the local, individual participants and their behavior, but as a model for *other* participants. When we test the efficacy of a teacher professional development program, for example, we are reporting our belief that the effects found can be replicated, under similar conditions, in some population of teachers. For primarily quantitative data, generalizability is established by the sampling frame—the methods by which the author makes the case that the sample represents the population of interest—the operational definition of the measure, and the appropriateness of the analyses. Standard errors are used to find the probability that a measure adequately reflects the typical behavior of the population. For such studies, size really does matter: The sample size is inversely proportional to the standard error. The issues of the complexity of sampling frames and the analyses mentioned above are largely important due to their impact on generalizability.

For other studies, those that use more qualitative methods, or those that cannot make random assignment to conditions, generalizability is difficult to impossible to establish statistically. Instead, a concept from design research becomes useful: Transportability of results. Transportability has to do with the functionality of the innovation being studied. Curricula, for example, may have different ways of being applied depending on teacher knowledge, available technology, state and local level standards, and so on. How robust the curriculum is, and how adaptable it is when transported from one situation to another, is a critical consideration for studies of applicability (Lamberg & Middleton, 2009).

In "Challenges in Conducting Large-Scale Studies of Curricular Effectiveness: Data Collection and Analyses in the COSMIC Project," Tarr and Soria (this volume) address both of these issues adroitly in their multi-level study of the impact of curriculum type on student achievement. The authors had to take multiple measures of prior achievement from state-level tests, convert them to z-scores, then map the state z-scores to NAEP scores to model student achievement as a result of reform-oriented curricula versus more traditional curricula. Effects of teachers, due to lack of observational data, were modeled using paper and pencil scales of teacher beliefs as proxies. Moreover, because so many teacher variables had potential impact on student achievement, potentially obscuring the impact of curriculum type, the authors reduced these dimensions using Principle Components Analysis. In this study of 4,600 students across 135 teachers, the sheer number of variables measured, and their potential interactions necessitated a large scale to have enough power to detect any effect curriculum might have had. Through iterative multi-level models, reducing the dimensionality of the system each iteration, they found that curriculum DOES matter, but prior achievement, opportunity to learn, and teacher effects mediate curriculum significantly.

The scale and sampling frame for this study establishes good generalizability of the results in a statistical sense. However, the iterative methods used in this chapter allowed the authors to show that many key variables impact the transportability of different curricula from one situation to another. Curriculum matters, but not to the exclusion of factors of implementation.

Type and Complexity of Data Analysis

"Large" is also determined, to a great extent, by the methods used to answer the research question. Observational methods, for example, because of their inherent cost in terms of time and analytic complexity, may constitute only dozens of records, depending on whether or not single units are observed multiple times, or whether multiple units are observed once or twice.

Shih, Ing, and Tarr (this volume), in "Addressing Measurement Issues in Two Large-Scale Mathematics Classroom Observation Protocols," for example, critique two different observational protocols, designed to view the same classroom phenomena, regarding how they account for, and treat as parameters sources of error variation. Their analysis highlights the need to run comparative analyses of reliability across competing or even seemingly complementary methods. One issue appears to be particularly important: Protocols aiming to determine general features of practice may tend to ignore or gloss over important differences in content and curriculum, which are the *central* features of other protocols, while those protocols focusing on the within-effects different tasks and curricula may report results that do not generalize across those factors. They also provide methodological insight by showing that utilization of multiple raters may improve reliability of observational protocols more effectively than increasing the number of items on a scale.

Like observation, face-to-face interview methods, all things being equal, will not allow samples as large as phone or online interviews. In the world of survey methods, the ability to use computerized (including online) collection methods enables larger sample sizes and more complex methods of assigning items to individuals. These methods, of course, both depend on, and interact with, the kinds of research questions being asked. As Shih et al. show, questions about the generalizability of a known finding requires more data than questions about the possible ways in which teachers might implement a particular concept in their class (also see Lewis et al., this volume).

In "Turning to Online Courses to Expand Access: A Rigorous Study of the Impact of Online Algebra I for Eighth Graders," Jessica Heppen and her colleagues (Heppen, Clements, & Walters, this volume) provide an excellent example of how the unit of analysis, coupled with the research question, influences what we consider "large." They report an efficacy study of providing online access to Algebra I to rural eighth-grade schools, which, heretofore had limited access to the content (some of the surveyed schools only had four eighth graders, presumably making staffing and curriculum adoption impractical and/or cost-prohibitive). In their study, the unit of analysis is *schools*. Schools are the appropriate unit for studying curriculum access, as individual students are typically nested within available curriculum, and typically schools adopt a single set of materials (see also Kramer, Cai, & Merlino, this volume). Thirty five schools receiving online access to Algebra 1 were compared to 33 control schools. The authors report that providing such access can improve eighth-grade performance as well as improve the probability of subsequent advanced mathematics coursetaking as students move to high school.

Characteristics of the Measurement

Size may matter, but what is being "measured" matters as well. It is clear, for example, from the high degree of variability and low goodness of fit for participant scores in mathematics assessments, that a large amount of any person's score is error of measurement. Any effect, therefore, includes not only true differences in the variable of interest, but also a whole host of spurious effects (Shadish, Cook, & Campbell, 2002).

Seltiz (1976) discusses the different components that make up a typical effect in social research. These effects include: (1) Stable characteristics other than those intended to be measured (such as the person's motivation in mathematics impacting their effort on a test of performance); (2) Relatively transient factors such as health or fatigue; (3) Variation in the assessment situation, for example, taking a test in a testing center versus the classroom or interviewing a teacher in her room versus in the researcher's lab; (4) Variation in administration (different instructions given or tools made available); (5) Inadequate sampling of items; (6) Lack of clarity of measuring instruments; and (7) Variation due to mechanical factors, such as marking an incorrect box on a multiple choice test, of incorrect coding of an interview item.

Multiple-methods and mixed methods (e.g., Mujtaba, Reiss, Rodd, & Simon, this volume) provide both statistical confidence and qualitative depiction of typical or expected attitudes, practices, or student behaviors in context and help the researcher understand when one or more of these factors may play an important role in measurement.

Error of Measurement

Inadequate or inconsistent sampling of items from the conceptual domain under study reduces the degree to which we can have confidence in the results of any assessment utilizing those items. In "Using NAEP to Analyze Eighth-Grade Students' Ability to Reason Algebraically," Kloosterman et al. (this volume) perform a secondary analysis of NAEP items, classifying them by their mathematical content, and then analyzing student performance for that content *longitudinally*. Their study represents a heroic effort just getting access to, and classifying NAEP items, given the proprietary nature to which specific item content and wording is guarded by the National Center for Education Statistics. Their results show that US eighth students' performance on NAEP, both overall and for algebra-specific content, has improved steadily from 1990 to 2011. Analysis of different items, however, shows consistent difficulty in content associated with proportional reasoning and on equations and inequalities utilizing proportional concepts. As a nationally representative sample, their works illustrate how huge-scale data can simultaneously provide us with information regarding how we are improving (or not), in mathematics instruction, but also provide specific critique on areas where we may still be falling short despite overall improvement.

For studies that assess the structure of variables in a network model or that employ advanced regression methods, the critical relationship between the number of items used to measure a construct and its reliability becomes extremely important. Even if each item is an excellent measure of its individual construct, the degree to which the items, together, predict some larger class of understandings can be eroded through their incorporation into a subscale. This increases the error of estimate of the latent variable.

Ebby and Sirinides (this volume) studied the interaction among several key variables, heretofore studied separately, in "Conceptualizing Teachers' Capacity for Learning Trajectory-Oriented Formative Assessment in Mathematics". They report on the development of an instrument to measure several aspects of teachers' Mathematical Knowledge for Teaching, including their assessment of the validity of the mathematics students used to solve problems, their assessment of students' mathematical thinking, and their orientation towards thinking of students' work in a learning trajectory. Fourteen hundred teachers were assessed by 15 different raters in this study! Using structural equation modeling (SEM), the authors found that teachers utilize their assessment of the validity of the mathematics to help them diagnose students' mathematical thinking. Their understanding of children's mathematical thinking, in turn, impacts their understanding of the students' learning trajectory. Together, these three variables significantly impact teachers' instructional decision making.

Complexity of the Measure

Assessments that measure multiple constructs versus a single one run into the tendency to under-sample the domain for each sub-construct, increase fatigue due the length of the administration of the assessment, and subsequently increase the number of mechanical errors recorded. Mujtaba et al. (this volume) clearly illustrate this in "Methodological issues in mathematics education research when exploring issues around participation and engagement". The authors studied motivational variables and their individual and collective impact on students' intended choice of mathematics in post-compulsory education. Multi-level modeling allowed the authors to account for school-based variation, to focus analyses on individual determinants of future course choice. What scale afforded the authors was an opportunity to examine *multiple* variables in concert, without sacrificing predictive validity of any variable apart from the others. Intrinsic motivation in mathematics, beliefs about extrinsic material gain from studying mathematics and advice all were shown to be significant contributors to students' decisions.

P = "Publish"

Large-scale studies are prone to errors due to "fishing." Because, particularly for secondary data analysis, researchers have access to so many variables at once, the tendency to run analyses without clear hypotheses or theoretical justification is almost too easy. The probability values of these results may be very low, due to the effect of large sample size on the standard error of measurement. The literature is currently full of findings of dubious utility, because the probability that a correlation is zero due to random chance may be very small. But how large is the correlation? For experimental research, an effect may have low probability of occurrence by random chance. But how large is the effect? Large-scale studies, because of the relative stability that large samples provide for estimates, can give us indication of the size of effect of an intervention, and therefore its potential practical significance.

Orletsky et al. (this volume), in "A Review of Three Large-Scale Datasets Critiquing Item Design, Data Collection, and the Usefulness of Claims," compare and contrast three large-scale longitudinal studies (ELS, NAEP, & TIMSS), examining the potential threats to validity that are probable when performing secondary data analysis. In particular, because of the relationship between sample size and standard error of estimate, the tendency for large-scale "findings" to have low *p-values* may yield *many* spurious results. Heppen et al. (this volume) utilize the narrow standard errors of large-scale research methodologically in a clever and unique way by hypothesizing that a *lack of* statistically significant side effects of an intervention may be considered supportive evidence for its efficacy. When combined with *significant* performance outcomes, large sample sizes enable researchers to examine unintended consequences of interventions statistically.

Zhu (this volume), in "Homework and Mathematics Learning: What Can We Learn from the TIMSS Series Studies in the Last Two Decades?," utilized the TIMSS database to compare the mathematics homework practices of five east Asian nations with three Western nations. Overall, though there were key differences from nation to nation, homework practices were found to be highly similar. Well over 90 % of teachers surveyed assigned homework. Homework varied from about ½ h per day (US, Japan, England), to about 45 min per day (Singapore). Most homework consisted of worksheet problems. One key finding shows that across all the studied nations, the prevalence of classroom discussion of homework problems has steadily increased from 1995 to 2011. Without large samples capable of being disaggregated by nation, the stability of these findings would have been near impossible to establish.

Level of Data Analysis

Many of the chapters in this volume address this issue explicitly, so we do not go into depth here. Suffice it to say that learning studies where students can be randomly assigned to experimental conditions require fewer records than nested designs. HLM and other multi-level methods are only valuable if the appropriate number of participants is sampled *at each level*. Within group variation then becomes an issue, depending on the heterogeneity of students within classrooms, or classrooms within schools. The more within group variation, the more groups will be needed to establish the group effect (Hwang et al., this volume; Kramer et al., this volume; Lewis & Perry, this volume).

Summary

This monograph is timely in that the field of mathematics education is becoming more diverse in its methods, and the need to investigate the efficacy of policies, tools, and interventions on mathematics teaching and learning is becoming more and more acute. In particular, the diverse ways in which students from a variety of backgrounds and with a variety of interests can become more powerful, mathematically, is still an open question. While examples can be provided with investigations of a few students in a few classrooms, the generality of those examples across the tremendous diversity of conditions of implementation in the world must be established with studies of a scale large enough to detect and estimate the probabilities of interventions' effectiveness with populations of interest disaggregated.

The chapters in this book show that large scale studies can be both illuminative—uncovering patterns not yet seen in the literature, and critical—changing how we think about teaching, learning, policy, and practice. The authors examine topics as diverse as motivation, curriculum development, teacher professional development, equity, and comparative education. Organizationally, we divide the chapters into four thematic sections:

Section I: Curriculum Implementation
Section II: Teachers and Instruction
Section III: Learning and Dispositions
Section IV: Methodology

But, it must be noted that most of this work crosses lines of teaching, learning, policy, and practice. The studies in this book also cross the boundaries of the six types of research discussed in the IES/NSF *Common Guidelines for Education Research and Development* (2013). We have selected these authors because their research and commentary are complex, illuminating problems to look out for, methodologically, as well as insight for how to better create robust, generalizable information for the improvement of curriculum, teaching, and learning. We anticipate this volume will help researchers navigate this terrain, whether engaging in designing and conducting efficacy research on the one hand, or analyzing secondary data on the other.

References

Anderson, L. W., & Postlethwaite, T. N. (2007). *Program evaluation: Large-scale and small-scale studies*. International Academy of Education: International Institute for Education Planning. www.unesco.org/iiep/PDF/Edpol8.pdf.

Dossey, J. A., & Wu, M. L. (2013). Implications of international studies for national and local policy in mathematics education. In *Third international handbook of mathematics education* (pp. 1009–1042). New York: Springer.

Institute for Education Sciences, National Science Foundation. (2013). *Common guidelines for education research and development*. Washington, DC: IES. http://ies.ed.gov/pdf/CommonGuidelines.pdf.

Lakin, J. M., & Lai, E. R. (2012). Multigroup generalizability analysis of verbal, quantitative, and nonverbal ability tests for culturally and linguistically diverse students. *Educational and Psychological Measurement, 72*(1), 139–158.

Lamberg, T., & Middleton, J. A. (2009). Design research perspectives on transitioning from individual microgenetic interviews in a laboratory setting to a whole class teaching experiment. *Educational Researcher, 38*(4), 233–245.

Romberg, T. A., & Shafer, M. C. (2008). *The impact of reform instruction on student mathematics achievement: An example of a summative evaluation of a standards-based curriculum*. London: Routledge.

Roschelle, J., & Shechtman, N. (2013). SimCalc at scale: Three studies examine the integration of technology, curriculum, and professional development for advancing middle school mathematics. In S. P. Hegedus & J. Roschelle (Eds.), *The SimCalc vision and contributions* (pp. 125–143). Dordrecht, The Netherlands: Springer.

Roschelle, J., Tatar, D., Hedges, L., & Shechtman, N. (2010). *Two perspectives on the generalizability of lessons from scaling up SimCalc*. Paper presented at the Society for Research on Educational Effectiveness, Washington, DC.

Seltiz, C. (1976). *Research methods in social relations*. New York: Holt, Rinehart and Winston.

Shadish, W. R., Cook, T. D., & Campbell, D. T. (2002). *Experimental and quasi-experimental designs for generalized causal inference*. New York: Houghton Mifflin.

Thomas, S. L., Heck, R. H., & Bauer, K. W. (2005). Weighting and adjusting for design effects in secondary data analyses. *New Directions for Institutional Research, 2005*(127), 51–72.

Towne, L., & Shavelson, R. J. (Eds.). (2002). *Scientific research in education*. Washington, DC: National Academies Press.

Part I
Curriculum

A Lesson for the Common Core Standards Era from the NCTM Standards Era: The Importance of Considering School-Level Buy-in When Implementing and Evaluating Standards-Based Instructional Materials

Steven Kramer, Jinfa Cai, and F. Joseph Merlino

> *Those who cannot remember the past are condemned to repeat it.*
>
> George Santayana

In June 2010, the Council of Chief State School Officers and National Governor's Association promulgated the Common Core State Standards (CCSS) for mathematics and literacy (CCSSO/NGA, 2010). The new standards are expected to "stimulate significant and immediate revisions…in classroom curriculum materials (Council of Chief State School Officers, Brookhill, & Texas Instruments, 2011)." Similarly, the new *Next Generation Science Standards* may require educators to develop and implement new instructional materials (NSTA, 2012).

Today's new standards build on previous efforts, including earlier standards promulgated by the National Council of Teachers of Mathematics (NCTM, 1989, 2000, 2009a, 2009b). Soon after the publication of the first *Standards* document (NCTM, 1989), the National Science Foundation (NSF) funded development of a number of elementary, middle, and high school mathematics curricula (hereafter referred to as "NSF-funded curricula") designed to implement the *Standards*. Studies evaluating

Research reported in this paper was supported by funding from the National Science Foundation Award Numbers: 9731483 and 0314806. Any opinions expressed herein are those of the authors and do not necessarily represent the views of the National Science Foundation.

S. Kramer (✉) • F.J. Merlino
The 21st Century Partnership for STEM Education, Conshohocken, PA, USA
e-mail: skramer@21pstem.org

J. Cai
University of Delaware, Ewing Hall 523, Newark, DE 19716, USA
e-mail: jcai@udel.edu

© Springer International Publishing Switzerland 2015
J.A. Middleton et al. (eds.), *Large-Scale Studies in Mathematics Education*,
Research in Mathematics Education, DOI 10.1007/978-3-319-07716-1_2

the effectiveness of the NSF-funded curricula can provide important lessons for today's new curriculum reform efforts.

The current study investigates the effectiveness of NSF-funded middle school mathematics curricula implemented with the assistance of the Greater Philadelphia Secondary Mathematics Project (GPSMP). The GPSMP, operating between 1998 and 2003, was an NSF-funded Local Systemic Change (LSC) initiative. This study differs from previous studies by evaluating the mediating effects of a school-level measure of stakeholder buy-in. We found that the degree of principal and teacher buy-in had a large impact on curriculum effectiveness. These results have potentially important implications for today's efforts to implement new instructional materials, providing insights both about how to support implementation and about how to evaluate the effectiveness of those materials.

Background

The original NCTM *Standards* emphasized student reasoning as being central to learning mathematics. Mathematics curriculum materials that had been in widespread use prior to promulgation of the *Standards* were perceived as placing too much emphasis on procedural fluency at the cost of ignoring conceptual understanding and applications (Hiebert, 1999). Based on early field trials of new curricula designed to implement the *Standards*, developers cautioned that teachers could find it difficult to change their practice (Cai, Nie, & Moyer, 2010). The NSF attempted to address this issue by establishing the LSC Initiative. The LSC theory of action argued that providing teachers with extensive professional development in the context of implementing the new NSF-funded curricula would result in teachers having both the inclination and capacity to implement the curricula. Between 1995 and 2002, NSF funded 88 multi-year LSC mathematics and/or science projects in Grades K-12 (Banilower, Boyd, Pasley, & Weiss, 2006).

The passage of the No Child Left Behind Act in 2001 and the establishment of the What Works Clearinghouse in 2002 heralded a new wave of educational reform focusing on student assessment and "scientifically based research" to investigate the effects of educational innovations (Slavin, 2002). Researchers began investigating the effectiveness of NSF-funded mathematics curricula. Syntheses of this early research tended to report positive achievement effects on researcher-administered tests using open-ended problems, but near-zero achievement effects on standardized tests measuring basic mathematical skills (Cai, 2003; Kilpatrick, 2003; Slavin, Lake, & Groff, 2008; U.S. Department of Education, 2007a). These early studies of NSF-funded curricula generally used a quasi-experimental *intent-to-treat* analysis, comparing achievement growth in schools and/or classrooms implementing new curricula with achievement growth in matched comparison schools/classrooms that implemented business-as-usual curricula. As shown in Fig. 1, intent-to-treat views curriculum implementation as a black box, comparing the achievement of students assigned to Treatment classrooms to the achievement of students assigned to Comparison classrooms without regard to actual classroom instruction (see, e.g., Riordan & Noyce, 2001).

Fig. 1 Intent-to-treat
evaluation model

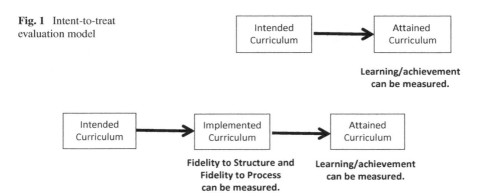

Fig. 2 Evaluation model with implementation fidelity

Recently, social science researchers have become concerned about information that is obscured by intent-to-treat studies. Evaluators have emphasized the importance of focusing not only on average effects, but also on mediating factors which can affect program outcomes when delivered under naturalistic conditions (Vuchinich, Flay, Aber, & Bickman, 2012). Among the potential mediating factors, evaluators have put particular emphasis on *fidelity of implementation* (Flay et al., 2005; U.S. Department of Education, 2007b). In an extensive review of the literature, O'Donnell (2008) defined fidelity of implementation as the degree to which an intervention is implemented as originally intended in the program design. Without a measure of fidelity of implementation, researchers may not be able to determine whether unsuccessful outcomes are due to an ineffective program or are due to failure to implement the program as intended (Dobson & Cook, 1980; Forgatch, Patterson, & DeGarmo, 2005; Hohmann & Shear, 2002; O'Donnell, 2008). As shown in Fig. 2, researchers focusing on fidelity of implementation differentiate between the intended curriculum embodied in curriculum materials, the implemented curriculum as seen in the classroom, and the attained curriculum as reflected in tests and other measures of student achievement (Cai, 2010). O'Donnell (2008) extended the concept of fidelity to include both *fidelity to structure* and *fidelity to process*. Fidelity to program structure means actually using the program materials as intended—and will be seen only in Treatment groups. Fidelity to process, in contrast, involves implementing processes congruent with the underlying program theory of action and might be seen in both Treatment and Control/Comparison groups.

Recent effectiveness studies have indeed confirmed an interaction between fidelity and treatment effects. In an evaluation of a supplemental elementary school math intervention aimed at increasing computational fluency, VanDerHeyden, McLaughlin, Algina, and Snyder (2012) found that a measure of fidelity to structure in Treatment classrooms predicted higher achievement on statewide test scores. Four recent studies evaluated inquiry-based middle school mathematics or science curricula while investigating fidelity to process (Cai, Wang, Moyer, Wang, & Nie, 2011; O'Donnell & Lynch, 2008; Romberg, Folgert, & Shafer, 2005; Tarr et al., 2008). All four found that Treatment classrooms with high fidelity to process showed more achievement growth than either Control classrooms or Treatment classrooms with low fidelity to process.

While many researchers and funders investigating program effectiveness have focused on fidelity of implementation, other researchers have taken a contrasting *mutual adaptation* or *co-construction* perspective that views fidelity of implementation itself as too simplistic a construct (e.g., Cho, 1998; Remillard, 2005). The mutual adaptation perspective emphasizes that any curriculum implementation necessarily involves teachers transforming the written curriculum, working with those materials to co-construct the enacted curriculum.

Researchers working on design experiments have used an evolutionary metaphor to describe this view. Some program changes are "lethal mutations" which decrease quality of learning, whereas other changes are "productive adaptations" which increase quality of learning (Brown & Campione, 1996; Design-Based Research Collective, 2003). Brown and Edelson (2001) described one such "productive adaptation" to the Global Warming Project (GWP), an inquiry-based middle school science curriculum that Brown had helped develop. They described how one teacher, Janet, implemented *The Sun's Rays*, an investigation that occurs approximately midway through the GWP.

> Rather than have her students follow the "recipe" for doing the lab, she decided to turn the activity into an opportunity for them to engage in experimental design. Instead of providing them with a set list of materials, she gave them access to a host of supplies which she gathered from her own supply closet and borrowed from other teachers. And rather than just connect the elements of the lab model to the actual phenomena they represented, she relied on the model throughout the lesson as a means to stimulate deep reflection and analysis of the results (p. 15).

The authors viewed Janet's extensive adaptations as consistent with the program philosophy and, if anything, an improvement on the original lesson materials.

The current study extends the "implementation fidelity" model in Fig. 2 by introducing *buy-in*, a concept taken from research into Comprehensive School Reform (Cross, 2004; Glennan, Bodilly, Galegher, & Kerr, 2004; Schwartzbeck, 2002). While buy-in is often discussed in the Comprehensive School Reform literature, the buy-in concept has seldom been formally defined. One exception is Turnbull (2002), who used a five-part operational definition for buy-in to a school reform model. Teachers bought in to the model if they understood the model, believed they could make the model work, were personally motivated to do so, and believed the model was good for their school and would help them become better teachers. Our definition of buy-in is consistent with Turnbull's, but more general, applying not only to comprehensive school reform, but to any school program, and to principals and other stake-holders as well as to teachers. Our definition is also influenced by a co-construction view of program implementation. We define buy-in as *the degree to which stakeholders understand the underlying program theory and embrace that theory*. Stakeholders who buy in to a program are less likely to introduce lethal mutations—and, to the degree their ability and situation allows, they are more likely to introduce productive adaptations. When applied to curriculum materials, buy-in reflects stakeholders' attitudes toward, beliefs about, and understandings of those

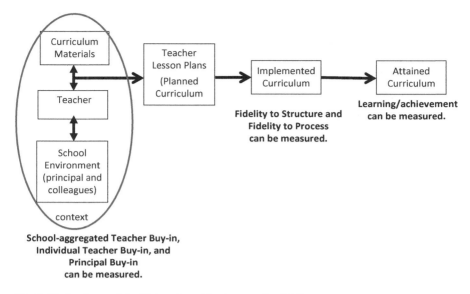

Fig. 3 Evaluation model with buy-in and implementation fidelity

materials. It is hypothesized that stakeholders with strong buy-in will be more likely to implement a program with fidelity to both structure and process.

Buy-in might be easier to measure than fidelity, especially fidelity to process. Measures of buy-in might be obtained via interviews or surveys of teachers and other stakeholders, or even indirectly via interviews or surveys of teacher leaders and mentors, whereas measures of fidelity to process might require teacher logs or classroom observations. Furthermore, professional development and other efforts to support program implementation might be more successful if such efforts seek to secure buy-in and help teachers and other stakeholders become active co-constructors of the curriculum, rather than seeking only to help them implement the program with fidelity to its structure and processes. Figure 3 (modified from Remillard, 2005) represents a curriculum evaluation model incorporating both buy-in and fidelity of implementation.

Buy-in has not often been addressed by curriculum effectiveness studies. In their evaluation of four elementary school math curricula, Agodini et al. (2010) asked teachers to state their interest in using their assigned curriculum again, if they were given a choice. The authors reported mean responses by curriculum, but did not attempt to analyze whether choosing "yes" correlated with higher achievement, perhaps because the dichotomous variable may have been too weak a measure of buy-in to achieve valid results. Similarly, in their evaluation of an intervention implementing supplemental math curriculum materials, VanDerHeyden et al. (2012) measured teacher-rated "acceptability" of the intervention by computing the school-level mean of teacher responses to 15 Likert-scale items measuring the program's perceived effectiveness, practicality, ease of implementation, potential risks, etc. The authors did not report whether or not schools with high average teacher buy-in achieved better results than did schools with low average teacher buy-in. An unpublished supplemental analysis did not detect any correlation between buy-in and achievement, but because buy-in variation among the schools was not large and the sample consisted of only seven Treatment schools, the data available would have very low power to detect any such correlation (A. VanDerHeyden, personal communication, 2013). Thus, the current study breaks new ground by quantitatively analyzing the relationship between buy-in and program effectiveness.

The current study investigates the effectiveness of one particular LSC, the GPSMP. The GPSMP operated between 1998 and 2003. The study was commissioned by NSF in 2003 to use retrospective data to analyze GPSMP effects at twenty middle schools that implemented one of two NSF-funded middle school curricula, either *Mathematics in Context* (MiC) or *Connected Mathematics* (CMP). This study differs from previous studies of NSF-funded curricula in that it investigates the effectiveness not only of the curricula themselves, but of the LSC theory of action, which combined curriculum adoption with extensive professional development for teachers. The study also differs from previous studies by using a measure of buy-in as a mediating variable.

Over its 5 years of operation, the GPSMP provided an average of 59 h of professional development to each of 249 middle school teachers at the 20 middle schools that participated in the retrospective study. GPSMP differed from some other LSCs in that mentors working for the project supplemented professional development by providing extensive assistance to mathematics teachers implementing NSF-funded curricula.

The LSC theory of action predicted that, at most middle schools, implementing an NSF-funded mathematics curriculum combined with extensive teacher professional development would lead to sufficiently high fidelity curriculum implementation so that positive impacts on student achievement might be expected. In contrast, anecdotal reports from GPSMP mentors indicated that this was not the case. The mentors reported that even when all schools participated in extensive professional development activities, there remained systematic differences among middle schools in quality of implementation. These differences appeared to be a function of initial teacher and principal assent or buy-in and district-level support for the new curriculum, a factor we have named "Will to Reform."

The mentors' focus on the importance of Will to Reform was supported by previous research investigating conditions that facilitate or inhibit the process of implementing a new curriculum. Important factors identified included the teachers' buy-in, as well as support from school principals and district superintendents (e.g., Fullan & Pomfret, 1977; Krainer & Peter-Koop, 2003; Little, 1993). The LSC capstone report (Banilower et al., 2006) also noted that, over time, LSC Principal Investigators came to feel that school principals had a larger impact on the quality and impact of program implementation than had been recognized in the original theory of action.

This paper has two goals. First, it evaluates the effects of a moderately large scale (20 middle schools) implementation of the LSC model (adopting a problem-based mathematics curriculum combined with extensive professional development) on student achievement, as measured by high-stakes, state-administered, standardized mathematics tests. Second, it focuses on Will to Reform, a school-level measure of buy-in to implementing the problem-based curriculum. It investigates how Will to Reform interacts with the LSC model to predict student achievement.

As shown in Fig. 3, teacher/curriculum interactions take place within a wider school context. While researchers from the 1980s (e.g., Goodlad, 1983) emphasized that teachers tend to work in isolation, more recent research has found that teachers see themselves as part of a larger coordinated system of instruction. For example, Kennedy (2004) reported that, when planning and implementing lessons, teachers often focused on their obligation to make sure students mastered the particular content the teachers who received their students the following year would expect them to have learned. Congruent with this "coordinated system" view, the GPSMP mentors described a school-wide gestalt Will to Reform. Consequently, it was reasonable to believe that this school-level measure of buy-in might mediate the effect of curriculum on student achievement. While it would have been worthwhile to evaluate the effects of buy-in measured at the teacher-level in addition to the effects of buy-in measured at the school level, the retrospective nature of our data made doing so impossible. Also due to the retrospective nature of the data, we do not have available any direct measure of implementation fidelity. Thus, the current study investigates how school-level buy-in variables interact with curriculum materials to predict student achievement, without considering the effects of any intervening variables. Nonetheless, establishing whether or not such school-level buy-in variables predict student achievement is an important first step towards studying the more complete model displayed in Fig. 3.

Method

Achievement Measures

Student achievement was measured using eighth-grade state mathematics tests: the New Jersey Grade Eight Proficiency Assessment (GEPA) and the Pennsylvania System of State Assessment (PSSA) test. There are several advantages to using these two tests as the measure of student achievement. High stakes state tests measure what Confrey et al. (2004) called "curriculum alignment with systemic factors"—i.e., the degree to which students achieved the learning goals laid out for them by local authorities. Between 1999 and 2004, the GEPA was designed to assess student mastery of the *New Jersey Core Curriculum Content Standards for Mathematics* (New Jersey Department of Education, 1996). Subsequent to 1998, the Grade 8 PSSA assessed mastery of the Pennsylvania *Academic Standards for Mathematics* (Pennsylvania Department of Education, 1999). By using students' scores on both GEPA and PSSA, we assessed mathematics content that was both important to local stakeholders and well-aligned with the state curriculum goals.

Will-to-Reform Scale

An independent educational research firm located in Philadelphia, PA was employed to gather information about program implementation within each of the districts participating in the GPSMP follow-on. Two qualitative researchers were assigned to the project. They interviewed school administrators, teachers, and math supervisors at each of the schools, talked to the mentors, and reviewed hundreds of pages of notes mentors had made about their interactions with GPSMP schools. They also collected quantitative data on the amount of professional development attended by teachers at the various districts.

At the same time, the GPSMP mentors were asked to rate teachers and school administrators on their Will to Reform. They did so blind to the work of the two independent qualitative investigators. In February, 2006, the two qualitative researchers, the mathematics mentor(s) for each school/district, and the GPSMP Principal Investigator held a meeting to compare ratings of each school on Will to Reform.[1] Will to Reform was determined as the sum of a school's gestalt rating on each of four subscales: Teacher Buy-in, Principal Support, District Coherence, and Superintendent Support. In advance of the meeting, the scales were defined in a fairly cursory manner, with a more detailed definition developed by group consensus at the meeting.

Teacher Buy-in:

1 = Low Buy-in. Some teachers avoid professional development, taking personal leave days since they would "rather miss a workshop day than a school day." Most teachers who do attend professional development go through the motions: they tend not to ask questions and tend not to be responsive. There is a subset of teachers who are vocal in criticizing the new program, including complaining to parents about it. In schools with adequate test scores, some teachers express the attitude, "If it ain't broke, don't fix it." In schools with lower test scores, teachers tend to see the problem as being located in "kids today," "parents today," "society today," or in poor elementary school teaching. No matter what the test scores, there is a tendency for teachers to believe that the curriculum and pedagogy "won't work with our type of kids."

3 = Medium Buy-in. Teachers tend to view themselves as professionals. They are willing to be team players and implement what the school asks of them. They make an effort to attend professional development, as long as they receive a stipend for doing so. They tend to believe that their old ways of teaching might be improved, and to be willing to give something new a shot.

---

[1] We attempted to keep raters as blind as possible to student test achievement data. Before they started their research, the qualitative researchers were instructed not to review such data. Further, we asked mentors not to discuss or compare standardized test data across districts when making their ratings. However, it is possible that during the time when mentors were working with the districts they had learned whether test scores were improving, and perhaps even developed some sense about which schools were seeing relatively weaker or relatively stronger improvement.

5 = High Buy-in. Teachers tend to be excited about the new program and are eager to attend related professional development regardless of the pay. There is already a school culture supporting learning for the staff as well as for the students. Teachers tend to participate vocally in professional development and in math meetings and are willing to have others come in and observe them and provide feedback. As a group, teachers are proactive in dealing with the community and the school board, organizing parent meetings and similar activities. They aren't just walking into curriculum change, but have been building up to it: looking at student data to diagnose problems, trying out new teaching techniques like cooperative learning, etc. In general, the curriculum fits with where the majority had already been going.

Although only three levels (low, medium, and high) were described, each rater could rate Teacher Buy-in as "2" (between low and medium) or "4" (between medium and high).

Principal Support:

1 = Individual does not support the program. He/she may give lip service to it in front of district officials, but in private will criticize it.

2 = Neutral and disengaged. Often, mentors had never met these individuals even after spending many hours working with teachers in the building.

3 = Generally supportive of the program, but not an advocate; allows it to happen, takes an interest, but not willing to go out and fight for it. If there is any flak from the community, the principal defers to math teachers, who are expected to be the experts and defend what they are doing.

4 = Not only supportive, but also an advocate. Talks about the program in public meetings, and runs "interference" defending teachers from any community criticism. Lets teachers know that he/she is strongly behind the new curriculum.

5 = Supportive and an advocate, and a mathematics instructional leader. The principal understands the mathematics and learning theory behind the curriculum. He/she uses this knowledge to inform discussions with teachers about classroom practice, to inform teacher observations, to decide what types of professional development activities are appropriate for the staff, etc.

District Coherence:

1 = Program is incoherent. There is a lot of conflict and/or disagreement among the school board, superintendent, principals, teachers, and community about exactly where the program should go or what should be done.

3 = Medium coherence. There is no overt or obvious conflict about mathematics among the school board, superintendent, principals, and teachers. Community disagreements tend to be dealt with in a spirit of communication, not conflict.

5 = High coherence. Everyone is "pulling in the same direction." Programs like ongoing professional development for new teachers and advanced professional development are in place. District support staffs take an active interest in the math program, in collecting data about mathematics achievement, etc.

Similar to Teacher Buy-in, although only three levels (incoherent, medium coherent, and high coherent) were described for district coherence, each rater could rate District Coherence as "2" (between incoherence and medium coherence) or "4" (between medium coherence and high coherence).

Superintendent Support:

1 = Individual does not support the program. In cases where this happened, it tended to be a superintendent who inherited a predecessor's program, and who was interested in setting a new direction.

2 = Neutral and disengaged. Again, these tended to be superintendents who inherited the math program, but who were not actively hostile to it.

3 = Generally supportive of the program, but not an advocate; allows it to happen, takes an interest, but not willing to go out and fight for it.

4 = Not only supportive, but also an advocate. Talks about the program in public meetings, and runs "interference" defending principals and teachers from any community criticism. Lets principals and teachers know that he/she is strongly behind the new curriculum.

5 = Supportive and an advocate, and a mathematics instructional leader. The superintendent understands the mathematics and learning theory behind the curriculum and can use this knowledge in explaining what the district is doing, and in making plans with principals and other instructional leaders.

To rate a school on each subscale, each member in the group first described any information and experiences relevant to that school. The description was intended to cover the period from initial implementation through the spring of 2004, so individuals were asked to describe information relative to the overall tenor of teacher buy-in, district coherence, and support of principals and superintendents about the implementation during this period. Then, the group as a whole developed a consensus rating for each factor for each particular school. Although the rating scales were developed using retrospective data, they were based on the input of independent observers who had interviewed relevant stakeholders and reviewed detailed field notes taken by the mentors, plus the observations of the mentors themselves, who had acted as participant-observers. (Recall that the average math teacher at these middle schools participated in 59 h of professional development, much of it either one-on-one with the mentor or in group sessions taught by the mentor.) Additional observations and information were provided by the GPSMP Principal Investigator who had worked closely with district administrators and principals throughout the 5-year GPSMP project.

Each Treatment school was assigned a composite Will-to-Reform score by summing the subscale scores. Since each of the four subscales was scored from 1 to 5, the composite Will-to-Reform score could theoretically vary from a minimum of 4 to a maximum of 20. In practice, there was wide variation among Treatment schools in Will to Reform, with an observed minimum score of 5 and an observed maximum score of 19. Across the 20 Treatment schools, the mean Will-to-Reform score was 11.5 and the standard deviation was 3.62. Figure 4 displays a dot-plot of the observed

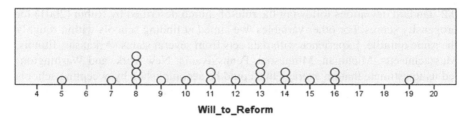

Fig. 4 Dot plot of observed Will-to-Reform scores

Will-to-Reform scores at the 20 Treatment schools. While necessarily imperfect due to the retrospective nature of the data, Will to Reform was a reasonable proxy measure for school-wide buy-in to the NSF-funded math curriculum that each Treatment school had implemented with GPSMP support.

Comparison Schools

School districts in suburban Pennsylvania and New Jersey are usually small, composed of one or two high schools and their feeder elementary and middle schools. For this reason, the participating GPSMP school districts each contained only one to four middle schools. Within each participating district, all middle schools adopted the chosen reform curriculum. Thus, we matched each GPSMP middle school to similar Comparison schools that were located in other, similar districts in the same state (either Pennsylvania or New Jersey).

Each GPSMP middle school was matched to a unique set of Comparison schools according to similar demographics (as reported by the National Center for Education Statistics 2004 data base) and test scores prior to GPSMP implementation. We chose to match using pre-determined "calipers" (maximum distance) on a set of covariates. While a number of studies have used propensity scores to match Treatment and Comparison groups (e.g., Stuart, 2007), for our study calipers had two advantages over propensity scores. Calipers allowed us to prioritize among covariates so that we matched most closely on baseline math scores and second most closely on baseline reading scores—the two covariates that were the best predictor of later-year math scores. Second, we found that while propensity scores match the entire set of Treatment schools so that on average they are similar to Comparison schools on each covariate, propensity scores might match an individual Treatment school to Comparison schools that are very different on specific covariates. The statistical models we used assumed that each Treatment school was individually matched to similar Comparison schools. Using calipers enabled us to accomplish this goal.

To select Comparison schools, we required a match within ±0.2 standard deviations on baseline scores in eighth grade mathematics and reading scores. We chose

0.2 standard deviations following the rule-of-thumb described by Rubin (2001) for propensity scores. For other variables, we aimed at finding schools within roughly the same quintile. Experience with data sets from several states (Arkansas, Illinois, Massachusetts, Michigan, Minnesota, Pennsylvania, New York, and Washington) led us to estimate that on average this could be accomplished by accepting schools within approximately ±17 % on Free and Reduced lunch and ±27 % on Percent White. We set the calipers for acceptable distance in Percent other races to be the same as calipers for Percent White. The actual matching proceeded along the following steps:

First, we first identified "Priority One" matches, defined as follows:

1. School-level Grade 8 math and reading scores in the "baseline year," which we defined as the school year prior to beginning GPSMP-supported professional development and/or curriculum implementation (1998 for all but one treatment school in Pennsylvania, and 1999 for New Jersey schools and the remaining Pennsylvania school), were within ±0.2 school-level standard deviations.
2. Within ±17 % Free and Reduced Lunch.
3. Within ±27 % for EACH of the following races: White, Black, Hispanic, Asian, and Native American.
4. Greater than 40 students enrolled in eighth grade in 2004.
5. School organization: Schools where students attended grades 6–8 but not earlier were matched to similar schools (either 6–8 or 6–12). Schools where students attended grades 5–8 were matched to similar schools (either 5–8 or K-8). Junior High schools (grades 7–8) were matched to other Junior High schools.

After identifying the set of Priority One Comparison schools, we sorted by three variables, in order: closeness of baseline math scores, closeness of baseline reading scores, and percent free/reduced lunch. For each Treatment school, we selected the top ten Priority One matches. (We used a predetermined algorithm to assign each Comparison school matching more than one Treatment school to one unique Treatment school.) If fewer than ten Priority One matches existed, we accepted all Priority One matches. In the few cases where this process yielded fewer than three Comparison schools, we used a predetermined algorithm to relax our criteria until we identified three acceptable Comparison schools. Table 1 compares Treatment to matched Comparison schools on baseline math and reading scores, as well as demographic variables. Because each Treatment school was paired with 3–10 Comparison schools, depending on how many good matches were available, for each variable we computed the average reported in Table 1 by first computing the mean for Comparison schools within each school-group, and then averaging across the 20 school-groups. Table 2 lists for each Treatment school the school's Will-to-Reform score, its math achievement growth between baseline year and 2004, the average math achievement growth of its matched comparison schools, and the number of Comparison schools identified by our matching algorithm. For each Treatment and Comparison school, math achievement growth was computed as within-state school level z-score in 2004 minus within-state school level z-score in baseline year.

Table 1 Average baseline ability and demographic characteristics of schools in the study

Variable	Treatment schools ($n=20$)	Comparison schools ($n=118$)
Baseline Math (in school-level standard deviations from state mean)	0.09	0.11
Baseline Reading (in school-level standard deviations from state mean)	0.18	0.17
Percent free/reduced lunch (%)	28	29
Percent White (%)	78	86
Percent Black (%)	9	8
Percent Hispanic (%)	10	4
Percent Asian (%)	3	2
Percent Native American (%)	<1	<1

Table 2 Mean growth for treatment and comparison schools

	Will-to-Reform score	Growth for treatment school	Mean growth comparison schools	Number of comparison schools
School 1	8	−0.03	−0.05	3
School 2	11	−0.52	0.27	10
School 3	7	−0.80	−0.23	10
School 4	19	0.54	−0.31	3
School 5	14	−0.63	0.03	10
School 6	15	0.11	0.11	10
School 7	16	1.12	−0.28	3
School 8	9	−0.82	0.27	3
School 9	8	−0.25	0.06	3
School 10	5	−0.51	0.46	3
School 11	8	−0.37	−0.20	10
School 12	16	0.05	−0.01	4
School 13	14	0.98	0.68	3
School 14	8	−1.09	0.18	4
School 15	12	0.32	0.37	3
School 16	11	−0.29	−0.20	10
School 17	13	−0.10	−0.09	10
School 18	13	0.12	−0.08	7
School 19	13	−0.18	−0.25	6
School 20	10	0.00	0.06	3
Mean		−0.12	0.04	

Statistical Model

Each school and all its Comparison schools were assigned to the same unique "school group." Further, all schools in each district plus all their Comparison schools were assigned to the same unique "district group." We used a growth model identical in form to Hierarchical Linear Models (HLMs) that track growth in individual achievement over time—but in our case, schools served as the "individuals" whose growth we were analyzing. Thus, the 4-level model measured observations, nested within schools, nested within school-groups, nested within district-groups. There were either six or seven observations per school, one for the mean math test score in the spring of each school-year from 1998 through 2004. (As noted above, for a few school groups the baseline year was 1999 instead of 1998.) We used an unstructured correlation matrix to model the six or seven observations within each school as being correlated with each other. We allowed the effects of year, of treatment, and of treatment-by-year to vary randomly between school-groups and between district-groups. (Because each school-group consisted of a Treatment school and all its matched Comparison schools and each district-group consisted of the Treatment schools within a district and all their matched Comparison schools, groups were defined by underlying similar characteristics that might lead to correlated results.) We treated State (Pennsylvania or New Jersey) as a fixed effect and allowed the fixed effect of "year" to vary between the two states. All statistical tests were run using SAS Proc Mixed. The "Satterthwaite" formula was used to estimate degrees of freedom.

Investigating the main effect of Treatment. To investigate the "main effects" of adopting an NSF-funded curriculum (either CMP or MiC) with GPSMP support, we used the model in Eq. 1:

$$
\begin{aligned}
\text{Math Test Score} &= \text{Baseline Score} + \beta_1 * \text{New Jersey} + \beta_2 * \text{Year} \\
&+ \beta_3 * \text{New Jersey} * \text{Year} + \beta_4 * \text{Treatment} + \beta_5 * \text{Treatment} * \text{Year} \quad (1) \\
&+ (\text{Error Terms})
\end{aligned}
$$

Definitions. Math Test Score: Mean score on the eighth-grade state math test (PSSA in Pennsylvania or GEPA in New Jersey) at a particular school in a particular year. For each year, these scores were standardized to a school-level z-score by subtracting the statewide average of school mean test scores and dividing by the statewide standard deviation of school mean test scores. This is analogous to what other large scale program evaluations have done (e.g., Garet et al., 2008) when they recentered student achievement data on each state's distribution by creating standard scores, except that we used schools, instead of students, as the unit-of-analysis.

Baseline Score: Model-estimated 1998 mean score for the Pennsylvania Comparison schools.

β_1: Difference between model-estimated 1998 mean score for Pennsylvania Comparison schools and model-estimated 1998 mean score for New Jersey Comparison schools.

β_2: Yearly growth rate in z-score for Pennsylvania Comparison schools. Because the dependent variable was a within-year z-score, this parameter would be significantly different from zero only if over time math test scores at the Comparison schools were systematically getting better or worse than test scores at other schools in Pennsylvania—an unlikely prospect.

β_3: Difference between the yearly growth rate in z-score for Pennsylvania Comparison schools and yearly growth rate in z-score for New Jersey schools. Like β_2, this parameter would ordinarily be near zero.

β_4: Model-estimated difference between Math Test Score at Treatment schools and Math Test Score at Comparison schools in the baseline year, 1998. Because each Treatment School was matched to Comparison schools using baseline test scores, by design this parameter was near zero.

β_5: This is the parameter of primary interest in the main-effects model. It is the difference in yearly achievement growth rate between Treatment and Comparison schools. A positive value would indicate that on average implementing *Mathematics in Context* or *Connected Mathematics* under the LSC model had a positive effect on achievement growth. A negative value would indicate that on average the program had a negative effect on achievement growth.

Error Terms: These were the error terms computed by the 4-level HLM. The following error terms were used: random differences among school groups in baseline score, yearly growth rate, baseline treatment effect, and treatment-by-growth interaction; random differences among district groups in baseline score, yearly growth rate, baseline treatment effect, and treatment-by-growth interaction; and seven correlated error terms for each year-within-school.

Investigating the effect of Will to Reform. We theorized that strong school-wide Will to Reform might catalyze the impact of NSF-funded middle school curricula, whereas low school-wide Will to Reform might interfere with the impact of the curricula. To test this theory, a valid and intuitively appealing approach would be to add for each Treatment school a recentered "Will-to-Reform" variable, i.e., the original Will-to-Reform score recentered around the middle value of 12 (halfway between 4 and 20).[2] For each comparison school, the recentered Will-to-Reform variable would be entered as zero. The new model would then add Will-to-Reform and Will-to-Reform*Year as fixed effects. The Will-to-Reform*Year slope would test whether the Will-to-Reform variable predicted how much achievement grew at Treatment schools, relative to achievement growth at other Treatment and Comparison schools.

This intuitively appealing approach had one potential drawback. Perhaps schools tended to have higher or lower Will to Reform because of some underlying background characteristic that was also associated with achievement growth. For example, perhaps baseline year achievement scores might predict Will to Reform and

---

[2] We recentered Will-to-Reform because otherwise the main effects for Treatment in Eq. 2 (reported in Table 4) would have been misleading. Table 4 would have reported Treatment effects at implementation schools where the Will-to-Reform was 0, a score below the minimum possible actual score of 4.

also predict achievement growth. In that case, Will to Reform might be correlated with achievement growth, but the correlation would be due to underlying school characteristics, not to the interaction between Will to Reform and the Treatment. One way to control for this possibility was to take advantage of each Treatment school's similarity to its matched comparison schools. In this model, each Treatment school's recentered Will-to-Reform score would be assigned both to the Treatment school and to its matched comparison schools. Then, four additional variables would be added to Eq. 1: Will-to-Reform, Will-to-Reform*Year, Will-to-Reform*Treatment, and Will-to-Reform*Treatment*Year. A positive slope for the Will-to-Reform*Treatment*Year interaction term would indicate that Treatment schools with high Will to Reform had a larger growth rate, relative to their matched Comparison schools, than did Treatment schools with low Will to Reform. A negative slope would indicate the opposite.[3]

Because both of these models were defensible, we ran each separately. Results of the two models were nearly identical. We report results from the second model, since that model theoretically did a better job controlling for possible spurious results. The model we used is described in Eq. 2.

$$
\begin{aligned}
\text{Math Test Score} = & \ \text{Baseline Score} + \beta_1 * \text{New Jersey} + \beta_2 * \text{Year} \\
& + \beta_3 * \text{New Jersey} * \text{Year} + \beta_4 * \text{Treatment} + \beta_5 * \text{Treatment} * \text{Year} \\
& + \beta_6 * \text{Will-to-Reform} + \beta_7 * \text{Will to Reform} * \text{Treatment} \\
& + \beta_8 * \text{Will-to-Reform} * \text{Year} + \beta_9 * \text{Will-to-Reform} * \text{Treatment} \\
& * \text{Year} + \left(\text{Error Terms} \right)
\end{aligned}
\tag{2}
$$

Definitions of new parameters

β_6: The effect of Will to Reform on predicted mean 1998 scores for Comparison schools. It is possible that initially high-achieving Treatment schools might have systematically lower or higher Will to Reform than low-achieving Treatment schools.

---

[3] It might appear that, by assigning each Treatment school's recentered Will-to-Reform score to its matched comparison schools, we are claiming that Will-to-Reform is a meaningful construct for the comparison schools, and further that the Will-to-Reform happens to be exactly the same at the matched comparisons as at the Treatment school. That is not what we have done. Will-to-Reform is our (retrospective and imperfect) measure of school-level buy-in at the Treatment school to their reform math curriculum. The matched comparison schools did not implement a reform math curriculum, so Will-to-Reform is not a meaningful concept for them. Within our HLM, by assigning the same value of Will-to-Reform to all members of a school-group we have made Will-to-Reform a variable that applies to school-groups, not to individual schools within a school-group. Conceptually, the HLM first estimates the growth over time at each school by computing slope for Year within that school. Then, the HLM estimates how Treatment affects the growth rate in each school-group by computing the slope for Treatment*Year within that school-group. Finally, the HLM estimates how Will-to-Reform impacts Treatment effects by computing *across school groups* the slope of Will-to-Reform*Treatment*Year. To be imprecise but conceptually correct, the model is treating Treatment*Year as a dependent variable with school-group as unit of analysis, and Will-to-Reform as the independent variable. In this way, parameter β_9 in Eq. 2 estimates whether the effect of Treatment in school-groups where the Treatment school had a high Will-to-Reform is different from the effect of Treatment in school-groups where the Treatment school had a low Will-to-Reform.

If this were the case, then our matching procedures would ensure each school's matched comparison schools would have similarly high or low baseline math scores.

β_7: The effect of Will to Reform on the difference in baseline math scores between a Treatment school and its matched Comparison schools. Our matching procedures were designed to ensure that this parameter would be close to zero, since in theory each Treatment school would have nearly the same baseline scores as its Comparison schools. Including this term in the model corrected for any remaining noise caused by imperfect matching.

β_8: The effect of a school-group's Will-to-Reform score on predicted growth rate at its Comparison schools. Some demographic characteristics were associated with a higher achievement growth rate. For example, between 1998 and 2004 in Pennsylvania, low-SES middle schools improved their eighth-grade math test scores more than did high-SES middle schools. If demographic characteristics also predicted the Will to Reform of a Treatment school, then it is possible that school-groups whose Treatment school had high Will to Reform might have systematically higher (or lower) achievement growth rates than school-groups whose Treatment school had low Will to Reform.

β_9: This is the parameter of primary interest in the Will-to-Reform model. It measures the degree to which Will to Reform was associated with an increased or decreased difference in growth rate between a Treatment school and its matched Comparison schools. A positive slope would indicate that implementing *Mathematics in Context* or *Connected Mathematics* under the LSC model had a more positive effect in high Will-to-Reform schools than in low Will-to-Reform schools. A negative slope would indicate the opposite.

Results

Overall Treatment Effects

None of the parameters in Eq. 1 differed significantly from zero. Most importantly, the slope of Treatment*Year was not significantly different from zero ($t = -0.44$, $p = 0.6714$), with a mean treatment effect of only -0.012 school-level standard deviations per year. Thus, on average, mathematics achievement growth at the 20 treatment schools was not statistically different from math achievement growth at matched similar schools.

A rough 95 % confidence interval indicates that each year the Treatment schools' growth rate differed from that at Comparison schools between -0.064 school-level standard deviations per year and $+0.041$ school-level standard deviations per year. Over 6 years, this difference in growth rate would predict a 95 % confidence that the total effect of Treatment by 2004 would be in the confidence interval $(-0.38, +0.25)$ school-level standard deviations, i.e., very near zero. Table 3 reports all fixed effects of the Main Effects model.

Table 3 SAS proc mixed solution for fixed effects, modeling treatment effects on yearly growth

Effect	STATE	Standard				
		Estimate	Error	DF	t Value	Pr > |t|
Intercept		0.238	0.253	4.23	0.94	0.399
STATE	NJ	−0.241	0.515	8.81	−0.47	0.652
STATE	PA	0				
YEAR*STATE	NJ	0.004	0.021	36.5	0.20	0.843
YEAR*STATE	PA	−0.006	0.007	17.8	−0.76	0.455
YEAR		0				
TREAT		−0.030	0.030	108	−0.99	0.326
YEAR*TREAT		−0.012	0.026	9.95	−0.44	0.671

Buy-in Effects

To investigate the interaction between Will to Reform and the effects of the two NSF-funded curricula, we used a HLM that in essence compared the value-added of high Will-to-Reform Treatment schools to the value-added of low Will-to-Reform Treatment schools. That is, we compared the degree to which growth in eighth grade math scores from the baseline year (1998 or 1999) through 2004 of high versus low Will-to-Reform schools exceeded growth at their matched Comparison schools.

Figure 5 provides a simplified visual display of this analysis. The Composite Will-to-Reform scale was a sum of four scales scored from 1 to 5, so possible scores ran from 4 to 20. As the figure shows, GPSMP Treatment schools scored over a wide range of possible Will-to-Reform scores, from a minimum of 5 to a maximum of 19 on the composite scale. The figure displays the "Value Added" at each Treatment school—i.e., how much math achievement growth from the baseline through 2004 at the Treatment school exceeded growth at its matched Comparison schools—as a function of Will to Reform. Figure 5 clearly shows that students in the treatment schools with high values of Will to Reform had higher growth from the baseline to 2004 on state test scores than those students in the treatment schools with low values of Will to Reform. By the end of 6 years of treatment, some high Will-to-Reform schools showed an increase in state test scores of more than 1 school-level standard deviation (about 0.4 student-level standard deviations) in comparison to their matched schools while some low Will-to-Reform schools showed a decrease of more than 1 school-level standard deviation.

The actual HLM, described in Eq. 2, calculated each school's growth rate based on data from all available years, providing a more accurate and stable estimate than would have been possible using just the baseline and 2004 test scores used to create Fig. 5. The analysis showed a statistically significant slope for only one parameter: β_9, the Will-to-Reform*Treatment*Year interaction ($t=4.51$, $p<.0001$), with a mean effect size of 0.021 standard deviations per Will-to-Reform point per year (95 % confidence interval between 0.012 and 0.030). That is, the higher a Treatment

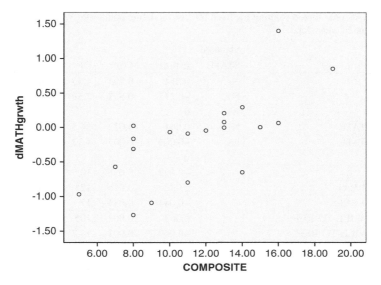

Fig. 5 Math achievement growth at treatment school from base year through 2004, minus math achievement growth at matched comparison schools (dMATHgrowth) as a function of composite Will-to-Reform score (COMPOSITE)

school's score on Will to Reform, the more achievement grew relative to that of matched Comparison schools. For detailed fixed effects from this analysis, see Table 4 in the appendix.

In practical terms, how much impact did the slope of 0.021 school-level standard deviations per Will-to-Reform point per year have on the relative effectiveness of Treatment schools? Our model's estimates for the three growth parameters in Eq. 2 were $\beta_5 = 0.002$, $\beta_8 = 0.000$, and $\beta_9 = 0.021$ where β_5 is the predicted growth rate difference between a Treatment school and its matched Comparison schools if the Treatment school had a middle value of 12 on the Will-to-Reform scale, β_8 is the (unsurprisingly zero) impact of a Treatment school's Will-to-Reform score on achievement growth at its matched Comparison schools, and β_9 is the impact of Will to Reform on achievement growth at the Treatment school. Thus, the predicted impact of the GPSMP Treatment at a school with the lowest observed Will-to-Reform score of 5 (7 less than 12) would be $0.002 + (-7*0.021) = -0.145$ school-level standard deviations per year, or -0.87 school-level standard deviations over 6 years. That is, by 2004, a school that implemented a Reform curriculum but had the lowest Will to Reform would be expected to be performing about 0.87 school-level standard deviations *below* its matched comparison schools. Assuming a normal distribution, this would be enough to bring a school from the 50th percentile in math scores statewide in 1998 down to the 19th percentile in 2004. In contrast, the predicted impact of GPSMP Treatment at a school with the highest observed Will-to-Reform score of 19 (7 more than 12) was $0.002 + (+7*0.021) = 0.149$ school-level

Table 4 SAS proc mixed solution for fixed effects, modeling composite Will-to-Reform effects on treatment-by-year slope

Effect	STATE	Standard Estimate	Error	DF	t Value	Pr > |t|
Intercept		0.214	0.293	4.55	0.73	0.500
STATE	NJ	−0.215	0.563	7.59	−0.38	0.712
STATE	PA	0				
YEAR*STATE	NJ	0.004	0.022	24.8	0.17	0.867
YEAR*STATE	PA	−0.004	0.009	4.12	−0.39	0.716
YEAR		0				
TREAT		−0.029	0.031	107	−0.94	0.348
YEAR*TREAT		0.002	0.016	125	0.10	0.920
WILL-TO-REF[a]		−0.044	0.054	17	−0.82	0.426
TREAT* WILL-TO-REF[a]		0.007	0.008	105	0.88	0.384
YEAR* WILL-TO-REF[a]		−0.0002	0.002	11.7	−0.09	0.930
YEAR*TREAT*WILL-TO-REF[a]		0.021	0.005	124	4.51	<.0001

[a]Recentered composite Will to Reform

standard deviations per year, or 0.89 school-level standard deviations over 6 years. That is, by 2004, a school that implemented a Reform curriculum but had the highest Will to Reform would be expected to be performing about 0.89 school-level standard deviations *above* its matched comparison schools. Assuming a normal distribution, this would be enough to bring a school from the 50th percentile in math scores statewide in 1998 up to the 81st percentile in 2004.

Movements of roughly this magnitude were in fact visible in the data set. For example, there were three middle schools in the data set (designated School 10, School 7, and School 4) that at the start of the GPSMP program in 1998 had the same PSSA score, at the 22nd percentile of all middle schools in Pennsylvania. School 10 (with the lowest observed Composite Will-to-Reform score of 5) moved from the 22nd percentile in 1998 down to the 18th percentile in 2004. In contrast, School 7 (tied for the second-highest observed Composite Will-to-Reform score of 16) moved from the 22nd percentile in 1998 up to the 69th percentile in 2004. School 4 (with the highest observed Composite Will-to-Reform score of 19) moved from the 22nd percentile in 1998 up to the 46th percentile in 2004.

Equation 2 did not control for reading achievement because both MiC and CMP incorporate extensive reading and thus might potentially improve eighth-grade reading as well as math scores. Nonetheless, we conducted a secondary analysis of Will-to-Reform effects on mathematics achievement while controlling for each school's reading score each year. After controlling for eighth-grade reading scores, the interaction between "Will to Reform" and the effects of GPSMP on mathematics achievement growth remained statistically significant ($t = 2.86$, $p < 0.005$). The point estimate for Will-to-Reform Effects was 0.011 school-level standard deviations per Will-to-Reform point per year (see Table 5). Thus, even after controlling for reading growth that might have been partly caused by the new

Table 5 SAS proc mixed solution for fixed effects, modeling composite Will-to-Reform effects on treatment-by-year slope, after controlling for school-level reading achievement

Effect	STATE	Standard Estimate	Error	DF	t Value	Pr > $|t|$
Intercept		0.073	0.157	4.86	0.47	0.659
YREAD[a]		0.610	0.024	842	25.74	<.0001
STATE	NJ	−0.031	0.297	7.48	−0.10	0.920
STATE	PA	0				
YEAR*STATE	NJ	−0.014	0.019	20.3	−0.76	0.454
YEAR*STATE	PA	0.001	0.008	5.11	0.13	0.905
YEAR		0				
TREAT		−0.050	0.035	109	−1.44	0.153
YEAR*TREAT		−0.00007	0.014	121	−0.00	0.996
COMPCTR[b]		−0.021	0.027	16.1	−0.78	0.447
YEAR*COMPCTR[b]		0.002136	0.002189	13.2	0.98	0.347
TREAT*COMPCTR[b]		0.006141	0.009673	105	0.63	0.527
YEAR*TREAT*COMPCTR[b]		0.01125	0.003929	121	2.86	0.005

[a]Current year mean eighth-grade reading score for the school
[b]Recentered composite Will to Reform

math program, math achievement at a school with the highest Will-to-Reform score would grow roughly 7*0.011 standard deviations faster than its comparison schools each year, or .462 standard deviations over 6 years—enough to bring a school from the 50th percentile in math scores statewide in 1998 up to the 68th percentile in 2004.

Effects of Will-to-Reform Subcomponents

While there was a significant interaction between composite Will To Reform and mathematics achievement growth, we were also interested in how each of the four Will-To-Reform subcomponents affected achievement growth. To that end, we ran four separate analyses and found that two of the components (Principal Support and Teacher Buy-in) were by themselves significant predictors of curriculum effectiveness. That is, when we replaced the Will-to-Reform variable in Eq. 2 with each of the individual subscale variables in turn, we could confirm the statistical significance of Principal-Support*Treatment*Year ($p = 0.0007$) and Teacher-Buy-in*Treatment*Year ($p = 0.0074$) (See Tables 6 and 7).

On their respective five-point scales, Principal-Support*Treatment*Year had a slope of 0.05 school-level standard deviations, and Teacher-Buy-in*Treatment*Year had a slope of 0.04 school-level standard deviations. Over 6 years, a school with principal buy-in of 5 would be expected to outperform a school with principal buy-in of 1 by $(5-1)*0.05*6 = 1.2$ school-level standard deviations. Over the same

Table 6 SAS proc mixed solution for fixed effects, modeling principal buy-in effects on treatment-by-year slope

Effect	STATE	Standard Estimate	Error	DF	t Value	Pr > |t|
Intercept		0.145	0.263	5.33	0.55	0.603
STATE	NJ	−0.093	0.521	9.58	−0.18	0.864
STATE	PA	0				
YEAR*STATE	NJ	0.002	0.021	31.9	0.07	0.943
YEAR*STATE	PA	−0.005	0.008	20.7	−0.63	0.534
YEAR		0				
TREAT		−0.026	0.032	107	−0.80	0.424
YEAR*TREAT		0.005	0.021	8.29	0.25	0.809
PRINCC[a]		−0.190	0.122	13.3	−1.55	0.145
TREAT*PRINCC[a]		0.018	0.027	109	0.67	0.504
YEAR*PRINCC[a]		−0.001	0.007	27.1	−0.13	0.901
YEAR*TREAT*PRINCC[a]		0.052	0.015	76.5	3.54	0.0007

[a]Zero-centered Principal buy-in

Table 7 SAS proc mixed solution for fixed effects, modeling teacher buy-in effects on treatment-by-year slope

Effect	STATE	Standard Estimate	Error	DF	t Value	Pr > |t|
Intercept		0.163	0.259	4.79	0.63	0.558
STATE	NJ	−0.067	0.527	9.49	−0.13	0.902
STATE	PA	0				
YEAR*STATE	NJ	−0.007	0.020	136	−0.32	0.747
YEAR*STATE	PA	0.0004	0.007	120	0.06	0.956
YEAR		0				
TREAT		−0.030	0.031	108	−0.94	0.351
YEAR*TREAT		−0.003	0.018	8.21	−0.16	0.876
TCHRBUYINC[a]		−0.145	0.109	14.2	−1.34	0.203
TREAT*TCHRBUYINC[a]		0.004	0.022	108	0.19	0.850
YEAR*TCHRBUYINC[a]		0.009	0.0057	121	1.84	0.068
YEAR*TREAT*TCHRBUYINC[a]		0.036	0.012	24.9	2.92	0.0074

[a]Zero-centered aggregate teacher buy-in

period, a school with teacher buy-in of 5 would be expected to outperform a school with teacher buy-in of 1 by $(5-1)*0.04*6 = 0.96$ school-level standard deviations.

It is important to note that the two school-level components of Will to Reform were not completely independent constructs. In fact, Principal Support and Teacher Buy-in were significantly correlated with each other ($r = 0.686$, $p < 0.01$). None of the other correlations among the four components of Will to Reform were statistically significant (See Table 8).

Table 8 Correlations between components of Will-to-Reform scale

	Teacher buy in	District coherence	Superintendent support
Principal support	0.686[a]	0.291	0.093
Teacher buy-in	0.286	0.094	
District coherence	0.329		

[a]Correlation is significant at the 0.01 level (2-tailed)

Neither of the two district-level variables was, by itself, a significant predictor for mathematics achievement growth (for Superintendent-Support*Treatment*Year, $p=0.3184$, and for District-Coherence *Treatment*Year, $p=0.0791$). The lack of statistical significance for district-level Will-to-Reform subcomponents may be an artifact of the small number of Treatment districts in the sample (only 9 district-groups, vs. 20 school-groups). Nonetheless, we cannot at this time confirm the independent importance of district-level Will-to-Reform subcomponents on the effectiveness of NSF-funded middle school mathematics curricula.

Discussion

In their comprehensive review of experimental and quasi-experimental studies that investigated the outcomes of mathematics programs for middle and high schools, Slavin et al. (2008) found a "lack of evidence that it matters very much which textbook schools choose (p. 42)." In particular, they reported a mean effect size of 0.00 standard deviations for 24 studies of NSF-funded curricula. At first blush, our findings appear to support the contention that choice of textbook doesn't matter. In our quasi-experimental study of 20 middle schools that adopted an NSF-funded math curriculum, the main effect was a statistically non-significant negative 0.012 school-level standard deviations per year.

However, when we added to our model Will to Reform, a measure of school-level buy-in to the new curriculum, we found that choice of textbook appears to have mattered very much indeed. Middle schools with very high scores on the Will-to-Reform scale saw dramatic improvements in mathematics achievement after adopting *Connected Mathematics* or *Mathematics in Context* with professional development support provided by the GPSMP. Middle schools with very low scores on the Will-to-Reform scale saw just as dramatic drops in mathematics achievement after adopting one of the new curricula—even though they too received significant professional development support from the GPSMP.

Our study also confirmed the importance of both the Teacher Buy-in and the Principal Support components of Will to Reform. The district-level components of Will to Reform—Superintendent Support and District Coherence—could not be confirmed as being independently important. It should be noted, however, that our sample consisted of only nine districts. A better test of district-level components would require a larger study incorporating a larger number of districts.

When considered in light of a co-construction view of program implementation (see Fig. 3), our results are consistent with a second finding reported by Slavin et al. (2008): reforms to instructional process strategies can have a strong positive effect on mathematics achievement. In our view, the implemented curriculum is a result not of the curriculum materials alone, but of an interaction between the teacher and the curriculum materials, as mediated by such factors as school context and teacher buy-in. That is, instructional processes, which actually affect learning, can only be predicted when curriculum materials and teachers' reactions to them are considered together.

This study is only a first step towards using the evaluation model displayed in Fig. 3 to study the effects of curriculum materials. Our study was limited by the retrospective nature of the data available. Will to Reform and its subscales were less than ideal measures of school-level buy-in. They were subject to potential limitations such as observer bias. Further, because we developed ratings by consensus, we did not have any measures of construct reliability. Moreover, our study did not have any teacher-level measure of buy-in, which might have been a more accurate predictor of program implementation than the school-level measures we used. Neither did we have available any direct measures of fidelity to implementation structure or fidelity to implementation process. To confirm the evaluation model and gain a deeper understanding of the interaction between buy-in, implementation fidelity, and student outcomes, future studies will need to correct these problems. Ideally, such studies would also include qualitative data documenting whether curriculum materials actually undergo lethal mutations in classrooms with low buy-in and productive adaptations in classrooms with high buy-in.

Future work to develop better measures of buy-in will need to consider trade-offs between the detail needed to obtain valid measures and the expense of collecting data. Would a Likert-type questionnaire for teachers, similar to that used by VanDerHeyden et al. (2012), have produced similar results to ours? Could a yes/no question about wanting to use the curriculum again, similar to that used by Agodini et al. (2010), have been sufficient?

In addition to replicating our findings using better measures of buy-in combined with measures of other variables in our model, it is also important to investigate whether our findings are applicable to other settings. Would a similar process occur in high school? In elementary school? Does buy-in predict results for subject matter other than mathematics? Compared to the reforms we implemented, many reforms (e.g., Saxon mathematics and Success for All language arts) are much more scripted. For such curricula, would strong buy-in lead to productive adaptations and positive results? Would weak buy-in lead to lethal mutations and negative results?

The retrospective nature of our study, in addition to limiting what independent variables we could study, limited us in several other ways. Only school-level, not student-level, data were available. A more detailed data set using student instead of school as unit-of-analysis would provide more precise estimates of program effects and would make it possible to investigate differential impacts on differing subgroups of students. Also, we had available only one measure of curriculum effect, the high-stakes eighth-grade mathematics tests administered by the local state

(Pennsylvania or New Jersey). A more diverse set of dependent variables would have been desirable. Past research has found that both *Math in Context* and *Connected Mathematics* tend to have a more positive impact on measures like the Balanced Assessment of Mathematics that are explicitly designed to test student problem-solving skill (Kilpatrick, 2003; Romberg et al., 2005; Tarr et al., 2008). Additionally, this was a quasi-experiment. While quasi-experiments can provide important and valid findings—especially when, as we did, they use large data bases and careful matching techniques—randomized control trials are less prone to error and provide more certain results.

Nonetheless, our results have potentially important implications for current and future implementations of instructional materials such as those designed to implement the newer *Common Core State Standards* or the *Next Generation Science Standards*. Researchers evaluating new instructional materials should strive to test the full model displayed in Fig. 3, including measures of principal buy-in, of school-wide teacher buy-in, and of individual teacher buy-in, as well as measures of structural fidelity and of process fidelity to the implementation. Further, quantitative data should be supplemented with qualitative data reporting how curriculum materials are adapted when those materials are actually used in the classroom.

Implementers of new instructional materials would be wise to attend to the role of principals and teachers as co-constructors of the planned and implemented curriculum—either by selecting materials that are a good match for local staff, or else by working closely with staff to ensure buy-in and minds-on implementation. Results of the current study support the hypothesis that doing so might encourage productive adaptations that improve student learning, while failing to do so might encourage lethal mutations that retard student learning.

References

Agodini, R., Harris, B., Thomas, M., Murphy, R., Gallagher, L., & Pendleton, A. (2010). *Achievement effects of four elementary school math curricula: Findings for first and second graders.* Washington, DC: Department of Education. Retrieved Jan 12, 2013, from http://ies.ed.gov/ncee/pubs/20094052/index.asp

Banilower, E. R., Boyd, S. E., Pasley, J. K., & Weiss, I. R. (2006). *Lessons from a decade of mathematics and science reform: A capstone report for the local systemic change through teacher enhancement initiative.* Chapel Hill, NC: Horizon Research, Inc.

Brown, A. L., & Campione, J. C. (1996). Psychological theory and the design of innovative learning environments: On procedures, principles, and systems. In R. Glaser (Ed.), *Innovations in learning: New environments for education* (pp. 289–325). Mahwah, NJ: Erlbaum.

Brown, M. W., & Edelson, D. C. (2001, April). *Teaching by design: Curriculum design as a lens on instructional practice.* Paper presented at the Annual Meeting of the American Educational Research Association, Seattle, WA.

Cai, J. (2003). What research tells us about teaching mathematics through problem solving. In F. Lester (Ed.), *Research and issues in teaching mathematics through problem solving* (pp. 241–254). Reston, VA: National Council of Teachers of Mathematics.

Cai, J. (2010). Evaluation of mathematics education programs. *International Encyclopedia of Education, 3*, 653–659.

Cai, J., Nie, B., & Moyer, J. C. (2010). The teaching of equation solving: Approaches in *Standards-based* and traditional curricula in the United States. *Pedagogies: An International Journal, 5*(3), 170–186.

Cai, J., Wang, N., Moyer, J. C., Wang, C., & Nie, B. (2011). Longitudinal investigation of the curriculum effect: An analysis of student learning outcomes from the LieCal Project. *International Journal of Educational Research, 50*(2), 117–136.

Cho, J. (1998, April). *Rethinking curriculum implementation: Paradigms, models, and teachers' work.* Paper presented at the annual meeting of the American Educational Research Association, San Diego, CA.

Confrey, J., Castillo-Chavez, C., Grouws, D., Mahoney, C., Saari, D., Schmidt, W., et al. (2004). *On evaluating curricular effectiveness: Judging the quality of K-12 mathematics evaluations.* Washington, DC: National Academies Press.

Council of Chief State School Officers and National Governors Association. (2010). *Common core state standards for mathematics.* Washington, DC: Council of Chief State School Officers and National Governors Association.

Council of Chief State School Officers, Brookhill Foundation, & Texas Instruments. (2011). *Common Core State Standards (CCSS) mathematics curriculum materials analysis project.* Washington, DC: Authors. Retrieved Jan 23, 2013, from https://www.k12.wa.us/CoreStandards/pubdocs/CCSSOMathAnalysisProj.pdf

Cross, C. T. (2004). *Putting the pieces together: Lessons from comprehensive school reform research.* Washington, DC: National Clearinghouse for Comprehensive School Reform.

Design-Based Research Collective. (2003). Design-based research: An emerging paradigm for educational inquiry. *Educational Researcher, 32*(1), 5–8.

Dobson, L. D., & Cook, T. J. (1980). Avoiding type III error in program evaluation: Results from a field experiment. *Evaluation and Program Planning, 3*, 269–276.

Flay, B., Biglan, A., Boruch, R., Castro, F., Gottfredson, D., Kellam, S., et al. (2005). Standards of evidence: Criteria for efficacy, effectiveness and dissemination. *Prevention Science, 6*(3), 151–175.

Forgatch, M. S., Patterson, G. R., & DeGarmo, D. S. (2005). Evaluating fidelity: Predictive validity for a measure of competent adherence to the Oregon model of parent management training. *Behavior Therapy, 36*, 3–13.

Fullan, M., & Pomfret, A. (1977). Research on curriculum and instruction implementation. *Review of Educational Research, 47*, 335–397.

Garet, M. S., Cronen, S., Eaton, M., Kurki, A., Ludwig, M., Jones, W., et al. (2008). *The impact of two professional development interventions on early reading instruction and achievement (NCEE 2008-4030).* Washington, DC: National Center for Education Evaluation and Regional Assistance, Institute of Education Sciences, U.S. Department of Education.

Glennan, T. K., Bodilly, S. J., Galegher, J. R., & Kerr, K. A. (2004). *Expanding the reach of education reform: Perspectives from leaders in the scale-up of educational interventions.* Santa Monica, CA: Rand. Retrieved Jan 12, 2013, from http://www.rand.org/pubs/monographs/MG248.html

Goodlad, J. I. (1983). A study of schooling: Some implications for school improvement. *Phi Delta Kappan, 64*(8), 552–558.

Hiebert, J. (1999). Relationships between research and the NCTM Standards. *Journal for Research in Mathematics Education, 30*, 3–19.

Hohmann, A. A., & Shear, M. K. (2002). Community-based intervention research: Coping with the "noise" of real life in study design. *American Journal of Psychiatry, 159*, 201–207.

Kennedy, M. M. (2004). Reform ideals and teachers' practical intentions. *Education Policy Analysis Archives, 12*(13). Retrieved Jan 2, 2013, from http://epaa.asu.edu/epaaa/v12n13/

Kilpatrick, J. (2003). What works? In S. L. Senk & D. R. Thompson (Eds.), *NSF funded school mathematics curricula: What they are? What do students learn?* (pp. 471–488). Mahwah, NJ: Erlbaum.

Krainer, K., & Peter-Koop, A. (2003). The role of the principal in mathematics teacher development. In A. Peter-Koop et al. (Eds.), *Collaboration in teacher education* (pp. 169–190). Dordrecht: Kluwer Academic.

Little, J. W. (1993). Teachers' professional development in a climate of educational reform. *Educational Evaluation and Policy Analysis, 15*, 129–151.

National Council of Teachers of Mathematics. (1989). *Curriculum and evaluation standards for school mathematics.* Reston, VA: National Council of Teachers of Mathematics.

National Council of Teachers of Mathematics. (2000). *Principles and standards for school mathematics.* Reston, VA: The Council.

National Council of Teachers of Mathematics. (2009a). *Curriculum focal points for prekindergarten through grade 8 mathematics: A quest for coherence.* Reston, VA: The Council.

National Council of Teachers of Mathematics. (2009b). *Focus in high school mathematics: fostering reasoning and sense making for all students.* Reston, VA: The Council.

National Science Teachers Association. (2012). *Recommendations on next generation science standards first public draft.* Arlington, VA: NSTA. Retrieved Feb 6, 2013, from http://www.nsta.org/about/standardsupdate/recommendations2.aspx

New Jersey Department of Education. (1996). *New Jersey core curriculum content standards for mathematics.* Trenton, NJ: Author. Retrieved Aug 7, 2007, from http://www.edsolution.org/CustomizedProducts/data/standards-frameworks/standards/09mathintro.html

O'Donnell, C. L. (2008). Defining, conceptualizing, and measuring fidelity of implementation and its relationship to outcomes in KI-12 curriculum intervention research. *Review of Educational Research, 78*(1), 33–84.

O'Donnell, C. L. & Lynch, S. J. (2008, March). *Fidelity of implementation to instructional strategies as a moderator of science curriculum unit effectiveness.* Paper presented at the annual meeting of the American Educational Research Association, New York. Retrieved Jan 12, 2013, from http://www.gwu.edu/~scale-up/documents/AERA%20O%27Donnell%20Lynch%202008%20-%20Fidelity%20of%20Implementation%20as%20a%20Moderator.pdf

Pennsylvania Department of Education. (1999). *Academic standards for mathematics.* Harrisburg. PA: Author. Retrieved Aug 7, 2007, from http://www.pde.state.pa.us/k12/lib/k12/MathStan.doc

Remillard, J. T. (2005). Examining key concepts in research on teachers' use of mathematics curricula. *Review of Educational Research, 75*(21), 1–246.

Riordan, J. E., & Noyce, P. E. (2001). The impact of two standards-based mathematics curricula on student achievement in Massachusetts. *Journal for Research in Mathematics Education, 32*(4), 368–398.

Romberg, T. A., Folgert, L., & Shafer, M. C. (2005).*Differences in student performances for three treatment groups.* (Mathematics in context longitudinal/cross-sectional study monograph 7). Madison, WI: University of Wisconsin, Wisconsin Center for Education Research. Retrieved May 28, 2014, from http://micimpact.wceruw.org/working_papers/Monograph%207%20Final.pdf

Rubin, D. (2001). Using propensity scores to help design observational studies: Application to the tobacco litigation. *Health Services & Outcomes Research Methodology, 2*, 169–188.

Schwartzbeck, T. D. (2002). *Choosing a model and types of models: How to find what works for your school.* Washington, DC: National Clearinghouse for Comprehensive School Reform.

Slavin, R. E. (2002). Evidence-based educational policies: Transforming educational practice and research. *Educational Researcher, 31*(7), 15–21.

Slavin, R. E., Lake, C. & Groff, C. (2008). *Effective programs in middle and high school mathematics: A best-evidence synthesis.* Baltimore, MD: Johns Hopkins University Center for Data Driven Reform in Education (CDDRE) Best Evidence Encyclopedia. Retrieved May 29, 2013, from http://www.bestevidence.org/word/mhs_math_Sep_8_2008.pdf

Stuart, E. A. (2007). Estimating causal effects using school-level data sets. *Educational Researcher, 36*, 187–198.

Tarr, J. E., Reys, R., Reys, B., Chávez, Ó., Shih, J., & Osterlind, S. (2008). The impact of middle school mathematics curricula on student achievement and the classroom learning environment. *Journal for Research in Mathematics Education, 39*(3), 247–280.

Turnbull, B. (2002). Teacher participation and buy-in: Implications for school reform initiatives. *Learning Environments Research, 5*(3), 235–252.

U.S. Department of Education. (2007a). *What works clearinghouse intervention report: connected mathematics project.* Washington, DC: Author. Retrieved Nov 1, 2007, from http://ies.ed.gov/ncee/wwc/pdf/WWC_CMP_040907.pdf

U.S. Department of Education. (2007b). *Mathematics and science specific 84.305A RFA.* Washington, DC: Author. Retrieved Dec 2, 2007, from http://ies.ed.gov/ncer/funding/math_science/index.asp

VanDerHeyden, A., McLaughlin, T., Algina, J., & Snyder, P. (2012). Randomized evaluation of a supplemental grade-wide mathematics intervention. *American Educational Research Journal, 49*(6), 1251–1284.

Vuchinich, S., Flay, B. R., Aber, L., & Bickman, L. (2012). Person mobility in the design and analysis of cluster-randomized cohort prevention trials. *Prevention Science, 13*(3), 300–313.

Longitudinally Investigating the Impact of Curricula and Classroom Emphases on the Algebra Learning of Students of Different Ethnicities

Stephen Hwang, Jinfa Cai, Jeffrey Shih, John C. Moyer, Ning Wang, and Bikai Nie

Classrooms in the United States are becoming increasingly ethnically diverse. However, disparities in the mathematics achievement of different ethnic groups remain a persistent challenge (Lubienski & Crockett, 2007). Although there was about a 10 % reduction in the eighth-grade White-Hispanic mathematics achievement gap on the 2011 National Assessment of Educational Progress (NAEP), since 2009 there have been no reductions in any of the other White-ethnic mathematics gaps at either grades 4 or 8 (NCES, 2012). Since teaching and learning are cultural activities, students with different ethnic and cultural backgrounds may respond differently to the same curriculum. Given the development and implementation of curricula based on the *Standards* documents developed by the National Council of Teachers of Mathematics (NCTM, 1989, 2000), a key question about curriculum reform is: How does the use of a *Standards*-based curriculum impact the learning of students of

An early version of this paper was presented at the 2012 PME-NA, Kalamazoo, MI: Western Michigan University. The research reported here is supported by a grant from the National Science Foundation (ESI-0454739 and DRL-1008536). Any opinions expressed herein are those of the authors and do not necessarily represent the views of the National Science Foundation.

S. Hwang • J. Cai (✉) • B. Nie
University of Delaware, Ewing Hall 523, Newark, DE 19716, USA
e-mail: hwangste@udel.edu; jcai@udel.edu

J. Shih
Department of Teaching & Learning, University of Nevada-Las Vegas,
4505 Maryland Parkway, Las Vegas, NV 89154-3005, USA

J.C. Moyer
Marquette University, 1250 W. Wisconsin Ave., Milwaukee, WI 53233, USA

N. Wang
Widener University, Chester, PA 19013, USA

© Springer International Publishing Switzerland 2015
J.A. Middleton et al. (eds.), *Large-Scale Studies in Mathematics Education*,
Research in Mathematics Education, DOI 10.1007/978-3-319-07716-1_3

45

color as compared to White students? The purpose of this study is to use the data from a longitudinal project to explore this research question. We begin by describing the larger longitudinal project of which this study is a part.

Background

The LieCal Project

In our project, Longitudinal Investigation of the Effect of Curriculum on Algebra Learning (LieCal), we used a longitudinal design to examine the similarities and differences between a *Standards*-based curriculum called the Connected Mathematics Program (CMP) and more traditional curricula (non-CMP). The CMP curriculum has been more broadly implemented than any other *Standards*-based curriculum at the middle school level. In the 2002–2003 school year, CMP was used in nearly 2,500 school districts in the United States. It has been used in all 50 states and some foreign countries (Rivette, Grant, Ludema, & Rickard, 2003; Show-Me Center, 2002). Thus, it provided us with a useful context to study student achievement in *Standards*-based and more traditional mathematics curricula. We investigated not only the ways and circumstances under which the CMP and non-CMP curricula affected student achievement gains, but also the characteristics of these reform and traditional curricula that hindered or contributed to the gains.

The LieCal Project was designed to provide: (a) a profile of the intended treatment of algebra in the CMP curriculum with a contrasting profile of the intended treatment of algebra in non-CMP curricula; (b) a profile of classroom experiences that CMP students and teachers have, with a contrasting profile of experiences in non-CMP classrooms; and (c) a profile of student algebra-related performance resulting from the use of the CMP curriculum, with a contrasting profile of student algebra-related performance resulting from the use of non-CMP curricula. One aspect of the LieCal analysis was an examination of potentially differential effects of curriculum and procedural and conceptual emphases in the classroom on the achievement of students of color. The longitudinal growth curve analysis of the LieCal data produced mixed results with respect to the achievement of students of color (Cai, Wang, Moyer, Wang, & Nie, 2011). Although the CMP curriculum contributed to significantly higher growth than the non-CMP curricula for all ethnic groups on more conceptually oriented measures (e.g., open-ended tasks), the situation was more complex for achievement on more procedurally oriented measures. African-American CMP students had a smaller growth rate on computation and equation solving than students from other ethnic groups and African-Americans using a non-CMP curriculum. However, the CMP program had a positive impact on Hispanic students' growth in these areas. In this paper, we present results from a cross-sectional analysis of student growth within each grade level. This analysis allows us to probe effects that are significant at individual grades, but which were not uncovered in our longitudinal analysis.

Algebra Readiness

The LieCal Project focused on the effects of the CMP and non-CMP curricula on middle-school students' learning of algebra. Middle school algebra lays the foundation for the acquisition of tools for representing and analyzing quantitative relationships, for solving problems, and for stating and proving generalizations (Bednarz, Kieran, & Lee, 1996; Carpenter, Franke, & Levi, 2003; Kaput, 1999; Mathematical Sciences Education Board, 1998; RAND Mathematics Study Panel, 2003). Thus, algebra readiness has been characterized as the most important "gatekeeper" to success in school mathematics (Pelavin & Kane, 1990), which has itself been considered a broader gatekeeper to educational and economic opportunities (Moses, Kamii, Swap, & Howard, 1989; Nasir & Cobb, 2002).

In particular, success in algebra and geometry has been shown to help narrow the disparity between minority and non-minority participation in post-secondary opportunities (Loveless, 2008). Research shows that completion of an Algebra II course correlates significantly with success in college and with earnings from employment. The National Mathematics Advisory Panel (2008) found that students who complete Algebra II are more than twice as likely to graduate from college as students with less mathematical preparation. Furthermore, the African-American and Hispanic students who complete Algebra II reduce the gap between their college graduation rate and that of the general student population by 50 %. However, success in high school algebra is dependent upon mathematics experiences in the middle grades, and middle school is a critical turning point for students' development of algebraic thinking (College Board, 2000).

Conceptual and Procedural Emphases

In a *Standards*-based curriculum like CMP, there is a greater emphasis on conceptual understanding and problem solving than on procedural knowledge. Students are expected to learn algorithms and master basic skills as they engage in explorations of worthwhile problems. However, a persistent concern about *Standards*-based curricula is that the development of students' higher-order thinking skills comes at the expense of fluency in computational procedures and symbolic manipulation. In addition, it is not clear whether this potential trade-off might play out differently for students from different ethnic backgrounds. Some reports have suggested that Hispanic and African-American students using the CMP curriculum may in fact show greater achievement gains than students from other backgrounds (Rivette et al., 2003). Our previous longitudinal analysis of the LieCal data using growth curve modeling showed that, over the three middle school years, CMP students' gains in conceptual understanding did not come at the expense of procedural skills (Cai et al., 2011). The use of either the CMP or a non-CMP curriculum improved the mathematics achievement of all students, including students of color. Moreover, the use of CMP contributed to significantly higher problem-solving growth for all ethnic groups (Cai et al., 2011). However, African-American students experienced greater gain in symbol manipulation when they used a traditional curriculum. Additional research is needed to assess whether and how the use of a *Standards-based* curriculum such as CMP can improve both the problem-solving and symbol-manipulation achievement of all students while helping to close achievement gaps (Lubienski & Gutiérrez, 2008; Schoenfeld, 2002).

At the same time, it is not sufficient to examine the achievement outcomes of students in different ethnic groups without also considering the ways that those outcomes are shaped through students' experiences in school (Lubienski & Bowen, 2000). For example, since the effectiveness of a curriculum depends critically on how it is implemented by teachers in real classrooms, studies of the effectiveness of *Standards*-based curricula must examine how teachers actually use the curricula (Kilpatrick, 2003; NRC, 2004; Wilson & Floden, 2001). Indeed, Tarr and his colleagues (2008) have found that the nature of the learning environment moderates the effects of *Standards*-based curricula. In particular, they noted that such curricula are associated with a positive impact on student achievement only when they are implemented in *Standards*-based learning environments. However, it is not clear whether the use of *Standards*-based curricula within *Standards*-based learning environments influences the achievement of students from different ethnic groups in the same way (Lubienski, 2000). Lubienski (2002) explicitly questions whether "some students enter the mathematics classroom better positioned than others to learn in the ways envisioned in the *Standards*" (p. 109) and, thus, whether such pedagogies might exacerbate rather than mitigate achievement differences. Research in mathematics education has not yet adequately addressed such questions.

Thus, to determine the effects of curriculum on learning, and in particular, on the learning of groups of students with different ethnic backgrounds, it is essential to take into account the classroom experiences of the teachers and students who are using the different curricula. In this paper, we consider features of classroom instruction related to conceptual and procedural emphases when we examine the impact of curricula on students' learning of algebra. In particular, we examine the extent to which teachers emphasize concepts and procedures in their classroom instruction. As was reported by Moyer, Cai, Nie, and Wang (2011), CMP teachers placed more emphasis on conceptual understanding in their instruction, whereas non-CMP teachers placed more emphasis on procedural knowledge. We seek to better understand how these emphases play out in the achievement of different groups of students using the CMP and non-CMP curricula. In this paper, we take a cross-sectional approach and examine the achievement of students of color at each grade level while controlling for the conceptual and procedural emphases in classroom instruction.

Method

Sample

The LieCal project was conducted in 14 middle schools of an urban school district serving a diverse student population. When the project began, 27 of the 51 middle schools in the district had adopted the CMP curriculum, and the remaining 24 had adopted more traditional curricula. Seven schools were randomly selected from the 27 schools that had adopted the CMP curriculum. After the seven CMP schools were selected, seven non-CMP schools were chosen based on comparable demographics. In sixth grade, 695 CMP students in 25 classes and 589 non-CMP students in 22 classes participated in the study. We followed these 1,284 students as they progressed from grades 6 to 8. Approximately 85 % of the participants were minority students: 64 % African-American, 16 % Hispanic, 4 % Asian, and 1 % Native American. Male and female students were almost evenly distributed.

Assessing Students' Learning

Learning algebra involves honing procedural skills with computation and equation-solving, fostering a deep understanding of fundamental algebraic concepts and the connections between them, and developing the ability to use algebra to solve problems. Thus, to assess students' learning of algebra, it is important to consider their conceptual understanding, their symbol manipulation skills, and their ability to solve problems. We used state standardized test scores in mathematics and reading as measures of prior achievement. The state tests were administered in the fall of the students' sixth-, seventh-, and eighth-grade years (2005, 2006, and 2007). We used LieCal-developed multiple-choice and open-ended assessment tests as dependent measures of procedural knowledge and conceptual understanding in algebra, respectively. The two LieCal-developed tests were administered on 2 consecutive days of testing during the students' regular classroom periods. In all, they were administered four times, once as a baseline in the fall of 2005, and again each spring (2006, 2007, and 2008).

The LieCal multiple-choice tests assessed whether students had learned the basic knowledge required to perform competently in introductory algebra. We chose to use multiple-choice items because of their potential for broad content coverage and objective scoring, their highly reliable format, and their low cost of scoring. Each of the four parallel versions of the multiple-choice test (F05, Sp06, Sp07, and Sp08) comprised 32 questions that assessed five mathematical components: translation, integration, planning, computation (or execution), and equation solving. The first four of these components are based on Mayer's (1987) model for analyzing cognitive components in solving word problems. Translation and integration involve the representing phase of problem solving, while planning and computation (execution) involve the searching phase of problem solving. To represent a problem, a student must be able to put the elements of a problem together into a coherent whole and translate them into an internal representation, such as an equation. In the searching phase of problem solving, the student must first plan the solution, and then find and execute an adequate algorithm. The LieCal multiple-choice tests included six items for each of Mayer's four cognitive components, and eight items to assess equation solving. For this paper, we report on the results from the translation, computation, and equation-solving components of the multiple-choice tasks.

The LieCal open-ended tests assessed students' conceptual understanding and problem-solving skills. In the open-ended tasks, students were asked to provide explanations of their solutions as part of their responses to the problems. The tasks used in these tests were adopted from various projects including Balanced Assessment, the QUASAR Project (Lane et al., 1995), and a cross-national study (Cai, 2000). In the fall of 2005, the LieCal sixth graders were given a baseline open-ended assessment with six tasks. Since only a small number of open-ended tasks can be administered in a testing period, and since grading students' responses to such items is labor-intensive, we distributed the non-baseline tasks over three forms (five items in each form) and used a matrix sampling design to administer them. Thus, starting in the spring of 2006, each third of the students was administered one of the three forms. The forms were rotated in the two subsequent administrations so that eventually each student received all three forms.

From one testing administration to another, 10 of the 32 multiple-choice items were identical, while the other 22 items were new, but parallel. The ten identical items were composed of two items from each of the five components. They served as linking items in the analysis. In a similar way, at least two identical open-ended tasks served as linking items from one form to another and one testing administration to another. We used standard scores to report and analyze the student achievement data. A two-parameter partial-credit Item Response Theory (IRT) model was used to scale student assessment data on each of the five components in the multiple-choice tasks as well as on the open-ended tasks (Hambleton, Swaminathan, & Rogers, 1991; Lord, 1980). Because IRT models simultaneously compute item difficulty and student ability, the use of linking items made it possible to place assessment results on the same scale even if students responded to different tasks at different times. Additional details and examples of the items and tasks used in the LieCal assessments can be found in Cai et al. (2011).

The multiple-choice items were scored electronically, either right or wrong. The open-ended tasks were scored by middle school mathematics teachers, who were trained using holistic scoring rubrics that had been developed previously by the investigators. Two teachers scored each response. On average, perfect agreement between each pair of raters was nearly 80 %, and agreement within 1 point out of 6 points (on average) was over 95 % across tasks. Differences in scoring were arbitrated through discussion.

Conceptual and Procedural Emphases as Classroom-Level Variables

Mathematical proficiency includes both conceptual and procedural aspects (NRC, 2001), and teachers can shape instruction in ways that emphasize either or both aspects. We used conceptual and procedural emphases as classroom variables when examining the impact of curriculum on students' learning. To do so, we estimated the levels of conceptual and procedural emphases in the CMP and non-CMP classrooms using data from 620 lesson observations of the LieCal teachers, which we conducted while the students were in grades 6, 7, and 8. Each class was observed four times per year, during two consecutive lessons in the fall and two in the spring. Further details about the observations are documented in Moyer et al. (2011).

One component of the observation was a set of 21 items using a 5-point Likert scale to rate the nature of instruction for each lesson. Of the 21 items, four were designed to assess the extent to which a teacher's lesson had a conceptual emphasis. For example, observers rated a lesson's conceptual emphasis using the following item: "The teacher's questioning strategies were likely to enhance the development of student conceptual understanding/problem solving." Another four items were designed to determine the extent to which a teacher's lesson had a procedural emphasis. For example, observers rated a lesson's procedural emphasis using this item: "Students had opportunities to learn procedures (by teacher demonstration, class discussion, or some other means) before they practiced them." Factor analysis of the LieCal observation data confirmed that the four procedural-emphasis items loaded on a single factor, as did the four conceptual-emphasis items. Since students changed their classrooms and teachers as they moved from grade 6 to grade 7 and from grade 7 to grade 8, each student could have a different procedural (or conceptual) score each year for 3 years. However, within each grade all students in the same classroom were assigned the same procedural (and conceptual) score.

Quantitative Data Analysis

To examine student growth within each school year while controlling for multiple factors such as gender, ethnicity, and classroom conceptual and procedural emphases, we used hierarchical linear modeling (HLM). Although we originally created three-level hierarchical models (students nested within teachers nested within schools) with the mathematics achievement measures as outcome measures, the school sample sizes and the relatively small intraclass correlation coefficients ultimately supported the use of two-level models (students nested within teachers). For each dependent variable and grade level, an unconditional model of the form:

$$Y_{ij} = p_{0j} + e_{ij}$$

$$p_{0j} = b_{00} + r_{0j}$$

was fitted, where Yij is the achievement of child i in classroom j, p_0j is the mean mathematics achievement of classroom j, b_{00} is the grand mean, eij is the random student effect, and r_0j is a random classroom effect. The unconditional models were tested to determine the intraclass correlation coefficients for each model.

After the unconditional models were fitted, two sets of conditional cross-sectional HLM analyses were conducted. The first set of these models was composed of cross-sectional hierarchical linear models that included student-level variables and a curriculum variable. These models used four of the LieCal-developed student achievement measures: open-ended, translation, computation, and equation solving. Each HLM model used data from one of the four dependent achievement measures in one of three middle grades, together with an independent prior achievement measure, namely the results of the state mathematics testing in the fall of the corresponding year. So, each model examined a single type of learning within a specific grade level. Since we had four achievement measures at each of three grade levels, there were 12 cross-sectional models in this first group.

The next set of models built on the first group of models by adding two classroom-level variables: the conceptual emphasis of the classroom and the procedural emphasis of the classroom. These cross-sectional HLM models were of the following form:

Level-1 Model

$$Y_{ij} = p_{0j} + p_{1j}\left(\text{Prior achievement}_{ijk} - \overline{X}^{-1}\right) + p_{2j}\left(\text{Gender}_{ijk} - \overline{X}^{-2}\right)$$
$$+ p_{3j}\left(\text{African American}_{ijk} - \overline{X}^{-3}\right) + p_{4j}\left(\text{Hispanic}_{ijk} - \overline{X}^{-4}\right)$$
$$+ p_{5j}\left(\text{Other ethnicity}_{ijk} - \overline{X}^{-5}\right) + r_{ijk}$$

Level-2 Model

$$p_{0j} = b_{00} + b_{01}\,\text{CMP} + b_{02}\,\text{Conceptual emphasis}_j + b_{03}\,\text{Procedural emphasis}_j + r_{0j}$$

In these models, pij is the student level slope capturing the effect of achievement due to classroom-level variable i with teacher j, b_0j is the school-level slope capturing the effect of achievement due to classroom-level variable j. Interactions between conceptual emphasis, procedural emphasis, and curricula were tested, but found to be not significant.

Results

We first present data from the state standardized tests of mathematics and reading that were used as measures of prior achievement. We then present the results of our HLM analyses in two parts. First, we report on the cross-sectional HLM models that included student-level and curriculum variables. Then, we examine the impact of including the classroom-level conceptual and procedural emphasis variables in the models.

State Standardized Tests

Table 1 shows the mean scores each year for CMP and non-CMP African-American, Hispanic, and White students on the state standardized tests of mathematics and reading.

The reading scores show a notable difference across the three student groups. For African-American and White students, growth over the middle grades in reading scores was comparable across curricula, ranging from an increase of 22.97 to 27.74 points. However, the Hispanic CMP students' reading scores grew more than any other student group, regardless of curriculum (36.69 points). In contrast, the Hispanic non-CMP students' reading scores grew by only 14.15 points.

Student-Level and Curriculum Cross-Sectional HLM Models

Analysis of combined CMP and non-CMP data. Table 2 shows the standardized results from an examination of the performance of African-American and Hispanic students relative to White students, when controlling for prior achievement, gender, and curriculum (but not conceptual and procedural classroom emphases).

In the sixth grade, an achievement gap was seen between African-American students and White students on all four student achievement measures, and between Hispanic students and White students on the open-ended, computation, and equation-solving measures. The gaps on the open-ended and equation-solving measures remained in the seventh grade for both groups. However, performance on the compu-

Table 1 Mean scores on state standardized mathematics and reading tests

	Fall 2005		Fall 2006		Fall 2007	
	Math	Reading	Math	Reading	Math	Reading
CMP						
African American	460.96	460.47	486.62	469.01	486.31	483.44
Hispanic	463.12	451.17	495.60	467.66	501.46	487.86
White	502.41	505.61	528.65	519.44	539.74	531.30
Non-CMP						
African American	464.05	459.03	494.86	472.83	492.17	485.01
Hispanic	477.49	459.92	496.13	468.03	500.42	474.07
White	497.79	508.89	536.55	519.24	538.72	536.63

Table 2 Effect of ethnicity on standardized mathematics achievement

	Grade 6		Grade 7		Grade 8	
	African American	Hispanic	African American	Hispanic	African American	Hispanic
Open-ended	−0.50***	−0.21*	−0.26**	−0.22*	−0.28***	−0.13*
Translation	−0.24**	−	−	−	−	−
Computation	−0.37***	−0.22*	−	−	−	−
Equation solving	−0.35**	−0.23*	−0.24**	−0.22*	−	−

$*p < .05$, $**p < .01$, $***p < .001$

Table 3 Effect of ethnicity on standardized mathematics achievement for CMP students

	Grade 6		Grade 7		Grade 8	
	African American	Hispanic	African American	Hispanic	African American	Hispanic
Open-ended	−0.40**	−	−0.27**	−0.31*	−0.36**	−0.28**
Translation	−	−	−	−	−	−
Computation	−0.43**	−	−	−	−0.22*	−
Equation solving	−0.35*	−	−0.44**	−0.45**	−0.23**	−0.22*

$*p < .05$, $**p < .01$, $***p < .001$

Table 4 Effect of ethnicity on standardized mathematics achievement for non-CMP students

	Grade 6		Grade 7		Grade 8	
	African American	Hispanic	African American	Hispanic	African American	Hispanic
Open-ended	−0.91***	−	−0.23*	−	−0.26**	−
Translation	−0.37*	−	−	−	−	−
Computation	−0.27**	−0.35**	−	−	−	−
Equation solving	−	−	−	−	−	−

$*p < .05$, $**p < .01$, $***p < .001$

tation and translation measures had equalized across the groups. In the eighth grade, the only gap that remained was on the open-ended items. The overall trend was a gradual decline or elimination of the achievement gap among the ethnic groups.

Analyses of separated CMP and non-CMP data. To better understand whether the use of the CMP curriculum reduced achievement gaps, we conducted separate parallel analyses for CMP and non-CMP students. The results are shown in Tables 3 and 4.

Recall that in the analysis of the combined CMP and non-CMP student data, achievement gaps for the translation and computation measures occurred only in the sixth grade, where there were three such gaps: White students outperformed African-American students on both measures, and White students outperformed

Hispanic students on computation. Although all three gaps persisted for the non-CMP students when we separated the CMP and non-CMP data, they did not all persist for the CMP students. That is, of these three gaps, the only one that persisted for the CMP students at sixth grade was the African-American gap in computation. In grades 7 and 8, the parity on computation and translation achievement that we observed in the combined CMP and non-CMP data was preserved in the separate analyses, except for the appearance of a gap between CMP eighth-grade African-American students and White students on computation items.

Mirroring the results from the combined data, our analyses of the separated data showed that White students outperformed African-American students on open-ended items across all three grades regardless of curriculum. For students using CMP, White students also outperformed Hispanic students on these items in grades 7 and 8 (but not grade 6). However, for non-CMP Hispanic students, there were no achievement gaps on the open-ended items.

With respect to the equation-solving items in the combined analysis, White students outperformed African-American and Hispanic students in grades 6 and 7, with no achievement gap in grade 8. The parallel CMP and non-CMP analyses indicate that these gaps were attributable to the CMP students; there were no achievement gaps found for equation-solving items among the non-CMP students. For CMP students, White students outperformed African-American students in all three grades, and White students outperformed Hispanic students in grades 7 and 8. The equation-solving gaps are most pronounced in seventh grade; their magnitude appears to decline in eighth grade.

Student-Level, Classroom-Level, and Curriculum HLM Models

Analysis of combined CMP and non-CMP data. We built on the results of Table 2 with the addition of the conceptual emphasis and procedural emphasis classroom-level variables. Our goal in adding these variables to the analysis was to begin to probe the complexity that underlies conclusions we might otherwise draw from one-dimensional comparisons of students in different ethnic groups. With respect to the analysis of the combined sample of CMP and non-CMP students, however, controlling for the classroom-level variables did not greatly perturb the results save for the disappearance of the gap in Hispanic students' performance on open-ended tasks in the eighth grade.

Analyses of separated CMP and non-CMP data. We again conducted parallel analyses for the CMP and non-CMP students, this time controlling for the conceptual and procedural emphasis classroom-level variables. The results are presented in Tables 5 and 6. Compared to the results of the models without controlling for the conceptual and procedural emphasis variables, some differences were apparent. For the CMP students, two achievement gaps were no longer statistically significant with the addition of the classroom variables: eighth-grade African-American

Table 5 Effect of ethnicity on standardized mathematics achievement for CMP students controlling for conceptual and procedural emphases

	Grade 6		Grade 7		Grade 8	
	African American	Hispanic	African American	Hispanic	African American	Hispanic
Open-ended	−0.40**	−	−0.28**	−0.30**	−0.36**	−0.29**
Translation	−	−	−	−	−	−
Computation	−0.37**	−	−	−	−	−
Equation solving	−0.35*	−	−0.45**	−0.44**	−0.21**	−

$*p < .05, **p < .01, ***p < .001$

Table 6 Effect of ethnicity on standardized mathematics achievement for non-CMP students controlling for conceptual and procedural emphases

	Grade 6		Grade 7		Grade 8	
	African American	Hispanic	African American	Hispanic	African American	Hispanic
Open-ended	−0.90***	−0.33*	−0.27*	−	−0.23**	−
Translation	−	−	−	−	−	−
Computation	−	−0.33**	−	−	−	−
Equation solving	−	−	−	−	−	−

$*p < .05, **p < .01, ***p < .001$

students on computation items, and eighth-grade Hispanic students on equation-solving items. For the non-CMP students, the performance gaps of sixth-grade African-American students on translation and computation items ceased to be significant. However, a performance gap appeared for sixth-grade non-CMP Hispanic students on open-ended items.

To summarize, in the analysis of the combined student groups (CMP and non-CMP) that included the classroom conceptual and procedural emphasis variables, by the end of eighth grade the performance of Hispanic students was not significantly different from White students on all four achievement measures. Similarly, the performances of eighth-grade African-American and White students were not significantly different except on the open-ended items; there was no achievement gap by the end of eighth grade between African-American and White students on translation, computation, and equation-solving items. When analyzed as separate groups, through the middle grades the CMP and non-CMP students of color, particularly African-American students, generally showed achievement gaps on open-ended items compared to White students using the same curriculum. Within the CMP student group, there were also persistent achievement gaps for African-American students on equation-solving items.

Discussion

In examining how *Standards*-based curricula such as CMP affect the mathematics learning of students of color, it is important to use nuanced analyses to look beyond one-dimensional comparisons (Lubienski, 2008). The longitudinal growth curve analysis of the LieCal data provided mixed conclusions regarding the use of the CMP curriculum with students of color (Cai et al., 2011). Although, over the course of the middle grades, African-American and Hispanic students had growth rates similar to students not in their ethnic groups on the open-ended and translation measures, African-American CMP students had smaller growth rates on the computation and equation-solving measures. However, Hispanic CMP students did not exhibit this pattern. The cross-sectional HLM analysis in this paper provides additional detail not captured in the longitudinal analysis.

Overall, when the CMP and non-CMP students are combined, the results of the cross-sectional analysis show a trend of decreasing gaps in achievement within each year. Whereas Hispanic and African-American students score significantly lower than White students on most or all of the measures at the end of sixth grade, by the end of eighth grade, only the open-ended measure still reflects a gap for that year. Moreover, when the differences between conceptual and procedural emphases in the classroom are controlled, the only difference that remains for eighth grade is in African-American students' performance on the open-ended tasks. Our cross-sectional analysis also pinpointed the longitudinal analysis' finding regarding the African-American students' slower growth rate on computation tasks as being largely limited to the sixth grade.

When the cross-sectional analysis is limited to the CMP students, the open-ended measure reflects a persistent gap between White students and students of color. Similarly, for African-American students in the CMP group, equation solving remains an area of challenge throughout the middle grades. Even when classroom conceptual and procedural emphasis variables are included, these gaps remain. Indeed, the open-ended performance gaps in the CMP analysis do not vanish or even consistently decrease over the course of the middle grades, as many of the other performance gaps do. This result suggests that there may be a need to seek opportunities within the CMP curriculum to develop open-ended problem-solving skills more robustly to better serve students of different ethnic backgrounds.

It is interesting to note how the influence of classroom emphasis variables on equation solving, translation, and computation played out differently for different student groups. On the one hand, the profile of Hispanic CMP students' equation-solving performance was somewhat different from the African-American CMP students'. Although the African-American CMP students' equation-solving gaps at grades 6–8 persisted when our analyses controlled for differences in the conceptual and procedural emphasis variables, the Hispanic CMP students' deficit in grade 8 disappeared. This result implies that differences in classroom conceptual and procedural emphases, not curriculum, appear to account for the Hispanic CMP students' performance gap in equation solving in the eighth grade.

For the translation and computation measures, on the other hand, the pattern is reversed, occurring at sixth grade (not eighth) for African-American non-CMP students (rather than Hispanic CMP students). More specifically, when controlling for classroom emphasis, the achievement gap on the translation and computation measures disappeared for the sixth grade African-American non-CMP students, but the gap in their Hispanic counterparts' performance on computation remained significant. This result implies that differences in classroom conceptual and procedural emphases, not curriculum, appear to account for the African-American non-CMP students' performance gap in translation and computation in the sixth grade. These differences in the effects of classroom emphasis on Hispanic and African-American students' performance merit further exploration.

The differences in student growth on the state standardized reading test suggest an additional avenue for analysis. Whereas the African-American and White students' reading scores increased at similar rates, the increase in the Hispanic CMP students' reading scores was markedly higher than for either the Hispanic non-CMP students or the other student groups. This may reflect the development of English language learners within the Hispanic student group. Indeed, the orientation in CMP toward instructional contexts that actively involve students in using language and discourse may be particularly supportive of English language learners (Moschkovich, 2002, 2006). However, it is not immediately clear how the difference in improvement on reading performance for the CMP Hispanic students might influence performance on the mathematical measures used in this study. Indeed, there remains a performance gap for the eighth-grade Hispanic CMP students on open-ended tasks, which would seem to be the type of task most amenable to increased performance due to improved reading skills.

In conclusion, the longitudinal and cross-sectional analyses paint complementary pictures of the effects of the CMP curriculum for students of color. Though African-American students' computation skills appeared to grow more slowly across grades 6–8, the effect of this difference seems to have been primarily limited to grade 6. However, the persistent gaps between African-American students and White students on the open-ended and equation-solving measures, even when classroom emphases are taken into account, invite further investigation.

References

Bednarz, N., Kieran, C., & Lee, L. (Eds.). (1996). *Approaches to algebra: Perspectives for research and teaching*. Dordrecht: Kluwer.

Cai, J. (2000). Mathematical thinking involved in U.S. and Chinese students' solving process-constrained and process-open problems. *Mathematical Thinking and Learning, 2*, 309–340.

Cai, J., Wang, N., Moyer, J. C., Wang, C., & Nie, B. (2011). Longitudinal investigation of the curriculum effect: An analysis of student learning outcomes from the LieCal project. *International Journal of Educational Research, 50*(2), 117–136.

Carpenter, T. P., Franke, M. L., & Levi, L. (2003). *Thinking mathematically: Arithmetic and algebra in elementary school*. Portsmouth, NH: Heinemann.

College Board. (2000). *Equity 2000: A systemic education reform model*. Washington, DC: Author.

Hambleton, R. K., Swaminathan, H., & Rogers, H. J. (1991). *Fundamentals of item response theory*. Newbury Park, CA: Sage.

Kaput, J. (1999). Teaching and learning a new algebra. In E. Fennema & T. A. Romberg (Eds.), *Mathematical classrooms that promote understanding* (pp. 133–155). Mahwah, NJ: Lawrence Erlbaum.

Kilpatrick, J. (2003). What works? In S. L. Senk & D. R. Thompson (Eds.), *Standards-based school mathematics curricula: What are they? What do students learn?* (pp. 471–488). Mahwah, NJ: Lawrence Erlbaum.

Lane, S., Silver, E. A., Ankenmann, R. D., Cai, J., Finseth, C., Liu, M., et al. (1995). *QUASAR cognitive assessment instrument (QCAI)*. Pittsburgh, PA: University of Pittsburgh, Learning Research and Development Center.

Lord, F. M. (1980). *Applications of item response theory to practical testing problems*. Mahwah, NJ: Lawrence Erlbaum.

Loveless, T. (2008). *The misplaced math student. The 2008 Brown Center report on American education: How well are American students learning?* Washington, DC: Brookings.

Lubienski, S. T. (2000). Problem solving as a means toward mathematics for all: An exploratory look through a class lens. *Journal for Research in Mathematics Education, 31*, 454–482.

Lubienski, S. T., & Bowen, A. (2000). Who's counting? A survey of mathematics education research 1982–1998. *Journal for Research in Mathematics Education, 31*(5), 626–633.

Lubienski, S. T. (2002). Research, reform, and equity in U.S. mathematics education. *Mathematical Thinking and Learning, 4*, 103–125.

Lubienski, S. T. (2008). On "gap gazing" in mathematics education: The need for gaps analyses. *Journal for Research in Mathematics Education, 39*, 350–356.

Lubienski, S. T., & Crockett, M. (2007). NAEP mathematics achievement and race/ethnicity. In P. Kloosterman & F. Lester (Eds.), *Results from the ninth mathematics assessment of NAEP* (pp. 227–260). Reston, VA: National Council of Teachers of Mathematics.

Lubienski, S. T., & Gutiérrez, R. (2008). Bridging the gaps in perspectives on equity in mathematics education. *Journal for Research in Mathematics Education, 39*, 365–371.

Mathematical Sciences Education Board. (1998). *The nature and role of algebra in the K-14 curriculum: Proceedings of a national symposium*. Washington, DC: National Research Council.

Mayer, R. E. (1987). *Educational psychology: A cognitive approach*. Boston: Little & Brown.

Moschkovich, J. (2002). A situated and sociocultural perspective on bilingual mathematics learners. *Mathematical Thinking and Learning, 4*, 189–212.

Moschkovich, J. (2006). *Statement for the National Mathematics Panel*. Retrieved from http://math.arizona.edu/~cemela/english/content/workingpapers/Moschkovich_MathPanel.pdf

Moses, R., Kamii, M., Swap, S. M., & Howard, J. (1989). The algebra project: Organizing in the spirit of Ella. *Harvard Educational Review, 59*, 423–444.

Moyer, J. C., Cai, J., Nie, B., & Wang, N. (2011). Impact of curriculum reform: Evidence of change in classroom instruction in the United States. *International Journal of Educational Research, 50*(2), 87–99.

Nasir, N. S., & Cobb, P. (2002). Diversity, equity, and mathematical learning. *Mathematical Thinking and Learning, 4*, 91–102.

National Center for Education Statistics (NCES). (2012). *Mathematics 2011: National assessment of education progress at grades 4 and 8*. Washington, DC: Author.

National Council of Teachers of Mathematics. (1989). *Curriculum and evaluation standards for school mathematics*. Reston, VA: Author.

National Council of Teachers of Mathematics. (2000). *Principles and standards for school mathematics*. Reston, VA: Author.

National Mathematics Advisory Panel. (2008). *Foundations for success: The final report of the National Mathematics Advisory Panel*. Washington, DC: U.S. Department of Education.

National Research Council. (2004). *On evaluating curricular effectiveness: Judging the quality of k-12 mathematics evaluations*. Washington, DC: National Academy Press.

National Research Council. (2001). *Adding it up: Helping children learn mathematics*. Washington, DC: National Academy Press.

Pelavin, S. H., & Kane, M. (1990). *Changing the odds: Factors increasing access to college.* New York: College Board.

RAND Mathematics Study Panel. (2003). *Mathematical proficiency for all students: Toward a strategic research and development program in mathematics education.* MR-1643-OERI.

Rivette, K., Grant, Y., Ludema, H., & Rickard, A. (2003). *Connected mathematics project: Research and evaluation summary 2003 edition.* Upper Saddle River, NJ: Pearson Prentice Hall.

Schoenfeld, A. H. (2002). Making mathematics work for all children: Issues of standards, testing, and equity. *Educational Researcher, 31,* 13–25.

Show-Me Center (2002). *CMP implementation map.* Retrieved from http://www.math.msu.edu/cmp/Overview/ImplementationMap.htm

Tarr, J. E., Reys, R. E., Reys, B. J., Chávez, O., Shih, J., & Osterlind, S. J. (2008). The impact of middle-grades mathematics curricula and the classroom learning environment on student achievement. *Journal for Research in Mathematics Education, 39,* 247–280.

Wilson, S. M., & Floden, R. E. (2001). Hedging bets: Standards-based reform in classrooms. In S. Fuhrman (Ed.), *From the capitol to the classroom: Standards-based reform in the states—One hundredth yearbook of the National Society for the Study of Education, Part 2* (pp. 193–216). Chicago: University of Chicago Press.

Exploring the Impact of Knowledge of Multiple Strategies on Students' Learning About Proportions

Rozy Vig, Jon R. Star, Danielle N. Dupuis, Amy E. Lein, and Asha K. Jitendra

Proportional reasoning is widely considered a major goal of mathematics education in the middle grades, where problems involving the use of proportional reasoning are most frequently encountered (Common Core State Standards Initiative, 2010; National Council of Teachers of Mathematics, 2000; National Research Council, 2001). The core of proportional reasoning, which involves multiplicative thinking, is foundational for more advanced mathematics (e.g., algebra, geometry, trigonometry, and calculus) encountered in high school and college (National Mathematics Advisory Panel, 2008). The development of proportional reasoning among students is a complex process that progresses gradually over many years (Lamon, 1999; Lesh, Post, & Behr, 1988). In spite of the centrality and promise of proportional reasoning in the middle grades, students experience great difficulty with this content domain (Lamon, 2007; Lobato, Ellis, & Zbiek, 2010; NRC, 2001). As an illustration consider a simple missing value proportion problem, $2/25 = n/500$. According to the National Assessment of Educational Progress (2009), 52 % of eighth-grade students failed to choose the correct answer of $n = 40$ from among a list of multiple-choice options.

In response to such student difficulties, a great deal of research has explored the teaching and learning of proportions (Behr, Harel, Post, & Lesh, 1992; Boyer, Levine, & Huttenlocher, 2008; Fujimura, 2001; Fuson & Abrahamson, 2005; Lamon, 2007; Lesh et al., 1988; Litwiller & Bright, 2002; Pitta-Pantazi & Christou, 2011; Van Dooren, De Bock, Hessels, Janssens, & Verschaffel, 2005). Most prominently, the Rational Number Project (e.g., Behr et al., 1992; Cramer, Post, & Currier, 1993; Harel & Behr, 1989; Lesh, Behr, & Post, 1987) has exerted a major influence on scholarship, curriculum, and policy around the teaching and learning of fractions,

R. Vig • J.R. Star (✉)
Harvard Graduate School of Education, 442 Gutman Library, 6 Appian Way,
Cambridge, MA 02476, USA
e-mail: jon_star@harvard.edu

D.N. Dupuis • A.E. Lein • A.K. Jitendra
University of Minnesota, Minneapolis, MN, USA

© Springer International Publishing Switzerland 2015
J.A. Middleton et al. (eds.), *Large-Scale Studies in Mathematics Education*,
Research in Mathematics Education, DOI 10.1007/978-3-319-07716-1_4

ratios, and proportions. While the peak of research on rational numbers may have been in the 1980s and early 1990s, work on proportional reasoning continues. More recently scholars have explored teacher knowledge of proportional reasoning (see, for example, Berk, Taber, Gorowara, & Poetzl, 2009), the role of multiple representations and/or technology in supporting students' understanding of proportional reasoning (see, for example, Fujimura, 2001), and the broader application of proportional reasoning to STEM curricula (see, for example, Bakker, Groenveld, Wijers, Akkerman, & Gravemeijer, 2014).

In this chapter, we revisit an issue that first emerged in the work of the Rational Number Project but has not been carefully explored in some time—namely, the strategies that students use when solving simple proportion problems. Our interest is in learning more about how students approach proportion problems, whether these approaches may have changed since this issue was last explored over 20 years ago, and whether strategy use has an impact on students' future learning about proportion.

Theoretical Background

Proportional reasoning refers to the ability to understand (interpret, construct, and use) relationships in which two quantities (ratio or rates) covary and to see how changes in one quantity are multiplicatively related to change in the other quantity. The presence of a multiplicative relationship between quantities and also within quantities is considered a defining feature of a problem that requires proportional reasoning (Behr et al., 1992). Typically, a proportion is defined as a statement of equality between two ratios. An example and commonly seen task relating to proportions in the elementary and middle school mathematics curriculum is to find the value of z that makes a proportion such as $3/9 = 6/z$ a true statement.

Of the many strategies that could be used to solve this kind of proportion problem, three (see Table 1) have been discussed at length in the literature (e.g., Post, Behr, & Lesh, 1988). The first is known as the cross-multiplication strategy (or CM), which involves multiplication across a problem's diagonals. For the problem $3/9 = 6/z$, CM could be used to rewrite the proportion as $3z = 9(6)$, and solve for z to

Table 1 Strategies for solving simple proportion problems

Cross-multiplication strategy	Equivalent fractions strategy	Unit rate strategy
Solve for z: $\dfrac{3}{9} = \dfrac{6}{z}$	Solve for z: $\dfrac{3}{9} = \dfrac{6}{z}$	Solve for z: $\dfrac{3}{9} = \dfrac{6}{z}$
$3 \cdot z = 9 \cdot 6$	$\dfrac{3}{9} \cdot \left[\dfrac{2}{2}\right] = \dfrac{6}{z} z = 18$	$3 \cdot [3] = 9$
$3z = 54$	$9 \cdot [2] = 18$	$6 \cdot [3] = 18$
$z = 54 \div 3$		
$z = 18$	$z = 18$	$z = 18$

yield an answer of $z = 18$. The second strategy is referred to here as the equivalent fractions strategy (or EF); EF involves examining the two ratios in a proportion and using their equivalence to solve for an unknown.[1] For the problem $3/9 = 6/z$, EF could be used to determine 2 as the multiplicative constant needed to arrive at an equivalent fraction; since 3 times 2 is 6, it follows that 9 times 2 is 18, so $z = 18$. Finally, we refer to a third strategy as the unit rate strategy (or UR); UR involves examining the multiplicative relationship *within* the quantity, determining the scalar multiple within a ratio or rate and using it to arrive at the missing value.[2] In the problem $3/9 = 6/z$, one could employ UR by noticing that 3/9 has a scalar multiple of 3, meaning that the denominator is three times as large as the numerator. The value of z can then be determined by multiplying 6 by 3 to arrive at 18. Note that the unit rate strategy (as we define it) does not require the explicit identification of a unit rate (in this case, 1/3)—only that the idea of a unit rate is implicitly used to determine the missing value of the variable.

It is worth noting that, while each of these strategies, if executed correctly, can yield the correct answer, one can argue that certain strategies may be easier than other strategies for particular problems. For example, for the problem $4/5 = 8/x$, EF might be considered easier than UR, since (especially for elementary and middle school students) using the multiplicative relationship between 4 and 8 to determine a solution (4 times 2 is 8, so 5 times 2 is 10; $x = 10$) is easier than using the multiplicative relationship between 4 and 5 (4 times 1.25 is 5, so 8 times 1.25 is 10). Conversely, for the problem $5/15 = 9/x$, UR (5 times 3 is 15, so 9 times 3 is 27; $x = 27$) is arguably easier than EF.

The existing literature on how students approach simple proportion problems such as the ones above suggests that students tend to rely heavily on the cross-multiplication strategy (Cramer & Post, 1993; Stanley, McGowan, & Hull, 2003). The consensus among many mathematics educators is that such a reliance on CM is problematic, primarily because of the belief that students often do not understand what they are doing when they perform the CM algorithm (e.g., Lesh et al., 1988). Furthermore, some have characterized CM as a conceptually opaque or even conceptually vacuous algorithm, in that multiplication across a diagonal is generally not considered a valid mathematical operation, and the algorithm does not make clear why it is permissible to perform this action for CM (e.g., Lesh et al., 1988).

In response to these types of concerns, many mathematics educators and researchers have advocated (1) delaying or even eliminating formal instruction in cross multiplication as a strategy for solving proportion problems and (2) teaching more intuitive strategies for proportion problems first (Cramer & Post, 1993; Ercole, Frantz, & Ashline, 2011; Lesh et al., 1988; Stanley et al., 2003). This de-emphasis on cross multiplication and advocacy of strategies such as EF has been steadily

---

[1] The equivalent fraction strategy is sometimes referred to as the factor-of-change method and involves attending to the multiplicative relationship between two ratios (Ercole et al., 2011).

[2] The unit rate strategy is sometimes referred to as the factor-of-change method (or scalar method) and involves the attending to the multiplicative relationship within each ratio (Ercole et al., 2011).

increasing since the Rational Number Project initially proposed it in the late 1980s. The Rational Number Project reports evidence in support for these recommendations; for example, when instruction is delayed on cross multiplication, students tend to use the unit rate strategy most frequently (Cramer & Post, 1993; Post et al., 1988).

The extent to which the two suggestions above have been implemented into practice is unclear. As noted above, the majority of work that explored students' strategies for solving simple proportion problems occurred in the 1980s and early 1990s, as part of the Rational Number Project. Given the significant changes that have occurred in US elementary and middle school mathematics curricula in the past 20 years, we were interested in revisiting the issue of how students approach these types of problems today. Despite apparent consensus for the two suggestions above, we are not aware of any recent studies that document changes since the 1980s and 1990s in how students approach simple proportion problems. As a result, it is worth noting that we began this study expecting to find that students continue to rely heavily or exclusively on cross multiplication for solving simple proportion problems.

In this chapter, we consider the following questions. First, do students continue to rely upon CM as a primary strategy for solving proportion problems, or have the past decades of de-emphasis of CM and advocacy of strategies such as UR and EF had an impact? Second, if students no longer rely as exclusively on CM, what potential impact might this have on their learning about proportional reasoning more generally? These questions were explored within the context of a larger research project investigating the impact of a curriculum unit on ratio, proportion, and percent word problems on student learning, as described below.

Method

As part of a study evaluating a 6-week curriculum unit on ratio, proportion, and percent problem solving, students were administered a pretest that evaluated their knowledge of strategies for solving simple proportion problems. Elsewhere we report the results of the larger study (Jitendra, Star, Dupuis, & Rodriguez, 2013); here our interest is in the strategy profile of students as demonstrated on the pretest and the relationship between students' strategy profiles and their future learning from the intervention.

Participants

Participants were 430 seventh-grade students drawn from 17 classrooms at three middle schools in two suburban school districts. Of the 430 students, 208 (48 %) were male, 200 (47 %) were eligible to receive free or reduced priced lunch, 37 (9 %) were English language learners, and 50 (12 %) received special education services. There were 216 (50 %) Caucasian students, 124 (29 %) African American students, 57 (13 %) Hispanic students, 23 (5 %) Asian students, and 9 (2 %) American Indian students. The mean student age was 12.5 years (SD=0.4 years). Within the larger study, all 430 students were in classrooms that implemented the intervention.

The two districts used either *Math Thematics Book 2* (Billstein & Williamson, 2008), a reform-oriented curriculum developed with funding from the National Science Foundation or *Math Course 2* (Larson, Boswell, Kanold, and Stiff (2007)), a more "traditional" mathematics curriculum.

Intervention

The complete details of the 6-week intervention are described elsewhere (Jitendra et al., 2013). In brief, the intervention contained 21 scripted lessons where students were introduced to the concepts of ratio, proportion, and percent and were taught strategies for how to solve ratio, proportion, and percent word problems. Each lesson required students to make use of schematic diagrams, multiple solution strategies, and metacognitive strategies. In prior studies, this intervention had been found to be effective (e.g., Jitendra et al., 2009), and the present study was designed to build on and extend existing work on the intervention's efficacy.

Measures

Students completed a 45 min pretest before the intervention and then a 45 min posttest at the conclusion of the intervention. The common questions on the pretest and posttest were taken or adapted from state, national, and international standardized tests and had been used in prior studies investigating the efficacy of the intervention.

The posttest was designed to measure students' learning from the intervention; all problems related to ratio, proportion, and percent problem solving. The posttest contained 3 open-response questions and 21 multiple-choice questions. As an example, one of the posttest multiple-choice problems was, "A machine uses 2.4 L of gasoline for every 30 h of operation. How many liters of gasoline will the machine use in 100 h?" Possible responses were 7.2, 8.0, and 8.4 L.

With respect to the pretest, of interest here are four problems that appeared only on the pretest and were designed to assess students' knowledge (prior to the intervention) for solving simple proportion problems (see Table 2). Within the 45 min pretest, students were given 15 min to complete these four problems. Problems 3 and 4 were considered as "prompted" items, in that students were shown one strategy (CM) and prompted to try to solve the simple proportion problem in a different way. The assumption guiding the inclusion of these two problems was that students knew and would rely upon CM; the items sought to determine whether students knew any other strategies for approaching this type of problem. We predicted that students would have trouble on problems 3 and 4, due to a lack of knowledge of

Table 2 Pretest problems on strategy use

Item	Type
1. Solve for x. Show your work and circle your answer. $$\frac{3}{5} = \frac{x}{15}$$	Unprompted
Describe how you solved the problem in 1–2 sentences	
2. Solve for y. Show your work and circle your answer $$\frac{2}{8} = \frac{3}{y}$$	Unprompted
Describe how you solved the problem in 1–2 sentences	
3. Miguel was asked to solve the problem for x: $$\frac{2}{7} = \frac{x}{21}$$	Prompted
Here is his solution: / Please solve this same problem again but in a different way. Show your work below	
$$\frac{2}{7} = \frac{x}{21}$$ $$7x = 2 \cdot 21$$ $$7x = 42$$ $$\frac{7x}{7} = \frac{42}{7}$$ $$x = 6$$	
Describe how you solved the problem in 1–2 sentences	
4. Ayana was asked to solve the problem for y: $$\frac{5}{20} = \frac{2}{y}$$	Prompted
Here is her solution: / Please solve this same problem again but in a different way. Show your work below	
$$\frac{5}{20} = \frac{2}{y}$$ $$5y = 40$$ $$\frac{5y}{5} = \frac{40}{5}$$ $$y = 8$$	
Describe how you solved the problem in 1–2 sentences	

strategies other than CM. Problem 3 was designed to suggest the use of EF, given the relationship between the two denominators of the problem (7 and 21). Problem 4 was designed to suggest the use of UR, given the relationship between the numerator and denominator of the given ratio (5 and 20).

In contrast, problems 1 and 2 were unprompted in that students could use whatever strategy they wanted for solving the proportion problems. We expected most students to use CM for problems 1 and 2. Problem 1 was designed to suggest the use of EF, while problem 2 was designed to suggest the use of UR. Problems 1 and 2 appeared on a single page of the pretest and problems 3 and 4 were placed on the next page. Students were instructed not to work backwards and to complete the test in the order that the problems were presented.

Strategy Coding

Posttests and the four pretest problems of interest were scored by two independent coders, who met to resolve all disagreements. For the scoring of the four pretest problems, students' strategies were coded based on which of the three strategies described above were used. Students' written mathematical work (in the "show your work below" box of each problem), as well as students' description "in 1–2 sentences" were used in determining a strategy code.

A student's strategy was coded as EF when the student indicated the (horizontal) relationship between the two denominators and used this relationship to find the value of the unknown. For example, for problem 1, $3/5 = x/15$, one student wrote, "5 goes into 15 three times, but I need to times the numerator by 3 too—which is 9". The UR code was given when a student's work indicated awareness of the (vertical) relationship between the numerator and denominator of one of the ratios in the proportion and used this relationship to determine the unknown. For example, for problem 2, $2/8 = 3/y$, one student wrote, "$y = 12$, because you divide $2/8 = 4$ and then you do 3 times $4 = 12$". The CM code was given when a student multiplied across the diagonals of the problem. For example, for problem 1, one student wrote, "Well what I did was multiply 3 times 15 and I got 45 so what I did was times 9 times 5 and I get 45." Arithmetic errors in executing the strategy were not taken into consideration in the strategy coding, as we were primarily interested in capturing student strategy and not the correctness of the solution.

In addition to codes for CM, UR, and EF, we also coded for the presence of common erroneous ways that UR and EF could be applied—when additive rather than multiplicative reasoning was used. For EF, we used the code "mal-EF" to indicate when a student made use of an additive relationship between denominators to determine the unknown. For example, on problem 1, $3/5 = x/15$, one student noted that 15 was 10 more than 5, noting, "I added ten on the bottom. So I added ten on the top." Similarly, we used the code "mal-UR" to note when a student used an additive relationship between numerator and denominator to find the unknown. For example, again on problem 1, one student noticed that 5 was 2 less than 3 and wrote, "I just thought of the pattern and just subtracted 2 from 15, which was 13." We did not use a mal-CM code, as we found no instances in which students applied the CM strategy in an erroneous way.

In addition, an OC or "other correct" code was used to indicate a correct strategy other than CM, UR, or EF. For example, on problem 2, $2/8 = 3/y$, one student multiplied each ratio by the reciprocal of $3/y$ which results in the equivalent equation $2y/24 = 1$. The student went on to explain, "after I multiplied by the reciprocals I got $y = 12$." The code OI or "other incorrect" was used when students had a decipherable strategy that involved steps that were not mathematically permissible. For example, again on problem 2 ($2/8 = 3/y$), one student multiplied the two numerators to arrive at a new numerator (2 times $3 = 6$), then multiplied all three of the given numbers in the problem to arrive at a new denominator (2 times 3 times $8 = 48$), and then reduced the resulting fraction (6/48) to arrive at the missing value ($6/48 = 3/24$ so $y = 24$). Note that the "incorrect" in OI refers to the steps of the strategy, rather than to the correctness of the answer, just as the "correct" in OC refers to the steps of the strategy rather than the correctness of the answer. Finally, we coded as "none" any instances where students arrived at an answer without showing any work or left the problem blank. In addition, for problems 3 and 4 (the prompted items where the problem is solved using CM and students are asked to use a different strategy), students received a code of "none" when they used CM to solve these problems—in essence, copying over the strategy that was already provided in the problem statement.

Results

Due to missing data, below we report the results based on the 423 students who completed the pretest and the 414 students who completed both the pretest and the posttest. We begin by reporting on students' strategy use at pretest. We then examine the relationship between strategy profiles at pretest and students' scores at posttest.

Strategy Use at Pretest

Recall that we expected (based on the literature) that students would rely on CM as their preferred strategy for these problems, and that the pretest problems were designed with the assumption that many students knew CM already. As shown in Table 3, these expectations were completely off base. Only 27 students (6 %) used CM on problems 1 and 2 on the pretest. (Recall that CM was illustrated on the prompted problems 3 and 4 and thus students could not use CM on these problems.) Of these 27 students, only 8 (2 %) used CM for both problems 1 and 2. Students' use of UR was small (12 % of students), but it is noteworthy that there were almost twice as many students who used UR as used CM.

To our surprise, EF was very widely used by students in our sample. Seventy-seven percent of students showed knowledge of the EF strategy. Almost all of these students used EF on both problems 1 and 3 (60 % of the total sample), with a few

Table 3 Pretest strategy use and posttest mean scores

	n	%	M	SD
Used EF	327	77	19.93	5.12
Used UR	52	12	22.27	4.66
Used CM	27	6	18.42	5.54
Did not use EF, UR, or CM	77	19	13.55	5.31
Used exactly one (EF, UR, or CM)	277	67	19.34	5.16
Used multiple strategies (at least two of EF, UR, or CM)	60	15	21.93	4.74

students using EF for only problem 1 (9 % of the total sample) or only problem 3 (8 % of the total sample). Note that problems 1 and 3 were the ones that were designed to be optimal for EF—where the problem numbers made EF easily applicable. Clearly EF was the preferred strategy for most students; furthermore, when the numbers in the problem indicated that EF would be possible, the majority of students consistently used EF.

Students' reliance on EF can also be seen in the prevalence of mal-EF—the strategy where students try to use EF but erroneously reason additively rather than multiplicatively. On problems where the relationship between denominators in the simple proportion problem was obviously multiplicative (such as problems 1 and 3), students overwhelmingly used EF correctly; only 3 students (1 %) used mal-EF on problem 1, for example. But on problems where the relationship between denominators was not overtly multiplicative, many students attempted to determine the additive relationship between denominators to solve for the unknown: 31 % of students used mal-EF on problem 2. Similarly, while only 1 student used mal-EF on problem 3, 36 students (9 %) used mal-EF on problem 4.

Students' interest in using EF whenever possible (even if this meant using a mal-adaptive version of EF, mal-EF) is further illustrated by examining all students who used EF on at least one problem, to see which of these students also used mal-EF on at least one problem. Almost half of EF users (43 %) used mal-EF on at least one problem. In addition, recall that our assessment was also designed to examine students' knowledge of multiple strategies, but the predominance of EF was the clear take-away. Most students (66 %) only used one strategy (of the three strategies of interest here—CM, UR, and EF) on the four pretest questions. For almost all (63 % of all students, or 95 % of one-strategy students) of these students, this one strategy was EF.

An additional goal of the pretest was to explore students' knowledge of which strategies were most appropriate for a given problem. As noted above, some problems were designed to potentially elicit EF while others hoped to elicit UR. Our original aim was to determine not only whether students knew strategies other than CM but also whether they were able to select the most appropriate strategy for a given problem. Because students rarely used CM, and because EF was so widely used, it was no longer of interest (or even feasible) to look for which students knew the most appropriate strategy for a given problem.

Relationship Between Strategy Profile and Posttest Performance

In addition to exploring students' strategy profiles on the pretest, a second focus of the present analysis is the nature of the relationship between students' strategies at pretest and their performance on the posttest. As such, a series of independent-samples *t*-tests and a one-way ANOVA were conducted. There are three main findings.

First, students with knowledge of at least one strategy (EF, UR, or CM) on the pretest scored higher on the posttest than students who did not exhibit knowledge of EF, UR, or CM at pretest, $t(412) = 9.52$, $p < .001$. Given the predominance of EF, one might interpret this result as suggesting that students who knew EF outscored those who did not know EF. A direct examination of this possibility indicated that it was indeed the case: Students who knew EF scored higher than those who did not know EF, $t(335) = 2.03$, $p = .043$.

Second, although only a few students used CM on the pretest, these students performed *no worse* on the posttest than those students who knew EF. There was no difference between posttest scores of students who knew CM ($M = 18.42$) and those who did not use CM but did use EF and/or UR ($M = 19.92$), $t(335) = 1.42$, $p = .157$. Not only did the literature's prediction about students' overreliance on CM not hold in our sample, but those students who did use CM learned as much as those who did not use CM. It is also interesting to note that students who used UR did better on the posttest ($M = 22.27$) than those who did not use UR but did use EF and/or CM ($M = 19.36$), $t(335) = 3.77$, $p < .001$.

Finally, students who used more than one strategy on the pretest (typically, EF plus one other strategy) outperformed students who only knew one strategy, who in turn scored higher than those who did not use any strategies on the pretest (see Table 3), $F(2, 411) = 52.89$, $p < .001$. Although only a few students used more than one strategy on the pretest (15 %), these students did quite well on the posttest.

Discussion

Our aims in this study were to explore students' strategies for solving simple proportion problems and to determine whether and how knowledge of one or more strategies impacted students' learning from our intervention. There were four main results. First and surprisingly, students relied quite heavily at pretest on EF. Our review of the literature suggested that either CM would be used/known by most students, or that (when instruction on CM was delayed) UR would be the most common strategy (Cramer & Post, 1993; Post et al., 1988). However, the majority of students in our study either knew only EF or knew EF in addition to one or more other strategies.

To better understand why so many students in this district were using EF, we informally talked to teachers and also examined the math texts that were in use at the elementary and middle schools in the district. Although (judging from the textbooks and teachers' reports) students had not received any prior instruction in how to solve simple proportion problems, we found that the text's treatment of equivalent fractions in fourth grade provided the foundations for the EF strategy. Our results suggest that many students were able to recall their work with equivalent fractions in fourth grade as they attempted to solve unfamiliar simple proportion problems in our study in the seventh grade. Furthermore and somewhat anecdotally, these districts were geographically relatively close to the University of Minnesota, home of several key members of the Rational Number Project, and apparently received professional development for many years that was consistent with the Rational Number Project suggestions about delaying formal instruction on CM. Regardless of the reasons, we find it noteworthy that claims made in the past about students' overreliance on cross multiplication may now (in some districts) be a bit dated. Perhaps due to the greater diversity and types of curricula in use in elementary schools, EF now appears to be the strategy of choice, at least for students in the districts that were included in the present study.

Second, while EF was the preferred strategy for students at pretest, results indicate that the widespread use of EF brought its own set of challenges. A central concern noted in the literature about CM is that students often do not know conceptually what they are doing and thus seem to be blindly following the CM algorithm. Another related concern is that CM fails to emphasize the proportionality that is central to thinking about and solving simple proportion problems. Many scholars view EF as improving on both of these concerns: EF may be better connected to conceptual knowledge (related to fractions and ratios), and EF appears to foreground proportionality. However, our results suggest that, for many students, EF brings challenges of its own. In particular, many students in our sample overgeneralized EF—in the interest of applying EF as often as possible, many students erroneously used EF additively rather than multiplicatively. It is certainly encouraging that (a) these students seemed to spontaneously apply a strategy that they learned for working with equivalent fractions to proportion problems and (b) these students appear to see the similarities between proportion problems and equivalent fractions. However, the frequency of overgeneralization—where students attempted to apply EF where the problem numbers made it difficult to do so, and then erroneously modified EF so that was applied additively, is problematic.

Third, the results of the current study show that prior knowledge of one or more solution methods can have a positive impact on students' ability to learn from an instructional intervention for proportional reasoning. This result is consistent with a growing body of research in mathematics education and psychology that suggests that students' learning is enhanced when they have the opportunity to learn multiple methods and compare and contrast them (e.g., Rittle-Johnson & Star, 2007). Finally, students who used CM performed no worse on the posttest than those who did not use CM but did use EF and/or UR.

Taken as a whole, these results suggest that much has changed in the many years since the Rational Number Project began investigating students' strategies for simple proportion problems. If the two districts in the present study are indicative of national trends, we do not see the same reliance on cross multiplication as earlier studies might have predicted—equivalent fractions was clearly the strategy of choice. While some mathematics educators might find the prevalence of EF to be an encouraging sign, it is also the case that students' difficulties with solving simple proportion problems persist. Clearly more work is needed to better understand the nature of students' difficulties with solving simple proportion problems—decreasing reliance on cross multiplication as a default strategy may not have been sufficient to significantly advance student understanding of this important mathematical topic.

References

Bakker, A., Groenveld, D., Wijers, M., Akkerman, S., & Gravemeijer, K. (2014). Proportional reasoning in the laboratory: An intervention study in vocational education. *Educational Studies in Mathematics, 86*(2), 211–221.

Behr, M., Harel, G., Post, T., & Lesh, R. (1992). Rational number, ratio and proportion. In D. Grouws (Ed.), *Handbook of research on mathematics teaching and learning* (pp. 296–333). New York: Macmillan.

Berk, D., Taber, S. B., Gorowara, C., & Poetzl, C. (2009). Developing prospective elementary teachers' flexibility in the domain of proportional reasoning. *Mathematical Thinking and Learning, 11*(3), 113–135.

Billstein, R., & Williamson, J. (2008). *Math thematics: Book 2* (new edition). Evanston, IL: McDougal Littell.

Boyer, T., Levine, S. C., & Huttenlocher, J. (2008). Development of proportional reasoning: Where young children go wrong. *Developmental Psychology, 44*(5), 1478–1490.

Common Core State Standards Initiative. (2010). *Common Core State Standards for Mathematics*. Retrieved from http://www.corestandards.org/assets/CCSSI_Math%20Standards.pdf

Cramer, K., & Post, T. (1993). Connecting research to teaching proportional reasoning. *Mathematics Teacher, 86*(5), 404–407.

Cramer, K., Post, T., & Currier, S. (1993). Learning and teaching ratio and proportion: Research implications. In D. Owens (Ed.), *Research ideas for the classroom* (pp. 159–178). New York: Macmillan.

Ercole, L. K., Frantz, M., & Ashline, G. (2011). Multiple ways to solve proportions. *Mathematics Teaching in the Middle School, 16*(8), 482–490.

Fujimura, N. (2001). Facilitating children's proportional reasoning: A model of reasoning processes and effects of intervention on strategy change. *Journal of Educational Psychology, 93*(3), 589–603.

Fuson, K. C., & Abrahamson, D. (2005). Understanding ratio and proportion as an example of the apprehending zone and conceptual-phase problem-solving models. In J. Campbell (Ed.), *Handbook of mathematical cognition* (pp. 213–234). New York: Psychology Press.

Harel, G., & Behr, M. (1989). Structure and hierarchy of missing value proportion problems and their representations. *Journal of Mathematical Behavior, 8*(1), 77–119.

Jitendra, A. K., Star, J. R., Dupuis, D., & Rodriguez, M. (2013). Effectiveness of schema-based instruction for improving seventh-grade students' proportional reasoning: A randomized experiment. *Journal of Research on Educational Effectiveness, 6*(2), 114–136.

Jitendra, A., Star, J. R., Starosta, K., Leh, J., Sood, S., Caskie, G., et al. (2009). Improving seventh grade students' learning of ratio and proportion: The role of schema-based instruction and self-monitoring. *Contemporary Educational Psychology, 34*(9), 250–264.

Lamon, S. J. (1999). *Teaching fractions and ratios for understanding*. Hillsdale, NJ: Lawrence Erlbaum.

Lamon, S. J. (2007). Rational numbers and proportional reasoning: Towards a theoretical framework for research. In F. Lester (Ed.), *Second handbook of research on mathematics teaching and learning* (pp. 629–666). Reston, VA: National Council of Teachers of Mathematics.

Larson, R., Boswell, L., Kanold, T. D., & Stiff, L. (2007). *Math course 2*. Evanston, IL: McDougal Littell.

Lesh, R., Behr, M., & Post, T. (1987). Rational number relations and proportions. In C. Janiver (Ed.), *Problems of representations in the teaching and learning of mathematics* (pp. 41–58). Hillsdale, NJ: Lawrence Erlbaum.

Lesh, R., Post, T., & Behr, M. (1988). Proportional reasoning. In J. Hiebert & M. Behr (Eds.), *Number concepts and operations in the middle grades* (pp. 93–118). Reston, VA: National Council of Teachers of Mathematics.

Litwiller, B., & Bright, G. (2002). *Making sense of fractions, ratios, and proportions*. Reston, VA: National Council of Teachers of Mathematics.

Lobato, J., Ellis, A., & Zbiek, R. (2010). *Developing essential understanding of ratios, proportions, and proportional reasoning for teaching mathematics: Grades 6-8*. Reston, VA: National Council of Teachers of Mathematics.

National Assessment of Educational Progress. (2009). U.S. Department of Education, Institute of Education Sciences, National Center for Education Statistics, Mathematics Assessment. Retrieved from http://nces.ed.gov/nationsreportcard/itmrlsx/search.aspx?subject=mathematics

National Council of Teachers of Mathematics. (2000). *Principles and standards for school mathematics*. Reston, VA: Author.

National Mathematics Advisory Panel. (2008). *Foundations for success: The final report of the National Mathematics Advisory Panel*. Washington, DC: U.S. Department of Education.

National Research Council. (2001). Adding it up: Helping children learn mathematics. In J. Kilpatrick, J. Swafford, & B. Findell (Eds.), *Mathematics learning study committee, center for education, division of behavioral and social sciences and education*. Washington, DC: National Academy Press.

Pitta-Pantazi, D., & Christou, C. (2011). The structure of prospective kindergarten teachers' proportional reasoning. *Journal of Mathematics Teacher Education, 14*(2), 149–169.

Post, T., Behr, M., & Lesh, R. (1988). Proportionality and the development of pre-algebra understandings. In A. Coxford & A. Shulte (Eds.), *The idea of algebra, K-12* (pp. 78–90). Reston, VA: National Council of Teachers of Mathematics.

Rittle-Johnson, B., & Star, J. R. (2007). Does comparing solution methods facilitate conceptual and procedural knowledge? An experimental study on learning to solve equations. *Journal of Educational Psychology, 99*(3), 561–574.

Stanley, D., McGowan, D., & Hull, S. H. (2003). Pitfalls of overreliance on cross multiplication as a method to find missing values. *Texas Mathematics Teacher, 11*, 9–11.

Van Dooren, W., De Bock, D., Hessels, A., Janssens, D., & Verschaffel, L. (2005). Not everything is proportional: Effects of age and problem type on propensities for overgeneralization. *Cognition and Instruction, 23*(1), 57–86.

Challenges in Conducting Large-Scale Studies of Curricular Effectiveness: Data Collection and Analyses in the COSMIC Project

James E. Tarr and Victor Soria

The COSMIC Project

Funded by the National Science Foundation, Comparing Options in Secondary Mathematics: Investigating Curriculum (COSMIC) is a 3-year longitudinal comparative study of integrated curricula and subject-specific curricula on mathematical learning in high schools that offer parallel curricular paths. In the subject-specific path, students experience a traditional course sequence, beginning with Algebra 1, followed by Geometry, and a course in Algebra 2. In the integrated path, students experience multiple branches of mathematics (algebra, geometry, statistics) within yearlong courses, Integrated I, Integrated II, and Integrated III. The dual path requirement ensured representation of both curriculum types and controlled for the number of instructional days, length of class periods, professional development opportunities, and so on. In all participating schools, the integrated curriculum was the Core-Plus Mathematics Program, an integrated comprehensive program with

The National Science Foundation under Grant No. (REC-0532214) supported the research reported in this chapter. The study was conducted as part of the Comparing Options in Secondary Mathematics: Investigating Curriculum (COSMIC) project, http://cosmic.missouri.edu. Any opinions, findings, and conclusions or recommendations expressed in this chapter are those of the authors and do not necessarily reflect the views of the National Science Foundation. Some of the results discussed herein were presented at the 2010, 2012, and 2013 annual meetings of the American Educational Research Foundation and/or published previously (e.g., Chávez, Tarr, Grouws, & Soria, 2015; Grouws et al., 2013; Tarr, Grouws, Chávez, & Soria, 2013). The authors wish to thank Robert Ankenmann, Angela Bowzer, Oscar Chavez, Dean Frerichs, Douglas A. Grouws, Melissa McNaught, Ira Papick, Greg Petroski, Robert Reys, Daniel Ross, Ruthmae Sears, and R. Didem Taylan for their assistance during various stages of the research.

J.E. Tarr (✉) • V. Soria
Department of Learning, Teaching & Curriculum, University of Missouri,
303 Townsend Hall, Columbia, MO 65211-2400, USA
e-mail: TarrJ@missouri.edu

© Springer International Publishing Switzerland 2015
J.A. Middleton et al. (eds.), *Large-Scale Studies in Mathematics Education*,
Research in Mathematics Education, DOI 10.1007/978-3-319-07716-1_5

the greatest market share. The subject-specific curriculum option included popular textbooks produced by several different publishers (e.g., Glencoe, McDougal Littell, Prentice Hall) that were remarkably similar in topic coverage and lesson structure.

The primary goal of the COSMIC project is to investigate whether there are differential curricular effects on secondary school students' mathematics learning using multiple measures of student achievement: two project-developed tests and one standardized measure. More specifically, based on content analyses of both curricula, we developed a *fair test* of common objectives and a test of *mathematical reasoning and problem solving*, each primarily comprised of constructed-response, rubric-scored items. Employing two-parameter Item Response Theory (IRT) using *item difficulty* and *item discrimination* indices, we generated scale scores for each student on each project-developed test.[1] We also assessed student learning using a *standardized measure*, the Iowa Test of Educational Development: Mathematical Concepts and Problem Solving, a 40-item multiple-choice test. To draw causal inferences between curricular programs and student outcomes, we gauged the fidelity of implementation of curricular materials using several techniques, including classroom observations, opportunity to learn (OTL) data, and teacher surveys.[2]

Our research design was quasi-experimental because neither teachers nor students were randomly assigned curriculum. Nonetheless, schools asserted that students were not tracked into one particular path, integrated or subject-specific, based on prior achievement, gender, race/ethnicity, or other student characteristics. The study included over 4,600 students taught by 135 teachers within 15 schools in five states over 3 years. Given the inherent nested data structure (students within classrooms, classrooms within schools), we constructed three-level hierarchical linear models (HLM) of three distinct measures of student learning.

Results of the COSMIC Project

The COSMIC project yielded results of cross-sectional and longitudinal analyses, reported previously but briefly summarized here. In cross-sectional analyses of student learning outcomes in Algebra 1 or Integrated I, after controlling for a variety of student-level characteristics and teacher-level factors, Grouws et al. (2013) reported a differential curricular effect in favor of the integrated program. Specifically, students in the integrated program scored significantly higher than students in the subject-specific program on all three outcome measures: the *fair test* of common objectives (Test A), assessment of mathematical reasoning and problem solving (Test B), and on the standardized measured—Iowa Tests of Educational Development Problem Solving and Concepts, Level 15 (ITED-15). Additionally, teacher

---

[1] For a robust description of the test development process, see Chávez, Papick, Ross, and Grouws (2011).

[2] For detailed descriptions of the conceptualization and development of multiple measures of implementation fidelity, see Tarr, McNaught, and Grouws (2012).

experience and OTL were significant predictors of student outcomes on Test A and the ITED-15. Student demographic data (e.g., gender, race/ethnicity, Individualized Educational Program [IEP]) added precisions to the models but were not a central focus of our study. Interestingly, there was a significant cross-level interaction on Test A and Test B—on both assessments, students with higher prior mathematics achievement benefited more from the integrated program than did students from the subject-specific program.

In Year 2 cross-sectional analyses of student learning in Geometry or Integrated II, Tarr, Grouws, Chávez, and Soria (2013) detected a differential effect in favor of the integrated program but on the standardized measure (ITED-16) only; curriculum type was not a significant predictor of student achievement on the fair test of common objectives (Test C) or the assessment of mathematical reasoning and problem solving (Test D). Teacher factor scores for OTL were a significant predictor of student scores on all three outcome measures, and factor scores measuring the Classroom Learning Environment (CLE) were significantly associated with higher performance on Test C and Test D. Similar to Grouws et al. (2013), Tarr et al. detected a significant cross-level interaction on Test C—students with higher prior achievement were better served by the integrated program than the subject-specific program.

In Year 3 cross-sectional analyses of student learning in Algebra 2 or Integrated III, Chávez, Tarr, Grouws, and Soria (2015) reported a differential curricular effect in favor of the integrated program on the fair test of common objectives (Test E) only. Across curricular programs, students scored comparably on the standardized measure (ITED-17). No test of mathematical reasoning and problem solving was administered in Year 3. In addition to curriculum type, several teacher-level variables were significant predictors of student achievement on Test E. For example, Orientation, a measure of teachers' beliefs about reform-oriented practices had a significant positive effect; that is, students whose teachers espoused practices commonly associated with the NCTM *Standards* scored significantly higher on Test E. Surprisingly, professional development in the past 12 months was significantly associated with lower scores on Test E, as was the class-level mean percent of students eligible for free and reduced lunch (%FRL). Significant effects of Orientation and %FRL were similarly found in analyses of ITED-17 scores.

In a longitudinal analysis of student scores on the ITED across Years 1–3, Tarr, Harwell, Grouws, Chávez, and Soria (2013) modeled the intercept (i.e., mean ITED-15 scale scores) and linear slopes (i.e., growth rate of ITED scale scores across years). In the intercept model, curriculum type was a statistically significant predictor of ITED-15 scores, with students in an integrated curriculum scoring more than 8 points higher than those in a subject-specific curriculum. Overall, IEP was the largest fixed effect indicating that, with other predictors held constant, students with an IEP program scored on average 13 points lower on the ITED-15 compared to students not having an IEP. Prior mathematics achievement was likewise a significant predictor of ITED-15 scores, as was the case for OTL, with classrooms with greater opportunities for students to learn associated with higher ITED-15 scores. Finally, Minority students on average scored more than 7 points lower on the ITED-15 than White students. None of the predictors were statistically significant moderators of linear growth rates. Stated alternatively, students in the integrated curriculum scored significantly higher on the ITED-15 and experienced similar growth rates across the years, thereby maintaining their advantage over students in the subject-specific curriculum after 3 years of high school mathematics.

This brief summary of results obscures many of the issues and complexities encountered by the COSMIC research team during data collection and analyses. In the following sections, we discuss key challenges related to our collection and analyses of student data, teacher data, and the construction of multilevel models of student outcome measures.

Issues Related to Collection and Analysis of Student Data

To add precision in the detection of curricular effects, our models included teacher-level and student-level variables that were hypothesized to explain variation in student outcomes. In particular, based on previous investigations of curricular effectiveness (e.g., Harwell et al., 2007; Post et al., 2008), we expected students' prior mathematics achievement to be strongly associated with student performance. However, the lack of a common measure of prior achievement presented method-ological challenges. Additionally, several issues arose with respect to composition of the student sample including (a) students missing a prior mathematics achieve-ment score, (b) the absence of free-and-reduced lunch (FRL)-status at the student level, (c) students who migrated across curricular paths, and (d) students who did not participate in all 3 years of the study. In this section, we discuss the nature of these problems, how they were ultimately resolved, and justify our decisions.

Lack of a Common Measure of Prior Achievement

Our quasi-experimental design necessitated that comparability be established by matching samples or making statistical adjustments using, among other factors, prior achievement measures. However, a pre-test administered to all students is rarely feasible in large-scale studies of curricular effectiveness across multiple states such as ours and, even it was feasible, there were several concerns about it including student motivation, asking for additional testing dates, and purchasing and scoring the exams. Consequently, we opted for a reasonable alternative, namely the utiliza-tion of scores on state-mandated grade 8 tests, typically administered in the spring of the academic year immediately preceding Year 1 of the study. These high-stakes tests generally purport to measure student achievement in mathematics at a common point in time (grade 8), and so they provided useful information in characterizing student knowledge prior to curricular treatments in the COSMIC project. Nevertheless, state tests are usually not nationally normed and are scored using dif-ferent scales. Moreover, because participating school districts were located in five US states, it was important to acknowledge and subsequently adjust for differences in student achievement *across* states, as average National Assessment of Educational Progress (NAEP) scores vary considerably across states. Consider results of the 2011 NAEP: An "average" grade-8 mathematics score of 299 in Massachusetts, the highest performing US state, is comparable to the 75th percentile in New Mexico which ranks 44th in state NAEP, while an "average" grade-8 mathematics score in New Mexico is comparable to the 25th percentile in Massachusetts. Therefore, efforts to map students' prior achievement onto a common scale needed to account for achievement patterns across US states in our sample.

For the vast majority of students in COSMIC, grade 8 scores on state-mandated tests were not nationally normed. In these cases, we converted students' scores in each state to *z*-scores before mapping these scores onto an NAEP scale score (see National Center for Education Statistics, 2007). We called the resulting score the COSMIC Prior Achievement (CPA) Score. Figure 1 depicts the process of transforming student prior achievement data into CPA Scores, and we offer the following illustrative

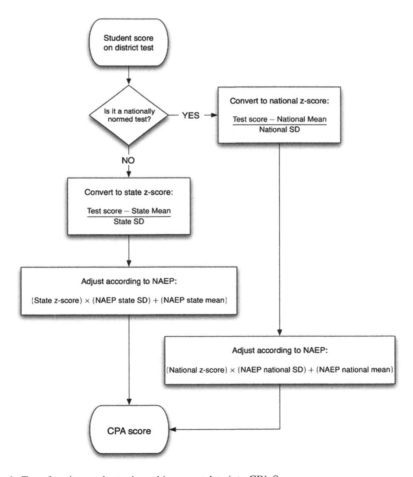

Fig. 1 Transforming student prior achievement data into CPA Scores

contrasting examples. Consider the first example in which, in grade 8, some students in District B were assessed using a nationally normed mathematics achievement test. In these cases, we simply converted their scores to a national *z*-score, which we then mapped onto the NAEP scale for grade 8. Therefore, a grade 8 student in State X scoring at the mean ($z=0$) was assigned to the mean NAEP scale score for State X while a student scoring 1 standard deviation above the mean was assigned a NAEP scale score that corresponded to the mean NAEP scale score plus 1 standard deviation.

Consider a second illustrative example in which Student A has a scale score of 709 on the 2005 grade 8 test mandated in State Y (Fig. 2). Because the assessment for State Y is *not* a nationally normed test, we converted this student's scale score to

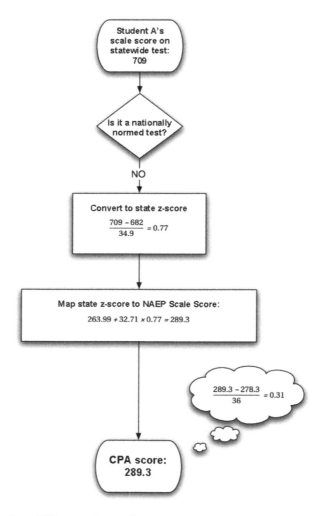

Fig. 2 Generation of CPA score for sample student

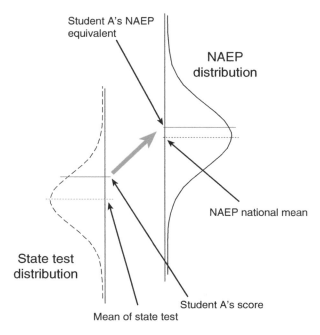

Fig. 3 Mapping Student A's state scale score onto the NAEP scale

a state z-score using descriptive statistics for the 2005 State Y test: a mean score of 682 and a standard deviation of 35. As depicted in Fig. 3, this state z-score was then mapped onto the NAEP Scale Score: State Y had an average NAEP scale score of 264 and a standard deviation of 35, yielding a CPA score of 289. Thus, although Student A scores 0.77 standard deviations above the mean relative to grade 8 students in State Y, Student A scored approximately 0.28 standard deviations above the mean relative to grade 8 students in the USA (see Fig. 3).

The resulting mapping student scores onto the CPA scale yielded an approximately normal distribution, and this property was characteristic of student samples in all 3 years. In subsequent analyses, CPA was used as a student-level variable and yielded the greatest effect sizes in cross-sectional analyses.

Composition of Student Sample

Students Missing Prior Achievement Measure

With one exception, all school districts were cooperative in supplying student records, including scores on state-mandated testing programs (subsequently converted to CPA scores) and demographic data including gender, race/ethnicity, and whether the student qualified for Free/Reduced Lunch or special services (IEP).

However, District R was surprisingly non-compliant in sharing prior achievement data despite having pledged to do so in principled (IRB) agreements. In said district, an administrator did not respond to repeated requests for student records over the span of several months. When the school district official finally responded to our request, he offered promises that the data would be forthcoming but did not follow through. When pressed, the administrator indicated that student achievement scores were difficult to compile or were incomplete due to the transient nature of its student population. However, the standoff ended only after we expressed our plans to send a member of our project team to work side-by-side with his staff to scrounge student records. A member of our project team was not dispatched to collect student records because the district finally acquiesced to our request and provided prior achievement scores for almost 75 % of students; about 25 % of students were missing this key indicator of prior knowledge.

The absence of a prior achievement measure for District R and, to a much lesser extent, in other school districts, called into question how to handle such student cases in cross-sectional and longitudinal analyses. Should we impute missing prior achievement scores or exclude these student cases? Student cases without a prior achievement measure could be discarded without introducing bias provided they are *missing completely at random*. Similarly, such cases *missing at random* can be excluded provided the regression controls for the variables that affect the probability of "missingness." Some argue that it is generally impossible to prove that data are missing at random because "we cannot be sure whether data are missing at random, or whether the missingness depends on unobserved predictors or the missing data themselves" (Gelman & Hill, 2006, p. 531). Although there are many available techniques for imputing missing data (e.g., mean imputation), we ultimately excluded student cases in which no prior achievement measure was available. Because of the critical importance of controlling for differences (Frank, 2000; What Works Clearinghouse, 2008) in mathematics proficiency, students who did not have a measure of prior mathematics achievement were excluded from the sample. The critical role of prior achievement as a covariate was evident in results of cross-sectional analyses. Specifically, CPA was the strongest predictor of student outcomes with effect sizes ranging from 0.53 (Test B) to 0.59 (ITED-15) in Year 1, from 0.53 (ITED-16) to 0.59 (Test D) in Year 2, and from 0.51 (ITED-17) to 0.53 (Test E) in Year 3. In principle, given the strong predictive power of students' prior achievement, we could have imputed CPA scores using scores on multiple measures of student learning but doing so would have violated fundamental assumptions upon which our models depended. Accordingly, we instead justified the missing-at-random assumption by including as many student-level predictors in our model, including gender, race/ethnicity, IEP-status, among others, thereby reducing the existence of possible "lurking variables" that might be attributed to missingness. Interestingly, District R dropped the integrated program after Year 2. Because no student from District R completed 3 years of the integrated program, they were excluded from longitudinal sample, thereby lessening the problem of missing prior achievement scores.

Missing FRL-Status at the Student Level

In our original research design, the effects of socioeconomic status (SES) were to be modeled at the student level using FRL-status as a proxy for SES. However, citing privacy concerns, District W was reluctant to provide information regarding Free/Reduced Lunch (FRL)-status at the individual student level despite our assurances (and IRB consensual agreements) that all project data would remain confidential and secure. Notwithstanding the concerns regarding student confidentiality, District W willingly provided FRL information but at the class level, not student level. For example, for Teacher A, district administrators indicated that 7 of 29 students in second period Algebra 1 qualified for FRL as did 9 of 24 students in third period Algebra 1. In the absence of such information at the student level, we were faced with a decision regarding how to model the effects of FRL. Given that more than 80 % of our student sample was not missing FRL-status, one idea was to randomly assign FRL-status to students in each class. Thus, in the case of Teacher A, 7 of 29 students would be selected at random and coded as qualifying for FRL. Alternatively, we entertained the notion of using other demographic data and logistic regression to predict which 7 of Teacher A's 29 students qualified for FRL. Because of the error likely introduced, we rejected both of these ideas and instead we opted to model FRL at the class level. Thus, for each teacher in the study, we computed the percentage of students qualifying for FRL and used the measure as a teacher-level variable; in doing so, we eliminated FRL as a student-level variable. Additionally, we modeled the effects of FRL at the school level as well.

Our decision regarding FRL data yielded some interesting findings. Most notably, in the absence of FRL at the student level, the effects of race/ethnicity might have been *overestimated*; similarly, the effects of FRL at the teacher level might have been *underestimated*. In particular, in cross-sectional analyses, we detected achievement gaps between African American and White students as well as between Hispanic and White students. However, at the teacher level, %FRL was largely not a significant predictor of student outcomes. Because some student groups (especially African American, American Indian/Native American, and Hispanic) are significantly more likely to qualify for the FRL program (National Center for Education Statistics, 2010), it is possible that race was confounded with SES at the student level. If so, then the true predictive value of FRL would be greatly diminished at the teacher- and school levels, and this appears to be the case in our study—after controlling for race/ethnicity at the student level, FRL did not yield significant predictive power at the teacher- or school level.

Students Who Migrated Across Curricular Paths

The offering of dual (parallel) curricular paths—integrated and subject specific—within participating high schools was inarguably a key strength of our research design. However, some school districts offered dual curricular options only as a compromise, to assuage various stakeholders including students, teachers, and

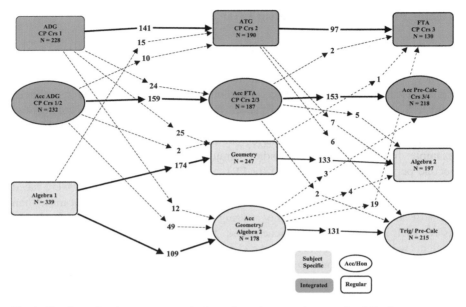

Fig. 4 The flow of students across curricular paths and courses for one school district

parents. Given the availability of distinct curricular options, it was expected that, once a student selected a particular curriculum type, he would continue in the same pathway for the three-course sequence. In other words, we expected students who began in Integrated I in Year 1 of the study would progress into Integrated II and Integrated III in Years 2 and 3, respectively. Although most students "stuck" with their initial selection of curriculum type, there was some migration across 3 years of high school mathematics, and this was most prominent in District T.

As depicted in Fig. 4, 61 of 460 students switched out of the integrated pathway after Year 1, enrolling in Geometry instead of Integrated II in Year 2. Similar migration was evident as 15 of 339 Algebra 1 students changed pathways after Year 1, enrolling in Integrated II in Year 2 instead of Geometry. Further migration continued into Year 3 as 15 of 377 students who completed Integrated II enrolled in Algebra 2 or Precalculus while 4 of 325 moved from the subject-specific option into Integrated III in Year 3. The reasons behind such patterns in enrollment were beyond the scope of our study and therefore not examined. Although curriculum "migrants" were included in cross-sectional analyses, these cases were excluded from our longitudinal analyses because they did not complete a 3-year sequence of mathematics courses in a single curriculum path. We considered students who migrated across paths to be "hybrid" cases, essentially cross-pollinated by two curricular options that were the central focus of our longitudinal study. Although curricular migration was most evident in District T, it was relatively non-prevalent overall as fewer than 2.67 % of students crossed pathways.

Students Who Participated Fewer than 3 Years

For the longitudinal study, our target population was students who completed 3 years of an integrated or subject-specific mathematics program. In most school districts, students moved through the study as a single cohort, completing all 3 years of mathematics in the same high school, typically beginning as grade 9 students enrolled in Integrated I or Algebra 1. In a few school districts, grade 9 students were housed in junior high (or middle) school buildings and subsequently changed buildings when they entered high school in Year 2 of the study. In one large school district (District W), at least some Year 1 participants studied in high schools that were not a part of the study in Year 2. Moreover, at least some students in Year 2 were previously enrolled in a junior high school that was not a part of the study in Year 1.

Because of these circumstances, not every student participated in all 3 years of the study, as shown in Table 1. Specifically, the nature of the Year 1 sample is represented by the first 4 rows of the table: Students participating in Years 1, 2, and 3 (row 1), Year 1 only (row 2), Years 1 and 2 only (row 3), and Years 1 and 3 only (row 4). The composition of student samples of Years 2 and 3 is similarly discerned by examining the entries for the second and third columns, respectively: Students who participated in Years 2 and 3 only (row 5), Year 2 only (row 6), and Year 3 only (row 7). It is worth noting that there are a variety of reasons why students provided only 1 year of outcome data. Students may have failed the course, enrolled in another school within the district that did not participate in the study, moved out of the school district completely, or been absent on testing dates in subsequent years, to name a few.

For cross-sectional analyses, all students who provided prior achievement data (CPA) and data on outcome measures were included in the sample regardless of whether they continued in the study. However, the longitudinal sample required that students complete 3 years of high school mathematics, as represented by rows 1, 4, 5, and 7 of Table 1. Whereas most students in the longitudinal study are represented by row 1, we decided to explore whether or not it made a difference to include students who provided fewer than 3 years of data. To that end, we coded students who did not provide 3 years of data as Missing (i.e., missing outcome scores in at least 1 year). The inclusion of students missing at least one ITED score increased the sample size but had essentially no effect on the intercept and linear slopes in our models. Specifically, in the longitudinal model Missing was not a significant predictor of

Table 1 Composition of the student sample in cross-sectional studies

Year 1	Year 2	Year 3
X	X	X
X		
X	X	
X		X
	X	X
	X	
		X

student achievement—the magnitude of t-values was -0.28 for intercept and 0.19 on linear slope related to time. Because our analyses yielded essentially the same results *with* and *without* the Missing predictor, we decided to include all students who completed their third year of high school mathematics, Integrated III or Algebra 2.

Issues Related to Collection and Analysis of Teacher Data

Our research design necessitated the documentation of curriculum implementation and the classroom learning environment. To that end, the COSMIC project team made a total of 326 classroom visits to 109 teachers during the first 2 years of data collection. Given the composition of our sample in most cases, classroom observations necessitated travel outside of the state of Missouri by 3–6 members of the research team. Although there were many merits in conducting classroom visits, these travel excursions were nonetheless pricy because they included the costs of airfare, parking, car rental, lodging, and per diem. Despite efforts to minimize travel expenses, the COSMIC project did not have sufficient funds to conduct classroom observations in Year 3, thereby relegating us to rely on surveys of teachers' curriculum use, demographics, and beliefs about teaching and learning mathematics. As a research team, we were genuinely concerned about the loss of observational data, particularly given the significant predictive power of Classroom Learning Environment (CLE) factor scores on Test C and Test D. Teacher scores on three CLE subscales—Focus on Sense Making (.880), Reasoning About Mathematics (.855), and Students' Thinking in Instruction (.835)—loaded heaviest on the CLE factor, and these scores were generated from classroom observations. Without observational data, how could measure the unique contributions of *curricular effects* and the effects of *classroom instruction*?

To address the lack of observational data in Year 3, we opted for a measure of teacher beliefs about reform-oriented teaching practices. Data on teacher beliefs were collected using 32 five-point Likert scale items on the Initial Teacher Survey, administered to all teachers across 3 years of the COSMIC project. Principal Components Analysis of teachers' belief responses extracted three factors that we named (1) reform-oriented practices, (2) didactic approaches, and (3) self-efficacy. The factor analysis yielded a scale score for each individual teacher on each of the three factors. Drawing on previous studies of teacher beliefs (Elis, Malloy, Meece, & Sylvester, 2007), we use "reform-oriented practices" to broadly refer to a teacher's use of instructional practices that align with NCTM Standards (1989, 1991, 2000). We named the factor score for teacher beliefs about reform-oriented practices, Orientation. Results of the Year-3 study indicate Orientation was a significant predictor of student scores on both Test E and ITED-17, with positive effect sizes of $g=0.15$ and $g=0.14$, respectively. In the absence of observational data, the Orientation variable provided an approximation to *unobserved* teaching practices, thereby enabling us to ascertain the unique contribution of curricular effects above and beyond the effects of classroom instruction.

In addition to the challenge of separating curricular and instructional effects, in Year 3 we could not use scale scores on the seven teacher factors used in Year 1 and Year 2 cross-sectional analyses. These factor scores offered a *relative* measure of the teacher attribute for each of the seven factors. Because we were unable to collect observational data in Year 3, we used a proxy for each teacher factor and included them in our models. More specifically, Opportunity to Learn (OTL) Index (measuring percentage of textbook lessons taught) was used as a proxy for the OTL factor (based on multiple teacher variables). Similarly, the Textbook Content Taught (TCT) Index (measuring the extent to which teachers, when teaching textbook content, followed their textbook, supplemented their textbook lessons, or used alternative curricular materials) was used as a proxy for the Fidelity factor. Likewise, PD12 (the number of hours of professional development in the last 12 months) was used as a proxy for the PD factor (comprised of multiple teacher variables). It is important to note that each of these proxies were *absolute* (not relative) measures. Thus, the OTL Index, in principle, ranged from 0 (indicating no textbook lessons were taught) to 100 (indicating all textbook lessons were taught). By way of contrast, the OTL factor scores were z-scores that ranged from about -2.5 to $+2.5$ and represented an individual teacher's relative position with respect to all teachers. Our use of proxies in Year 3 analyses resulted in different interpretation of slope coefficients in our models. For example, in Year 3 a one-increment increase in OTL Index was tantamount to 1 % greater coverage of textbook lessons; in Year 1 and Year 2, a one-increment increase in OTL factor scores represented a 1 standard deviation increase in how much OTL the individual teacher afforded students compared to all teachers. Our use of proxy measures likewise made it somewhat cumbersome to make direct comparisons of cross-sectional results across 3 years.

Issues Related to Modeling Student Outcomes

Prior to conducting quantitative analyses, the COSMIC project team engaged in discussions about what variables should be included to model the effects of curriculum type and curriculum implementation and on students' mathematical learning. Given the data structure, we knew our models needed to take into account student demographics (e.g., IEP-status, prior achievement) as well as teacher characteristics (e.g., experience, professional development). Even though these variables were not central to the study, they potentially held explanatory power and, without their inclusion, our models might be misspecified. Of course, our models also needed to include variables that directly addressed our research questions: curriculum type and curriculum enactment, that latter of which included multiple measures of implementation fidelity. But is it possible to have too much data where results would be incomprehensible, and how much is too much? In total, we had 27 teacher variables—too many variables to model, particularly given the inherent interdependencies among them. For example, fidelity of implementation of curricular materials and teacher satisfaction with the textbook are conceptually related, not independent—as textbook satisfaction increases, it is reasonable to expect greater adherence to the textbook. We used Principal Components Analysis (PCA) to responsibly and substantially reduce the number of teacher-level variables, resulting in a more manageable and coherent data set comprised of 7 teacher factors: Opportunity to Learn (OTL), Classroom Learning Environment (CLE), Fidelity of Implementation (FIDELITY), Technology and Collaboration (TECH & COLLAB), Professional Development (PD), Knowledge of Standards (STANDARDS), and Experience. Our factor analysis enabled us to detect interdependencies as well as to ascertain which teacher variables did not "perform" well and were therefore tenuous. The teacher factor scores were included in cross-sectional analyses in Years 1 and 2. However, they were excluded in Year 3 because the lack of observational data, thereby greatly diminishing the number of teacher variables.

To test our research hypothesis on whether particular factors were associated with student learning, we developed a "full" model that included all pertinent variables. To some, the full model tells the entire story: Some variables (e.g., CPA) offer significant predictive power while others simply do not. However, a contrary position is that parsimonious "reduced" models (containing only significant predictors) tell the story more coherently. In our view, the story is best told by including both the "full" and "reduced" models and consistent with this view, we reported the "full" and reduced "final" models in all three cross-sectional reports. However, there are multiple approaches to achieving a final model. For example, instead of removing all the statistically nonsignificant variables in one block, we chose to remove variables using an iterative process. We justify this decision because one-step removal could potentially eliminate variables that may be significant but are being suppressed by the inclusion of all the variables.

In cross-sectional analyses, each dependent measure warranted its own parsimonious model to describe the influence of curriculum on student learning. As an example of the iterative process used to develop parsimonious models, we provide a complete overview of the models of ITED-16 scores in Table 2. Whereas the initial (full) model and reduced (final) model were reported by Tarr et al. (2013), the intermediary models were not reported due to space limitations. The first model contains the full set of 22 variables: 6 student-level, 14 classroom-level (including four interaction terms), 1 school-level, and 1 cross-level interaction. Initially, 8 variables had p-values of less than or equal to .05 but the dichotomous Curriculum variable (0=subject-specific, 1=integrated) was not among them. Mindful of our central objective to investigate curricular effects, we retained Curriculum in the second iteration but removed two nonsignificant interaction variables (Curriculum×CLE, Curriculum×Fidelity) and the school-level predictor. The third iteration excludes another nonsignificant interaction term (Curriculum×FRL), leaving 18 variables in the model including Curriculum (with $p=0.106$). Interesting, by removing Curriculum×FRL, the interaction Curriculum×OTL changed in significance and this might have not been observed without using a one-step reduction. The fourth iteration removes a block of six nonsignificant variables at the teacher level—5 teacher factor scores and Time_LD (% to class period devoted to lesson development), leaving 12 variables in the model. At this point, Curriculum is a

Table 2 Parsimonious HLM fixed effects models of models scores on test ITED-16

Fixed effect	First Coeff	First p	Second Coeff	Second p	Third Coeff	Third p	Fourth Coeff	Fourth p	Fifth Coeff	Fifth p	Sixth Coeff	Sixth p	Seventh Coeff	Seventh p
Intercept	280.654	0	280.844	0	279.369	0	277.65	0	277.46	0	278.78	0	278.801***	0
Level 1 (student)														
CPA	0.661	0	0.662	0	0.66	0	0.662	0	0.689	0	0.689	0	0.692***	0
Gender	-3.266	0.001	-3.261	0.001	-3.278	0.001	-3.279	0.001	-3.267	0.001	-3.27	0.001	-3.247***	0.001
Afr-American	-12.451	0	-12.443	0	-12.164	0	-12.157	0	-11.965	0	-12.04	0	-12.242***	0
Hispanic	-5.963	0.001	-5.96	0.001	-5.967	0.001	-5.975	0.001	-5.943	0.001	-6.018	0.001	-5.958**	0.001
Other	-6.598	0.008	-6.588	0.008	-6.483	0.009	-6.483	0.009	-6.462	0.01	-6.569	0.008	-6.538**	0.009
IEP	-4.733	0.024	-4.694	0.025	-4.699	0.025	-4.691	0.025	-5.271	0.012	-5.271	0.012	-5.266*	0.012
Level 2 (teacher)														
Curriculum	4.043	0.346	4.101	0.305	6.218	0.106	9.304	0.001	9.94	0.001	10.456	0.001	10.311**	0.001
%FRL	-0.156	0.269	-0.078	0.479	-0.173	0.072	-0.179	0.058	-0.2	0.037	-0.171	0.072		
CLE	1.469	0.427	1.881	0.082	1.992	0.072	1.866	0.081						
FIDELITY	-0.97	0.528	-0.995	0.373	-0.895	0.422								
OTL	8.887	0	9.066	0	9.298	0	10.097	0	9.847	0	7.939	0	9.184***	0
TECH & COLLAB	2.35	0.131	2.386	0.114	1.667	0.258								
STANDARDS	0.3	0.82	0.293	0.815	-0.262	0.831								
Time_LD	0.035	0.526	0.031	0.57	0.027	0.626								
Experience	-1.132	0.679	-1.44	0.59	-0.642	0.807								
PD	1.331	0.212	1.484	0.155	1.142	0.264								
Curriculum×FRL	-0.27	0.108	-0.269	0.108										
Curriculum×CLE	0.656	0.767												
Curriculum×FID	-1.473	0.61												
Curriculum×OTL	-8.237	0.017	-7.83	0	-5.263	0.054	-4.951	0.065	-4.376	0.112				
Level 3 (school)														
FRL	0.128	0.441												
Interaction (cross-level)														
Curr×CPA	0.062	0.126	0.063	0.121	0.072	0.073	0.066	0.099						

*p < .05, **p < .01, ***p < .001

significant predictor ($p = 0.001$). We sought to further reduce the model, and in the fifth iteration we excluded CLE and Curriculum × CPA whose p-values were 0.081 and 0.099, respectively. In the sixth model, we excluded Curriculum × OTL because its p-value had risen to 0.112. In the seventh iteration, we excluded FRL because its p-value was 0.072. This final model includes only significant predictors, therefore making it simpler to interpret. However, which of the seven models best "tells the story" of student achievement on ITED-16? It depends. One could argue the full model tells the story in its entirety; yet, with the inclusion of 22 variables, one might counter-argue that the full model is bloated and results are somewhat incomprehensible. Perhaps the "true" story lies somewhere in between the full and final models. For example, the fourth model contains 12 variables, all of which have associated p-values less than 0.10. However, is the customary $p = 0.05$ too strict as the threshold of significance? The rigidity of your beliefs about significance thresholds determines which of the seven models in Table 2 is most appropriate.

Conclusion

Relationships between on curriculum, instruction, and student learning have long been the focus of mathematics education research. In recent decades, curriculum reform initiatives and innovative curricular materials have been introduced to address the historical lagging mathematics achievement of US students. As improving student performance in STEM fields has become a national priority, federal funding has supported several large-scale, rigorous studies of curricular effectiveness. Funded by the National Science Foundation, the COSMIC project sought to ascertain relationships between curriculum type, curriculum enactment, and student learning. However, attributing student learning to a particular curricular program is both a massive, complex, and expensive undertaking.

In the COSMIC project, our research team confronted numerous challenges in the collection and analysis of student and teacher data—some anticipated, others unanticipated—each of which needed to be resolved. We anticipated the lack of a common measure of students' prior achievement but we did not expect one school district's reluctance to provide student scores on their state-mandated achievement test. We expected the problem of missing data but did not anticipate one school district's refusal to provide FRL-status at the student level. And while we anticipated some students would not complete 3 years of high school mathematics, we did not expect some students to migrate across curricular paths. Furthermore, although observational enhanced an already robust set of teacher data in Years 1 and 2, the lack of classroom visits in Year 3 introduced challenges that we simply did not anticipate. Finally, we grappled with several issues in modeling the effects of student and teacher characteristics, curriculum type, and curriculum implementation, including *whether* and *how* to reduce the models. With this chapter, we hope our discussion of these key challenges provides further insights into how the COSMIC project studied curricular effectiveness in secondary mathematics and informs the data collection and analyses of future large-scale studies in mathematics.

References

Chávez, Ó., Papick, I., Ross, D. J., & Grouws, D. A. (2011). Developing fair tests for mathematics curriculum comparison studies: The role of content analyses. *Mathematics Education Research Journal, 23*(4), 397–416. doi:10.1007/s13394-011-0023-2.

Chávez, Ó., Tarr, J. E., Grouws, D. A., & Soria, V. (2015). Content organization, curriculum enactment, and student learning in the third-year high school mathematics curriculum: Results from the COSMIC project. *International Journal of Mathematics and Science Educational Research, 13*(1), 97–120. doi:10.1007/s10763-013-9443-7.

Elis, M. E., Malloy, C. E., Meece, J. L., & Sylvester, P. R. (2007). Convergence of observer ratings and student perceptions of reform teaching practices in sixth-grade mathematics classrooms. *Learning Environments Research, 10*(1), 1–15.

Frank, K. A. (2000). Impact of a confounding variable on a regression coefficient. *Sociological Methods and Research, 29*, 147–194.

Gelman, A., & Hill, J. (2006). *Data analysis using regression and multilevel/hierarchical models*. Cambridge, MA: Cambridge University Press.

Grouws, D. A., Tarr, J. E., Chávez, O., Sears, R., Soria, V., & Taylan, R. D. (2013). Curriculum and implementation effects on high-school students' mathematics learning from curricula representing subject-specific and integrated content organizations. *Journal for Research in Mathematics Education, 44*(2), 416–463.

Harwell, M. R., Post, T. R., Maeda, Y., Davis, J. D., Cutler, A. L., & Kahan, J. A. (2007). Standards-based mathematics curricula and secondary students' performance on standardized achievement tests. *Journal for Research in Mathematics Education, 38*(1), 71–101.

National Center for Education Statistics. (2007). *Mapping 2005 state proficiency standards onto the NAEP scales (NCES 2007-482)*. U.S. Department of Education. Washington, DC: Author.

National Center for Education Statistics. (2010). *Status and trends in the education of racial and ethnic minorities (NCES 2010-015)*. U.S. Department of Education. Washington, DC: Author.

National Council of Teachers of Mathematics. (1989). *Curriculum and evaluation standards for school mathematics*. Reston, VA: Author.

National Council of Teachers of Mathematics. (1991). *Professional standards for teaching mathematics*. Reston, VA: Author.

National Council of Teachers of Mathematics. (2000). *Principles and standards for school mathematics*. Reston, VA: Author.

Post, T. R., Harwell, M. R., Davis, J. D., Maeda, Y., Cutler, A., & Andersen, E. (2008). Standards-based mathematics curricula and middle-grades students' performance on standardized achievement tests. *Journal for Research in Mathematics Education, 39*(2), 184–212.

Tarr, J. E., Grouws, D. A., Chávez, O., & Soria, V. (2013). The effect of content organization and curriculum implementation on students' mathematics learning in second-year high school courses. *Journal for Research in Mathematics Education, 44*(4), 683–729.

Tarr, J. E., Harwell, M., Grouws, D. A., Chávez, Ó., & Soria, V. (2013, April). *Longitudinal effects of curriculum organization and implementation on high school students' mathematics learning*. Paper presented at the Annual Meeting of the American Educational Research Association, San Francisco.

Tarr, J. E., McNaught, M. D., & Grouws, D. A. (2012). The development of multiple measures of curriculum implementation in secondary mathematics classrooms: Insights from a three-year curriculum evaluation study. In I. Weiss, D. Heck, K. Chval, & S. Zeibarth (Eds.), *Approaches to studying the enacted curriculum* (pp. 89–115). Greenwich, CT: Information Age Publishing.

What Works Clearinghouse. (2008). *Procedures and standards handbook* (version 2). Retrieved from U.S. Department of Education website: http://nces.ed.gov/nationsreportcard/pdf/main2005/2007468.pdf

Part II
Teaching

Turning to Online Courses to Expand Access: A Rigorous Study of the Impact of Online Algebra I for Eighth Graders

Jessica B. Heppen, Margaret Clements, and Kirk Walters

Overview

A body of research shows that Algebra I operates as a gateway to more advanced mathematics courses in high school and college, and that students who succeed in Algebra I in middle school have more success in math throughout high school and college than students who take Algebra I later (e.g., Nord et al., 2011; Smith, 1996; Spielhagen, 2006; Stevenson, Schiller, & Schneider, 1994). Based on this research, policymakers have persistently called for broadening access to Algebra I in eighth grade (e.g., U.S. Department of Education, 1997). More recently, the 2008 report by the National Mathematics Advisory Panel recommended that "all prepared students [should] have access to an authentic algebra course" and that schools and districts should prepare more students to enroll in such a course by eighth grade (2008, p. 23). This recommendation is evident in the content and sequencing of topics of the Common Core State Standards for Mathematics (CCSSM), which are

This chapter is based on a report prepared for the National Center for Education Evaluation and Regional Assistance, Institute of Education Sciences, under contract ED-06C0-0025 with Regional Educational Laboratory Northeast and Islands administered by the Education Development Center, Inc. The citation for the full report is Heppen, J.B., Walters, K., Clements, M., Faria, A., Tobey, C., Sorensen, N., and Culp, K. (2011). *Access to Algebra I: The Effects of Online Mathematics for Grade 8 Students* (NCEE 2011-4021). Washington, DC: National Center for Education Evaluation and Regional Assistance, Institute of Education Sciences, U.S. Department of Education.

J.B. Heppen (✉) • K. Walters
American Institutes for Research, 1000 Thomas Jefferson Street, Washington, DC, NW 20007, USA
e-mail: jheppen@air.org

M. Clements
Education Development Center, 96 Morton Street, 7th Floor, New York, NY 10014, USA

© Springer International Publishing Switzerland 2015
J.A. Middleton et al. (eds.), *Large-Scale Studies in Mathematics Education*, Research in Mathematics Education, DOI 10.1007/978-3-319-07716-1_6

configured to effectively prepare all K–7 students for eighth-grade Algebra I (National Governors Association Center for Best Practices & Council of Chief State School Officers, 2010).

A push to prepare more students to take Algebra I in eighth grade creates the need for more Algebra I teachers and classes in middle schools. Some districts and schools are able to respond to this need while others are not. National grade 8 Algebra I enrollments increased from 16 % in the 1990s to 31 % in 2007 (Loveless, 2008), suggesting that many middle schools have found ways to increase students' access to "formal" Algebra I courses or equivalent content. With the large-scale adoption of the CCSSM now taking effect, this access is likely to continue to increase—perhaps even dramatically—in the coming years.

However, there are gaps in course access for eighth-grade students in schools across the country and particularly in rural areas. An analysis of data from the Early Childhood Longitudinal Study (ECLS-K; U.S. Department of Education, 2009a) indicated that nationally, approximately 25 % of students who scored in the highest quartile on the study's fifth-grade math assessment were not enrolled in a formal Algebra I course in eighth grade (Walston & Carlivati McCarroll, 2010). Additional analysis of ECLS-K data indicated that, compared to urban or suburban schools, a larger proportion of high-achieving students do not take Algebra I in eighth grade in rural schools. Thirty-nine percent of students attending rural schools who scored in the highest quartile on the fifth-grade assessment were not enrolled in Algebra I in eighth grade. A similar picture emerges when examining school-level access to Algebra I for eighth graders. Nationally, 16 % of all schools serving eighth graders report they do not offer Algebra I to eighth graders; in rural areas the rate is 24 % (compared to 21 % of urban schools and 9 % of suburban schools; U.S. Department of Education, 2009a).

In schools that do not offer Algebra I, curriculum offerings may be limited by a number of constraints including staffing, space, and enrollment. These issues are particularly prevalent in small or rural schools, where student populations are low and attracting qualified and experienced teachers is difficult (Hammer, Hughes, McClure, Reeves, & Salgado, 2005; Jimerson, 2006). In such schools, there is often a paucity of funds to hire teachers with specialized content knowledge to teach relatively small classes that are not offered to all students. As technology capacity grows in schools around the country, online courses are increasingly seen as a viable means for expanding curricular offerings and expanding access to key courses, especially in small and rural schools (Hanum, Irvin, Banks, & Farmer, 2009; Picciano & Seaman, 2009; Schwartzbeck, Prince, Redfield, Morris, & Hammer, 2003).

This chapter describes a study that focused on broadening eighth-grade students' access to Algebra I through an online course. The online course was offered to eighth graders, mostly from rural schools, who were considered academically ready for Algebra I (i.e., were "algebra-ready"), but who attended a middle-grade school that did not typically offer a formal Algebra I course. The primary goal was to determine whether using an online course to broaden access to Algebra I in eighth grade could improve algebra-ready students' knowledge of algebra in the short term, open doors to more advanced course sequences in the longer term, or both.

The secondary goal was to determine whether there were any unintended consequences (side effects) of offering online Algebra I to students identified as algebra-ready by their schools. For one, did taking an online Algebra I course in the eighth grade have an impact on these students' general mathematics achievement at the end of eighth grade? Also, offering the online Algebra I course to the higher-achieving students could lead to academic tracking that wouldn't otherwise exist and this could have unintended consequences for the eighth graders in the school who remain in the general mathematics course—through, for example, peer effects, changes in course emphasis, or smaller class sizes.

This study was designed to rigorously assess the effects—positive and negative—of offering an online Algebra I course on all eighth graders in middle schools with no access or limited access to Algebra I. This includes schools in which there were no opportunities for academically ready students to take Algebra I, as well as schools in which a limited number of academically ready students could take Algebra I at a nearby school if schedules and transportation needs could be accommodated. This study sought to produce useful information for education decision makers considering investing in an online course as a means of broadening access to Algebra I in grade 8, particularly in rural schools.

The randomized controlled trial was conducted in 68 mostly rural middle schools in Maine and Vermont that offered limited or no access to Algebra I to eighth graders. Half of the schools were randomly assigned to offer an online Algebra I course to students they considered academically ready for the course; the other half of the schools conducted business as usual during the study year (2008–2009). A total of 211 algebra-ready students took the online course as part of the study, and while only 43 % of them completed the entire course, over 82 % of them completed more than half the course. As described in this chapter, implementation of the course played out as expected in some ways, and not as expected in others.

Algebra-ready students in schools that offered the course were compared with their counterparts in control schools. These students in control schools typically took a general mathematics class for eighth graders—however, we found that these eighth-grade math classes had a substantial focus on algebraic content (in nearly all of the control schools, the eighth-grade math class had a focus on algebraic content of 50 % or more). Moreover, although the expected counterfactual was the absence of access to a formal Algebra I course, 20 % of the eligible, algebra-ready students in control schools took either a traditional, classroom-based or online Algebra I course in their middle school or traveled to the local high school to take the course.

The results showed that offering Algebra I as an online course to students considered academically ready by their schools is an effective way to broaden access in schools that do not typically offer Algebra I to eighth graders, or do so only on a limited basis. Taking the course significantly improved students' algebra achievement at the end of eighth grade, and significantly increased their likelihood of taking advanced math courses in high school (based on ninth- and tenth-grade coursetaking patterns). The course also had no discernible side effects for the algebra-ready students on their general math achievement, or on achievement and coursetaking outcomes for the remaining eighth graders, those who were not ready to take Algebra I.

In this chapter, we summarize the background and rationale for the study, the research design, sample, and measures. We describe the online course and how it was implemented, and summarize the findings. We conclude with a summary of implications of the findings, and future directions for rigorous research on technology-based math interventions.

Background and Rationale

The background for this study is based both on research on the benefits of taking Algebra I prior to high school and the lack of research on the effectiveness of online courses, despite the rapid increase in adoption of online courses for many reasons, including expanding curriculum offerings.

Significance of Algebra I

Algebra I is a gatekeeper course because it is a prerequisite for the high-school mathematics and science courses considered essential, if not required, for getting into college. High-school mathematics courses are ordered sequentially; students must successfully complete Algebra I before taking subsequent mathematics courses (Smith, 1996; Wagner & Kieran, 1989). If students succeed in Algebra I, they typically take Geometry, Algebra II, and then more advanced courses, such as Trigonometry, Precalculus, and Calculus.[1] Several research studies have shown that success in Algebra I is highly correlated with enrollment in more advanced math and science courses (Atanda, 1999; Kilpatrick, Swafford, & Findell, 2001; Lacampagne, Blair, & Kaput, 1995; Nord et al., 2011).

Previous research, mainly correlational, suggests that having access to Algebra I in eighth grade benefits at least some students. Using data from the 1988 National Educational Longitudinal Study, Stevenson et al. (1994) examined the relationship between students' mathematics and science opportunities in grade 8 and their later opportunities in mathematics and science in high school (see also Schneider,

---

[1] The Algebra I → Geometry → Algebra II sequence is known as the traditional mathematics coursetaking pathway (Common Core Standards Initiative, 2010; National Mathematics Advisory Panel, 2008). Some schools reverse the order of Algebra II and Geometry, yielding an Algebra I → Algebra II → Geometry sequence; this ordering is less common than the traditional sequence. Other schools offer an integrated course pathway that combines the content of Algebra I, Geometry, and Algebra II into integrated courses, which typically have generic names, such as Mathematics 1, Mathematics 2, and Mathematics 3 (National Governors Association Center for Best Practices, Council of Chief State School Officers, 2010; National Mathematics Advisory Panel, 2008).

Swanson, & Riegle-Crumb, 1998). They defined three course sequences, each of which hinged on students taking Algebra I:

- *Advanced*: Completion of both Geometry and Algebra II or any higher level course by grade 10
- *Intermediate*: Completion of either Geometry or Algebra II by grade 10
- *Low*: Completion of neither Geometry nor Algebra II by grade 10

According to the study, 42 % of students who took Algebra I in grade 8 participated in an advanced course sequence in high school. In contrast, only 12 % of students who did not take Algebra I in grade 8 participated in an advanced course sequence. The study's authors conclude that math course opportunities in eighth grade are related to students' subsequent opportunities to take, and succeed in, advanced course sequences in high school. Another study using the same data found that 60 % of students who took Calculus by grade 12 had taken algebra in eighth grade (National Center for Education Statistics, U.S. Department of Education 1996).

Research also suggests that students who take Algebra I in middle school subsequently have higher math skills that are sustained over time. Smith (1996) used data from the High School and Beyond study to estimate the relationship between middle school algebra and later mathematics outcomes, controlling for differences in student background (social and demographic background, aptitude, and academic emphasis or interest in mathematics). Students who took Algebra I in middle school completed an average of 1 more year of mathematics courses than students who took Algebra I in high school (2.3 versus 1.3 years). They also outscored their counterparts who took Algebra I in high school on the High School and Beyond mathematics assessment in both grades 10 and 12. Similarly, Spielhagen (2006) used data from a large urban district to compare high school and college outcomes of students with and without access to Algebra I in grade 8 and concluded that eighth graders who were provided access to Algebra I followed a more advanced coursetaking sequence in high school than those with similar academic abilities who took Algebra I in grade 9.

In turn, taking advanced mathematics courses in high school (typically defined as completing courses above Algebra II) is related to college and future success. For example, Horn and Nuñez (2000) found that three-quarters of students who participated in an advanced coursetaking sequence in high school enrolled in 4-year colleges. Adelman (1999, 2006) found that the odds of completing college are twice as high for students who take a sequence of advanced mathematics courses in high school. Enrollment in higher-level mathematics and science courses in high school is also related to future educational and employment opportunities (Gamoran & Hannigan, 2000; U.S. Department of Education, 1997). Rose and Betts (2001) showed that students who take higher-level mathematics classes in high school have higher earnings 10 years after high-school graduation, even after controlling for background characteristics and eventual educational attainment (see also Jabon et al., 2010).

Thus the established links between success in Algebra I and subsequent high-school mathematics coursetaking and later postsecondary outcomes are clear (though not causal). However, less clear is whether having more algebra content in earlier grades similarly benefits coursetaking and later outcomes for students. As mentioned previously, the CCSSM is structured to prepare K–7 students for an eighth-grade Algebra I course. The CCSSM accomplishes this by incorporating algebraic concepts into earlier grades, starting in kindergarten with patterns, and building throughout elementary and middle school (National Governors Association Center for Best Practices & Council of Chief State School Officers, 2010). By eighth grade, students who do not take a formal Algebra I course are likely to have access to a substantial amount of algebraic content in their regular grade 8 mathematics class. Future research will be able to determine the degree to which this deepening and expansion of algebra in earlier grades benefits student learning over time. Nevertheless, some students in schools that offer substantial algebra throughout middle school are still ready for a formal Algebra I course, and presumably, successfully completing a formal Algebra I course before entering high school will enable those students to participate in more advanced coursetaking sequences than they otherwise would.

Use of Online Courses to Expand Offerings

In schools that do not offer particular classes because of a lack of resources such as space and available teachers, online courses are one way to provide courses to interested or eligible students. Offering coursework virtually is a strategy that schools use to expand the curricula available to their students (National Education Association, 2006), particularly in small schools and isolated communities that do not have access to critical courses in science, technology, engineering, and mathematics (Picciano & Seaman, 2009; Tucker, 2007).

The increasing popularity of online courses is driven by both technological advancements and the flexibility with which online courses can provide access to content and instruction (U.S. Department of Education, 2009b). Online courses allow schools to take advantage of a broader pool of qualified teachers, which can enable students to take courses that are otherwise not offered or taught by qualified teachers.

Some researchers and education stakeholders suggest that online courses may present an affordable option for expanding students' access to courses to schools with limited funds by reducing costs for teaching staff or school facilities (Anderson, Augenblick, DeCesare, & Conrad, 2006; Greaves, Hayes, Wilson, Gielniak, & Peterson, 2010; Moe & Chubb, 2009; Smith & Mitry, 2008). However, studies determining whether this is actually the case for online learning in K–12 settings have yet to be conducted (Office of Educational Technology, U.S. Department of Education, 2012).

The use of online courses in K–12 settings has been on the rise over the past decade. According to the National Center for Education Statistics (NCES), 37 % of school districts used technology-based distance learning during 2004–2005 (Zandberg & Lewis, 2008). By 2009–2010, NCES reported that the proportion had grown to 55 % (Queen & Lewis, 2011). As of 2007, 28 states had virtual high-school programs, enabling students to take online courses in addition to their school-based courses to fill curriculum gaps (for example, Advanced Placement [AP] courses) or providing opportunities for credit recovery (Tucker, 2007). While the exact number of students involved in online learning is not available, the Sloan Consortium conducted a national survey of K–12 public school districts in 2007– 2008 and estimated that the total number of K–12 public school students engaged in online courses exceeded one million—a 43 % increase from the 700,000 students reported in 2005–2006 (Picciano & Seaman, 2009).

In the majority of districts using technology-based distance learning in the NCES studies, fully online courses are typically offered to students otherwise enrolled in traditional brick-and-mortar schools, as opposed to students taking all of their courses online. When asked about their reasons for providing online courses, over half of all responding districts listed "providing courses not otherwise available at the school" as one of the most important.

Surveys of K–12 public schools have suggested that rural districts and schools are especially interested in online learning. In the Sloan Consortium surveys, respondents from small rural school districts reported that they use online courses to provide opportunities they would not otherwise be able to offer (Picciano & Seaman, 2009). Two other surveys examined the prevalence of rural schools' use of distance learning, a broader category that overlaps with online learning. In these surveys, a majority of rural educators reported using distance education to expand access to advanced coursework for students (Hanum et al., 2009; Schwartzbeck et al., 2003).

Prior Research on Online Course Effectiveness

The study summarized in this chapter was the first to rigorously test the efficacy of using an online course to broaden access to a course to which students' access was otherwise limited. What little rigorous research exists has focused on comparing online learning with traditional face-to-face learning, mostly at the postsecondary level. Though the *Access to Algebra I* study was not designed to compare online Algebra I to traditional face-to-face versions of the course, this literature is relevant because it provides information on the utility of online courses as an educational experience compared with traditional face-to-face coursework. A meta-analysis of 99 studies conducted primarily in postsecondary settings found that online instruction yields positive effects relative to face-to-face instruction (U.S. Department of Education, 2009b).[2]

One prior study included in the meta-analysis examined the effects among eighth- and ninth-grade students of an online Algebra I course relative to a face-to-face Algebra I course (O'Dwyer, Carey, & Kleiman, 2007). In this quasi-

---

[2] Studies that qualified for inclusion in the meta-analysis used an experimental or quasi-experimental design (if quasi-experimental, the study must have included statistical controls for prior achievement). They also reported data sufficient for calculating effect sizes per the What Works Clearinghouse (2008) guidelines.

experimental study, students in 18 classrooms participating in the Louisiana Algebra I Online Project were compared with students in comparison classrooms on outcomes including an end-of-year Algebra I assessment. At the end of the school year, the difference in scores between students in the two types of classrooms on an algebra posttest was not statistically significant,[3] suggesting that online Algebra I can produce outcomes that are similar to the outcomes in traditional Algebra I courses. This study provided foundational information on potential effects of online Algebra I, and the current study, while neither designed nor intended to directly compare the effectiveness of online versus face-to-face Algebra I courses, builds on O'Dwyer et al.'s findings.

Study Design and Methodological Considerations

Given this research and policy context, this study rigorously tested the impact of expanding access to Algebra I to eighth-grade students by offering an online course in schools that do not typically offer Algebra I in grade 8. It is the first randomized controlled trial testing the impact of providing an online Algebra I course on students' mathematics achievement and coursetaking trajectories over time. Furthermore, the study was designed to investigate whether changing mathematics instruction in this way resulted in any unintended consequences.

Goals and Research Questions

The primary goal of the study was to measure the effects of offering an online Algebra I course to eighth graders considered algebra-ready in schools that do not typically offer the course. The related research questions asked whether access to online Algebra I improves these students' knowledge of algebra in the short term and whether it opens doors to more advanced mathematics course sequences in the longer term. Specifically:

1. What is the impact of offering an online Algebra I course to algebra-ready students on their algebra achievement at the end of grade 8?
2. How does offering an online Algebra I course to algebra-ready students affect their likelihood of participating in an advanced course sequence in high school?

The secondary goal of the study was to estimate whether offering online Algebra I to students considered ready for algebra resulted in potential unintended consequences (or side effects) for any of the eighth-grade students. The motivations for exploring the secondary research questions were twofold. First, offering the online

---

[3] The p-value associated with the coefficient representing the difference in posttest scores between students in online and face-to-face classes was 0.093.

Algebra I course may affect algebra-ready students in unintended ways, for example, by adversely affecting their general math achievement. Second, providing online Algebra I to algebra-ready students may have unintended consequences for non-algebra-ready students—the students who remain in the general mathematics course. When the algebra-ready students are removed from the general eighth-grade math class, outcomes for the remaining students may be affected because of peer effects, smaller class sizes, a change in course emphasis (for example, less algebra), or other reasons.

The study design, described in this section, allowed us to address these issues in a secondary set of research questions:

1. What is the effect of providing online Algebra I to algebra-ready students on their general mathematics achievement at the end of grade 8?
2. What is the effect of providing online Algebra I to algebra-ready students on the following outcomes for *non*-algebra-ready students?

   • End-of-eighth grade algebra achievement
   • End-of-eighth grade general math achievement
   • Planned high-school math coursetaking

By answering the primary and secondary research questions, this study examined what happens to the entire population of eighth graders—including potential benefits and possible negative consequences—when a school uses an online course as a way to offer Algebra I to students considered academically ready for the course. The study thus sought to inform decision makers who are considering investing in an online course as a means to broaden access to Algebra I in grade 8.

Study Design

This study was a randomized experimental trial with random assignment of schools to condition. Schools in Maine and Vermont that did not offer a full section of Algebra I to eighth graders (as of the 2007–2008 school year) were eligible for the study. Sixty-eight eligible schools were randomly assigned to one of two study groups. Schools in the treatment condition received the online algebra course for the 2008–2009 school year; schools in the control condition did not receive the online Algebra I course during the 2008–2009 school year, and offered their usual math curriculum and instruction for all students.[4]

In all cases, students were identified as academically ready for Algebra I in eighth grade in spring 2008 (as rising seventh graders), *prior to random assignment*. This way, knowledge of whether a school was or was not going to be offering the online course could not affect decisions about which students would be eligible.

---

[4] Schools in the control group received the online course for the 2009–2010 school year. All schools (treatment and control) were provided the online course for 2 consecutive years.

This process produced two groups of algebra-ready students (those in treatment schools and those in control schools) that were statistically similar on all measured characteristics at baseline (as described in the "Sample" section below).

For the sample of algebra-ready students, this study was also longitudinal in design. The premise of "pushing down" Algebra I to grade 8 is that it prepares students for more rigorous coursetaking in mathematics through high school, which better prepares them to succeed in advanced course sequences that prepare them for college. Therefore, of critical interest from a policy perspective is the extent to which offering an online algebra course to students who would otherwise not have been able to take Algebra I in grade 8 has a sustained impact on their mathematics coursetaking in high school. For this reason, researchers tracked the algebra-ready students who attended participating schools into high school to collect mathematics coursetaking information (grade 9 courses and grades and grade 10 planned courses) at the end of grade 9. These data were used to categorize students as participating in an advanced mathematics course sequence in high school.[5]

To estimate the impacts of online Algebra I on relevant outcomes for the algebra-ready students, we compared outcomes for students in treatment schools with those in control schools at the end of grade 8 (spring 2009) and at the end of grade 9 (spring 2010). Owing to random assignment and the fact that algebra-ready students were identified in all schools prior to random assignment, the comparison of algebra-ready students in treatment and control schools (arrow 1 in Fig. 1) provides an unbiased estimate of the effects of online Algebra I on these students. In this way, the study revealed the overtime effects on students' mathematics achievement of using an online course to broaden access to Algebra I for students who otherwise would have no access or only limited access to a formal Algebra I course until high school.

Because schools were the unit of assignment, the students in treatment and control schools who were not academically ready for Algebra I can also be compared to obtain an unbiased estimate of the effect (or "side effect") of the offering of online Algebra I to algebra-ready students on the students who remained behind in the general eighth-grade math classes (see arrow #2 in Fig. 1). As shown in Fig. 1, algebra-ready and non-algebra-ready students were not compared to each other; each group of students was only compared to their counterparts in schools in the other condition.

Research studies often have multiple research questions which may produce mixed findings. Mixed findings, while interesting, can make it difficult to draw conclusions about the effectiveness of the intervention being tested. Therefore, prior to conducting the study, we determined the combination of results with which we would consider the online Algebra I course a successful means for broadening

---

[5] We did not follow non-algebra-ready students into high school because of cost constraints (there were approximately three times as many non-algebra-ready students as eligible students) and because assessing the impact of the online Algebra I course on algebra-ready students' subsequent high-school coursetaking was most critical and relevant for the study.

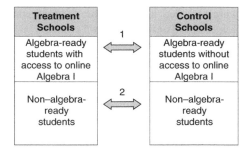

Fig. 1 Framework for estimating impacts of online Algebra I on algebra-ready students and non-algebra-ready students. *Notes*: Algebra-ready students were those considered by their schools—prior to random assignment—to be eligible to take Algebra I if the course could be offered. Non-algebra-ready students were the eighth graders in the participating schools in fall 2008 that had not been on the schools' list of eligible students

access. Specifically, we a priori determined that we would consider the online course a success if we saw both:

- A statistically significant positive impact on either algebra-ready students' algebra scores at the end of eighth grade *or* high-school coursetaking.
- The absence of statistically significant negative side effects on eligible or non-algebra-ready students.[6]

Only this combination of results would provide evidence that there are benefits of adopting an online Algebra I course for eligible, algebra-ready students without significant negative consequences to them or the non-eligible students in their schools. Ultimately, the intended outcome of the study was to provide rigorous evidence to help districts and schools determine whether adopting an online Algebra I course is a good choice in middle schools where access is limited.

Analytic Methods

All analyses were conducted separately on the algebra-ready and non-algebra-ready student samples. The samples used for analyses are intent-to-treat samples, meaning that all students with consent who were identified as eligible for Algebra I before

---

[6]For the secondary questions, the study was not designed to determine whether the groups are statistically equivalent. A lack of statistical significance for an impact estimate does not mean that the impact being estimated equals zero. Rather, it means that the estimate cannot reliably be distinguished from zero, an outcome that may reflect the small magnitude of the impact estimate, the limited statistical power of the study, or both. For the secondary questions, lack of statistical significance was defined as a difference with a p-value greater than 0.05, at 80 % power.

random assignment were included in the eligible student sample, whether or not they enrolled in or stuck with the online Algebra I course.

Given the nested structure of the data (the clustering of students within schools), we used multilevel models with students nested within schools to estimate the impacts of online Algebra I on the study outcomes. Analyses of continuous outcome measures (including algebra and general math posttest scores) used hierarchical linear modeling (Raudenbush & Bryk, 2002); analyses of coursetaking sequences used hierarchical generalized linear models that assumed a Bernoulli sampling distribution and logit link function (McCullagh & Nelder, 1989; Raudenbush & Bryk, 2002). To increase the precision of the estimates in these analyses, we used a set of baseline characteristics as covariates, including school factors (state and school size) and student factors (baseline math achievement scores from the prior year's state assessment, and demographics including gender, eligibility for free or reduced-price lunch, and special education status).

If an impact estimate was statistically significant, it is possible to conclude with some confidence that the online Algebra I course had an effect on the outcome being assessed. If an impact estimate was not statistically significant, the nonzero estimate may be a product of chance. To maintain the probability of falsely detecting a statistically significant result ($p < 0.05$) if there were no true impact on either of the two primary outcomes (algebra achievement and high-school coursetaking for algebra-ready students), we adjusted the statistical significance level for each of the two primary outcomes to 2.5 % (Bonferroni correction).

The study had low rates of missing data. To handle missing data, we used multiple imputation by chained equations. Multiple imputation models were specified on the basis of the analysis of predictors of missingness; they included student and school covariates and interaction terms between student covariates.

Study Sample

In this section, we describe the schools and students who participated in the study.

School Recruitment

The target population included schools in Maine and Vermont that served students in grade 8 and did not offer a stand-alone Algebra I class in 2007–2008, when recruitment for the study took place.[7] These two criteria—serving grade 8 students

---

[7] A stand-alone class was defined as one full section of Algebra I taken by at least 20–25 % of grade 8 students in the school, with a dedicated teacher. This proportion was derived from the percentage of grade 8 students in the northeast US who took Algebra I as of 2007, which was 25 % (U.S. Department of Education, 2007).

and not offering stand-alone Algebra I—plus the willingness to comply with the requirements of the study were the only eligibility criteria for participation.

Based on our knowledge of the educational landscape in these states, we knew in advance that there would likely be substantial amounts of algebraic content provided to eighth graders in schools that participated. Rather, we sought the participation of schools in which a stand-alone, "formal" Algebra I class was not offered to at least 20–25 % of eighth-grade students, and where expanding access could potentially yield benefits for some students. (During the course of the study, we collected information about the amount and type of algebra taught in math classes in study schools, to clarify the "treatment contrast" for the online course.) Schools eligible to participate in the study offered algebraic content to at least some students in several different ways. Eligible schools included those in which some students took Algebra I in the local high school (when scheduling and transportation allowed), and those in which some students sat in the "back" of the eighth-grade math class with an Algebra I textbook. Schools that offered part or all of a full Algebra I course to some but not all of their algebra-ready students were also eligible for the study, as were schools that typically delivered Algebra I content to some students by providing accelerated material in the context of the regular grade 8 mathematics curriculum.

The study was conducted in two states in the Northeast region, Maine and Vermont. Maine was chosen because in addition to relatively low overall enrollments in Algebra I among grade 8 students (20–25 % in 2007), the state had a strong technology initiative that could support the infrastructure needed to offer an online course in schools. The Maine Learning Technology Initiative provided all grade 8 students and teachers in Maine with their own laptop computer for use throughout the school year, both in and out of school. Students were thus familiar with using computers as part of their daily educational activities (Berry & Wintle, 2009; Silvernail & Gritter, 2007). The technology infrastructure in Maine helped drive interest in online courses.

Vermont was selected because it shares demographic and geographic characteristics with Maine that were factors in the selection of Maine for the study. Specifically, Vermont has the second-highest proportion of rural schools in the United States, after Maine (Johnson & Strange, 2007) and these schools serve students who, although racially and ethnically homogenous, are diverse in socioeconomic status. Like rural schools in Maine, rural schools in Vermont find it challenging to offer a full range of courses to students who might benefit from them. Although at the time of the study Vermont did not have the laptop initiative that Maine had, the state had both the necessary technological capacity and an interest in exploring ways to use technology to improve education.

During recruitment, we informed schools of the study requirements if they were randomized to the treatment group. First, students taking the online Algebra I course would take the course as their grade 8 mathematics course, not as a supplemental course to the general grade 8 mathematics class. Second, each participating student would have access at school to a computer with a high-speed Internet connection. Access to a computer was necessary for students to access the course and communicate with the online teacher.[8] Third, schools were required to assign a class period during the school day during which students would access the online Algebra I. The online class had to occur with the same frequency and duration as the regular grade 8 mathematics classes, but schools could schedule the period whenever they wanted, create multiple periods during the school day, and determine the location of the online course (for example, the regular classroom or the library). Fourth, schools would provide a staff member to serve as an on-site proctor to support the students taking the online course.

Based on power calculations conducted while planning the study, we aimed to recruit a minimum of 60 schools to participate to achieve a minimum detectable effect size of 0.25 standard deviations (for continuous outcomes like math achievement scores). A total of 68 schools agreed with the terms for participation and signed on to participate.

Description of Participating Schools

The 68 participating schools were spread across Maine and Vermont. Sixty-two were rural schools. The grade 8 enrollments for the schools ranged from fewer than 4 to nearly 150 students, with an average enrollment of 32 students. Fifty-two schools (76 %) served grades pre-K–8 or K–8; 10 schools (14 %) served middle grades (grades 5–8, 6–8, or 7–8); and the remaining six schools served other grade spans including K–12, 7–12, and 3–8.

On average, the participating schools had the following characteristics, according to the Common Core of Data as of the 2007–2008 school year:

- The average total enrollment for study schools was 186 students.
- 94 % of the schools were Title I schools.
- 48 % of the students were eligible for free or reduced-price lunch.
- 95 % of the students were white.
- 53 % of the students scored at or above the proficiency level on the state mathematics assessment in the 2007–2008 school year.

As mentioned earlier, all schools were required to identify algebra-ready students who would be offered the online course if the school was randomized to the treatment group. The eligible, algebra-ready students were rising eighth graders whom schools perceived as having the requisite skills at the end of grade 7 to take Algebra I in 2008–2009.[9] Before random assignment, the 68 participating schools

---

[8] Technology requirements and system specifications were provided to all schools during recruitment and again to the schools in the treatment group before the beginning of the 2008–2009 school year.

[9] Schools made decisions about which students were ready for algebra on the basis of teacher perceptions of preparedness; grades in mathematics classes through grade 7; scores on state assessments; and scores on other assessments, such as algebra readiness tests. The research team did not

identified a total of 468 eligible students, or an average of 6.9 students per school. These students represented 23 % of the rising grade 8 students in participating schools as of spring 2008.

To randomly assign schools to condition, we stratified by size and state and assigned half of the schools within each block to the treatment group and the other half to control. A total of 35 schools were assigned to the treatment group and 33 schools were assigned to the control group. All 68 schools cooperated with data collection and fully participated in the study during the 2008–2009 school year. This participation included implementation of the online Algebra I course in all 35 treatment schools.

Description of Students in Participating Schools

As described above, the study included two samples of eighth-grade students: algebra-ready students and non-algebra-ready students. At the start of the 2008–2009 school year, 445 eligible, algebra-ready students were enrolled in the 68 participating schools. (A total of 23 students had moved over the summer.) They represented 22 % of the eighth-grade students in participating schools. The other eighth graders who were attending the study schools totaled 1,554 students and comprised the non-eligible student sample. Table 1 provides a breakdown of the number of schools and students in the study.

To describe the study sample and check for baseline equivalence of the treatment and control groups, we collected information about students' background characteristics from administrative records from the state for Maine participants and from the supervisory unions (school districts) for Vermont participants. These data included race and ethnicity, gender, eligibility for free or reduced-price lunch, and eligibility for special education services. The administrative data also included prior achievement on state math assessments taken the year before the study, when students were in seventh grade. The Maine state assessment is given in the spring of each year; the scores collected were from spring 2008. The Vermont state assessment is given in October each year; the scores collected were from fall 2007. Because Maine and Vermont use different tests, it was necessary to translate scores into a common metric. All scores were standardized to Z-scores using the mean and the standard deviation of the test scores within each state, including only schools participating in the study (algebra-ready and non-algebra-ready samples

---

impose a definition or set of criteria for algebra readiness on the participating schools for two main reasons. First, there were no common instruments across all schools that were administered prior to random assignment that were specifically measures of algebra readiness. Second, the study aimed to test the effectiveness of offering an online Algebra I course in a real-world context, where local decision-making about student eligibility for the course would be the norm. We found that the students identified as algebra-ready had, as expected, significantly higher prior math achievement scores than those who were not.

Table 1 Number of schools and students per condition as of fall 2008

Item	Total	Treatment	Control	p-value
Number of schools	68	35	33	a
Number of grade 8 algebra-ready students	445	218	227	0.670
Number of grade 8 non-algebra-ready students	1,554	782	772	0.800
Total number of grade 8 students	1,999	1,000	999	0.982
Average number of algebra-ready students per school (standard deviation)	6.54 (5.23)	6.23 (5.21)	6.73 (5.32)	0.698
Average number of grade 8 students per school (standard deviation)	31.94 (37.01)	31.00 (40.93)	32.94 (32.96)	0.830

Note: Sample includes 68 schools (35 treatment, 33 control) and 1,999 students (445 eligible students, 1,554 non-eligible students). Algebra-ready students were identified before random assignment in June 2008. Tests of significance were conducted using two-tailed χ^2 and independent sample t-tests
Source: Records obtained from each school before random assignment (June 2008) and school rosters examined in fall 2008
[a]Not applicable, because schools were allocated to treatment and control using a block randomized procedure

combined). Student records were available for more than 97 % of the students attending study schools.

Table 2 summarizes the characteristics of students in the participating schools as of fall 2008. It shows baseline characteristics for the two student samples overall (across conditions) and by condition.

Based on pre-random assignment data from the prior school year, we found no significant differences between the treatment and control groups on any measured demographic characteristics and prior mathematics achievement between algebra-ready students in treatment schools and control schools. The same was true for non-algebra-ready students in treatment and control schools, indicating that the random assignment of schools to condition successfully produced groups of students that were statistically comparable.

As expected, there were some differences between the algebra-ready and non-algebra-ready student samples, providing empirical evidence that schools identified higher-achieving students as algebra-ready. In particular,

- 32 % of algebra-ready and 46 % of non-algebra-ready students in the study participated in the National School Lunch program (i.e., received free or reduced-price lunch).
- 3 % of algebra-ready and 17 % of non-algebra-ready students received special education services.
- On average, algebra-ready students scored nearly 1 standard deviation above the sample mean on their grade 7 state mathematics test ($z = 0.95$). The average among non-algebra-ready students was 0.24 standard deviations below the mean.

Table 2 Baseline student characteristics of algebra-ready and non-algebra-ready student samples

Characteristic	Overall	Treatment	Control	p-value
Algebra-ready students				
Percent with free or reduced-price lunch ($n=436$)	32	34	30	0.596
Percent receiving special education services ($n=437$)	3	3	4	0.622[a]
Percent limited English proficient ($n=437$)	3	5	2	0.646[a]
Percent female ($n=440$)	49	49	49	0.833
Percent racial/ethnic minority ($n=440$)	7	8	5	0.975
Mean grade 7 score on state mathematics assessment (standardized)[b] ($n=437$)	0.95 (0.69)	0.97 (0.59)	0.94 (0.77)	0.584
Non-algebra-ready students				
Percent with free or reduced-price lunch ($n=1,403$)	46	46	47	0.883
Percent receiving special education services ($n=1,419$)	17	19	16	0.255
Percent limited English proficient ($n=1,419$)	3	4	2	0.927[a]
Percent female ($n=1,439$)	50	49	50	0.731
Percent racial/ethnic minority ($n=1,438$)	5	7	4	0.596[a]
Mean grade 7 score on state mathematics assessment (standardized)[b] ($n=1,403$)	−0.24 (0.86)	−0.25 (0.84)	−0.22 (0.89)	0.609

Source: Maine state department of education and Vermont supervisory unions; study records
Sample includes 68 schools (35 treatment, 33 control); Full samples included 440 algebra-ready students (218 treatment, 222 control); and 1,445 non-algebra-ready students (744 treatment, 701 control); 4 control schools had no non-eligible students. Student sample sizes vary for each row, based on the amount of missing data for each student characteristic
Values are unadjusted. Differences in student characteristics by condition were tested using a model that accounts for the clustered data structure and blocking used for randomization. Figures in parentheses are standard deviations
[a]The model did not converge to produce estimates when controlling for five state by size dummy blocking variables. Reported p-value represents a model that controls for state and two dummy indicators for medium and large schools rather than their interactions
[b]State mathematics scores were standardized by using the mean and standard deviation of the test scores within each state, including only schools participating in the study. Data were missing for fewer than four algebra-ready students; data were also missing for 42 non-algebra-ready students (23 from treatment schools and 19 from control schools)

- In fall 2008, the average pretest score for algebra-ready students was 349.9 (SD=23.3); for non-algebra-ready students, the average pretest score was 312.6 (SD=27.2).

Significance tests confirmed that algebra-ready students scored higher on measures of prior math achievement than non-algebra-ready students. They were also less likely to receive free or reduced-price lunch or special education services. The two samples of students were similar on other demographic characteristics, including gender, race/ethnicity, and English proficiency.

Measures

We collected a range of data for this study from fall 2008 through spring 2010, including the already-described background characteristics data for the schools and students participating in the study to describe the sample, check for baseline equivalence, and use as covariates in impact analyses. In this section we describe the data collected to measure key aspects of implementation and to measure the study outcomes.

Implementation Measures

To measure implementation of the online course in treatment schools, we used data archived from the online course itself and weekly logs completed by the in-class online course proctors. We also conducted site visits to each treatment school. To measure aspects of mathematics instruction including content in both treatment and control schools, we used teacher surveys and collected classroom materials.

Online course activity data: The online course management system used to deliver the course (Moodle) automatically logs and stores data documenting online course activity of students, online teachers, and proctors. For all online course sections, Moodle records the date and time that users logged into the system, as well as the sender, recipient, and content of all messages. The system also records the content accessed by students, quiz and exam grades, and records of teachers' and proctors' review of student grades. These logs were the primary source of information used to assess online teachers' monitoring of and communication with students. The study team "observed" direct online interactions between online teachers and their students using these archived data. For each of the ten online course sections, we randomly selected 1 school day each month over an 8-month period (October–May) and downloaded all online activity over a 24-h period for a total of 80 observations.

Proctor logs: Each week on-site proctors recorded the amount of time they spent performing specific types of activities as part of their proctor role. Data from the logs were used to monitor implementation across treatment schools. All proctors completed at least half of their weekly logs.

Site visits: Study team members visited every treatment school once during the 2008–2009 school year to assess the extent to which the online course was being implemented as intended. At each school, we noted the physical location of students taking the online course and proctor activities while students accessed the course.

Teacher survey: We administered a web-based teacher survey in spring 2009 to all grade 8 mathematics teachers in study schools and to online Algebra I teachers. The survey served two main purposes. First, it provided data on characteristics of the teachers (for example, degree earned, years of teaching experience). Second, it provided data on the organization and delivery of grade 8 mathematics instruction in treatment and control schools, as well as in the online Algebra I classes. For example, the survey collected information about the number of students enrolled in each section of grade 8 mathematics and the different types of learning opportunities provided to students, including differentiated and accelerated learning opportunities (such as Algebra I). The response rate on the teacher survey was 95 %.

Classroom materials: To describe the content coverage in the general grade 8 mathematics classes in study schools, the study team worked closely with participating teachers to collect detailed sets of instructional materials, including pacing guides and course syllabi, textbook information/tables of contents, classroom assignments, and exams. When collecting these instructional materials, the study team used a structured protocol to capture how much of the curriculum/textbook each teacher completed and the associated mathematics topics. Math content experts staffed on the study analyzed the annotated materials and calculated the percentage of class time spent on algebra topics, using a code of 1–4 (1 = 25 % of time spent on algebra, 2 = 50 % of time spent on algebra, 3 = 75 % of time spent on algebra, 4 = 100 % of time spent on algebra). The study team obtained classroom instructional materials for 90 % of the eighth-grade math classes in study schools.

Outcome Measures

Algebra and general math posttests: The posttests were computer-adaptive assessments administered to all grade 8 students in participating schools in May and June 2009. The posttest was delivered as a 40-item test that included 20 items from a general mathematics item bank and 20 items from an algebra item bank.

The items given to each student were targeted to ability level, depending on their answers to previous questions. The general math item bank included approximately 1,000 items distributed across six domains (number, computation and estimation, measurement, geometry, probability and statistics, and algebraic concepts) ranging in difficulty from the fifth- to eighth-grade level. Scores were generated by a linear transformation of the underlying Rasch scale and are reported on a scale of 200–400. The algebra item bank contained approximately 300 items, and scores are reported on a scale of 400–500.

Although students took both tests in the same sitting, the algebra posttest scores represent the primary achievement measure at the end of grade 8. (Scores on the general mathematics posttest were a secondary measure of achievement at the end of grade 8.) Response rates on the algebra and general mathematics posttests were 99 % for the algebra-ready students and 94 % for the non-algebra-ready students.[10]

High-school math coursetaking: To measure high-school math coursetaking for the algebra-ready students, we collected actual course enrollment and grades for grade 9 and planned coursetaking for grade 10 from the high schools that students attended the next year (2009–2010).

Specifically, we collected course titles and grades for the mathematics courses taken in ninth grade and course titles for the math courses in which students planned to enroll for tenth grade. We collected this information from the high schools the algebra-ready students attended in grade 9, and we obtained usable coursetaking data for 427 algebra-ready students (97 % of the sample).

For the non-algebra-ready students, we simply collected *planned* high-school math coursetaking information.[11] We obtained this information from the participating middle schools. We collected the name of the planned ninth-grade math course for each student at the end of eighth grade, in spring 2009. These data were made available to the study team for 93 % of the non-algebra-ready students. (Planned courses and the high-school students planned to attend were also provided by the middle schools for 97 % of the algebra-ready students.)

Coding of high-school coursetaking data was based on methods used by the NCES for the National Assessment of Educational Progress and Education Longitudinal Study transcript studies. Transcript coding protocols guided the extraction of course identifiers. Mathematics education experts coded the course titles, using the Classification of Secondary School Courses, which is based on information available in school catalogs and other information sources (U.S. Department of Education, 2007).

The creation of the coursetaking indicators based on these codes was guided by previous research on typical high-school course sequencing and definitions of "advanced," "intermediate," or "low" high-school course sequences by Schneider et al. (1998) and Stevenson et al. (1994). In US high schools, the typical sequence is Algebra I → Geometry → Algebra II → Precalculus/Trigonometry → Calculus. Advanced, intermediate, and low sequences are defined by where students are in this pipeline during each year of high school. The study team drew on this research to define two coursetaking sequences for the study: "advanced" for algebra-ready students and "intermediate" for non-algebra-ready students.

---

[10] However, scores based on less than 5 min of testing were determined to be invalid by the test developer and thus were dropped and treated as missing. For the algebra posttest, there were fewer than four such cases in the algebra-ready student sample and 118 in the non-algebra-ready sample (73 in treatment and 45 in control).

[11] Data on high-school mathematics coursetaking were not collected for the non-algebra-ready students because of cost constraints and the determination that assessing the impact of online Algebra I in grade 8 on subsequent coursetaking was most critical and relevant for the already-ready students.

For algebra-ready students (who were followed into high school), the study team coded whether their actual grade 9 courses and grades and planned grade 10 courses were indicative of an *advanced* sequence, defined by Schneider et al. (1998) as the successful completion of Geometry and Algebra II by grade 10.

Algebra-ready students were coded as participating in an advanced course sequence if they met the following criteria:

• Completed a full-year course above Algebra I or equivalent in grade 9.
• Earned an end-of-year grade of C or above in the grade 9 course (if more than one grade 9 course was taken, the grade had to be C or higher in the most advanced course taken to meet this criterion).
• Enrolled in Algebra II (or the next course in the sequence) for grade 10.

Students who did not meet all three criteria were coded as not participating in an advanced course sequence.

For non-algebra-ready students, we assumed that they would not follow an advanced sequence in high school, since they had not been identified as "ready for algebra" as rising eighth graders. We therefore coded whether their planned grade 9 courses were indicative of an *intermediate* sequence, defined by Schneider et al. (1998) as the successful completion of Algebra I in grade 9 and Geometry in grade 10. Non-algebra-ready students were assigned codes for planned grade 9 courses according to whether or not the course for grade 9 was at or above Algebra I.

The Online Algebra I Course: Course Content, Online Teachers, and On-Site Proctors

To identify the online course for the study, we established a set of criteria that matched the study's design to examine the impact of offering an online Algebra I course to broaden grade 8 students' access. One criterion was that the content of the online Algebra I course represent what is typically taught in a high school-level Algebra I course. Along these same lines and to increase the external validity of the study, we selected an online course provider with an existing Algebra I course that was being used by secondary schools in the United States. A second criterion was that the online course provider would hire, train, and supervise the online teachers. This criterion was considered essential given the study's focus on rural schools, which typically do not have the resources to provide an Algebra I teacher or course to their grade 8 students.

We identified 11 online course providers that offered an Algebra I course and interviewed those who were willing to consider participating in the study. At the conclusion of the interviews, Class.com was the only online course provider whose online Algebra I course met the study's criteria and who would agree to participate in the randomized study and to operate within the study's parameters.

In a traditional face-to-face course, the classroom teacher determines the course content covered, as well as the pace of instruction. In most online courses, the expectation is that students will work primarily on their own and that their progress in the course will be supervised or monitored by an online teacher and, in some cases, by an on-site school staff member. The online Algebra I course implemented for the study included these three components: the online course curriculum, online teachers, and school staff who served as on-site "proctors." These three components are described in the following sections.

Course Content

While the structure and content of the Class.com online Algebra I course was similar to that of many Algebra I courses in terms of the curriculum and assignments, the mode of instruction represents a significant departure from what students typically encounter in a face-to-face course. The Class.com course used an *asynchronous* online instructional model, which means that students and online teachers were not online at the same time. The role of the online teacher was to supervise students' learning and progress through the course. All communication between the online teachers and their students took place through asynchronous (non-instant) messages sent through the online course management system.

The Class.com course was divided into two parts, Algebra IA and Algebra IB, with each part designed to be equivalent to a semester in a traditional middle- or high-school Algebra I course. Algebra IA had 5 units, which addressed symbols and number properties, functions and equations, equations and problem solving, inequalities and absolute value, and polynomials. Algebra IB had four core units, which focused on functions and relations, systems of equations and inequalities, the simplification of rational and radical expressions, and quadratic equations. Two additional units focused on statistics and probability.

Even though students worked primarily independently while logged into the course, the stated expectation for the study was that students would follow a schedule for completing each topic, and that they would complete the course by the end of the academic year. According to Class.com, students needed 32–34 weeks (160–170 days) of 40–50 min of instruction each day to complete Algebra IA and IB. During the study year (2008–2009), both Maine and Vermont mandated a 175-day school year for grade 8 students, which theoretically provided sufficient time for students to complete both parts of the Algebra I course.

Each of the topics in the online course was designed to be completed in a single class period. The course material for each topic was presented to students in the form of an electronic, interactive textbook. The following are examples of types of course material and instruction activities that students encountered daily.[12]

- All topics began with static text for students to read, with rollover definitions available for important terms that appeared in bold.
- Computer-scored fill-in-the-blank "your-turn" problems embedded within lessons and end of lesson practice problem sets were a primary component of the online course. Students could receive immediate feedback on whether their answers were correct or incorrect by selecting the "Check My Answers" option. However, it was the students' responsibility to complete the problems, check their answers, and follow-up if they did not understand something, as the online course management system did not record whether students attempted the practice problems or not. Students' progress through the course was entirely based on their performance on quizzes and exams, not on daily problem sets.
- For some of the topics, online teachers developed "chalktalks" for their students and posted them through the online course management system. These were mini-lessons consisting of a short video with an audio voiceover describing the solution steps of the problem presented.
- Interactive activities were a part of the course but were available to students less frequently than other noninteractive activities. The complexity and instructional purpose of the interactive activities varied, and included interactive demonstrations, guided questions, and open-ended prompts.
- Quizzes (both practice and graded) were included at the end of each lesson, and exams (both practice and graded) were given at the end of each unit. Quizzes and exams consisted of item sets randomly generated by the course management system from the Class.com item bank.

For each topic, students typically encountered an average of seven web pages of static and interactive text, one or more chalktalks, eight "your-turn" problems, and ten practice set problems. Topics ended with a summary of key ideas.

Online Teachers

Class.com hired eight teachers from its network of mathematics teachers to serve as online teachers for the study. The eight online teachers taught a total of ten online course sections in the 35 treatment schools; the average number of students per section was approximately 20 students. All of the online teachers were certified to teach math and met both states' "highly qualified teacher" criteria. Class.com provided a 2-day training workshop before the beginning of the school year (attended by all online teachers) and an optional 1-day workshop in January 2009 (attended by six of

---

[12] The examples are drawn from the Slope-Intercept topic of the Other Forms of Linear Equations lesson in the Linear Equations unit.

the online teachers). The purpose of the summer training was to familiarize the teachers with the structure and operation of the online course, demonstrate how to operate the courseware and use the embedded communication tools, and suggest methods for guiding student progress. The January workshop reviewed the topics presented in the summer workshop. Class.com also provided a senior mathematics specialist to oversee the administration of the online course.

On-Site Proctors

Participating schools agreed to assign students taking the online Algebra I course to a regular class period that met with the same frequency and duration as the regular eighth-grade math classes and to provide a school staff member who would be available to students during the designated class periods. The role of the proctor was to ensure that students had access to the required technology, proctor exams, supervise students' behavior, serve as a personal contact for students and parents, and serve as the liaison between the online teacher and the school or parents. Because their role did not include providing math instruction, proctors did not need to be certified math teachers. The schools selected the staff members who served as proctors.

The on-site monitors also participated in a training workshop prior to the beginning of the 2008–2009 school year. The training session covered the structure of the online course and provided hands-on training on operating the courseware, including viewing students' progress through the course, accessing their assessment scores, and using the embedded communication tools. The training also suggested methods for helping students keep track of their own progress.

Implementation Findings

In this section, we describe the implementation of the online course in treatment schools, and the content of the math classes taken by algebra-ready and non-algebra-ready students in control schools. As described in the Measures section, we collected several types of implementation data to describe the implementation of the online Algebra I course, including archived data collected by the online course management system, weekly proctor logs, and in-person observations on site visits. To describe the content of the math classes, we collected and analyzed classroom materials.

Online Course Activity

We used the archived online course data to conduct "virtual observations" of the online course activity including the interactions between the online teachers and students. These data included information about when and how often students and teachers logged into the course, various types of online activity, and all online communication between the online teachers and students. As noted above, we randomly selected one school day each month (October through May) for each of the ten online course sections and then downloaded and coded all online activity that occurred during a 24-h period. The result was a total of 80 observations (eight per section). In this section, we describe what we learned from these observations.

Student log-ins: Students were expected to log into the online course at least once per day. However, the archived course data showed that only 75 % of students taking the course logged into on average during each of the 24-h observation periods. This is in line with anecdotal reports from Class.com that students missed a number of days, particularly in the second half of the year, because of grade- or school-wide activities.[13]

Online teacher activities: According to Class.com, the role of online teachers was to grade written assignments, review students' scores on quizzes and exams, coach and motivate students, conduct online discussions, and demonstrate concepts and processes. During the 2-day training workshop for the online teachers, Class.com demonstrated the means through which teachers could use the online course management system to monitor students' progress in the course and to communicate with them asynchronously.

During the training, Class.com indicated that online teachers would be expected to monitor students' progress on a daily basis and communicate with them on most days. To assess the regularity with which online teachers actually logged in to the course to monitor student activity, we analyzed the archived online course activity data to tally how often each section's online teacher logged on to the course management system at least once during each of the 24-h observation periods, as well as the online activities in which they engaged while online.

We found that the online teachers:

- Logged into the online course at least once during almost all of the observation periods 96 %)
- Monitored students' login activity in only 70 % of the observed sessions
- Examined student grades and course progress in only 43 % of the observation periods

Teacher–student communications: Class.com was explicit about its expectation that online teachers should communicate with students on a daily basis through the messaging feature of the online course management system; this included sending messages to students or reading and responding to messages from students within 24-h. To determine the frequency of teachers' communication with individual students, we first counted the number of teacher-to-student messages. Then, we coded the content of the message as providing administrative feedback (for example, grades, the pace at which a student was progressing through the course), mathematics content (for example, encouraging understanding, reflection or critical

---

[13] We were not able to compare this to attendance in control schools, because we did not have mathematics class-specific attendance data for students in control schools.

thinking; providing constructive feedback; and using incorrect answers as learning opportunities), or some other content (such as greetings).

We found that teachers did send at least one message in almost all of the observations. However, only 27 % of students taking the course were recipients of these messages. About half of the observations included an instance of teachers sending a message containing administrative feedback and teachers directed these messages to 13 % of their students on average. Teachers sent even fewer messages that contained mathematics content; these messages occurred in 18 % of the observations overall and were directed to 13 % of the students in each section, on average.

Online teachers were quite responsive to the messages they received from students. At least one student asked their online teacher a question through the course messaging system in 81 % of the observed sessions, and the online teachers almost always replied to students' messages within 24-h. As was the case with the teacher-initiated messages, only some of the students participated in these communications; on average, only 10 % of the students in a section sent a message to the online teacher during one of the observed periods.

On-Site Proctors

Even though the proctor role as defined for the study did not include providing math instruction, in 28 of the 35 treatment schools (80 %), the proctor was the eighth-grade mathematics teacher. In most of these cases, the proctor was teaching the regular eighth-grade math class at the same time they were responsible for supervising the online students. In 24 of these schools, students taking the online course sat in a designated area of the same classroom where the proctor/mathematics teacher was simultaneously teaching the regular eighth-grade math class to other students. In other treatment schools where the proctor was the eighth-grade mathematics teacher, students sat in a location near the classroom while the teacher taught the regular eighth-grade math class (e.g., an unoccupied neighboring classroom, the hall). In treatment schools in which the proctor was not the eighth-grade math teacher, other teachers in the school, the principal, or an education technology specialist served as the proctor. These proctors had other simultaneous responsibilities in addition to monitoring the online students such as lesson-planning and grading, performing administrative tasks, or supervising a computer lab.

Although the role of the proctor as defined for the study did not include providing mathematics instruction, the proctors were a source of instructional support for online students. This was not surprising given that so many of them (80 %) were eighth-grade math teachers, but was not initially expected at the outset. Proctors' weekly logs revealed that they spent an average of about 50 min a week throughout the year answering students' algebra-related questions and 10–14 min a week answering non-algebra math questions. They also assisted students with technical issues for about 8–10 min per week. According to the logs, the proctors communicated with the online teachers only for about 5–6 min per week.

Online Course Completion Rates

As described above, the online course was a full-year course. Students could move slightly ahead or spend extra time covering topics as necessary, but the course was not intended to be completely self-paced and the expectation was that students would finish the course by the end of the academic year.

We found that the pace at which students actually progressed varied, as did rates of course completion. Student "completed" a unit by passing the respective end-of-unit test with a score of 60 % or higher. We defined passing for the study as 60 % or higher because the standard criterion for "passing" varied across participating schools, with some setting the passing criterion at 60 % and others at 70 %. The 60 % threshold was chosen so that students would not be held to higher standards by the course than was typical for their school.

We found that about 43 % of the 211 algebra-ready students who took the course completed all 9 units of Algebra IA and IB, and another 39 % completed all of IA and some of IB (6, 7, or 8 units). Another way of looking at this is that algebra-ready students, on average, completed 7.5 of the 9 Algebra I units, or 85 % of the online course.

Interestingly, while we found variation in course completion rates both within and between course sections, student background characteristics including gender, race/ethnicity, limited English proficiency, eligibility for free or reduced-price lunch special education services, or—surprisingly—prior year scores on the state mathematics achievement tests did not predict course completion.

Course Content in Control Schools: Treatment Contrast

We analyzed classroom materials to capture the content of the general math courses offered in control schools, as these courses as taken by the algebra-ready students in control schools served as the contrast to the online Algebra I course in treatment schools. The classroom materials we collected included the name of the textbook used and any of the following that were available: course syllabi, curricular pacing guides, annotated tables of contents of mathematics textbooks, and course exams. Mathematics content experts on the study team coded the general grade 8 class materials, indicating the degree to which they focused on algebraic content: 25, 50, 75, or 100 %. We used just four categories because detailed pacing information was available from too few schools to estimate more precisely (by, for example, the number of weeks spent on algebra).

We found that in control schools, over 90 % of the eighth-grade math classes had a curricular focus on algebraic content of 50 % or higher. More than one-third (35 %) of the control schools had a focus on algebra of 75 %, and 16 % had an algebraic focus of 100 %. As noted earlier, we had expected the general math courses in control schools to include a substantial amount of algebraic content, based on our review of state content standards and recruitment discussions with state and local educators, and this was confirmed. However, it is important to note that without a separate comparison group of schools that was unaware of the study, there is no way to know whether the amount of algebra in the study control schools represents what is typical or whether the amount of algebra offered was affected by participation in the study.

Our analysis of classroom materials also found that in seven control schools, most (94 %) of the algebra-ready students took a separate Algebra I course at their middle school or the local high school. In total, 45 algebra-ready students in control schools took a formal Algebra I course, representing 20.3 % of the total sample of algebra-ready students in control schools. Some of these students took a face-to-face Algebra I course and others took an online course from a different provider (i.e., not Class.com).

Summary of the Implementation Findings

- Most schools opted *not* to place students taking the online course in a separate space or at a different time from the regular eighth-grade math class.

    - In most (80 %) of the treatment schools, the regular eighth-grade math teacher served as the proctor.
    - In 69 % of the schools, students taking the online course sat in the same classroom as students taking the regular eighth-grade math class.

- The types and amount of interaction between students taking the online course and their teachers and proctors deviated from our expectations.

    - Online teachers spent less time communicating directly with students than expected. The online teachers logged in to the course at least once a day to monitor students' activity or progress, but they communicated directly with only about 25 % of the students every day. Communications containing math content were infrequent; however, when a student contacted the online teacher directly, the teacher almost always (96 % of the time) replied within 24-h.
    - In-class proctors spent more time providing math content support than expected. Although the proctors' role did not require providing math instruction, they spent an average of about 60–75 min per week answering algebra-related or other non-algebra-related math questions for students taking the online course.

- Rates of course completion in the online course varied.

    - The average number of course units completed by students who took the course was 7.5, or 85 % of the online course.
    - 43 % of the online algebra students completed the entire course, including all 9 core units of the full-year course (5 units in Algebra IA and 4 in IB).
    - 82 % of the online algebra students completed the first half of the course (Algebra IA) and part of Algebra IB.

Impact Findings

In this section, we first present the results of analyses conducted to test the primary research questions regarding the direct effects of offering online Algebra I to algebra-ready eighth graders on their algebra achievement at the end of grade 8 and their likelihood of participating in an advanced math course sequence in high school. Second, we present results of analyses conducted to test the secondary research questions, regarding potential "side effects" of offering online Algebra I to algebra-ready eighth graders on their general math achievement and outcomes for their non-algebra-ready peers.

Impacts on Algebra-Ready Students' Algebra Scores and High-School Coursetaking

Algebra score at the end of grade 8: We used a two-level hierarchical model with students nested within schools to estimate the impact of online Algebra I on algebra-ready students' algebra assessment scores at the end of eighth grade. To improve the precision of the impact estimates, we included students' prior state mathematics test scores and background characteristics (gender, eligibility for free or reduced-price lunch, and special education status) as covariates in the model. School-level covariates included blocking variables (state and school size dummy variables). Except for the treatment status indicator, all covariates were centered on the grand mean.

We found that algebra-ready students in schools randomly assigned to offer the online Algebra I course scored higher on the algebra posttest than their counterparts in schools that did not receive the course (Table 3). The average algebra score for algebra-ready students in treatment schools was 5.53 scale score points higher than the average score for eligible students in control schools (effect size = 0.40). This difference is equivalent to moving students from the 50th to the 66th percentile in achievement by taking the online course.[14]

---

[14] As described above, 20 % of algebra-ready students in control schools took a formal Algebra I course; some of these students took a traditional face-to-face version of the course and others took an alternate online version of the course. The study was not designed to compare the outcomes of algebra-ready students in treatment schools with those of subgroups of students in control schools and it would be inappropriate to conduct significance tests for these comparisons. However, to address questions regarding how the algebra scores of algebra-ready students in treatment schools compared to those of algebra-ready students in control schools who took a formal Algebra I course in grade 8 and algebra-ready students in control schools who did not take a formal Algebra I course in grade 8, we report the observed means for these three groups below. It is important to note that the study was not designed to test for the statistical significance of these differences and that the means reported below are based on the original and not imputed data, are not model-adjusted, and should be interpreted with caution. The observed Promise Assessment posttest mean scores were 447.91 (SD = 15.12) for algebra-ready students in treatment schools, 444.00 (SD = 11.16) for algebra-ready students in control schools who took a formal Algebra I course, and 440.98 (SD = 12.56) for algebra-ready students in control schools who took their schools' eighth-grade general mathematics course.

Table 3 Impact of online Algebra I on algebra scores of algebra-ready students in treatment and control schools

Mean in treatment schools (standard deviation)	Mean in control schools (standard deviation)	Estimated impact (standard error)	p-value	Effect size
447.17 (15.04)	441.64 (12.29)	5.53[a] (1.57)	0.001	0.40

Source: Algebra scores on study-administered computer-adaptive algebra posttest

Note: Sample includes 68 schools (35 treatment, 33 control) and 440 algebra-ready students (218 treatment, 222 control). The treatment group and control group means are the model-adjusted mean scores for algebra-ready students, controlling for all covariates in the impact model. The effect size was calculated using a pooled standard deviation of the outcome for algebra-ready students in treatment and control schools that incorporates both within and between imputation variance (SD = 13.78)

[a]Two-tailed statistical significance. Because of a multiple comparison adjustment that accounts for two primary analyses, a p-value less than 0.025 is considered statistically significant

High-school coursetaking: We used a two-level hierarchical generalized linear model, appropriate for binary outcomes, to estimate the effect of online Algebra I in eighth grade on the likelihood of participating in an advanced mathematics course sequence in high school for algebra-ready students. The model assumed a Bernoulli sampling distribution and logit link function (McCullagh & Nelder, 1989; Raudenbush & Bryk, 2002) and controlled for the same student- and school-level covariates as the model used to test the impact of online Algebra I on algebra-ready students' algebra scores.

The outcome measure for this analysis was participation in an advanced course sequence, based on the ninth-grade math courses taken, grades earned, and the course planned for grade 10. Students were considered advanced if they took a course above Algebra I in ninth grade, passed their ninth grade course with a grade of C or higher,[15] and enrolled in Algebra II or higher for tenth grade.

The results indicate that algebra-ready students from schools randomly assigned to offer the online Algebra I course were significantly more likely to follow an advanced mathematics course sequence than their counterparts in schools that did not offer the course (Table 4). Specifically, the average probability of participating in an advanced course sequence was 0.26 for algebra-ready students from control schools and 0.51 for algebra-ready students from treatment schools. The online course yielded a difference in the probability of participating in an advanced course sequence of 0.25, meaning that algebra-ready students from treatment schools were

---

[15] If students took more than one mathematics course in grade 9, they had to have earned a grade of C or better on the more advanced grade 9 course to meet this criterion.

Table 4 Predicted probability of algebra-ready students participating in an advanced math course sequence in high school

Treatment school (standard error)	Control school (standard error)	Difference in probability attributed to online Algebra I	p-value
0.51 (0.07)	0.26 (0.05)	0.25[a]	0.007

Source: Coursetaking data collected from high-school algebra-ready study students attended in 2009–2010

Note: Sample includes 68 schools (35 treatment, 33 control) and 440 students (218 treatment, 222 control). Coursetaking patterns were coded as representing successful completion of a course above Algebra I in grade 9 and enrollment in Algebra II or a higher course in grade 10) or not. The probabilities are the average model-predicted probabilities, controlling for all covariates specified for the model

[a]Two-tailed statistical significance. Because of a multiple comparison adjustment that accounts for two primary analyses, a p-value less than 0.025 is considered statistically significant

nearly twice as likely to participate in an advanced mathematics course sequence as algebra-ready students in control schools.[16]

Summary of main analyses: Offering online Algebra I to algebra-ready eighth graders yielded benefits for both their algebra achievement at the end of grade 8 and their coursetaking patterns in high school. As intended, offering Algebra I as an online course in eighth-grade-enabled eligible students to learn a substantial amount of algebra while still in middle school and made them more likely to bypass Algebra I in high school, thus opening doors to more advanced math courses as they moved through the pipeline. These results were consistent across a number of alternative model specifications that we tested as sensitivity analyses.

Impacts on Algebra-Ready Students' General Math Achievement and Non-Algebra-Ready Students' Outcomes

The primary focus of the impact of access to online Algebra I for algebra-ready students was on their algebra achievement at the end of eighth grade and subsequent high-school coursetaking. A secondary outcome was their achievement at the end of

---

[16]To address questions regarding how the percentage of algebra-ready students in treatment schools participating in an advanced mathematics course sequence compared to the percentage of algebra-ready students in control schools who took a formal Algebra I course in grade 8 and the percentage of students in control schools who did not take a formal Algebra I course in grade 8, we report the observed percentages for these three groups below. Again, it is important to note that the study was not designed to test for the statistical significance of these differences and that the percentages reported are based on the original and not imputed data, are not model-adjusted, and should be interpreted with caution. The observed percentage of students participating in an advanced mathematics course sequence was 54 % for algebra-ready students in treatment schools, 42 % for algebra-ready students in control schools who took a formal Algebra I course, and 24 % for algebra-ready students in control schools who took their schools' eighth-grade general mathematics course.

Table 5 Impact of online Algebra I on algebra-ready students' general math scores at the end of grade 8

Treatment schools (standard deviation)	Control schools (standard deviation)	Estimated impact (standard error)	p-value	Effect size
361.42 (24.79)	357.82 (25.43)	3.60 (2.80)	0.204	0.14

Source: General mathematics scores on study-administered Promise Assessment posttest
Note: Sample includes 68 schools (35 treatment, 33 control) and 440 students (218 treatment, 222 control). The treatment and control group means are the model-adjusted mean scores for algebra-ready students, controlling for all covariates in the impact model. Result is not statistically significant. The effect size was calculated using a pooled standard deviation of the outcome for eligible students in treatment and control schools that incorporates both within and between imputation variance (SD = 25.22)

eighth grade on a general math test. A significant negative effect on this outcome could signal a potential downside of offering an online Algebra I course to eligible students.

To test whether access to online Algebra I affected eligible students' general mathematics scores, we used the same two-level model with general mathematics scores as the outcome. The results revealed no significant difference by condition (effect size = 0.14) (Table 5).[17]

We conducted three additional secondary analyses to determine the impact of offering Algebra I online to eligible students on *non*-algebra-ready students' outcomes. The three outcomes of interest were algebra and general mathematics scores at the end of eighth grade and planned ninth grade courses. For these analyses, non-algebra-ready students in treatment schools were compared with non-algebra-ready students in control schools.

The analytic sample for these analyses included 1,445 non-algebra-ready students enrolled in the 68 participating middle schools (744 in treatment schools and 701 in control schools; 4 control schools had no non-eligible students).

To estimate impacts on non-algebra-ready students' algebra and general math posttest scores, we used a two-level hierarchical model with students nested within schools. (The models were identical to those described earlier for algebra-ready students for the same achievement outcomes.)

The results showed no significant differences in algebra or general mathematics posttest scores between non-algebra-ready students in schools that offered online Algebra I and their non-algebra-ready counterparts in control schools (Table 6). The impact of online Algebra I translates to an effect size of 0.06 on algebra scores and 0.02 on general mathematics scores, neither of which is statistically significant.

To estimate the effect of offering online Algebra I (to *algebra-ready* eighth graders) on non-algebra-ready students' probability of enrolling in Algebra I in grade 9 (as per an intermediate course sequence), we first coded planned ninth grade courses for non-algebra-ready students as "1" for intermediate (a course at or above Algebra I)

---

[17] The lack of a significant difference does not definitively show that general math scores for algebra-ready students in treatment and control schools were equivalent. It simply implies that the difference was not large enough to be distinguished from chance, given the size of the sample.

Table 6 Impact of online Algebra I on non-algebra-ready eligible students' algebra and general mathematics scores at the end of grade 8

Subject area	Treatment schools (standard deviation)	Control schools (standard deviation)	Estimated impact (standard error)	p-value	Effect size
Algebra	430.76 (15.36)	429.80 (15.64)	0.96 (1.25)	0.443	0.06
General mathematics	324.86 (28.42)	324.21 (30.04)	0.65 (2.41)	0.789	0.02

Source: Algebra and general mathematics scores on study-administered posttests
Note: Sample includes 68 schools (35 treatment, 33 control) and 1,445 non-eligible students (744 treatment, 701 control); 4 control schools had no non-algebra-ready students. Estimates were averaged across 10 multiply imputed datasets. The treatment and control group means are the model-adjusted mean scores for non-algebra-ready students, controlling for all covariates in the impact model. Results are not statistically significant. Effect sizes were calculated using a pooled standard deviation of the outcome for non-algebra-ready students in treatment and control schools that incorporates both within and between imputation variance (SD = 15.50 for Algebra and 29.39 for General Mathematics)

Table 7 Predicted probability of non-algebra-ready students enrolling in intermediate mathematics course sequence in grade 9

Treatment schools (standard error)	Control schools (standard error)	Difference in probability attributed to online Algebra I	p-value
0.89 (0.04)	0.79 (0.06)	0.10	0.099

Source: Planned courses indicated by study students at the end of grade 8
Note: Sample includes 68 schools (35 treatment, 33 control) and 1,445 non-eligible students (744 treatment, 701 control); 4 control schools had no non-algebra-ready students. Estimates were averaged across 10 multiply imputed datasets
Probabilities are the average model-predicted probabilities, controlling for all covariates specified for the model. Result is not statistically significant

and "0" for not intermediate (a course below Algebra I, such as Pre-algebra). Next, we used a two-level hierarchical model for binary outcomes. We found that the differences between students in treatment and control school were not statistically significant (Table 7).

Summary of secondary analyses: Offering online Algebra I to algebra-ready eighth graders yielded no significant side effects on their general math achievement scores or on any of the measured outcomes for non-algebra-ready students in the same schools. As with the primary analyses, these results were consistent across a number of alternative model specifications that we tested as sensitivity analyses. The secondary analyses were important to test because adopting the online Algebra I course in treatment schools represented a departure from business-as-usual math instruction, where algebra-ready and non-algebra-ready students were typically (though not strictly) heterogeneously mixed in general eighth-grade math classes.

First, our finding that taking online Algebra I did not negatively affect algebra-ready students' general math achievement responds to potential concerns that students who take Algebra I in eighth grade—in this case as an online course—will miss out on general math content. Grade 8 state math assessments are not algebra assessments, and some schools may worry that offering online Algebra I will lead to lower proficiency rates. The study findings suggest that students who take online Algebra I in eighth grade do not score significantly differently on a general math assessment than they would have had their schools not offered the online course.

Second, our finding that offering online Algebra I to algebra-ready students did not negatively affect outcomes for non-algebra-ready students responds to concerns about potential side effects of introducing tracking in schools that typically have heterogeneously mixed math classes. Preselecting some students as academically ready for algebra and then offering the online Algebra I course to those students meant that the higher-achieving students were not members of the general math class with the rest of the eighth graders in the school. It was conceivable that the result might be a watering down of the content of the regular eighth-grade math course, which in turn could yield lower achievement for students in those classes. Our study was carefully designed to test for such side effects and the results were clear that they were not present.

Conclusions and Future Directions

The combination of findings in this study showed that the implementation of an online Algebra I course to broaden students' access to Algebra I could effectively increase student performance on algebra content and enhance their probability of taking advanced coursework in high school. A successful intervention in this context was defined as one that yielded positive impacts on either end-of-eighth grade algebra achievement *or* subsequent high-school coursetaking for algebra-ready students, with no significant negative side effects on their general mathematics scores or on any achievement or coursetaking outcomes for non-algebra-ready students. The results showed that algebra-ready students with access to online Algebra I in grade 8 outperformed their counterparts in control schools on an end-of-year algebra assessment and were more likely to follow an advanced course sequence in high school. There were no obvious side effects of the course on algebra-ready students' end-of-year general mathematics achievement or on any of the non-algebra-ready students' measured outcomes. Thus, the results suggest that offering an online course to algebra-ready students in eighth grade is an effective way to broaden access to the specific course, and later, to more challenging mathematics course opportunities, for students in schools that do not typically offer Algebra I to eighth graders.

The study was designed to provide information to educators who are looking for ways to offer a key gateway course (Algebra I) to their grade 8 students who are ready for it, but for various reasons cannot typically offer full access to the course in a standard or traditional way. The goal for the intervention was to not only have an impact on algebra-ready students' short-term algebra knowledge but also influence a sequence of mathematics opportunities and outcomes over time. The hypothesis associated with the primary questions for this study was that offering an online Algebra I course would benefit the eligible students' outcomes in contrast to the mathematics instruction they would have received in absence of the online course.

It may seem obvious that students with access to an online Algebra I course in grade 8 should learn more algebra and take more advanced courses earlier in high school than those that do not. For multiple reasons, however, the results observed in the primary and secondary analyses were not necessarily obvious and addressed gaps in the research base. First, before this study, there was no prior rigorous evidence that an online version of a formal Algebra I course could be offered to grade 8 students by schools that do not typically offer the course, in terms of technology and content support. Second, though the logistical implementation of the course went as planned, just under half (43 %) of the algebra-ready students who enrolled in the course fully completed it, meaning that many of the eligible students in the treatment group were not exposed to the entire course. At the same time, algebra-ready students in control schools were exposed to a substantial amount of algebraic content in the context of their general mathematics classes, and one out of five algebra-ready students in control schools actually *did* take a formal Algebra I course either in a traditional classroom setting or in an alternate online program. Despite these circumstances, this study still demonstrated that the intervention as implemented was more effective in promoting students' success in mathematics than existing practices in these schools.

Limitations of the Study and Future Research Directions

This study was conducted with a sample of schools in Maine and Vermont that met the eligibility criteria for participation and agreed to take part in a random assignment study. Many of these schools were small (48 % had grade 8 enrollments of less than 17 students), and 90 % were in rural areas. While we know that 24 % of rural middle-grade schools do not offer Algebra I (U.S. Department of Education, 2009a), it is not clear whether the study schools represent rural schools located in other parts of the region or country. Nor do we know the extent to which the results observed in these schools generalize to other schools interested in using online courses to expand access to Algebra I to grade 8 students.

The online course that we evaluated in this study was Class.com's Algebra I course, which was similar in content and focus to the offerings of other providers. However, it is not clear that similar results would have been observed had another course provider been chosen. Moreover, the results observed in this study cannot necessarily be generalized to more recently developed online courses, including online courses that have been developed to reflect the content and structure of the CCSSM.

For these and other reasons, replication of this study is necessary to gain a better understanding of the potential impacts of using an online course to expand access to Algebra I to grade 8 students. This is particularly important as the proportion of students enrolling in grade 8 Algebra I increases with the wide-scale adoption and implementation of the CCSSM. In particular, future studies should examine longer-term effects of access to online Algebra I in grade 8—through high school, college, and even beyond. This study included a 1-year follow-up to track students from grade 8 into high school. A longer study is needed to assess whether access to online

Algebra I in grade 8 continues to have an impact on participation in advanced mathematics coursetaking through the end of high school.

As the use of online courses continues to increase in US schools, future research should continue to study their effects on student coursetaking patterns and achievement in key content areas. Further investigation of the effectiveness of online courses should contrast the offering of them with various relevant business-as-usual situations. These include school settings where students lack access to specific courses (where the control group does not take the course) as well as school settings where particular courses are oversubscribed or taught by under-qualified or uncertified teachers (where the control group would take a standard face-to-face version of the online course). More research is also needed regarding the ideal roles for online teachers and on-site proctors, both for fully online courses and in blended learning models. Our implementation findings suggest that future research should further examine issues related to communication between online teachers and students, both in general and specifically about subject matter content, as well as the nature and type of support—instructional and otherwise—needed by students from their on-site proctors.

Schools around the country, particularly those in rural areas, are in search of innovative ways to expand their course offerings. To address this need, this study focused on the use of an online course to provide access to Algebra I in schools that do not typically offer the course in grade 8. It did not compare the effects of taking online Algebra I versus a standard face-to-face version of the course in grade 8, and the results should not be interpreted to indicate that offering online Algebra I is better than (or as good as) offering a face-to-face Algebra I course to eighth graders. In addition, given that the study compared the offering of an online Algebra I course to a lack of (or limited) access to Algebra I in grade 8, it is not possible to isolate the portion of the observed effects that is due to the fact that the course was online. As noted in the earlier description of the intervention under test, the content of the course (Algebra I) cannot be untangled from the mode of instruction (online). Thus it is possible that broadening access to any type of formal Algebra I course to algebra-ready grade 8 students would yield similar effects. In this way, this study provides some of the first experimental evidence about the effects of offering Algebra I in middle school, contributing to an existing body of research that is growing but fraught with challenges related to selection. Continued rigorous research into the effects of expanding options for more students to take Algebra I in middle school will further contribute to a growing understanding of how best to increase equity and access to mathematics opportunities in high need schools.

References

Adelman, C. (1999). *Answers in the toolbox: Academic intensity, attendance patterns, and bachelor's degree attainment*. Washington, DC: U.S. Department of Education.

Adelman, C. (2006). *The toolbox revisited: Paths to degree completion from high school through college*. Washington, DC: U.S. Department of Education.

Anderson, A., Augenblick, J., DeCesare, D., & Conrad, J. (2006). *Costs and funding of virtual schools: An examination of the costs to start, operate, and grow virtual schools and a discussion of funding options for states interested in supporting virtual school programs.* Report prepared for the BellSouth Foundation. Denver, CO: Augenblick, Palaich, and Associates.

Atanda, R. (1999). *Do gatekeeper courses expand education options?* (NCES 1999-303). Washington, DC: National Center for Education Statistics, U.S. Department of Education.

Berry, A. M., & Wintle, S. E. (2009). *Using laptops to facilitate middle school science learning: The results of hard fun.* Gorham, ME: University of Southern Maine, Maine Education Policy Research Institute.

Common Core State Standards Initiative. (2010). Common Core State Standards for Mathematics. Washington, DC: National Governors Association Center for Best Practices and the Council of Chief State School Officers.

Gamoran, A., & Hannigan, E. C. (2000). Algebra for everyone? Benefits of college-preparatory mathematics for students with diverse abilities in early secondary school. *Educational Evaluation and Policy Analysis, 22,* 241–254.

Greaves, T., Hayes, J., Wilson, L., Gielniak, M., & Peterson, R. (2010). *The technology factor: Nine keys to student achievement and cost-effectiveness. Project RED.* Shelton, CT: MDR.

Hammer, P. C., Hughes, G., McClure, C., Reeves, C., & Salgado, D. (2005). *Rural teacher recruitment and retention practices: A review of the research literature, national survey of rural superintendents, and case studies of programs in Virginia.* Charleston, WV: Edvantia.

Hanum, W. H., Irvin, M. J., Banks, J. B., & Farmer, T. W. (2009). Distance education use in rural schools. *Journal of Research in Rural Education, 24*(3).

Horn, L., & Nuñez, A. M. (2000). *Mapping the road to college: First generation students' math track, planning strategies and context of support* (NCES 2000-153). Washington, DC: National Center for Education Statistics, U.S. Department of Education.

Jabon, D., Narasimhan, L., Boller, J., Sally, P., Baldwin, J., & Slaughter, R. (2010). The Chicago algebra initiative. *Notices of the American Mathematical Society, 57,* 865–867.

Jimerson, L. (2006). *Breaking the fall: Cushioning the impact of rural declining enrollment.* Washington, DC: The Rural School and Community Trust.

Johnson, J., & Strange, M. (2007). *Why rural matters 2007: The realities of rural education growth.* Arlington, VA: The Rural School and Community Trust.

Kilpatrick, J., Swafford, J., & Findell, B. (Eds.). (2001). *Adding it up: Helping children learn mathematics.* Washington, DC: Division of Behavioral and Social Sciences and Education, Center for Education, National Research Council.

Lacampagne, C. B., Blair, W. D., & Kaput, J. J. (Eds) (1995). *The algebra initiative colloquium papers.* Paper presented at a conference on reform in algebra. Washington, DC: U.S. Department of Education, Office of Educational Research and Improvement, National Institute on Student Achievement, Curriculum, and Assessment.

Loveless, T. (2008). *The misplaced math student: Lost in eighth-grade algebra.* Washington, DC: The Brookings Institute.

McCullagh, P., & Nelder, J. A. (1989). *Generalized linear models.* London: Chapman & Hall.

Moe, T., & Chubb, J. E. (2009). *Liberating learning: Technology, politics, and the future of American education.* San Francisco: Jossey-Bass.

National Governors Association Center for Best Practices & Council of Chief State School Officers. (2010). *Common core state standards for mathematics.* Washington, DC: Authors.

National Mathematics Advisory Panel. (2008). *Foundations for success: The final report of the National Mathematics Advisory Panel.* Washington, DC: U.S. Department of Education.

National Education Association. (2006). *Guide to teaching online courses.* Washington, DC: Author.

Nord, C., Roey, S., Perkins, R., Lyons, M., Lemanski, N., Brown, J., et al. (2011). *The nation's report card: America's high school graduates* (NCES 2011-462). Washington, DC: National Center for Education Statistics, U.S. Department of Education, U.S. Government Printing Office.

O'Dwyer, L. M., Carey, R., & Kleiman, G. (2007). A study of the effectiveness of the Louisiana algebra I online course. *Journal of Research on Technology in Education, 39,* 289–306.

Picciano, A., & Seaman, J. (2009). *K–12 online learning: A 2008 follow-up survey of the U.S. school district administrators.* Needham, MA: The Sloan Consortium.

Queen, B., & Lewis, L. (2011). *Distance education courses for public elementary and secondary school students: 2009–10* (NCES 2012-008). Washington, DC: National Center for Education Statistics, U.S. Department of Education, U.S. Government Printing Office.

Raudenbush, S. W., & Bryk, A. S. (2002). *Hierarchical linear models: Applications and data analysis methods* (2nd ed.). Thousand Oaks, CA: Sage.

Rose, H., & Betts, J. (2001). *Math matters: The links between high school curriculum, college graduation, and earnings.* San Francisco: Public Policy Institute of California.

Schneider, B., Swanson, C. B., & Riegle-Crumb, C. (1998). Opportunities for learning: Course sequences and positional advantages. *Social Psychology of Education, 2*, 25–53.

Schwartzbeck, T., Prince, C., Redfield, D., Morris, H., & Hammer, P. C. (2003). *How are rural school districts meeting the teacher quality requirements of no child left behind?* Charleston, WV: Appalachia Education Laboratory.

Silvernail, D. L., & Gritter, A. K. (2007). *Maine's middle school laptop program: Creating better writers.* Gorham, ME: Maine Education Policy Research Institute, University of Southern Maine.

Smith, D., & Mitry, D. (2008). Investigation of higher education: The real costs and quality of online programs. *Journal of Education for Business, 83*(3), 147–152.

Smith, J. B. (1996). Does an extra year make any difference? The impact of early algebra on long-term gains in mathematical attainment. *Educational Evaluation and Policy Analysis, 18*, 141–153.

Spielhagen, F. R. (2006). Closing the achievement gap in math: The long-term effects of eighth-grade algebra. *Journal of Advanced Academics, 18*, 34–59.

Stevenson, D. L., Schiller, K. S., & Schneider, B. (1994). Sequences of opportunities for learning. *Sociology of Education, 67*, 184–198.

Tucker, B. (2007). *Laboratories of reform: Virtual high schools and innovation in public education.* Washington, DC: Education Sector.

National Center for Education Statistics, U.S. Department of Education. (1996). *Pursuing excellence: A study of U.S. eighth-grade mathematics and science teaching, learning, curriculum, and achievement in international context.* Washington, DC: National Center for Education Statistics, Institute for Education Sciences, U.S. Department of Education.

U.S. Department of Education. (1997, October). *Mathematics equals opportunity* (White paper prepared for U.S. Secretary of Education Richard W. Riley).

U.S. Department of Education. (2007). *America's high school graduates: Results from the 2005 NAEP high school transcript study* (NCES 2007-467). Washington, DC: National Center for Education Statistics, Institute of Education Sciences, U.S. Department of Education.

U.S. Department of Education. (2009a). *Early childhood longitudinal study, kindergarten class of 1998–99 (ECLS-K) kindergarten through eighth grade full sample public-use data and documentation* [DVD] (NCES 2009-005). Washington, DC: National Center for Education Statistics, Institute of Education Sciences, U.S. Department of Education.

U.S. Department of Education. (2009b). *Evaluation of evidence-based practices in online learning: A meta-analysis and review of online learning studies.* Washington, DC: Office of Planning, Evaluation, and Policy Development, U.S. Department of Education.

Office of Educational Technology, U.S. Department of Education. (2012). *Understanding the implications of online learning for educational productivity.* Washington, DC: Office of Educational Technology, U.S. Department of Education.

Wagner, S., & Kieran, C. (Eds.). (1989). *Research issues in the learning and teaching of algebra.* Reston, VA/Hillsdale, NJ: National Council of Teachers of Mathematics/Erlbaum.

Walston, J., & Carlivati McCarroll, J. (2010). *Eighth-grade algebra: Findings from the eighth-grade round of the early childhood longitudinal study, kindergarten class of 1998–99 (ECLS-K)* (NCES 2010-016). Washington, DC: National Center for Education Statistics, Institute of Education Sciences, U.S. Department of Education.

What Works Clearinghouse. (2008). Procedures and standards handbook (version 2). Retrieved from U.S. Department of Education website: http://nces.ed.gov/nationsreportcard/pdf/main2005/2007468.pdf.

Zandberg, I., & Lewis, L. (2008). *Technology-based distance education courses for public elementary and secondary school students: 2002–03 and 2004–05* (NCES 2008-08). Washington, DC: National Center for Education Statistics, Institute of Education Sciences, U.S. Department of Education.

A Randomized Trial of Lesson Study with Mathematical Resource Kits: Analysis of Impact on Teachers' Beliefs and Learning Community

Catherine C. Lewis and Rebecca Reed Perry

Both theory and empirical findings suggest that improvement of teaching is not a one-shot activity: It requires ongoing effort by teachers, who must integrate improvements into the complex juggling act of classroom practice (e.g., Clarke & Hollingsworth, 2002; Lampert, 2001). To change teaching successfully often requires repeated cycles of classroom trial, reflection, feedback, and revised trial. For example, (Schorr & Koellner-Clark, 2003) chronicle the experience of a teacher who changed his classroom teaching with the intention of having students contribute to the construction of mathematical ideas in the classroom. He was initially pleased with his effort, but when he showed video of his classroom practice to colleagues, they saw the class discussion as lacking in mathematical rigor. Their reactions led him to reevaluate the quality of his class discussions and engage in further work to better establish mathematical focus while building student contributions.

Situations like those described by Schorr and Koellner-Clark (2003), in which teachers must engage in repeated cycles of experimentation and reflection to improve practice, are probably more the rule than the exception, even though models of professional learning impact often show a unidirectional arrow from professional development program to change in instruction (see review by Clarke & Hollingsworth, 2002). If repeated cycles of experimentation are typically needed to improve one's teaching, then what are the implications for the design of professional learning programs? For example, how might we design professional learning programs to catalyze development of the beliefs and dispositions needed to continue such challenging, ongoing work?

This chapter reports a randomized, controlled trial of lesson study supported by mathematical resources. The experimental treatment, described in more detail below, significantly increased the mathematical knowledge of both participating

C.C. Lewis (✉) • R.R. Perry
Mills College School of Education, 5000 MacArthur
Boulevard, Oakland, CA 94613, USA
e-mail: clewis@mills.edu

© Springer International Publishing Switzerland 2015
J.A. Middleton et al. (eds.), *Large-Scale Studies in Mathematics Education*,
Research in Mathematics Education, DOI 10.1007/978-3-319-07716-1_7

teachers and students, as reported elsewhere (Gersten, Taylor, Keys, Rolfhus, & Newman-Gonchar, 2014). This chapter focuses, however, on two "Intermediate Outcomes" shown in Fig. 1: teachers' beliefs and teacher learning community. We examine (1) the impact of the intervention (lesson study with mathematical resources) on teachers' beliefs and teacher learning community and (2) the role of teachers' beliefs and teacher learning community as mediators of teachers' and students' increases in mathematical knowledge.

Background on Lesson Study

As shown on the left side of Fig. 1, lesson study is collaborative, practice-based professional learning in which teachers study the academic content of the curriculum and plan, enact, observe, and analyze a live classroom lesson (Fernandez & Yoshida, 2004; Lewis & Hurd, 2011; Lewis & Tsuchida, 1997, 1998; Perry & Lewis, 2010; Stigler & Hiebert, 1999; Wang-Iverson & Yoshida, 2005). As shown in the center of Fig. 1, the practice-based cycles that comprise lesson study are hypothesized to improve instruction by simultaneously improving five basic inputs to instruction: teachers' knowledge; teachers' beliefs and dispositions; teacher learning community; learning resources and tools; and system features.

Although the term "lesson study" often evokes images of lesson planning, in fact lesson planning is just a small portion of lesson study. It may be useful to think of lesson study and other familiar professional learning approaches as overlapping circles in a Venn diagram to highlight the characteristics lesson study shares with other familiar professional learning approaches. For example, lesson study overlaps with many other professional learning approaches in the shared element of *teachers' study of content knowledge*, which previous research has identified as a feature of effective professional learning (Garet, Porter, Desimone, Birman, & Yoon, 2001) and which in lesson study occurs most heavily during the first parts of the lesson study cycle, when teachers engage in *kyouzai kenkyuu*, the study of content and curriculum materials (Takahashi, Watanabe, Yoshida, & Wang-Iverson, 2005). In Japan, the teacher's manual provides a key resource for *kyouzai kenkyuu*, since it includes discussion of both the curriculum content and of common student thinking and misconceptions (Lewis, Perry, & Friedkin, 2011). In the current study, mathematical resource kits were designed to substitute for the materials available to Japanese teachers as they conduct *kyouzai kenkyuu*.

A second element of lesson study is *observation of live practice*, and this element is shared, for example, with many professional learning programs that include coaching or mentoring (e.g., Campbell & Malkus, 2011), although lesson study focuses on observation of students rather than on critique of teaching. A third element of lesson study is *analysis of student thinking and student work*; again, many well-known professional learning approaches strongly emphasize this element (Carpenter, Fennema, Franke, Levi, & Empson, 1999). In lesson study, teachers observe students *as* they think and work, as well as analyzing student work products.

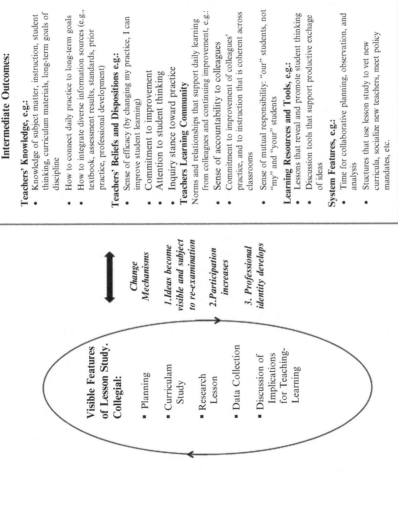

Fig. 1 Lesson study cycle

Intermediate Outcomes:

Teachers' Knowledge, e.g.:
- Knowledge of subject matter, instruction, student thinking, curriculum materials, long-term goals of discipline
- How to connect daily practice to long-term goals
- How to integrate diverse information sources (e.g., textbook, assessment results, standards, prior practice, professional development)

Teachers' Beliefs and Dispositions e.g.:
- Sense of efficacy (by changing my practice, I can improve student learning)
- Commitment to improvement
- Attention to student thinking
- Inquiry stance toward practice

Teachers Learning Community
Norms and relationships that support daily learning from colleagues and continuing improvement, e.g.:
- Sense of accountability to colleagues
- Commitment to improvement of colleagues' practice, and to instruction that is coherent across classrooms
- Sense of mutual responsibility: "our" students, not "my" and "your" students

Learning Resources and Tools, e.g.:
- Lessons that reveal and promote student thinking
- Discussion tools that support productive exchange of ideas

System Features, e.g.:
- Time for collaborative planning, observation, and analysis
- Stuctures that use lesson study to vet new curricula, socialize new teachers, meet policy mandates, etc.

Change Mechanisms
1. *Ideas become visible and subject to re-examination*
2. *Participation increases*
3. *Professional identity develops*

Visible Features of Lesson Study.
Collegial:
- Planning
- Curriculam Study
- Research Lesson
- Data Collection
- Discussion of Implications for Teaching-Learning

Improvement of Instruction (Research Lesson and Daily Practice)

Improvement of Student Learning

A fourth element of lesson study is *teacher-led inquiry*. Lesson study asks teachers to pose and investigate research questions about practice and to answer them through the study of live practice; this element is shared with approaches to professional learning such as action research, self-study, and inquiry (e.g., Noffke & Somekh, 2009). A fifth element of lesson study is *collaboration with colleagues* (e.g., McLaughlin & Talbert, 2006), which is intended to build ongoing instructional collaboration and to reshape school routines to better focus on improvement of classroom practice (Sherer & Spillane, 2011).

In summary, lesson study, which originated in Japan and has been practiced there for over a century, combines familiar elements including study of content and curriculum, study of student thinking and work, and observation of live instruction, with collaborative planning and analysis of instruction. These elements are brought together in cycles of teacher-led inquiry.

Method

The Study Design and Conditions

The tension between teacher "ownership" of an innovation and its faithful implementation is a quintessential dilemma of educational reform. Our intervention approached this dilemma by joining a teacher-led form of professional learning–lesson study–to research-based mathematics resources. We designed the mathematics resource kits because, as noted above, some US teacher's manuals do not provide sufficient information on mathematical content and student thinking to support the *kyouzai kenkyuu* during the first phase of lesson study (Lewis et al., 2011).

Using an electronic mailing list (lsnetwork@mailman.depaul.edu) and personal contacts, we recruited groups of US educators to engage in locally led lesson study, using resource kits centrally designed and distributed by our group. Since local educators took responsibility for recruiting lesson study teams and organizing and managing their own learning with the resource kits, this model departs substantially from centrally planned and "delivered" professional development in which educators are expected to faithfully implement a centrally designed set of instructional changes (Lewis, Perry, & Murata, 2006). The design of the experimental condition was intended to support local flexibility for users, by allowing each lesson study group to collaboratively make its own locally appropriate decisions about participants, use of the fractions resources we provided, and time allocation for lesson study and for specific components of the work.

More than 100 groups requested an opportunity to participate in the study. Four criteria were considered in selecting the sample of 39 sites: permission from local authorizing agencies and administrators; willingness to be randomly assigned to a study condition; site demographic characteristics (we sought diversity in region of the USA, urbanicity and student socioeconomic status); and the ability to participate within our study time frame. In the interest of supporting naturally occurring collaborative groups, we did not specify group membership, except to require that at least one group member be a classroom teacher within grades 2–5. Since some lesson study groups find it beneficial to collaborate across grades, groups could include educators at other levels or from non-classroom positions (e.g., mathematics coach). Educators who responded to our call for participation recruited local groups of 4–9 educators.

Triads of demographically matched sites (matched where possible on district and SES of students served) were created and one site from each triad was assigned by random draw to each of the three study conditions. Random assignment was not performed until after groups had completed teacher and student pre-assessments. *Condition 1* (C1) is the primary experimental treatment and consists of lesson study supported by a mathematical resource kit focused on fractions (described further in Lewis & Perry, 2014, and summarized in Fig. 2). Because we worried that the mathematical resource kit might be experienced by teachers as prescriptive, thereby undermining the sense of inquiry that should be integral to lesson study, we designed *Condition 2* (C2) as a control treatment in which teachers conducted lesson study *without* the fractions resource kit, on a topic of the group's *own* choosing other than fractions. We asked C2 groups to avoid fractions both for practical reasons (to avoid cross-condition "contamination," since many districts had groups in more than one condition) and for ethical reasons (we did not want to put local mathematics educators who had recruited more than one group in the position of withholding resources from some groups). C2 groups did receive sections 4 and 5 of the resource kit (see Fig. 2), the generic materials to support lesson study. These included tools and protocols to help groups set norms, anticipate student thinking, and plan, observe, and discuss a research lesson. Groups in *Condition 3* (C3) received no materials from our group, but participated in all study procedures (such as pre- and post-assessment) and received the study stipend upon documentation of expenses for professional development. C3 thus served as a control for selection factors (such as willingness to participate in lesson study) and for study procedures (such as study assessments and stipend). Groups in all three conditions were offered a $4,000 stipend upon documentation of expenses related to professional learning (e.g., substitutes, stipends for after-school work, course fees). The average length of participation (calendar days from student pretest to posttest) was roughly comparable across conditions: 91 days for Condition 1 groups; 80 days for Condition 2 groups; and 84 days for Condition 3 groups. No groups dropped out of the study, and only one teacher failed to complete the study.

Figure 2 summarizes the materials in the resource kit. Overall, the resource kit was designed to maintain the qualities of lesson study that are appealing to educators such as active investigation of a problem of practice, study of student thinking, and application of newly learned ideas in the classroom, while also providing ready access to mathematical resources. The basic flow of the toolkit, summarized in Fig. 2, begins with teachers solving mathematics tasks individually, sharing solutions, predicting how students might solve the tasks, and then examining actual student responses. Teachers then examine curriculum materials and research on

Introduction

Section 1: Mathematics Tasks to Solve and Discuss

Groups do and discuss problems such as <u>Problem 2</u>: "Find two fractions between $\frac{1}{2}$ and 1 and write them here," then study associated sample student work.

Section 2: Curriculum Inquiry: Different Models of Fractions

Groups examine different fraction models, the Japanese textbook and fractions curriculum trajectory, and classroom video of fractions instruction. Groups also solve a hands-on fractions task mirroring one shown on the video ("Mystery Strip").

Section 3: Choosing a Focus for Your Lesson Study Work

Groups choose either Path A or Path B. Path A centers around an introduction to fractions using the linear measurement context. Path A groups study materials based on the Japanese curriculum introduction to fractions (e.g., lesson plans; Japanese elementary Course of Study; Teaching Manuals). Path B groups study and read about another aspect of fractions, such as: 1) understanding that fractions are accumulations of unit fractions; or 2) understanding fractions on the number line, etc.

Section 4: Planning, Conducting, and Discussing the Research Lesson

Groups follow lesson study protocols, guidelines, and suggestions on reflection included in this section.

Section 5: Lesson Study Refresher: Overview and Suggestions for Getting Started

Groups new to lesson study may refer to this section for background information on how to conduct lesson study (e.g., setting norms in the group or choosing a research topic).

Fig. 2 Fractions resource kit: summary and examples of contents

fractions, with a particular focus on the linear measurement representation and how it may help students understand fractions as rational numbers. Many of the resources in this section are drawn from the Japanese curriculum, including a textbook, teacher's edition, lesson plan, and video of a Japanese educator introducing fractions to a US class, using the linear measurement representation of fractions. Finally, lesson study groups plan, conduct, observe, and discuss at least one research lesson on fractions, followed by reflection on what they learned during the lesson study cycle.

Measurement of Teachers' Beliefs and Teacher Learning Community

The central portion of Fig. 1 posits that change in teachers' beliefs and dispositions is one route by which lesson study produces changes in instruction. Changes in beliefs and dispositions may directly influence instruction; for example, change in beliefs about the value of student struggle may lead a teacher to give students more time to struggle with a challenging problem. Changes may also operate indirectly; for example, a strengthened sense of efficacy may allow a teacher to engage in the repeated cycles of experimentation needed to successfully implement a new teaching strategy that initially proves difficult.

A large body of research documents the ways that teachers' beliefs, dispositions, and identity both *influence* and *are influenced* by professional learning experiences (Clarke & Hollingsworth, 2002; Goldsmith, Doerr, & Lewis, 2014; Zech, Gause-Vega, Bray, Secules, & Goldman, 2000). (In the current work, we do not try to distinguish among beliefs, dispositions, and identity, but refer to them all using the term "beliefs.") Teachers' expectations for student achievement have been the focus of a number of studies that have shown, for example, that opportunities to closely observe students can increase teachers' expectations (Borko, Davinroy, Bliem, & Cumbo, 2000; Chazan, Ben-Chaim, & Gormas, 1998; Kazemi & Franke, 2004; Lin, 2001; Puchner & Taylor, 2006), as can collegial learning that focuses on student thinking (Lin, 2002; Tobin & Espinet, 1990). Since working with colleagues to closely observe students is a core feature of lesson study, we included survey items designed to measure teachers' expectations for student achievement.

Another type of belief that may be important to instructional improvement is interest in student thinking and in eliciting it during instruction because student thinking can provide crucial instructional feedback to teachers (e.g., Grandau, 2005; Kazemi & Franke, 2004; Remillard & Bryans, 2004; Seymour & Lehrer, 2006; Steinberg, Empson, & Carpenter, 2004). We included four survey items designed to measure teachers' sense of efficacy in eliciting and using students' mathematical thinking.

Inquiry stance toward practice and identity as a learner and teacher of mathematics is a third focus of the survey items we selected for inclusion. Prior research has shown that professional learning can impact teachers' inquiry stance, leading to increased use of analysis and knowledge-based reasoning and allowing a shift away from an evaluative stance (Sherin & Han, 2004; Sherin & van Es, 2009; Ticha & Hospesova, 2006; van Es & Sherin, 2008). Items in this area focused on teachers' interest in learning about mathematics and its teaching–learning. Prior research shows that teachers' identities as learners and teachers of mathematics can shape, for example, their learning from mathematics curriculum materials (Remillard & Bryans, 2004; Spillane, 2000). Because the resource kits we provided to the teachers in this study included curriculum and research materials, four survey items tapped teachers' perception that research and curricular materials (including those from other countries) are useful to teachers.

Another hypothesized route of lesson study influence on instruction (see Fig. 1) is through changes in the teacher learning community. Prior research indicates that professional development can, for example, engender collegial encouragement and support that enables teachers to try new types of teaching (Britt, Irwin, & Ritchie, 2001; Chazan et al., 1998; Manouchehri, 2001) and to see colleagues as a source of useful feedback and knowledge (Fisler & Firestone, 2006; Taylor, Anderson, Meyer, Wagner, & West, 2005; Thijs & van den Berg, 2002; Zech et al., 2000). The duration of the current study was somewhat brief to expect development of the teacher learning community, as it included just one cycle of lesson study. However, we included in the survey a number of items focused on workplace collaboration, including a number of items from established measures of school-site professional community (e.g., Michigan State University, 2003; CRC, 1994) and some new items designed to tap the perceived efficacy of working with colleagues to improve mathematics teaching ("Collegial Learning Effectiveness").

The pre- and post-teacher survey included the items shown in Appendix, which were interspersed for administration along with some additional items such as self-ratings of fractions knowledge (not reported here). Factor analysis and item content review were used to construct the scales, which are shown in Appendix along with scale alphas and item sources.

Measurement of Teachers' and Students' Fractions Knowledge

Since some of our analyses examine the impact of teachers' beliefs and teacher learning community on the development of fractions knowledge, we briefly describe the assessments of teachers' and students' fraction knowledge. The measure of teachers' fraction knowledge was a 33-item assessment, with 21 of the items drawn from Learning Mathematics for Teaching (2007) and the remaining items drawn from other sources including Diagnostic Teacher Assessments in Math and Science (Center for Research in Mathematics and Science Teacher Development, 2005); New Zealand Maths (Ward & Thomas, 2009); and mathematics education research and curriculum materials (Beckmann, 2005; Newton, 2008; Norton & McCloskey, 2008; Post, Harel, Behr, & Lesh, 1988; Schifter, 1998; Zhou, Peverly, & Xin, 2006).

Because the study included students in grades 2–5, three overlapping assessments of students' fraction knowledge were constructed, including between 17 items (for grade 2–3 students) and 41 items (for grade 5 students) drawn from sources including published mathematics education research (Hackenberg, Norton, Wilkins, & Steffe, 2009; Saxe, 2007; Van de Walle, 2007); the National Assessment of Educational Progress (Institute of Education Sciences/National Center for Education Statistics (IES/NCES), 2007); Japanese teachers' manuals and student texts (Hironaka & Sugiyama, 2006); and the California Standards Test (California Department of Education, 2005).

In addition to the survey and assessment data, we collected written reflections at the end of the lesson study cycle from teachers in both lesson study conditions in response to the following prompt:

> *Describe in some detail two or three things you learned from this lesson study cycle that you want to remember, and that you think will affect your future practice.* These might be things about fractions or mathematics, about teaching, about student learning, or about working with colleagues. (If you don't feel you learned anything from this cycle of lesson study, please note that and identify changes that might have made the lesson study work more productive for you.)

Data Collection

Teacher and student pre-assessments were mailed out to the sites, along with guidelines for administration. Once the completed assessments had been mailed back and received in our office, the site was randomly assigned to a condition and the appropriate study materials (for example, the resource kit) were mailed out to the site. Post-assessments were mailed out at the end of the study period. Participants in Conditions 1 and 2 (the lesson study conditions) were also asked to video record their lesson study meetings and research lessons, to collect materials from the lesson study cycle (such as student work and lesson plans), and to complete reflection forms at the end of each meeting and at the end of the lesson study cycle. Sites periodically mailed these data to our office. Due to budgetary constraints, we did not observe or attempt to measure changes in teachers' regular classroom instruction.

The 39 groups of educators included groups in 11 US states and the District of Columbia and in 27 school districts, totaling 213 teachers across the three study conditions. Table 1 provides demographic information on the teachers by study condition. The treatment and control conditions are generally comparable in teachers' years of experience and grade-level assignment. However, teachers in the treatment group were more likely to have a math degree or credential than control teachers ($\chi^2(2, N=213) = 10.39$, $p = .006$) and also had slightly more lesson study experience ($t(122) = 2.756$, $p = .007$), although the means for lesson study experience of all three conditions were in the range of 1–2 years. To control for baseline differences, these two teacher characteristics were included as covariates in subsequent analyses.

Table 1 Demographic data at study baseline

Indicator (dichotomous variable-D; continuous variable-C)	Percentage if dichotomous; mean (SD) if continuous				Tests of difference between condition 1 and other control groups combined		
	All groups ($N=213$)	Cond 1 ($N=73$)	Cond 2 ($N=73$)	Cond 3 ($N=67$)	X^2/t	df	p
Elementary grade teacher (D)	87 %	86 %	92 %	84 %	$X^2=2.23$	211	.329
Less than 5 years experience (D)	28 %	23 %	25 %	37 %	$X^2=4.07$	211	.130
More than 15 years experience (D)	25 %	27 %	30 %	18 %	$X^2=3.01$	211	.223
Math degree/ credential (D)	11 %	21 %	4 %	9 %	$X^2=10.39$	211	.006
Lesson study experience (C, scale 1–5)	2.27 (1.32)	2.63 (1.48)	2.10 (1.29)	2.06 (1.09)	$t=2.76$	122	.007

Although we suggested a time allocation of about 12–14 group meetings (including at least one classroom research lesson) for completion of the study requirements, groups organized their own meeting logistics, determining the total time, number of meetings, and meeting length. As a result, group participation time varied. Excluding time for assessments, estimated participation time for Condition 1 groups ranged from 7 to 42 h and time for Condition 2 groups ranged from 1.5 to 29 h. Meeting time was calculated from video records and self-reported meeting schedules to the extent they were available. Video records may err on the side of underestimation, since groups sometimes started the video camera late or let it run out before the meeting ended. Because teachers in Condition 3 engaged in various professional development activities (some individually, some in groups) a comparable participation figure is difficult to calculate. For example, teachers in one Condition 3 group jointly attended a regional mathematics conference, while other groups requested stipend funds to support future lesson study efforts. Variability in time devoted to lesson study (within the two lesson study conditions) is probably due to a range of factors. For example, some groups asked members to review materials as "homework," so that some of their time did not get picked up in the video record. Likewise, time spent planning the lesson or talking with group members outside of formal group meetings did not get captured. Hence, the time estimates should be considered imprecise. One factor we could identify that impacted participation time is that groups that decided to teach the research lesson more than once tended to have longer participation times.

Data Analysis

HLM analyses were conducted to assess the impact of the experimental condition on changes in teachers' beliefs and teacher learning community. For teacher outcomes, we used a two-level HLM model with teachers at Level 1 ($n=213$) and groups at Level 2 ($n=39$) to account for the nesting of teachers within lesson study groups. We chose three Level 1 covariates on the basis of baseline data and prior similar research: pretest value on the scale, lesson study experience, and possession of a mathematics degree or credential (Akiba, Chiu, Zhuang, & Mueller, 2008; Birman et al., 2009; Bloom, Richburg-Hayes, & Black, 2007; Desimone, Smith, & Ueno, 2006; Hill, 2010; Smith & Desimone, 2003). For each outcome measure, the Level 1 pretest value, the dummy indicator for possession of a math degree/credential, and lesson study experience (continuous variable) were included as grand-mean centered variables in the model (Raudenbush & Bryk, 2002). At Level 2, we included as an uncentered variable the group assignment to Condition 1 (lesson study with resource kit), assigned a value of 1, and a value of 0 otherwise. Our primary interest in this analysis was the estimate of the treatment effect on each of the six measures of belief and teacher learning community, captured by the Level 2 parameter γ_{01} in the fully conditional model shown below.

Level-1 Model

$$Y_{ij} = \beta_{0j} + \beta_{1j}\left(\text{pretest\_value}\right)_{ij} + \beta_{2j}\left(\text{math\_degree / credential}\right)_{ij}$$
$$+ \beta_{3j}\left(\text{lesson\_study\_experience}\right)_{ij} + r_{ij}.$$

Level-2 Model

$$\beta_{0j} = \gamma_{00} + \gamma_{01}\left(\text{toolkit\_assignment}\right)_{j} + u_{0j}$$
$$\beta_{1j} = \gamma_{10}$$
$$\beta_{2j} = \gamma_{20}$$
$$\beta_{3j} = \gamma_{30}$$

In addition to the HLM analyses to look at the impact of experimental assignment on the six outcome measures (beliefs and teacher learning community), we conducted additional HLM analyses designed to explore the impact of the six belief and teacher community measures on teachers' and students' development of fractions knowledge during the study period. Specifically, we investigated whether the six measures of teachers' beliefs and learning community predicted changes in teachers' and students' fractions knowledge.

Results

Table 2 shows the pre- and post-intervention scores and change scores for teachers' beliefs and teacher learning community for all three study conditions, as well as *t*-tests for the comparison of change rates between the experimental treatment (lesson study with mathematical resource kit) and the combined control conditions (lesson study only and locally chosen professional development). HLM analyses are shown in Table 3; they indicate a positive and statistically significant impact of the experimental treatment on two of the six scales: Collegial Learning Effectiveness, and Expectations for Student Achievement. In addition, the intervention shows a marginally significant impact on the Using and Promoting Student Thinking scale.

To avoid inflating the experiment-wise significance level, we limit significance testing to comparison of Condition 1, which is the full experimental treatment, with the remaining conditions. Tables 4 and 5 show the results of HLM analyses that examine changes in the teacher belief and teacher learning community measures as mediators of teachers' and students' change in fractions knowledge. Table 4 indicates that increases in collegial learning effectiveness and in professional community both significantly predict teachers' gain in fractions knowledge during the study period for the overall study sample. Likewise, Table 5 indicates that increase in teachers' collegial learning effectiveness significantly predicts students' gain in fractions knowledge during the study period.

The end-of-cycle written reflections provide insights into the kinds of experiences that increased teachers' beliefs in the effectiveness of learning with colleagues:

> This has made me think of how essential it is to observe other teachers and take as many ideas as possible to integrate in my classroom.
>
> I think this was my 7th or 8th cycle of working with lesson study and every time I am amazed at the amount of growth and learning that happens professionally for me.... The biggest impact for me is having more ears around the room listening to the students' conversations and what they are actually thinking. For example, during one of the lessons, a pair of students had recorded the correct fraction and written it the correct way, however, I overheard one partner say to the other, "1/2 means we have 1 m and 2 more." During a typical lesson and without "extra" ears around the room, the classroom teacher would have thought that pair of students knew the answer and the misconception would not have been noted.
>
> I found it so helpful to come together as a team, look closely at work that we had recently observed in action, and not all agree at what the student demonstrated. This made it clear to me that my "research" can be flawed if I am not listening and watching closely as my students talk and solve problems.
>
> I feel the collaboration piece is one of the greatest benefits for each of us. As I look back at each of the reflection notes, it is amazing how many things were discussed and how many different perspectives came out as we discussed any research or topics as part of the discussion.... The collaboration piece is also important *during* the lesson.... With more eyes there is more information, which we have found helps us create better lessons with more student learning. Even after the second lesson, our post lesson discussion has us thinking about what we still could improve and where to go from here. I definitely feel that the lack of collaboration is a weakness in our American schools. (italics not in original)

Likewise, the reflections highlight experiences that increased teachers' expectations for student achievement:

> As I watched the lesson unfold I saw how, with good intentions, we teachers stop the thinking of our students by providing too much scaffolding.... I saw students working

Table 2 Teachers' beliefs and teacher learning community, by time and condition

	C1-LS with resource kit (N=73)[a]			C2-lesson study only (N=73)			C3-control (N=67)			t-Test (211 df) for change in C1 vs. C2 and C3
	Pre	Post	Change	Pre	Post	Change	Pre	Post	Change	
Collegial learning effectiveness										
Mean	3.95	4.11	**0.17**[b]	3.96	3.94	−0.02	3.83	3.84	0.00	2.265[c]**
SD	0.65	0.58	0.51	0.56	0.66	0.54	0.67	0.61	0.55	
Expectations for student learning										
Mean	4.26	4.27	0.01	4.33	3.97	**−0.36**	4.31	4.09	**−0.22**	3.927***
SD	0.48	0.44	0.53	0.48	0.62	0.57	0.52	0.55	0.51	
Interest in mathematics and inquiry stance										
Mean	4.19	4.2	0.01	4.23	4.02	**−0.2**	4.13	4.12	−0.01	1.928+
SD	0.5	0.5	0.38	0.45	0.66	0.54	0.49	0.57	0.42	
Professional community										
Mean	3.71	3.54	**−0.18**	3.62	3.56	−0.06	3.58	3.4	**−0.18**	−0.689
SD	0.71	0.84	0.62	0.71	0.75	0.55	0.74	0.75	0.53	
Research relevance for practice										
Mean	4.05	4.18	**0.13**	4.2	4.01	**−0.18**	4.12	4.15	0.03	2.382*
SD	0.57	0.52	0.57	0.58	0.73	0.65	0.63	0.61	0.6	
Using and promoting st. thinking										
Mean	3.69	3.9	**0.21**	3.77	3.72	−0.05	3.87	3.84	−0.03	3.138**
SD	0.53	0.5	0.51	0.49	0.6	0.6	0.49	0.5	0.54	

[a]N for each study condition represents the number of educators in each condition and is consistent within condition across the 6 belief measures

[b]Bolded means in the "change" column for each of the study assignments indicate a statistically significant difference between pretest and posttest as follows. C1—Collegial learning effectiveness ($t(72)=2.751$, $p<.01$); Professional community ($t(72)=2.414$, $p<.05$); Research relevance for practice ($t(72)=1.945$, $p<.10$); Using and promoting student thinking ($t(72)=3.499$, $p<.01$). C2—Expectations for student achievement ($t(72)=5.421$, $p<.001$); Interest in mathematics and inquiry stance ($t(72)=3.253$, $p<.01$); Research relevance for practice ($t(72)=2.381$, $p<.05$). C3—Expectations for student achievement ($t(66)=3.461$, $p<.01$); Professional community ($t(66)=2.802$, $p<.01$)

[c]*$p<.10$, *$p<.05$, **$p<.01$, ***$p<.0

Table 3 HLM results: impact of lesson study with mathematical resource kits on teachers' beliefs and learning community

Fixed effects	Collegial learning effectiveness Coefficient	Expectations for student achievement Coefficient	Interest in mathematics and inquiry stance Coefficient	Professional community Coefficient	Research relevance for practice Coefficient	Using and promoting student thinking Coefficient
Intercept	3.907 (.039)***	4.027 (.046)***	4.086 (.051)***	3.502 (.054)***	4.084 (.057)***	3.773 (.044)***
Teacher predictors						
Pretest value	.604 (.058)***	.528 (.089)***	.750 (.067)***	.737 (.037)***	.496 (.072)***	.435 (.062)***
Math degree or credential	.085 (.098)	.111 (.114)	.168 (.069)*	−.082 (.141)	.133 (.092)	.202 (.077)*
Lesson study experience	.018 (.026)	.011 (.023)	.017 (.018)	.050 (.033)	−.023 (.031)	.019 (.023)
Group predictor—assignment to toolkit	.173 (.075)*	.251 (.069)***	.089 (.063)	−.066 (.131)	.082 (.086)	.146 (.075)+
Random effects	Variance component	Variance component	Variance component	Variance component	Variance component	Variance component
Unconditional model						
Variance between groups	0.033	0.022	0.029	0.187	0.005	0.033
Variance within groups	0.359	0.284	0.311	0.434	0.390	0.257
Full model						
Variance between groups	0.000	0.006	0.018	0.073	0.016	0.013
Variance within groups	0.232	0.229	0.180	0.235	0.310	0.219
Effect size—resource kit	**0.03**	**0.07**	**0.02**	**−0.01**	**0.02**	**0.03**

L1 sample=213 teachers; L2 sample=39 groups

***$p<.001$, **$p<.01$, *$p<.05$, +$p<.10$

Table 4 HLM results: changes in teachers' beliefs and professional community as predictors of teachers' knowledge gain

Fixed effects	Fractions knowledge (Z score)	Fractions knowledge (Z score)	Fractions knowledge (Z score)	Fractions knowledge (Z score)	Fractions knowledge (Z score)	Fractions knowledge (Z score)	Fractions knowledge (Z score)	Fractions knowledge (Z score)
Intercept	-.062 (.056)	-.059 (.055)	-.049 (.055)	-.049 (.057)	-.052 (.055)	-.063 (.055)	-.065 (.055)	-.058 (.055)
Teacher predictors								
Pretest value	.802 (.046)***	.806 (.046)***	.804 (.045)***	.802 (.047)***	.802 (.045)***	.800 (.046)***	.791 (.043)***	.802 (.046)***
Math degree or credential	.084 (.101)	.123 (.110)	.102 (.099)	.079 (.100)	.067 (.097)	.084 (.101)	.104 (.099)	.076 (.100)
Lesson study experience	-.032 (.030)	-.025 (.029)	-.029 (.028)	-.032 (.029)	-.032 (.030)	-.033 (.030)	-.039 (.030)	-.032 (.029)
Collegial learning effectiveness (change score)	–	–	.181 (.083)*	–	–	–	–	–
Expectations for student achievement (change score)	–	–	–	.124 (.084)	–	–	–	–
Using and promoting student thinking (change score)	–	–	–	–	.115 (.076)	–	–	–
Research relevance for practice (change score)	–	–	–	–	–	-.210 (.063)	–	–
Professional community (change score)	–	–	–	–	–	–	.186 (.061)**	–
Interest in math/inquiry stance (change score)	–	–	–	–	–	–	–	.093 (.116)
Group predictors								
Assignment to toolkit	.178 (.077)*	.173 (.075)*	.142 (.078)+	.141 (.082)+	.151 (.078)+	.183 (.077)*	.190 (.077)*	.168 (.076)*
Random effects								

(continued)

Table 4 (continued)

Fixed effects	Fractions knowledge (Z score)	Fractions knowledge (Z score)	Fractions knowledge (Z score)	Fractions knowledge (Z score)	Fractions knowledge (Z score)	Fractions knowledge (Z score)	Fractions knowledge (Z score)	Fractions knowledge (Z score)	Fractions knowledge (Z score)
Unconditional model									
Intercept (variance between groups)	0.031	0.031	0.031	0.031	0.031	0.031	0.031	0.031	0.031
Level 1 (variance within groups)	0.970	0.970	0.970	0.970	0.970	0.970	0.970	0.970	0.97
Full model									
Intercept (variance between groups)	0.004	0.002	0.002	0.007	0.003	0.004	0.004	0.005	0.004
Level 1 (variance within groups)	0.359	0.361	0.353	0.354	0.358	0.361	0.361	0.349	0.360
Effect size of assignment to resource kit[a]	**0.18**	**0.17**	**0.14**	**0.14**	**0.15**	**0.18**	**0.18**	**0.19**	**0.17**
Effect size of assignment to resource kit[b]	**0.02**	**0.02**	**0.02**	**0.02**	**0.02**	**0.02**	**0.02**	**0.02**	**0.02**

L1 sample = 213 teachers; L2 sample = 39 groups

[a]Effect sizes are standardized coefficients

[b]Computed effect sizes using formula from Song & Herman, 2010

***$p < .001$, **$p < .01$, *$p < .05$, +$p < .10$

Table 5 HLM results: changes in teachers' beliefs and professional community as predictors of students' knowledge gain

Fixed effects	Fractions knowledge (Z score)	Fractions knowledge (Z score)	Fractions knowledge (Z score)	Fractions knowledge (Z score)	Fractions knowledge (Z score)	Fractions knowledge (Z score)	Fractions knowledge (Z score)
	Coefficient	Coefficient	Coefficient	Coefficient	Coefficient	Coefficient	Coefficient
Intercept	-.186 (.050)***	-.160 (.051)***	-.185 (.049)***	-.182 (.048)**	-.188 (.051)**	-.182 (.049)**	-.186 (.051)**
Student predictors							
Pretest score	.682 (.029)***	.681 (.028)***	.682 (.029)***	.683 (.027)***	.682 (.029)***	.683 (.029)***	.682 (.029)***
Grade 23 test	-.056 (.099)	-.102 (.091)	-.058 (.095)	-.058 (.095)	-.037 (.103)	-.1058 (.096)	-.049 (.099)
Grade 5 test	-.085 (.127)	-.148 (.136)	-.089 (.139)	-.105 (.121)	-.064 (.135)	-.091 (.119)	-.075 (.131)
Teacher predictors							
Math degree or credential	-.193 (.069)**	-.133 (.081)**	-.197 (.062)**	-.287 (.082)**	-.186 (.063)**	-.195 (.060)**	-.196 (.069)**
Lesson study experience	.016 (.033)	.023 (.031)	.017 (.033)	.015 (.032)	.012 (.032)	.021 (.033)	.013 (.032)
Collegial learning effectiveness (change score)	–	.219 (.073)**	–	–	–	–	–
Expectations for student achievement (change score)	–	–	.011 (.072)	–	–	–	–
Using and promoting student thinking (change score)	–	–	–	.101 (.076)	–	–	–
Research relevance for practice (change score)	–	–	–	–	-.066 (.064)	–	–
Professional community (change score)	–	–	–	–	–	-.070 (.059)	–
Interest in math/inquiry stance (change score)	–	–	–	–	–	–	-.061 (.066)

(continued)

Table 5 (continued)

Fixed effects	Fractions knowledge (Z score)	Fractions knowledge (Z score)	Fractions knowledge (Z score)	Fractions knowledge (Z score)	Fractions knowledge (Z score)	Fractions knowledge (Z score)	Fractions knowledge (Z score)
	Coefficient	Coefficient	Coefficient	Coefficient	Coefficient	Coefficient	Coefficient
Group predictors							
Assignment to toolkit	.496 (.136)**	.405 (.125)**	.494 (.133)***	.477 (.130)**	.505 (.137)**	.489 (.136)**	.500 (.136)**
Random effects	Variance component	Variance component	Variance component	Variance component	Variance component	Variance component	Variance component
Unconditional model							
Intercept1/intercept2 (variance between groups)	0.250	0.250	0.250	0.250	0.250	0.250	0.250
Intercept (variance within groups between classes)	0.124	0.124	0.124	0.124	0.124	0.124	0.124
Level 1 (variance within classes)	0.603	0.603	0.603	0.603	0.603	0.603	0.603
Full model							
Intercept1/intercept2 (variance between groups)	0.044	0.047	0.044	0.035	0.048	0.040	0.049
Intercept (variance within groups between classes)	0.050	0.036	0.050	0.053	0.046	0.051	0.047
Level 1 (variance within classes)	0.290	0.290	0.290	0.290	0.290	0.290	0.290
Effect size of assignment to resource kit	**0.5**	**0.41**	**0.49**	**0.48**	**0.51**	**0.49**	**0.41**

themselves from an incorrect answer to recognizing the answer was wrong, puzzling over how to correct it only to have a teacher ask "yes–no" questions that stopped their problem solving and led them to the correct answer. I recognize this trait in myself and have committed myself to allowing the students time to struggle…

My students discovered on their own that the more you divide the whole the smaller the fractional parts…. Because of this discovery many students began to make awesome connections…. Once when comparing 3/4 and 5/6 Daniella claimed that she can compare the fractions just by looking at them. Other students thought it would be too difficult because the size of the parts and number of parts were different. Daniella used her understanding of unit fractions to compare the numbers. She said that 3/4 is 1/4 away from equaling 1 whole but 5/6 is only 1/6 away from equaling 1 whole, 1/4 is larger than 1/6, so 3/4 is less than 5/6. This came completely unprompted and it led to a student explaining and demonstrating using fraction strips and the other students agreeing and taking part in a cool discovery. This never happened before because I never put much time as a 5th grade teacher into my student understanding of unit fractions.

I love that one teacher did a 360 [complete turnaround] from her initial response to the math lesson, "My students cannot do this," to "I would love to see my students do this." That raising of the bar, while at the same time knowing the students well enough to plan for success, proved to be the best surprise of all.

Discussion

The HLM analyses indicate that participation in lesson study with mathematical resources significantly increased two of the six outcome measures related to teachers' beliefs and teacher learning community: Expectations for Student Achievement and Collegial Learning Effectiveness. The intervention also had a marginally significant effect on a third outcome: Using and Promoting Student Thinking. End-of-cycle reflections illuminate the specific experiences that enabled these changes in beliefs, such as hearing other teachers' perspectives and seeing students respond to a challenging mathematical task.

Although we generally combined the two control groups for analysis to avoid inflating the experiment-wise significance levels, examining the data for all three conditions provides insights into how lesson study with the specially designed mathematical resources (Condition 1) differed from more typical lesson study (Condition 2) and from locally chosen professional development other than lesson study (Condition 3). Going down the "change" columns in Table 2 for each of the three conditions suggests that the two control conditions may be more similar to each other than to the experimental condition (lesson study with the mathematical resource kit) in terms of impact on teachers' beliefs and professional learning community. Why would this be? The mathematical resources provided to Condition 1 teachers may have catalyzed more opportunities to change beliefs than the resources Condition 2 teachers located on their own. For example, the mathematical resources included fractions chapters from a Japanese teacher's edition, and previous research has shown major differences between USA and Japanese teacher's editions, such as more presentation of varied student thinking in the Japanese vs. the US teacher's edition (28 % vs. 1 % of statements) and more discussion of the rationale for tasks and instructional design (10 % vs. 0 %) (Lewis et al., 2011). So the resource kits may have catalyzed a more substantial collegial discussion than the materials (such as US teacher's editions) located by Condition 2 groups, making colleagues more valuable in making sense of the materials. A number of participants mentioned how helpful it was to see how colleagues solved the math tasks; it is likely that Condition 2 lesson study groups located, solved, and discussed fewer tasks than did teachers using the resource kit, since it included a number of mathematical tasks and specific prompts to solve them individually and then discuss.

Similarly, the resource kit's emphasis on linear measure models and unit fractions seem to have been useful in revealing students' mathematical potential. The linear measure model made it easy for students to compare 3/4 and 5/6 (as noted by the teacher quoted above) and to use this in classroom discussion, which in turn allowed the teacher to see students' potential to reason mathematically. After describing students' "awesome discovery" the teacher wrote: "This never happened before because I never put much time as a 5th grade teacher into... unit fractions." Condition 2 teachers, who sought out lesson materials on their own, may have had a harder time finding materials that supported such changes during the brief period of the study.

One interesting feature of the findings is the difference in results for the two scales related to learning from colleagues: Collegial Learning Effectiveness and Professional Community. These scales differ in two major ways. First, several of the items on the Professional Community scale focus on all colleagues at the school site, for example: "There is a lot of discussion among teachers at this school about how to teach." In contrast, the Collegial Learning Effectiveness scale refers to colleagues self-identified by the respondent, for example, "I have learned a great deal about mathematics teaching from colleagues." Since the lesson study we report was conducted by small groups of teachers, not by all teachers in a school, differences between the two measures would be expected if teachers' attitudes toward their lesson study colleagues do not necessarily extend to the broader set of all colleagues at their school site. A second difference between the scales is that the Professional Community scale focuses on the *frequency* of learning with colleagues, whereas the Collegial Learning Effectiveness Scale focuses on its *usefulness and impact* (for example, whether respondents believe they have learned about student thinking from colleagues). Finally, the Collegial Learning Effectiveness Scale is more heavily focused on mathematics (4 of 5 items) than the Professional Community Scale (2 of 6 items). Hence, Collegial Learning Effectiveness is better designed to pick up changes in usefulness of collegial learning among educators with whom the respondent collaborates, as opposed to frequency of collegial interaction within the school as a whole.

Another interesting aspect of Table 2 is that scores on many of the scales declined in the 3–4 months period between the baseline administration (usually in September) and the final administration (usually in January), especially in the two control groups. The September administration may have captured the most hopeful moment of the school year.

One limitation of this study is that several of the scales have a relatively small number of items, marginal scale reliability, and little or no prior evidence of predictive validity. Given the length of the fractions assessment, there was not sufficient time to administer a large number of survey items related to beliefs or teacher learning community. Many existing scales (with evidence of predictive validity) did not seem adequately aligned to the intervention at hand, lesson study by a small group of educators at a site (or across sites) on a particular topic within mathematics (fractions). More thorough investigation of the middle box of Fig. 2, the changes that occur during lesson study in teachers' beliefs and collegial relationships, is certainly warranted.

Conclusions

As far as we are aware, this is the first randomized, controlled trial of a lesson study intervention, and we believe that it contributes in several ways to the current research base on professional learning. First, it documents that in a brief period of about 3 months, self-organized groups of educators scattered across the USA, supported by mathematical resource kits, were able to conduct lesson study that significantly increased not only their own and students' knowledge of fractions but also their expectations for student achievement and the reported efficacy of working with colleagues–beliefs that may have enormous implications for *future* efforts to improve. Prior qualitative research has provided evidence of changes in teachers' beliefs and professional relationships during lesson study (e.g., Lewis, Perry, & Hurd, 2009; Murata, 2003), and the current study confirms these changes in a much larger sample using a randomized trial. The findings suggest the fruitfulness of taking a much closer look at the middle box of Fig. 1 to document the changes in beliefs and collegial work that may allow lesson study to produce changes in both teachers' and students' learning and to support teachers' continued learning from practice over time.

Given the arguments made at the outset that improvement of instruction is likely to require teachers to engage in repeated cycles of trial and revision, it is essential to identify the beliefs and collegial learning structures that allow teachers to keep up this effortful work over time. Our findings indicate that the intervention significantly increased teachers' perceptions of the usefulness of collegial work and their expectations for student achievement and that these changes significantly predicted increases in teachers' and students' mathematical knowledge over the study period. This was true for the intervention group and for the study sample as a whole.

Finally, the results of this study should encourage us to think in new ways about scale-up of instructional improvement. The intervention was "low-touch," in that local, self-managed groups of educators worked independently at a distance from us, without any centralized supervision. These groups organized their learning in ways that made sense locally, rather than adhering to centrally prescribed rules designed to achieve implementation fidelity. In this way, the intervention supported educators' own agency and leadership, while also allowing them to build their mathematical knowledge.

These results suggest a promising solution to the conundrum of faithful implementation of high-quality materials versus teachers' "ownership" of professional learning. Through a lesson study process supported by mathematical resources, teachers can participate in a process that values their ideas and leadership, while at the same time increasing their expectations for student achievement and the effectiveness of their collegial work, as well as their own mathematical knowledge and that of their students.

Appendix: Scales to Measure Teachers' Beliefs and Teacher Learning Community

Stem: "Please indicate how well each of the following statements describes your attitude" (Rated on a 5-point scale ranging from 1 ("strongly disagree") to 5 ("strongly agree.")).

Expectations for student achievement (7 items; Alpha = .63 on pretest; .64 on posttest)

No matter how hard I try, some students will not be able to learn aspects of mathematics (reverse coded) (McLaughlin & Talbert, 2001).

My expectations about how much students should learn are not as high as they used to be (reverse coded) (McLaughlin & Talbert, 2001).

Students who work hard and do well deserve more of my time than those who do not (reverse coded) (McLaughlin & Talbert, 2001).

The attitudes and habits students bring to my classes greatly reduce their chances for academic success (reverse coded) (McLaughlin & Talbert, 2001).

There is really very little I can do to ensure that most of my students achieve at a high level (reverse coded) (McLaughlin & Talbert, 2001).

Most of the students I teach are not capable of learning material I should be teaching them (reverse coded) (McLaughlin & Talbert, 2001).

By trying a different teaching method, I can significantly affect a student's achievement (CRC, 1994).

Using and promoting student thinking: (4 items; .63 at pretest and .68 at posttest)

I am able to figure out what students know about fractions (Project-developed).

I have some good strategies for making students' mathematical thinking visible (Project-developed).

I can help students "catch up" who come to me lacking in math skills (Adapted from CRC, 1994).

When students are confused about fractions, I am able to provide good examples and explanations (Project-developed).

Interest in mathematics and inquiry stance (*8 items*; Alpha = .74 on pretest; .84 on posttest)

I enjoy teaching mathematics (Horizon Research, 2000).

I like solving mathematics problems (Project-developed).

Student mathematical thinking is fascinating to me (Project-developed).

I think of myself as a researcher in the classroom (Project-developed).

I am always curious about student thinking (Adapted from MSU, 2003).

I actively look for opportunities to learn more mathematics (Project-developed).

I am interested in the mathematics taught at many grade levels (Project-developed).

I would like to learn more about fractions (Adapted from LMT, 2007).

Research relevance for practice (4 items; .64 at pretest and .66 at posttest)

Educational research often provides useful insights for teaching (Project-developed).

In general, curriculum materials from other countries are not useful (Project-developed).

Most research is not relevant to my needs as a teacher (Project-developed).

I find it interesting to read about a variety of educational programs and ideas (Project-developed).

Collegial learning effectiveness (5 items; .62 on pretest and .63 on posttest; based on items adapted from CRC, 1994 and Horizon Research Inc., 2000.)

I have learned a lot about student thinking by working with colleagues.

Working with colleagues on mathematical tasks is often unpleasant (reverse coded) (Project-developed).

I have good opportunities to learn about the mathematics taught at different grade levels (Adapted from CRC, 1994).

I have learned a great deal about mathematics teaching from colleagues.

I find it useful to solve mathematics problems with colleagues (Project-developed).

Professional Community (6 items; .80 at pretest and .82 at posttest)

My colleagues and I regularly share ideas and materials related to mathematics teaching.

Mathematics teachers in this school regularly observe each other teaching classes as part of sharing and improving instructional strategies.

I feel supported by other teachers to try out new ideas in teaching.

There is a lot of discussion among teachers at this school about how to teach (Adapted from CRC, 1994; MSU, 2003).

I plan and coordinate with other teachers (MSU, 2003).

I don't know how other teachers in this school teach (Adapted from CRC, 1994).

References

Akiba, M., Chiu, Y.-F., Zhuang, Y.-L., & Mueller, H. E. (2008). Standards-based mathematics reforms and mathematics achievement of American Indian/Alaska native eighth graders. *Education Policy Analysis Archives, 16*(20), 1–31.

Beckmann, S. (2005). *Mathematics for elementary teachers*. Boston: Pearson.

Birman, B. F., Boyle, A., Le Floch, K., Elledge, A., Holtzman, D., Song, M., et al. (2009). *State and local implementation of the no child left behind act: Volume VIII–teacher quality under NCLB: Final report*. Washington, DC: U.S. Department of Education.

Bloom, H. S., Richburg-Hayes, L., & Black, A. R. (2007). Using covariates to improve precision for studies that randomize schools to evaluate educational interventions. *Educational Evaluation and Policy Analysis, 29*(1), 30–59.

Borko, H., Davinroy, K. H., Bliem, C. L., & Cumbo, K. B. (2000). Exploring and supporting teacher change: Two third-grade teachers' experiences in a mathematics and literacy staff development project. *The Elementary School Journal, 100*(4), 273–306.

Britt, M. S., Irwin, K. C., & Ritchie, G. (2001). Professional conversations and professional growth. *Journal of Mathematics Teacher Education, 4*(1), 29–53.

California Department of Education. (2005). CST released test questions—standardized testing and reporting (STAR). From http://www.cde.ca.gov/ta/tg/sr/css05rtq.asp

Campbell, P. F., & Malkus, N. N. (2011). The impact of elementary mathematics coaches on student achievement. *The Elementary School Journal, 11*(3), 430–454.

Carpenter, T. P., Fennema, E., Franke, M. L., Levi, L., & Empson, S. B. (1999). *Children's mathematics: Cognitively guided instruction*. Porthsmouth, NH: Hienemann.

Center for Research in Mathematics and Science Teacher Development. (2005). Diagnostic teacher assessment in mathematics and science, elementary mathematics, rational numbers assessment, version 1.2 and middle school mathematics, number and computation assessment, version 2. From http://louisville.edu/education/research/centers/crmstd/diag_math_assess_elem_teachers.html

Chazan, D., Ben-Chaim, D., & Gormas, J. (1998). Shared teaching assignments in the service of mathematics reform: Situated professional development. *Teaching and Teacher Education, 14*(7), 687–702.

Clarke, D., & Hollingsworth, H. (2002). Elaborating a model of teacher professional growth. *Teaching and Teacher Education, 18*(8), 947–967.

CRC (Center for Research on the Context of Teaching). (1994). *Survey of elementary mathematics education in California: Teacher questionnaire*. Stanford, CA: Stanford University School of Education.

Desimone, L. M., Smith, T. M., & Ueno, K. (2006). Are teachers who need sustained, content-focused professional development getting it? An administrator's dilemma. *Educational Administration Quarterly, 42*(2), 179–215.

Fernandez, C., & Yoshida, M. (2004). *Lesson study: A case of a Japanese approach to improving instruction through school-based teacher development*. Mahwah, NJ: Lawrence Erlbaum.

Fisler, J. L., & Firestone, W. A. (2006). Teacher learning in a school-university partnership: Exploring the role of social trust and teaching efficacy beliefs. *Teachers College Record, 108*(6), 1155–1185.

Garet, M. S., Porter, A. C., Desimone, L., Birman, B. F., & Yoon, K. S. (2001). What makes professional development effective? Results from a national sample of teachers. *American Educational Research Journal, 38*(4), 304–313.

Gersten, R., Taylor, M. J., Keys, T. D., Rolfhus, E., & Newman-Gonchar, R. (2014). *Summary of research on the effectiveness of math professional development approaches*. (No. REL 2014-010). Washington, DC: US Department of Education, Institute of Education Sciences, National Center for Educational Evaluation and Regional Assistance, Regional Educational Laboratory Southeast.

Goldsmith, L. T., Doerr, H. M., & Lewis, C. C. (2014). Mathematics teachers' learning: A conceptual framework and synthesis of research. *Journal of Mathematics Teacher Education, 17*(1), 5–36.

Grandau, L. (2005). Learning from self-study: Gaining knowledge about how fourth graders move from relational description to algebraic generalization. *Harvard Educational Review, 75*(2), 202–221.

Hackenberg, A., Norton, A., Wilkins, J., & Steffe, L. (2009). *Testing hypotheses about students' operational development of fractions*. Paper presented at the NCTM Research Pre-Session.

Hill, H. C. (2010). The nature and predictors of elementary teachers' mathematical knowledge for teaching. *Journal for Research in Mathematics Education, 41*(5), 513–545.

Hironaka, H., & Sugiyama, Y. (2006). *Mathematics for elementary school, grades 1-6*. Tokyo: Tokyo Shoseki. English language version available at http://www.globaledresources.com.

Horizon Research, Inc. (2000). 2000 National Survey of Science and Mathematics Education Mathematics Teacher Questionnaire. http://www.horizon-research.com/horizonresearchwp/wp-content/uploads/2013/04/math_teacher.pdf.

Institute of Education Sciences/National Center for Education Statistics (IES/NCES). (2007). NAEP questions tool. Retrieved from http://nces.ed.gov/nationsreportcard/itmrlsx/landing.aspx

Kazemi, E., & Franke, M. L. (2004). Teacher learning in mathematics: Using student work to promote collective inquiry. *Journal of Mathematics Teacher Education, 7*(3), 203–235.

Lampert, M. (2001). *Teaching problems and the problems of teaching*. New Haven, CT: Yale University Press.

Learning Mathematics for Teaching. (2007). *Survey of teachers of mathematics, winter 2007, form LMT PR-07*. Ann Arbor, MI: University of Michigan, School of Education.

Lewis, C., & Hurd, J. (2011). *Lesson study step by step: How teacher learning communities improve instruction*. Portsmouth, NH: Heinneman.

Lewis, C., & Perry, R. (2014). Lesson study with mathematical resources: A sustainable model for locally led teacher professional learning. *Mathematics Teacher Education and Development, 16*(1), 22–42.

Lewis, C., & Perry, R. (under review). *Scale-up of instructional improvement through lesson study.* Unpublished manuscript.

Lewis, C., Perry, R., & Friedkin, S. (2011). Using Japanese curriculum materials to support lesson study outside Japan: Toward coherent curriculum. *Educational Studies in Japan: International Yearbook, 6,* 5–19.

Lewis, C., Perry, R., & Hurd, J. (2009). Improving mathematics instruction through lesson study: A theoretical model and North American case. *Journal of Mathematics Teacher Education, 12*(4), 285–304.

Lewis, C., Perry, R., & Murata, A. (2006). How should research contribute to instructional improvement? The case of lesson study. *Educational Researcher, 35*(3), 3–14.

Lewis, C., & Tsuchida, I. (1997). Planned educational change in Japan: The case of elementary science instruction. *Journal of Educational Policy, 12*(5), 313–331.

Lewis, C., & Tsuchida, I. (1998). A lesson is like a swiftly flowing river: Research lessons and the improvement of Japanese education. *American Educator*, Winter, 14–17 & 50–52.

Lin, X. (2001). Reflective adaptation of a technology artifact: A case study of classroom change. *Cognition and Instruction, 19*(4), 395–440.

Lin, P. (2002). On enhancing teachers' knowledge by constructing cases in the classrooms. *Journal of Mathematics Teacher Education, 5*(4), 317–349.

Manouchehri, A. (2001). Collegial interaction and reflective practice. *Action in Teacher Education, 22*(4), 86–97.

McLaughlin, M. W., & Talbert, J. E. (2001). *Professional communities and the work of high school teaching.* University of Chicago Press.

McLaughlin, M. W., & Talbert, J. E. (2006). *Building school-based teacher learning communities.* New York: Teachers College Press.

Michigan State University. (2003). *Teacher learning through professional development, 2003 SVMI teacher survey.* East Lansing, MI: Michigan State University.

Murata, A. (2003). *Teacher learning and lesson study: Developing efficacy through experiencing student learning.* Paper presented at the annual meeting of the School Science and Mathematics Association, Columbus, OH.

Newton, K. J. (2008). An extensive analysis of preservice elementary teachers' knowledge of fractions. *American Educational Research Journal, 45*(4), 1080–1110.

Noffke, S., & Somekh, B. (Eds.). (2009). *The SAGE handbook of educational action research.* London: Sage.

Norton, A. H., & McCloskey, A. V. (2008). Modeling students' mathematics using Steffe's fraction schemes. *Teaching Children Mathematics, 15*(1), 48–54.

Perry, R., & Lewis, C. (2010). Building demand for research through lesson study. In C. E. Coburn & M. K. Stein (Eds.), *Research and practice in education: Building alliances, bridging the divide* (pp. 131–145). Lanham, MD: Rowman & Littlefield.

Post, T. R., Harel, G., Behr, M. J., & Lesh, R. A. (1988). Intermediate teachers' knowledge of rational number concepts. In E. Fennema (Ed.), *Papers from the First Wisconsin Symposium for Research on Teaching and Learning Mathematics* (pp. 194–219). Madison, WI: Wisconsin Center for Education Research.

Puchner, L. D., & Taylor, A. R. (2006). Lesson study, collaboration and teacher efficacy: Stories from two school-based math lesson study groups. *Teaching and Teacher Education, 22*(7), 922–934.

Raudenbush, S., & Bryk, A. (2002). *Hierarchical linear models: Applications and data analysis methods.* Thousand Oaks, CA: Sage.

Remillard, J. T., & Bryans, M. B. (2004). Teachers' orientations toward mathematics curriculum materials: Implications for teacher learning. *Journal for Research in Mathematics Education, 35*(5), 352–388.

Saxe, G. B. (2007). Learning about fractions as points on a number line. In W. G. Martin, M. E. Strutchens, & P. C. Elliott (Eds.), *The Learning of Mathematics* (Vol. 69th Yearbook, pp. 221–237). Reston, VA: The National Council of Teachers of Mathematics.

Schifter, D. (1998). Learning mathematics for teaching: From the teachers' seminar to the classroom. *Journal of Mathematics Teacher Education, 1*(1), 55–87.

Schorr, R. Y., & Koellner-Clark, K. (2003). Using a modeling approach to analyze the ways in which teachers consider new ways to teach mathematics. *Mathematical Thinking and Learning, 5*(2), 191–210.

Seymour, J. R., & Lehrer, R. (2006). Tracing the evolution of pedagogical content knowledge as the development of interanimated discourses. *The Journal of the Learning Sciences, 15*(4), 549–582.

Sherer, J., & Spillane, J. (2011). Constancy and change in work practice in schools: The role of organizational routines. *Teachers College Record, 113*(3), 611–657.

Sherin, M. G., & Han, S. Y. (2004). Teacher learning in the context of a video club. *Teaching and Teacher Education, 20*(2), 163–183.

Sherin, M. G., & van Es, E. A. (2009). Effects of video club participation on teachers' professional vision. *Journal of Teacher Education, 60*, 20–37.

Smith, T., & Desimone, L. M. (2003). Do changes in patterns of participation in teachers' professional development reflect the goals of standards-based reforms? *Educational Horizons, 81*(3), 119–129.

Song, M., & Herman, R. (2010). Critical issues and common pitfalls in designing and conducting impact studies in education lessons learned from the what works clearinghouse (Phase I). *Educational Evaluation and Policy Analysis, 32*(3), 351–371.

Spillane, J. P. (2000). A fifth-grade teacher's reconstruction of mathematics and literacy teaching: Exploring interactions among identity, learning, and subject matter. *The Elementary School Journal, 100*(4), 307–330.

Steinberg, R. M., Empson, S. B., & Carpenter, T. P. (2004). Inquiry into children's mathematical thinking as a means to teacher change. *Journal of Mathematics Teacher Education, 7*(3), 237–267.

Stigler, J. W., & Hiebert, J. (1999). *The teaching gap: Best ideas from the world's teachers for improving education in the classroom.* New York: Summit Books.

Takahashi, A., Watanabe, T., Yoshida, M., & Wang-Iverson, P. (2005). Improving content and pedagogical knowledge through kyozaikenkyu. In P. Wang-Iverson & M. Yoshida (Eds.), *Building our understanding of lesson study* (pp. 77–84). Philadelphia: Research for Better Schools.

Taylor, A. R., Anderson, S., Meyer, K., Wagner, M. K., & West, C. (2005). Lesson study: A professional development model for mathematics reform. *Rural Educator, 26*(2), 17–22.

Thijs, A., & van den Berg, E. (2002). Peer coaching as part of a professional development program for science teachers in Botswana. *International Journal of Educational Development, 22*, 55–68.

Ticha, M., & Hospesova, A. (2006). Qualified pedagogical reflection as a way to improve mathematics education. *Journal of Mathematics Teacher Education, 9*(2), 129–156.

Tobin, K., & Espinet, M. (1990). Teachers as peer coaches in high school mathematics. *School Science and Mathematics, 90*(3), 232–244.

Van de Walle, J. A. (2007). *Elementary and middle school mathematics: Teaching developmentally* (6th ed.). Boston, MA: Pearson.

van Es, E. A., & Sherin, M. G. (2008). Mathematics teachers' "learning to notice" in the context of a video club. *Teaching and Teacher Education, 24*(2), 244–276.

Wang-Iverson, P., & Yoshida, M. (2005). *Building our understanding of lesson study.* Philadelphia: Research for Better Schools.

Ward, J., & Thomas, G. (2009). From presentation on teaching scenarios. Retrieved June 1, 2009, from http://www.nzmaths.co.nz/fractions

Zech, L. K., Gause-Vega, C. L., Bray, M. H., Secules, T., & Goldman, S. R. (2000). Content-based collaborative inquiry: A professional development model for sustaining educational reform. *Educational Psychologist, 35*(3), 207–217.

Zhou, Z., Peverly, S. T., & Xin, T. (2006). Knowing and teaching fractions: A cross-cultural study of American and Chinese mathematics teachers. *Contemporary Educational Psychology, 31*(4), 438–457.

Conceptualizing Teachers' Capacity for Learning Trajectory-Oriented Formative Assessment in Mathematics

Caroline B. Ebby and Philip M. Sirinides

TASK (Teachers' Assessment of Student Knowledge) is an online tool designed to measure teacher's capacity for learning trajectory-oriented formative assessment in mathematics, specifically focusing on their ability to analyze student work and make instructional decisions based on that work. Formative assessment has proved to be one of the most powerful current educational practices in terms of improving student learning (Black & Wiliam, 1998; Kluger & DeNisi, 1996). A meta-analysis of more than 250 studies on formative assessment indicates substantial evidence linking formative assessment with higher student achievement, with typical effect sizes ranging from an impressive 0.4–0.7 (Black & Wiliam, 1998). Yet numerous studies have concluded that teachers struggle to make effective use of student learning data (Datnow, Park, & Wohlstetter, 2007; Heritage, Kim, Vendlinski, & Herman, 2009; Kerr, Marsh, Ikemoto, Darilek, & Barney, 2006; Young, 2006).

While the term formative assessment is often used erroneously in educational contexts to refer to assessment instruments themselves, it is more accurately defined as a process whereby an assessment provides feedback to both the learner and the teacher and this feedback causes an adjustment in instruction (Bennett, 2014; Black & Wiliam, 1998; Shepard, 2008). Formative assessment is therefore fundamentally an interpretive process. Effective formative assessment—assessing student understanding relative to a standard or goal, providing feedback to the student in the form of instructional guidance, and continually working to diminish the gap between the student's performance and the instructional goal—requires that teachers are able to understand and analyze student thinking to develop an instructional response that will move the learner forward. TASK is an open-ended measure situated in the context of looking at student-generated work that can be used to measure these specific aspects of teacher knowledge and also explore the nature of that knowledge.

C.B. Ebby (✉) • P.M. Sirinides
Consortium for Policy Research in Education, Graduate School of Education,
University of Pennsylvania, 3440 Market St, Suite 560, Philadelphia, PA 19104, USA
e-mail: cbe@gse.upenn.edu

© Springer International Publishing Switzerland 2015
J.A. Middleton et al. (eds.), *Large-Scale Studies in Mathematics Education*,
Research in Mathematics Education, DOI 10.1007/978-3-319-07716-1_8

In this chapter, we begin by articulating the conceptual framework behind learning trajectory-oriented formative assessment and describing the instrument, scoring rubrics, and ongoing development of TASK. We then present the results of a large-scale field test of TASK, both in terms of the overall results and additional studies of the properties of the instrument. We also draw on the results of this field test to investigate the relationships between various dimensions of teachers' ability to analyze student work in mathematics and their instructional decision making.

Conceptual Framework: Learning Trajectory-Oriented Formative Assessment

At the foundation of formative assessment is a clear understanding of the gap between the learner's current state of understanding and the learning goal or standard. A well-designed assessment should illuminate the learner's current state so that the gap is evident. The assessment becomes formative only when (1) the information provides useful feedback to the learner, (2) the information provides useful feedback to the teacher, and (3) the teacher is able to provide an instructional response that will help the learner move closer to the goal. This is an iterative process, the cycle repeating until the gap is closed and new learning goals are established (Bennett, 2014; Black & Wiliam, 1998; Heritage, 2008).

Learning progressions, or "successively more sophisticated ways of thinking about a topic" (National Research Council, 2007, p. 219), have recently become prominent in mathematics educational research as well as in conceptualizations of assessment and instruction (Clements & Sarama, 2004; Confrey, 2008; Daro, Mosher, & Corcoran, 2011; Sztajn, Confrey, Wilson, & Edgington, 2012). *Learning trajectories*, as they are most often called in mathematics education, can provide a guiding framework as teachers assess where students are in the trajectory of learning those concepts and skills and then use that information to design and enact instructional responses that support students' movement along that trajectory towards the learning goal (Heritage, 2008). Learning trajectories can be described as a path through the complex terrain of a particular mathematical topic (Battista, 2011; Daro et al., 2011). While this path is not necessarily linear, knowledge of the key stages or levels that characterize this path can help teachers both determine where students are and what experiences are likely to help them move forward. In other words, knowledge of learning trajectories can enhance the formative assessment process.

In conceptualizing the knowledge that teachers need to implement effective formative assessment in the classroom, we draw upon a conception of teaching as a complex activity that is dependent on distinct but interconnected bodies of knowledge (Ball, Thames, & Phelps, 2008; Putnam & Borko, 2000; Shulman, 1987). Arguing that teachers draw on knowledge that is distinct from either knowledge of subject matter, Shulman defines *pedagogical content knowledge* (PCK) as "the ways of representing and formulating the subject matter that make it comprehensible to others" (p. 9) and frames it as the intersection between content and pedagogy. Building on this work to study the work that teachers do when teaching mathematics in the classroom setting, Ball and colleagues have further defined and delineated mathematical knowledge for teaching (MKT) by breaking down the domain of content knowledge into *common content knowledge, specialized content knowledge,* and *horizon content knowledge* and pedagogical content knowledge into *knowledge of content and students, knowledge of content and teaching,* and *knowledge of content and curriculum* (Ball et al., 2008).

More recently, Sztajn et al. (2012) bring together research on learning trajectories with research on teaching to propose the construct of *learning trajectory-based instruction* as "teaching that uses student learning trajectories as the basis for instruction (p. 147)." In addition to presenting a learning trajectory interpretation of the six MKT categories, they define a learning trajectory interpretation of formative assessment as the case where teachers are "guided by the logic of the learner" rather than only by disciplinary goals when eliciting student thinking and providing feedback to students. In developing the TASK instrument and analyzing the results of the field test, we draw on these frameworks to explore how teachers actually make sense of evidence of student thinking for their instruction.

The TASK Instrument

We designed the TASK instrument to capture and explore teacher knowledge in relation to learning trajectories in several core mathematical content areas. Open-ended prompts were designed to elicit the information teachers glean from student work and the instructional response they develop based on that evidence. While the MKT is an established measure of "mathematics knowledge for teaching," these multiple choice measures have not been as useful in capturing teacher reasoning or more subtle manifestations of teacher conceptual change (Goldsmith & Seago, 2007). Hill, Ball, and Schilling (2008) describe the challenges of using multiple choice measures to assess "knowledge of content and students," or teachers' knowledge of mathematical thinking and learning, including the fact that performance can be influenced by test-taking skills or mathematical content knowledge. They conclude that open-ended items may be a more effective way to assess the kind of reasoning skills about student thinking that are called for in classroom-based instructional practice. We have developed, field tested, and validated TASK to provide a contextualized measure of teachers' ability to (a) analyze students' mathematical thinking within a grade-specific content area in relation to research-based learning trajectories, and (b) formulate effective instructional responses.

We began the development of TASK by first determining what kinds of knowledge are brought to bear in the process of formative assessment. Formative assessment involves a critical shift from *scoring* student work to *interpreting evidence* of student thinking and considering that evidence in light of research on the development of understanding of mathematical content. With this in mind, we initially posited that the following six domains of knowledge are relevant for *learning trajectory-oriented formative assessment*:

1. *Content Knowledge*—At the most basic level, teachers need to be able to understand and correctly solve math problems that assess the content they are teaching.
2. *Concept Knowledge*—To assess student understanding, teachers must be able to identify and articulate the concept and related sub-concepts that a particular mathematics problem or item is assessing.
3. *Mathematical Validity*—Once a teacher administers an assessment to a student, he/she must be able to understand the logic or mathematical validity of the strategy that the student uses to solve the problem.[1]
4. *Analysis of Student Thinking (AST)*—To build on student thinking, teachers need to be able to go beyond determining whether or not a response is correct or incorrect to identify the underlying conceptual understanding or misconceptions that are present in student work.
5. *Learning Trajectory Orientation (LTO)*—After analyzing the strategy a student uses to solve a math problem, teachers need to be able to position that strategy along a learning trajectory for the respective math content. Thus, teachers must have a sense of what the developmental progress looks like for the particular math concept and where to place students along that continuum and be able to use this as a framework to interpret and respond to student thinking.
6. *Instructional Decision Making (IDM)*—Finally, teachers must choose an appropriate instructional response and be able to describe why that instructional intervention is designed to move students from their current level of understanding along the developmental trajectory towards greater understanding.

To further explore these domains, we constructed a performance assessment that requires teachers to draw upon and articulate these types of knowledge in the context of classroom practice. Specifically, we situated TASK in the activity of looking at and responding to a carefully designed set of typical student responses to a mathematics problem in a particular content area. The student responses characterize different levels of sophistication of student thinking as well as common misconceptions that are supported by mathematics education research. Through an online instrument, teachers are presented with the student work and then led through a series of questions designed to measure these six key domains of knowledge related to the specific mathematical concept that is being assessed.

Seven TASK instruments have been developed in the following mathematics content areas: (1) *addition*, for teachers in grades K-1; (2) *subtraction*, for teachers in grades 2–3; (3) *multiplicative reasoning*, for teachers in grades 3–5; (4) *fractions*, for teachers in grades 3–5; (5) *proportional reasoning* for teachers in grades 6–8; (6) *algebraic reasoning* for teachers in grades 7–12; and (7) *geometric reasoning*, for

---

[1] As Ball et al. (2008) point out, determining whether a students' thinking is mathematically sound requires a kind of knowledge that a person with strong knowledge of mathematics content who is not a teacher may not necessarily possess. It is therefore distinct from common content knowledge.

> Each carton holds 24 oranges. Kate's carton is 1/3 full. Paul's carton is 2/4 full. If they put all their oranges together, would Kate and Paul fill one whole carton?
> Solve the problem. Show your work.

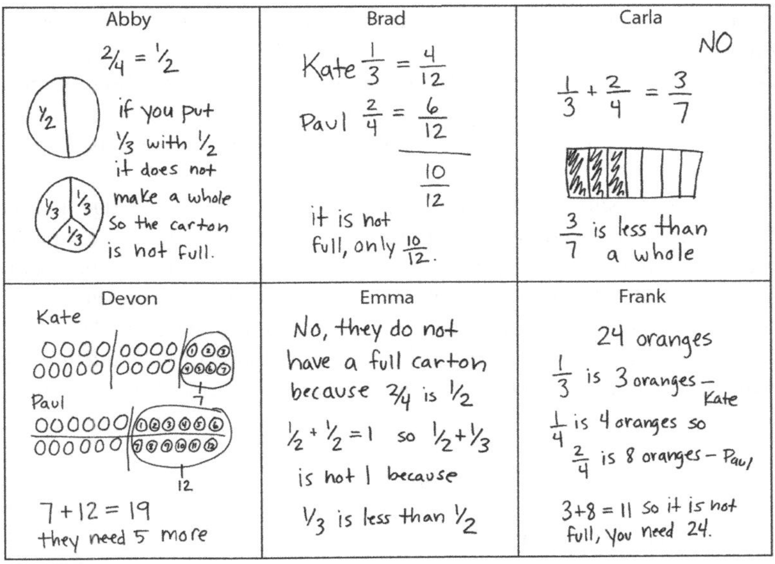

Abby	Brad	Carla
$\frac{2}{4} = \frac{1}{2}$ if you put $\frac{1}{3}$ with $\frac{1}{2}$ it does not make a whole So the carton is not full.	$Kate \frac{1}{3} = \frac{4}{12}$ $Paul \frac{2}{4} = \frac{6}{12}$ $\frac{10}{12}$ it is not full, only $\frac{10}{12}$.	NO $\frac{1}{3} + \frac{2}{4} = \frac{3}{7}$ $\frac{3}{7}$ is less than a whole
Devon	Emma	Frank
Kate 0000|0000|000 00000|0000|0000 Paul 000000|000000 000000|000000 $7 + 12 = 19$ they need 5 more	No, they do not have a full carton because $\frac{2}{4}$ is $\frac{1}{2}$ $\frac{1}{2} + \frac{1}{2} = 1$ so $\frac{1}{2} + \frac{1}{3}$ is not 1 because $\frac{1}{3}$ is less than $\frac{1}{2}$	24 oranges $\frac{1}{3}$ is 3 oranges — Kate $\frac{1}{4}$ is 4 oranges so $\frac{2}{4}$ is 8 oranges — Paul $3 + 8 = 11$ So it is not full, you need 24.

Fig. 1 Problem and designed student responses from the grades 3–5 fractions TASK

teachers in grades 9–12. These content areas represent core or fundamental mathematical ideas at the different grade level bands, and the TASKs are designed around key concepts in those domains (e.g., part/whole, equivalency, and magnitude for fractions). While the content areas are different across grade levels, all TASKs follow a consistent structure in both the prompts and the fact that the student work reflects key stages in the development of student thinking in the content area. The K-8 TASKs focus around six samples of student solutions; however, for algebra and geometry, since the problems have a higher level of complexity and longer student responses, there are only four samples of student work for the teacher to interpret.

An example of the different levels of sophistication of students' thinking and common strategies and misconceptions that are embedded in the student responses is presented in the fractions TASK for grades 3–5 in Fig. 1. The problem involves reasoning about whether two fractional quantities combine to make a whole. As shown in Fig. 1, Abby, Carla, and Devon's work reflect the use of visual models to make sense of parts and wholes, while Brad and Emma's work demonstrate more abstract reasoning about equivalence and addition. Carla, Devon, and Frank's work are less developed and contain misconceptions about partitioning, part/whole understanding, and the meaning of fractions. In this way, the student work represents some of the important landmarks that have been identified in current research on children's learning of fractions as well as an overall progression from concrete to more abstract understanding of fractional quantities (Confrey, 2008; Lamon, 2012; Steffe & Olive, 2010). Thus, TASK is designed to provide a realistic context from which to elicit information about what teachers pay attention to when they examine student strategies that they are likely to come across in their own classrooms.

Similarly for the other content areas, student work was constructed to represent key stages in the development of addition, subtraction, multiplication, proportional reasoning, and algebraic thinking, with student responses reflecting strategies of different levels of sophistication as well as strategies reflecting both procedural and conceptual errors.

The prompts are shown in relation to each dimension of knowledge and method of scoring in Table 1. Three of the response types are forced-choice or short answer and can be scored automatically while the rest are constructed responses scored by trained raters with a rubric or a combination of a coding scheme and rubric. The rubrics, described in the next section, are based on a four point ordinal scale to characterize the teachers' orientation towards the interpretation of the student work on a continuum that ranges from general to procedural to conceptual to developmental.

Table 1 TASK prompts and scoring

Domain of teacher knowledge	Prompt	Scoring	Scale
Content knowledge	Examine the math problem and state the correct answer	Automated	Correct/incorrect
Concept knowledge	Explain what a student at that grade-level needs to know and/or understand to solve the problem	Scored and coded by rater	Rubric score (1–4)
Mathematical validity	Examine the solutions of 4–6 typical students and determine if their solution processes are mathematically valid	Automated	Percent correct
Analysis of student thinking	Comment on four students' solution process in terms of what the work suggests about the student's understanding of the mathematics	Scored and coded by rater	Rubric score (1–4)
Learning trajectory orientation	Rank each student's solution of the level of sophistication of the mathematical thinking that is represented	Automated	Rubric score (1–4)
	Explain the rationale for the rankings given to each student	Scored by rater	Rubric score (1–4)
Instructional decision making	Suggest instructional next steps and explain the rationale for those next steps for two student solutions	Scored by rater	Rubric score (1–4)

As described above, the six samples of student work were constructed to represent both correct and incorrect solution strategies, common conceptual errors, as well as a range of sophistication of strategies. To prevent the instrument from becoming too time-consuming, respondents were asked to comment on a subset of four solution strategies, but then to rank and explain their ranking for all six. The four solutions represent beginning, transitional, and advanced strategies (with correct answers) as well as one solution that reflected a correct strategy with a conceptual error and incorrect answer. Likewise, respondents were only asked to describe instructional responses for two of the solutions: (a) a correct, but less sophisticated response to the problem and (b) a response with a conceptual weakness.

TASK was developed as an online instrument where teachers are sent an email link to complete the survey. Respondents move through several screens where the student work is shown as it is in Fig. 1 along with the respective prompts. Responses for mathematical validity and ranking are entered by clicking on radio buttons (see Fig. 2 below), while the open-ended responses for concept knowledge, analysis of student thinking, ranking-rationale, and instructional decision making are entered into text boxes. Respondents also have the option to expand their view of the student work by hovering the mouse over the image. A benefit of the online administration is that the system can target reminders to non-respondents to achieve a high response rate.

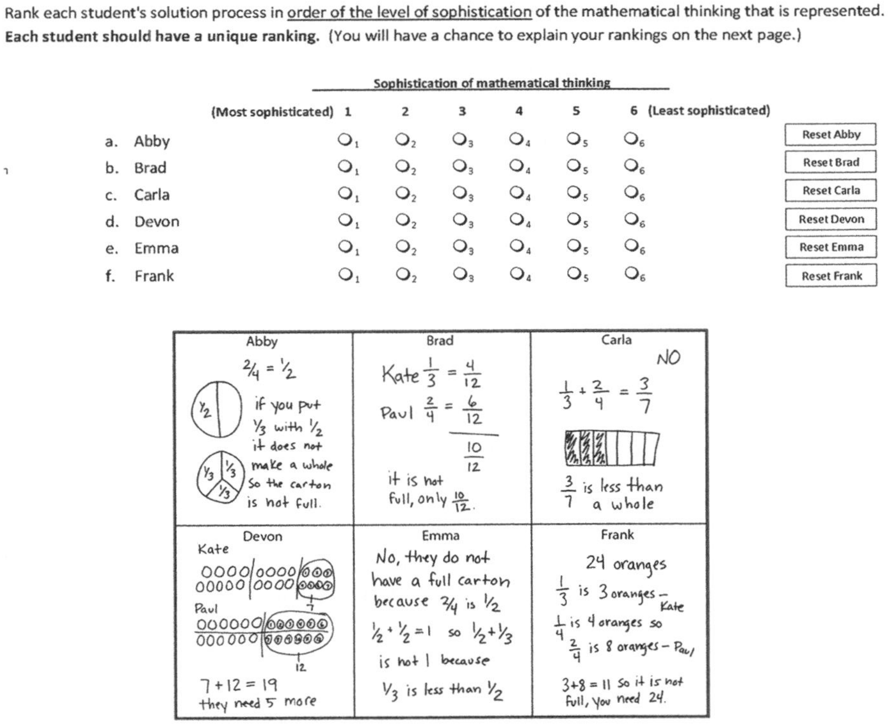

Rank each student's solution process in <u>order of the level of sophistication</u> of the mathematical thinking that is represented. **Each student should have a unique ranking.** (You will have a chance to explain your rankings on the next page.)

Fig. 2 Ranking screen of the online TASK

Scoring Rubric

The rubrics that raters used to score specific prompts about student work were based on a four-point ordinal scale to capture the overall orientation toward teaching or student understanding. We developed this rubric from the pilot data (described in the next section) through both an inductive and deductive process. First, a team of researchers read the entire set of responses to generate initial categories and codes to capture what teachers were referencing in their responses to each question. These codes were then grouped into larger categories, drawing on existing research in mathematics education to guide the analysis in terms of the degree to which the response reflected elements of a learning trajectory orientation. The distinction between procedures, or what students did, and concepts, or what students understood, became salient across all domains. The shift from procedural to more conceptual views of mathematics has long been promoted in mathematics reform literature (e.g., Hiebert, 1986; National Council of Teachers of Mathematics, 1988; National Research Council, 2001), and since learning trajectories by nature focus on conceptual development, a conceptual orientation toward student work was rated as higher than one that was only procedural. More recently, research on learning trajectories has promoted a developmental view, where students' conceptual knowledge develops in relation to instruction along a predictable path toward more complex and sophisticated thinking (Battista, 2011). Therefore, for a response to be at the highest level of the rubric, we determined that a teacher's focus on conceptual understanding must have evidence of drawing upon a developmental framework. We then had four ordinal categories (general, procedural, conceptual, and learning trajectory) that applied to each question on the TASK. The general rubric shown in Table 2 describes each of the TASK rubric categories. These categories are seen as cumulative where each level builds on the one before it; therefore, a conceptual response might also contain some procedural focus. Four domains were scored with more specific and detailed versions of this rubric: Concept Knowledge, Analysis of Student Thinking, Learning Trajectory Orientation, and Instructional Decision Making (Ebby, Sirinides, Supovitz, & Oettinger, 2013).

For Concept Knowledge and Analysis of Student Thinking, raters were asked to utilize a coding scheme organized in the form of a checklist, with descriptors under the main categories: general/superficial, procedural, conceptual, and learning trajectory. After the raters assigned the relevant codes, they used those results to help determine a rubric score. This technique also allows for tabulation of the specific concepts and procedures that are referenced by teachers which can be used to decompose patterns of teacher responses within each of the rubric categories.

The sample teacher responses shown below in Table 2 are taken from the pilot administration of the grades 6–8 proportions TASK. Teachers are describing a piece of student work where the strategy reflected a conceptual error stemming from additive thinking and led to an incorrect response. The sample teacher responses reflect different levels of analysis of that evidence: at the lowest level, the teacher evaluates the strategy but misses the nature of the conceptual error completely. At the procedural level, the teacher describes what the student did to get the incorrect answer, but

Table 2 TASK rubric levels and descriptions

Score	Category	Description	Sample response
4	Learning trajectory	Response draws on developmental learning trajectory to explain student understanding or develop an instructional response	Devon shows that he has some basic understanding of multiplicative reasoning when it comes to doubling both quantities of the rate. However, he then goes to additive reasoning to get to $20. He is not distinguishing the difference between multiplying and adding/subtracting in relation to proportionality
3	Conceptual	Response focuses on underlying concepts, strategy development, or construction of mathematical meaning	Devon has just a beginning understanding of a proportion as demonstrated by doubling both 12 and 15. He knew he needed to get to $20, but he didn't know how to use the proportion so he subtracted
2	Procedural	Response focuses on a particular strategy or procedure without reference to student conceptual understanding	Devon did not figure out the cost per can. He subtracted $10 so he also subtracted 10 cans but 1 can is not equivalent to $1
1	General	Response is general or superficially related to student work in terms of the mathematics content	Devon had a good strategy but did not perform the operations correctly

does not relate this to underlying conceptual understanding. At the conceptual level, the teacher recognized the proportional understanding in part of the student's solution strategy. The learning trajectory response is further distinguished by interpreting the students' use of doubling in terms of multiplicative reasoning and the conceptual error in terms of additive reasoning, an important distinction in established learning trajectories for proportional reasoning (e.g., Confrey, 2008; Lamon, 2012).

Ongoing TASK Development

TASK began with a pilot administration in the fall of 2011 with a convenience sample of 60 teachers and at least 10 responses at each grade band. The pilot data were used for two purposes. The first purpose was to begin development of the detailed scoring rubrics for each domain of the instrument and the second was to advance the design of the instrument. Both the actual responses and participant feedback contributed to our modifications of the instrument. Based on what we learned from this feedback, the instruments were substantially modified and scoring rubrics were developed.

In the spring of 2012, we administered TASKs in 6 content areas to a sample of about 1,800 teachers in grades K-10 from 5 public school districts in 5 states. Recruitment for this validation study used a stratified random sample of teachers by grade/subject; however, participation was voluntary. The five districts vary in terms of size, student demographics, and programs of math instruction. Overall, about 1,400 teachers completed the TASK,[2] for a 74 % response rate. Fifteen raters, including researchers with math content expertise and experienced mathematics teachers and coaches, were trained to code the open-ended responses for references to procedures and concepts and then make an overall judgment about a teachers' written response in relation to the four point rubric. TASKs were only scored by raters after they had established a reliability of at least 75 % direct agreement on all of the rubric scores with other reliable raters. Drawing from the results of the analysis of this field test data, much of which is described in the sections that follow, we have refined the TASK instrument to focus on the three most salient and robust domains: Analysis of Student Thinking (AST), Learning Trajectory Orientation (LTO), and Instructional Decision Making (IDM) while further streamlining the coding and analysis process.[3] In addition, we are developing multiple forms for repeated administrations as well as new TASKs in additional content areas.

Large-Scale Field Trial Results

Descriptive Statistics

Analysis of the field test data resulted in descriptive statistics for each of the domains on each TASK using unit weighting scoring as the average of scores within domain (Ebby et al., 2013). Across the domains examined on the TASK the majority of teacher responses were procedural, focusing on what the student did to solve the problem, rather than underlying conceptual understanding or sophistication of reasoning. Table 3 shows the breakdown of rubric scores for the domains of AST, LTO, and IDM. While these results are briefly summarized below, a more complete analysis of the descriptive results can be found in our interactive report (Supovitz, Ebby, & Sirinides, 2014).

Analysis of student thinking (AST). In this domain, the vast majority of teacher responses were procedural, focusing on what the student did to solve the problem rather than commenting on underlying conceptual understanding. Fewer than one fifth of the teachers surveyed, across all grade levels, interpreted the student solutions in terms of underlying conceptual understanding. The highest level of procedural responses were found in grades K-1 addition (93 %), while the highest level of conceptual and learning trajectory responses (19 %) were found in grades 3–5 fractions. Particularly striking is the fact that all of the responses for proportions in grades 6–8 were either general or procedural, with 21 % of the teachers providing

---

[2] Thousand two hundred and sixty-one fully completed TASKs in five content areas were analyzed from this field test. Responses to the geometry TASK have not yet been analyzed.

[3] For example, the latest version of the TASK for multiplicative reasoning includes some multiple choice questions to augment the open ended prompt for Instructional Decision Making.

Table 3 Percent of teacher responses by TASK domain, content, and score

Domain/content/grade	n	General	Procedural	Conceptual	Learning trajectory
AST					
Addition (K-1)	246	4	93	3	0
Subtraction (1–2)	185	13	76	11	0
Fractions (3–5)	376	7	73	18	1
Proportions (6–8)	291	21	79	0	0
Algebra (9–10)	163	9	88	3	0
LTO (rationale)					
Addition (K-1)	246	0	83	17	0
Subtraction (1–2)	185	4	69	22	5
Fractions (3–5)	376	9	76	14	1
Proportions (6–8)	291	14	78	7	1
Algebra (9–10)	163	15	57	26	2
IDM					
Addition (K-1)	246	13	79	8	0
Subtraction (1–2)	185	19	59	20	2
Fractions (3–5)	376	22	60	16	2
Proportions (6–8)	291	28	55	14	3
Algebra (9–10)	163	15	46	30	9

Note: 1,261 teacher responses to the TASK were collected in spring 2012

only general analyses of student work (e.g., "understands proportions" or "demonstrates strong reasoning."). The results highlight the widespread lack of a conceptual focus in teachers' analysis of student thinking around proportions among middle grades teachers.

Learning trajectory orientation (LTO). Again the vast majority of teachers explained their ranking of student work by pointing to procedural aspects of student work rather than what students understood or how that understanding was situated in a learning trajectory. It should be noted that teachers were somewhat more successful in choosing the ranking than they were in providing a reasoned rationale for that ranking, though fewer than half of teachers in grades K-8 were able to correctly order student strategies in terms of sophistication.

Instructional decision making (IDM). Across all grade levels, the majority of teachers' instructional suggestions for specific students focused on teaching a student a particular strategy or procedure rather than on developing mathematical meaning or understanding. The percentage of teachers who gave conceptual or learning trajectory responses was highest for algebra and lowest for addition in grades K-1.

Together, these results suggest that there is a great deal of room for growth in relation to teacher's ability to interpret and respond to conceptual understanding in student work, and even more so in relation to learning trajectories. We also used these results to provide information back to the participating districts in the form of reports that detailed the relative proportion of teachers at each grade level who responded at each level of the rubric in each domain. These reports were designed to allow districts to view both strengths and weaknesses in their teachers' capacity for learning trajectory-oriented formative assessment.

Instrument Properties

The design of the instrument and validation methods were directly influenced by the *Standards for Educational and Psychological Testing* (American Educational Research Association, American Psychological Association, & National Council on Measurement in Education, 1999), which provides strong guidelines for high-quality and technically sound assessments. Our methods of ongoing instrument validation were chosen to supply evidence that the resulting scores from this theoretically grounded instrument are reliable and valid for the purposes of evaluating teachers' capacity for learning trajectory-oriented formative assessment in mathematics. Unless otherwise noted, data for these analyses were collected from the large-scale field trial described above. The technical report (Ebby et al., 2013) provides more details about the measurement studies for the TASK.

To examine the validity of TASK scores as a measure of pedagogical content knowledge, we have analyzed its association with another similar established test, the measures of Mathematical Knowledge for Teaching (MKT) (Ball et al., 2008; Hill, Schilling, & Ball, 2004; Schilling, Blunk, & Hill, 2007). The MKT is a measure of the Common Content Knowledge and Specialized Content Knowledge that teachers need for effective mathematics instruction. The MKT is most aligned with the TASK domains of Content Knowledge and Mathematical Validity, but we expect that there would still be a positive, though smaller, relationship with the other domains, for which no validated measures exist. In the technical report (Ebby et al., 2013), we present descriptive statistics and correlation matrices for domain scales and the MKT separately for each TASK based on a sample of 486 teachers across the five districts. We find that the statistical associations of MKT and TASK domains reflect a low relationship and note that correlations are largest ($r=0.56$) for TASK domains with the most variance.[4] The positive direction and low magnitude of the statistics suggests that the constructs are related but distinct from MKT.

Collectively, results from a series of ongoing instrument validation studies are generating evidence that the instrument yields reliable and valid scores of teachers' learning trajectory-oriented formative assessment capacity in mathematics, is feasible for widespread use in a variety of settings, and provides useful reporting of results. Ongoing research studies will focus on how TASK can be used to measure change in teacher knowledge over time and whether it can predict student outcomes.

---

[4] We are mindful that score reliabilities for the TASK are still under investigation and that correlations may be underestimated in the presence of measurement error (i.e., attenuation) (Lavrakas, 2008).

Pathways Analyses

In this section, we highlight an investigation of the relationships between the various dimensions of teachers' ability to analyze student work in mathematics and their instructional decision making. More specifically, we investigate the relationships between: (1) mathematical validity; (2) analysis of student thinking; (3) learning trajectory orientation; and (4) instructional decision making.[5] The theoretical framework is based on the research literature in mathematics education and our hypothesis that analyzing student work for underlying conceptual understanding should contribute to a more sophisticated instructional response. Given the current focus on learning trajectories in mathematics education research, and standards, we also investigated whether the ability to place student work in a learning trajectory would have an effect on instructional decision making, and if so, how strong that relationship is compared to other dimensions.

The conceptual framework guiding the empirical study is summarized by the structural pathways in Fig. 3. This framework includes a series of relationships among independent and dependent constructs, which characterize the mechanism through which the analysis of student work influences instructional decision making. The analysis of student work in terms of Mathematical Validity (MV), Analysis of Student Thinking (AST) and Learning Trajectory Orientation (LTO) is theorized to affect instructional decision making (IDM). Additionally, AST is theorized to be

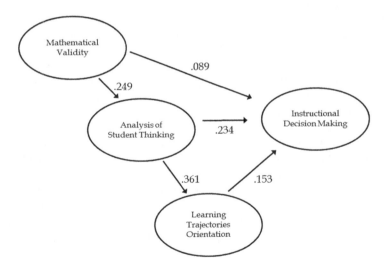

Fig. 3 Conceptual model of teachers' assessment of student knowledge and instructional decision making

---

[5] We do not include the domains of content knowledge or concept knowledge in this analysis as we do not expect them to have as strong of an influence on instructional decision making.

indirectly predictive of IDM through its effect on LTO. Finally, MV indirectly affects IDM through AST's direct and indirect paths to IDM. It is important to note that our method of analysis (described below) cannot be employed as a causal modeling approach because it cannot satisfy assumptions of directionality (Duncan & Hodge, 1963) or spurious correlation (Simon, 1957). As such, study findings do not adopt a causal interpretation, in the sense of confirming a presumed hypothesized network of causation. Rather this study sheds light on the tenability of the theorized causal model and results may be used as grounds for future research to further investigate causal mechanisms that are implied by this correlational study.

For this study, a statistical modeling approach was needed to meet several analytic goals. First, the study of indirect effects required the modeling of mediating variables. Path analysis (Wright, 1934) met this need because it offered a single framework for a system of multiple equations. Another analytic goal was the inclusion of latent variables in the model. This study examined relationships among theorized dimensions that pertain to learning trajectory-oriented formative assessment, which are not measured directly, but rather are measured by the TASK using a set of indicators and rubrics. Structural equation modeling (SEM) expands the path analysis framework to include a measurement model. In the measurement component of the SEM, latent variables are modeled as exogenous predictors of multiple observed items. The structural component of SEM specifies relationships among latent or observed variables. A benefit of using SEM is that both the measurement model and the structural model are estimated as one system of equations.

An empirical model was specified according to the structural pathways in the conceptual model (Fig. 3). Each of the four hypothesized domains were modeled as unobserved factors represented by the ten constituent rubric scores using all 1,261 complete TASK records. The latent factors were identified in the model by assigning each a variance of one, making the factor covariance interpretable as a factor correlation. The observed data were analyzed as continuous outcomes and the pathways between factors were freely estimated parameters. The model was estimated using Full Information Maximum likelihood using MPlus 7.1.

The full structural equation model defined by the structural pathways specified by our conceptual model was estimated and did not meet conventional thresholds for model fit, with a significant overall model chi square statistic and RMSEA=0.18 (recommended<0.10). Despite the marginal fit of the model, we find that all path coefficients in the structural model are statistically significant and consistent with the hypothesized direction of the relationships. The three antecedent dimensions accounted for 23 % of variation in instructional decision making and the estimated direct, total indirect, and total effects presented in Table 4 are all statistically significant at $p<0.05$. Table 5 presents the standardized estimated correlation matrix for the latent variables.

Estimated correlations between TASK domains are positive, as expected. Further, all correlations are low suggesting that the measured domains are not highly associated (Cohen, 1988). The direction and magnitude of the statistics across TASK instruments suggests that the domains we are measuring are distinct. Across the subject areas, we observe that the largest correlations are between

Table 4 Standardized direct, indirect, and total effects on instructional decision making

	Direct	Total indirect	Total
MV	0.089	0.072	0.161
AST	0.234	0.055	0.289
LTO	0.153	–	0.153

Note: All estimates are significant at $p < 0.05$

Table 5 Standardized estimated correlation matrix for the latent variables

	MV	AST	LTO	IDM
MV	1.00			
AST	0.224	1.00		
LTO	0.058	0.259	1.00	
IDM	0.118	0.179	0.114	1.00

the domains of Analysis of Student Thinking (AST) and Learning Trajectory Orientation (LTO). In addition, Analysis of Student Thinking (AST) is more strongly correlated with each of the other domains, particularly in grades K-5, suggesting that overlap exists between this domain and the other domains.

These results are interpreted as preliminary findings that will inform ongoing instrument development as well as alternative conceptual models that may improve the fit of the measurement and structural components using a second round of multi-district TASK data. The ability of a teacher to analyze student thinking in terms of conceptual understanding was the largest predictor of instructional decision making in both its direct and total effect. A teacher's ability to assess the mathematical validity of student work is also predictive of IDM with nearly half of the total effect being mediated by their analysis of students' thinking and learning trajectory orientation. Overall, these findings provide preliminary evidence that a teacher's depth of understanding of student thinking may have the largest total effect on instructional decision making in terms of the degree to which these decisions draw upon learning trajectories with a significant amount of that relationship being mediated by the teachers' learning trajectory orientation.

These results confirm and add to some of the existing findings of qualitative studies of the relationship between teachers' interpretation of student work and their ability to develop informed instructional responses. In studying teachers' use of interim test data, Goertz, Oláh, and Riggan (2009) found that teachers who interpreted student errors conceptually, rather than only procedurally, were more likely to generate substantive instructional responses. Similarly, in analyzing teacher logs, Riggan and Ebby (2013) found a clear linkage between the way teachers analyze their student work and the nature of the instructional responses they develop. Teachers who described student work in terms of conceptual understanding were more likely to state that they would reteach the content differently using strategies that were tailored to the individual student. Adding to this research base, our analysis of TASK suggests that the depth of teachers' interpretation of student work is moderately related to their tendency to develop a learning trajectory-oriented instructional response.

Building Capacity for Effective Mathematics Instruction

TASK was developed as a tool for researchers and evaluators to assess teacher capacity for learning trajectory-oriented formative assessment and the impact of initiatives that seek to develop that capacity. The development of TASK and the various ongoing studies described in this chapter has led to some key findings about the instrument itself, the current capacity of teachers to interpret student thinking in relation to learning trajectories, and the nature of the knowledge that teachers need for effective mathematics instruction. TASK was developed to explore and measure an understudied component of mathematical knowledge for teaching: teacher knowledge in the context of formative assessment. Taken together, our analyses highlights three key domains—analysis of student thinking, learning trajectory orientation, and instructional decision making—that advance the conceptualization of the teacher knowledge required for learning trajectory-oriented formative assessment.

Our results offer empirical evidence that teachers' tendency to analyze student thinking for underlying conceptual understanding is related to their ability to develop instructional responses that build on the current state of students' thinking to move them towards more sophisticated understanding. Yet the vast majority of teachers surveyed across grade levels analyzed student work procedurally, in terms of what students did to solve the problem, rather than in relation to underlying conceptual understanding. Given the current emphasis in mathematics education on rigor as a balance between conceptual and procedural understanding, this suggests that there is a great deal of room for growth in teacher capacity to identify, interpret, and respond to students' conceptual understanding. Furthermore, results point to understanding a learning trajectory orientation, or the ability to order different student strategies in terms of the sophistication of mathematical thinking, as rooted in analysis for conceptual understanding and an immediate predictor of instructional decision making.

The TASK instrument is thus an important step towards identifying more precisely the components of teacher knowledge that can influence and potentially improve classroom instruction. TASK also represents an important new tool for researchers in mathematics education that has the capability to gauge more than just content knowledge or general pedagogical content knowledge. The ability to measure teacher knowledge, capacity, and growth in relation to the understanding and use of learning trajectories will become increasingly important as states and districts work to train teachers to reach the goals of new and more rigorous standards.

References

American Educational Research Association, American Psychological Association, & National Council on Measurement in Education. (1999). *The standards for educational and psychological testing*. Washington, DC: Authors.

Ball, D. L., Thames, M. H., & Phelps, G. (2008). Content knowledge for teaching: What makes it special? *Journal of Teacher Education, 59*, 398–407.

Battista, M. T. (2011). Conceptualizations and issues related to learning progressions, learning trajectories, and levels of sophistication. *The Mathematics Enthusiast, 8,* 507–570.

Bennett, R. E. (2014). Formative assessment: A critical review. *Assessment in Education: Principles, Policy & Practice, 18*(1), 5–25.

Black, P. B., & Wiliam, D. (1998). Assessment and classroom learning. *Assessment in Education, 5*(1), 7–74.

Clements, D., & Sarama, J. (2004). Learning trajectories in mathematics education. *Mathematical Thinking and Learning, 6*(2), 81–89.

Cohen, J. (1988). *Statistical power analysis for the behavioral sciences* (2nd ed.). Hillsdale, NJ: Lawrence Erlbaum.

Confrey, J. (2008). *A synthesis of the research on rational number reasoning: A learning progression approach to synthesis.* Paper presented at The International Congress of Mathematics Instruction, Monterrey, Mexico.

Daro, P., Mosher, F. A., & Corcoran, T. (2011). *Learning trajectories in mathematics: A foundation for standards, curriculum, assessment, and instruction.* (Research report no. 68). Madison, WI: Consortium for Policy Research in Education.

Datnow, A., Park, V., & Wohlstetter, P. (2007). *Achieving with data: How high-performing school systems use data to improve instruction for elementary students.* Los Angeles: Center on Educational Governance, University of Southern California.

Duncan, O. D., & Hodge, R. W. (1963). Education and occupational mobility: A regression analysis. *The American Journal of Sociology, 68,* 629–644.

Ebby, C. B., Sirinides, P., Supovitz, J., & Oettinger, A. (2013). *Teacher analysis of student knowledge (TASK): A measure of learning trajectory-oriented formative assessment.* Technical report. Philadelphia: Consortium for Policy Research in Education.

Goertz, M., Oláh, L., & Riggan, M. (2009). *From testing to teaching: The use of interim assessments in classroom instruction.* CPRE research report #RR-65. Philadelphia: Consortium for Policy Research in Education.

Goldsmith, L. T., & Seago, N. (2007). *Tracking teachers' learning in professional development centered on classroom artifacts.* Paper presented at the Conference of the International Group for the Psychology of Mathematics Education, Seoul, Korea.

Heritage, M. (2008). *Learning progressions: Supporting instruction and formative assessment.* Los Angeles: National Center for the Research on Evaluation, Standards, and Student Testing (CRESST).

Heritage, M., Kim, J., Vendlinski, T., & Herman, J. (2009). From evidence to action: A seamless process in formative assessment? *Educational Measurement: Issues and Practices, 28*(3), 24–31.

Hiebert, J. (Ed.). (1986). *Conceptual and procedural knowledge: The case of mathematics.* Hillsdale, NJ: Lawrence Erlbaum.

Hill, H. C., Ball, D. L., & Schilling, S. G. (2008). Unpacking pedagogical content knowledge: Conceptualizing and measuring teachers' topic-specific knowledge of students. *Journal for Research in Mathematics Education, 39*(4), 372–400.

Hill, H. C., Schilling, S. G., & Ball, D. L. (2004). Developing measures of teachers' mathematics knowledge for teaching. *The Elementary School Journal, 105,* 11–30.

Kerr, K. A., Marsh, J. A., Ikemoto, G. S., Darilek, H., & Barney, H. (2006). Strategies to promote data use for instructional improvement: Actions, outcomes, and lessons from three urban districts. *American Journal of Education, 112*(4), 496–520.

Kluger, A. N., & DeNisi, A. (1996). The effects of feedback interventions on performance: A historical review, a meta-analysis, and a preliminary feedback intervention theory. *Psychological Bulletin, 119*(2), 254–284.

Lamon, S. J. (2012). *Teaching fractions and ratios for understanding: Essential content knowledge and instructional strategies for teachers* (3rd ed.). New York: Routledge. National Council of Teachers of Mathematics, 1988.

Lavrakas, P. J. (2008). *Encyclopedia of survey research methods: AM (Vol. 1).* Thousand Oaks, CA: Sage.

National Council of Teachers of Mathematics. (1988). *Curriculum and evaluation standards for school mathematics*. Reston, VA: NCTM.

National Research Council (Ed.). (2001). *Adding it up: Helping children learn mathematics*. Washington, DC: The National Academy Press.

National Research Council. (2007). *Taking science to school: Learning and teaching science in grades K-8*. Washington, DC: The National Academies Press.

Putnam, R. T., & Borko, H. (2000). What do new views of knowledge and thinking have to say about research on teacher learning? *Educational Researcher, 29*, 4–15.

Riggan, M., & Ebby, C. B. (2013). *Enhancing formative assessment in elementary mathematics: The ongoing assessment project* (Working paper). Philadelphia: Consortium for Policy Research in Education.

Schilling, S. G., Blunk, M., & Hill, H. C. (2007). Test validation and the MKT measures: Generalizations and conclusions. *Measurement: Interdisciplinary Research and Perspectives, 5*(2–3), 118–127.

Shepard, L. A. (2008). Formative assessment: Caveat emptor. In C. A. Dwyer (Ed.), *The future of assessment: Shaping teaching and learning* (pp. 279–303). New York: Lawrence Erlbaum.

Shulman, L. (1987). Knowledge and teaching: Foundations of the new reform. *Harvard Educational Review, 57*(1), 1–22.

Simon, H. A. (1957). *Models of man*. New York: Wiley.

Steffe, L., & Olive, J. (2010). *Children's fractional knowledge*. New York: Springer.

Supovitz, J., Ebby, C. B., & Sirinides, P. (2014). *Teacher analysis of student knowledge: A measure of learning trajectory-oriented formative assessment*. Retrieved from http://www.cpre.org/sites/default/files/researchreport/1446_taskreport.pdf

Sztajn, P., Confrey, J., Wilson, P. H., & Edgington, C. (2012). Learning trajectory based instruction: Towards a theory of teaching. *Educational Researcher, 41*(5), 147–156.

Wright, S. (1934). The method of path coefficients. *The Annals of Mathematical Statistics, 5*(3), 161–215.

Young, V. M. (2006). Teachers' use of data: Loose coupling, agenda setting, and team norms. *American Journal of Education, 112*(4), 521–548.

Part III
Learning

Using NAEP to Analyze Eighth-Grade Students' Ability to Reason Algebraically

Peter Kloosterman, Crystal Walcott, Nathaniel J.S. Brown, Doris Mohr, Arnulfo Pérez, Shenghi Dai, Michael Roach, Linda Dager Hall, and Hsueh-Chen Huang

The National Assessment of Educational Progress (NAEP) is the United States government program for tracking the performance of elementary, middle, and high school students. This chapter focuses on some of the issues that arise when NAEP mathematics data are used by researchers. After discussing NAEP mathematics assessments in general, we present analysis of eighth graders' algebraic reasoning as a sample of the type of research that can be done by looking at performance on specific NAEP items. We close by discussing logistical and methodological issues in the use of NAEP data. Some of these issues are unique to the mathematics data but many, as will be apparent, are issues that arise when using NAEP data from any content area.

P. Kloosterman (✉) • S. Dai • M. Roach • H.-C. Huang
School of Education, Indiana University, 201 N. Rose Ave., Bloomington, IN 47405, USA
e-mail: klooster@indiana.edu

C. Walcott
Indiana University Purdue University Columbus, Columbus, IN, USA

N.J.S. Brown
Boston College, Chestnut Hill, MA, USA

D. Mohr
University of Southern Indiana, Evansville, IN, USA

A. Pérez
The Ohio State University, Columbus, OH, USA

L.D. Hall
St. Patrick's Episcopal Day School, Washington, DC, USA

© Springer International Publishing Switzerland 2015
J.A. Middleton et al. (eds.), *Large-Scale Studies in Mathematics Education*,
Research in Mathematics Education, DOI 10.1007/978-3-319-07716-1_9

The NAEP Assessments

NAEP assessments take place in the areas of the arts, civics, economics, geography, mathematics, reading, science, technology and engineering literacy, US history, and writing. Economics and technology have only been assessed once but other content areas are assessed every 2–10 years. There is also a NAEP High School Transcript Study (HSTS) that provides high school transcripts of students who participated in one of the NAEP content area studies. The No Child Left Behind act (NCLB, 2001) requires that mathematics and reading be assessed every 2 years and, because these areas have been assessed since the early 1970s, there are more data sets available for them than for other areas.

The National Center for Education Statistics (NCES), which administers NAEP, provides reports on overall findings after each NAEP mathematics assessment (e.g., Braswell et al., 2001; National Center for Education Statistics, 2012). A major advantage of NAEP as opposed to state-level or college entrance exams (e.g., SAT) is that NAEP data come from a representative national sample of students and thus conclusions drawn from these data are valid for the United States as a whole. A student taking the mathematics portion of NAEP completes at most 30 of the 150 or more mathematics items in the pool at each grade level. The remaining items are completed by different individuals so when results from all students are combined, there is information on a wide variety of mathematics concepts and skills. In addition, each student who completes NAEP mathematics items, along with the teacher and principal of that student, completes a background questionnaire focusing on things like language used at home (student), years of experience (teacher), and percentage of students in the school receiving free or reduced-price lunches (principal).

There are currently two NAEP mathematics programs, Long-Term Trend (LTT) NAEP and Main NAEP. LTT NAEP began in the early 1970s and from 1978 through 2004 used the same items, so it was possible to track performance on those items for a long period of time (Kloosterman, 2010, 2011, 2014). This assessment collects data on 9-, 13-, and 17-year-old students and, while it was updated after 2004, the LTT mathematics NAEP is still predominantly an assessment of the basic skills that were taught in the twentieth century. In contrast, Main NAEP for mathematics, which originated in 1990, collects data on students in grades 4, 8, and 12, and has items that are continually updated to be representative of mathematics curricula in use at the time of each assessment. The 2009 Main NAEP mathematics assessment, for example, used 159 items at grade 8. Of these, 46 were in the algebra strand.

Although all NAEP assessments provide data broken down by demographic subgroup, Main NAEP collects data on enough students in every state to report results on a state-by-state basis. The confidence interval for the national sample is usually less than 1 scale point on either side of the mean and thus national scale scores are very accurate. State scores are less accurate with confidence intervals for states of 2–3 points on either side of the mean. After each administration, roughly one-fourth of the items are replaced so that there are enough items to track trends over time while allowing for updates to keep the assessment consistent with changes in curriculum. Most retired items are released to the public and available online (see http://nces.ed.gov/nationsreportcard/itmrlsx). One of the reasons that items are retired is that they no longer represent what is being taught in schools and thus the National Center for Education Statistics (NCES) cautions that released items are not necessarily representative of the NAEP assessment as a whole.[1]

NAEP Framework and Scoring

NAEP items are based on frameworks, or assessment blueprints, that are updated periodically to reflect curricular shifts. The NAEP mathematics frameworks for grades 4 and 8 have changed minimally since the 1990s, allowing for comparison of performance trends over time. The grade 12 framework changed significantly before the 2005, 2009, and 2013 assessments making performance trend analyses more complicated at this grade level.

The National Assessment Governing Board (NAGB) is charged with maintaining the framework documents. The current version of the framework (NAGB, 2012) sets assessment guidelines in four categories: mathematics content, mathematical complexity, item format, and assessment design. Each NAEP assessment item aligns to a content objective from one of five content strands: Number Properties and Operations; Measurement; Geometry; Data Analysis, Statistics, and Probability; and Algebra. The assessment framework calls for the greatest emphasis on number properties and operations at grade 4 and on algebra at grades 8 and 12. In addition, a mathematical complexity level of low, moderate, or high is reported for each item. NAGB defines mathematical complexity as the level of demand on thinking and takes into account what mathematics students are asked to do in a particular task. The framework specifies that about 50 % of assessment time should be spent on moderate level tasks with the remainder split between low and high complexity tasks.

The NAEP framework outlines three item formats: multiple choice, short-constructed response, and extended-constructed response. Designed to span all three mathematical complexity levels, multiple-choice items at grades 8 and 12 have five distractors and those at grade 4 contain four. In contrast to extended-constructed response items that often have multiple parts and require a written explanation in support of an answer, short-constructed response items require a brief response such as a numerical answer, a simple drawing, or a brief explanation. The NAEP framework calls for 50 % of the testing time to be devoted to multiple-choice items and the remaining 50 % of the time spent on short- or extended-constructed response items. The framework also calls for the use of calculators and manipulatives or tools on a limited number of items and includes recommendations for accommodations for students with disabilities.

The NAEP assessment is not designed to report individual student scores. Instead, the results represent the achievement of the national sample and its various

---

[1] Our analysis of the mathematics items released at grades 4 and 8 in recent years indicates that they are reasonably representative of the content and format of items found on the current assessment. Items released most recently are closer in content to current items than items released in the 1990s. Because of the major changes in the framework and content of the grade 12 assessment after 2000, grade 12 items released prior to 2005 are not representative of the current grade 12 assessment.

subgroups. NAEP results are reported in two ways: scale scores and achievement levels (NCES, 2012). Scale scores range from 0 to 500 for grades 4 and 8 and from 0 to 300 at grade 12. Average scale scores are determined through Item Response Theory (IRT) statistical procedures. NAEP reports a composite scale score for mathematics as a whole as well as scale scores for each of the five content strands at grades 4 and 8. Reporting at grade 12 is the same except that measurement and geometry are combined into a single strand. NAEP also reports results in terms of the percentage of students performing within the achievement levels of basic, proficient, and advanced. The formal definition of each level is complex, but in general, grade 8 students performing at the basic level are expected to have developed procedural and conceptual knowledge of objectives within the five content strands. Students falling within the proficient category are able to apply procedural and conceptual knowledge to problem-solving situations, while students within the advanced category are able to generalize and synthesize this knowledge across the five content strands. Examples of items at each proficiency level can be found on the NAEP website (http://nces.ed.gov/nationsreportcard/itemmaps/).

Access to NAEP Data

In addition to providing items no longer used in the assessments along with data on student performance for those items, the NAEP website (http://nces.ed.gov/nationsreportcard) provides a software tool for doing basic analyses of student performance. Called the *NAEP Data Explorer*, this tool allows a researcher to compare things such as performance of Hispanic grade 8 students who are eligible for free or reduced-price lunch to performance of Hispanic grade 8 students who are not eligible. Analyses can be done for the entire United States or an individual state although, because the number of private school students sampled in some states was small, state-level analyses are restricted to public school students only. There are, however, significant limitations to the types of analyses that can be done with the online tool. To complete more complex analyses, researchers must have the original data set, which requires a secure site license (see http://nces.ed.gov/statprog/instruct.asp). The analyses described in this chapter required use of the full data set because we needed performance information on items for each year they were administered and because we wanted to do analyses that went beyond the basic statistical analyses provided with the online *Data Explorer*.

What Can NAEP Tell Us About Students' Algebraic Reasoning Skills?

Since the early years of NAEP, mathematics educators have provided interpretive reports based on the specific items used for the mathematics assessment. Of the reports that focused on algebraic reasoning, for example, Chazan et al. (2007)

reported that gains in the algebra strand were greater than gains in any other content strand for grades 4 and 8 from the mid-1990s through 2003 but that large gaps in performance based on race/ethnicity persisted. These researchers also found that performance on items used at both grades 4 and 8 was always higher at grade 8 although the amount of difference between grades varied substantially by item. Looking at NAEP performance of grade 8 students, Sowder, Wearne, Martin, and Strutchens (2004) reported on 10 algebra items that were used in 1990, 1992, 1996, and 2000. Two of those items focused on patterns. For the first item, students were told that a pattern of As and Bs repeated in groups of 3 and then asked to fill in the missing letters in the sequence $A\ B\ \_\ A\ \_\ B\ \_\ \_\ \_$. Performance on this item increased from 50 % answering correctly in 1990 to 64 % answering correctly in 2000. The second item, although multiple choice, was much more difficult. Students had to identify the number of the term of the sequence 1/2, 2/3, 3/4, 5/6, ... that was equal to 0.95. Performance increased from 19 % answering correctly in 1990 (essentially the chance level) to 27 % answering correctly in 2000. Relative to other items, these were fairly substantial jumps. Performance also increased significantly on two of six items involving algebraic expressions or equations (5 and 6 % increases), and on both items involving graphing (7 and 14 %).

In a more general review of research on algebra learning, Kieran (2007) identified three broad categories of research: understanding letter-symbolic algebra, understanding multiple representations in algebra, and using algebra in the context of word problems. With respect to letter-symbolic algebra, Kieran notes that recent research often focuses on factors that make algebra difficult to learn. For example, MacGregor and Stacey (1997) found that when presented with algebraic expressions and equations, students sometimes ignore variables, replace them with constant values, or treat them simply as names rather than representations of a range of values. Obviously, such interpretations make it difficult to understand the meaning of the variables and expressions. With respect to multiple representations, Kieran referenced a study by Lobato, Ellis, and Munoz (2003) that found that students studying the slope-intercept form of a linear function often interpreted slope as the amount a line goes up rather than as a ratio. Lobato et al. recommended more instructional focus on covariation to help students see that slope is a relationship between two variables rather than just an increase in one. In her summary of research on algebra in the context of word problems, Kieran indicated that many studies show that when the option presents itself, students often want to use informal methods rather than formal algebraic methods (i.e., writing and solving equations). In brief, research on the learning of algebra documents that while students use logical processes when presented with algebra problems, they sometimes get incorrect answers because their logic is based on inappropriate assumptions or procedures that do not work in the context provided.

Looking across content areas, D'Ambrosio, Kastberg, and Lambdin (2007) argued that factors beyond item content affected student performance. In particular, they argued that item wording and item format impacted student performance and that the impact varied by demographic subgroup.

In the research reported here, we extended prior studies of algebraic reasoning and studies of what NAEP data say about algebra learning by analyzing student performance on grade 8 algebraic reasoning items used between 2003 and 2011.

More specifically, this study used the 2003–2011 data on items that require algebraic reasoning to address the questions of what grade 8 students in the United States know about the forms of algebraic reasoning outlined in previous NAEP work and by Kieran and how that knowledge changed between 2003 and 2011.

Method

The data set for this study was all grade 8 algebra-related Main NAEP mathematics items used between 2003 and 2011. Items were divided into the categories of (a) patterns, relations, and functions, (b) algebraic representations, (c) variables and expressions, and (d) equations and inequalities. Tables were constructed showing the items in each category along with performance on those items for each year they were administered.

We developed a coding scheme to facilitate identification of items relevant to given topics such as patterns or reading graphs. The scheme, which uses the five content strands identified in the 2011 NAEP framework (NAGB, 2010), includes a simplified list of topics featured in two monographs (Kloosterman & Lester, 2004, 2007) produced by earlier iterations of our project. We also added an additional subtopic—fractions, decimals, and percents—as a part of the number properties and operations strand because of the heavy focus on this topic in the monographs. All items coded to fractions, decimals, and percents were also coded to other objectives such as number sense or algebraic representations.

The NAEP secure data set identifies the content area for each item but not the subtopic. Information about the subtopic would have been helpful for our coding, but our system was not designed to recreate the official NAEP coding. Instead, it was designed to allow researchers to easily find items that might be related to topics of interest. To this end, we allowed for multiple codes. Two members of the project team coded each item but made no attempt at consensus, reasoning that an item might be of interest for the subtopic if either team member found it relevant. While this process identified all possible items for each subtopic, it led to inclusion of some items that were not an ideal fit for some of the subtopics. Thus, when the codings were used to place items into categories, authors eliminated released items that did not provide much insight into the primary topic and non-released items on which too little was known to verify that the items fit the category.

The algebra items that make up the four tables in this chapter resulted from the coding process. Exact format items that have been released to the public and the diagrams that accompanied some items can be seen by using the year, block, and item number given after the item description in the tables to identify the item in the NAEP online *Questions Tool* (http://nces.ed.gov/nationsreportcard/about/naeptools.asp). Only general information can be provided on non-released items such as Item 7 in Table 1.

Some items, including Item 15 in Block 8 of the grade 8 items released in 2011, appear in more than one table. This item (Item 10 in Table 1 and Item 22 in Table 2)

Table 1 Performance on items involving understanding and use of patterns, relations, and functions

Item description	Item type	Percent correct				
		2003	2005	2007	2009	2011
Patterns						
1 Which of the figures below should be the fourth figure in the pattern shown above? [2005-M4 #3]	MC	90	90			
2 Write the next two numbers in the number pattern. 1 6 4 9 7 12 10 ___. Write the rule that you used to find the two numbers you wrote [2009-M5 #11]	SCR	67	68	70	70	
3 The same rule is applied to each number in the pattern 1, 9, 25, 48, 81, …. What is the 6th number in the pattern? [2005-M12 #13]	MC	60	61			
4 According to the pattern suggested by the four examples above, how many consecutive odd integers are required to give a sum of 144? [2005-M3 #12]	MC	40	40			
5 In the sequence 35, 280, 2,240, ___, the ratio of each term to the term immediately following it is constant. What is the next term of this sequence after 2,240? [2009-M10 #9]	SCR	32	32	32	33	
6 Each figure in the pattern below is made of hexagons that measure 1 cm on each side. If the pattern of adding one hexagon to each figure is continued, what will be the perimeter of the 25th figure in the pattern? Show how you found your answer [2007-M7 #14]	SCR		18	19		
7 Extend and generalize a number pattern	ECR			12	14	
Graphs of relations and functions						
8 From the starting point on the grid below, a beetle moved in the following way. It moved 1 block up and then 2 blocks over, and then continued to repeat this pattern. Draw lines to show the path the beetle took to reach the right side of the grid [2005-M4 #10]	SCR	54	54			
9 Which of the following is the graph of the line with equation $y = -2x + 1$? [2007-M11 #11]	MC	22	23	25		
10 The linear graph below describes Josh's car trip from his grandmother's home directly to his home [2011-M8 #15]	ECR					
(a) Based on this graph, what is the distance from Josh's grandmother's home to his home?						90
(b) Based on this graph, how long did it take Josh to make the trip?						94
(c) What was Josh's average speed for the trip? Explain how you found your answer						54
(d) Explain why the graph ends at the *x*-axis						19
(a) through (d) all correct					12	11

(continued)

Table 1 (continued)

Item description	Item type	Percent correct				
		2003	2005	2007	2009	2011
Other						
11　For 2 min, Casey runs at a constant speed. Then she gradually increases her speed. Which of the following graphs could show how her speed changed over time? [2011-M9 #3]	MC				69	70
12　Which of the following equations represents the relationship between x and y shown in the table? [2005-M12 #17]	MC	51	54			
13　The number of gallons of water, y, in a tank after x hours may be modeled by the linear equation $y = 800 - 50x$. Which of the following statements about the tank is true? [2011-M12 #15]	MC			46		
14　Find change in y given x for linear equation	MC			47	48	
15　Tom went to the grocery store. The graph below shows Tom's distance from home during his trip. Tom stopped twice to rest on his trip to the store. What is the total amount of time that he spent resting? [2009-M10 #10]	MC		41	42	44	
16　An airplane climbs at a rate of 66.8 ft per minute. It descends at twice the rate that it climbs. Assuming it descends at a constant rate, how many feet will the airplane descend in 30 min? [2007-M11 #19]	MC	38	37	37		
17　In the equation $y = 4x$, if the value of x is increased by 2, what is the effect on the value of y? [2005-M3 #10]	MC	33	34			
18　Which of the following is an equation of a line that passes through the point (0, 5) and has a negative slope? [2011-M12 #7]	MC			26	29	31
19　Sarah has a part-time job at Better Burgers restaurant and is paid $5.50 for each hour she works. She has made the chart below to reflect her earnings. (a) Fill in the missing entries in the chart. (b) If Sarah works h hours, then, in terms of h, how much will she earn? [2007-M9 #10]	SCR	24	27	27		
20　In which of the following equations does the value of y increase by 6 units when x increases by 2 units? [2013-M3 #9]	MC	22	23	25		

Table 2 Performance on items involving understanding and use of algebraic representations

Item description	Item type	Percent correct				
		2003	2005	2007	2009	2011
Number lines						
1. On the number line above, what number would be located at point P? [2005-M4 #5]	SCR	89	89			
2. Jorge left some numbers off the number line below. Fill in the numbers that should go in A, B, and C [2009-M5 #14]	SCR	75	76	77	77	
3. Weather records in a city show the coldest recorded temperature was −20 °F and the hottest was 120 °F. Which of the following number line graphs represents the range of recorded actual temperatures in this city? [2007-M7 #5]	MC		59	59		
4. What number is represented by point A on the number line above? [2005-M12 #12]	MC	41	43			
5. Which of the graphs below is the set of all *whole* numbers less than 5? [2005-M12 #10]	MC	36	37			
Graphs						
6. The map above shows eight of the counties in a state. The largest city in the state can be found at location B-3. In which county could this city lie? [2005-M12 #14]	MC	85	86			
7. The graph above shows lettered points in an (x, y) coordinate system. Which lettered point has coordinates $(−3, 0)$? [2007-M11 #2]	MC	72	75	78		
8. According to the graph, between which of the following pairs of interest rates will the *increase* in the number of months to pay off a loan be greatest? [2011-M12 #9]	MC			71	70	70
9. Which point is the solution to both equations shown on the graph above? [2009-M10 #7]	MC		67	69	72	
10. If the points Q, R, and S shown above are three of the vertices of rectangle QRST, which of the following are the coordinates of T (not shown)? [2005-M3 #13]	MC	58	61			
11. In which of the following groups do all the ordered pairs lie on the line shown above? [2013-M6 #6]	MC		48	53	56	
12. For the figure above, which of the following points would be on the line that passes through points N and P? [2009-M10 #14]	MC		50	52	55	
13. From the starting point on the grid below, a beetle moved in the following way. It moved 1 block up and then 2 blocks over, and then continued to repeat this pattern. Draw lines to show the path the beetle took to reach the right side of the grid [2005-M4 #10]	SCR	54	54			

(continued)

Table 2 (continued)

Item description	Item type	Percent correct 2003	2005	2007	2009	2011
14. On the curve above, what is the *best* estimate of the value of x when $y=0$? [2005-M3 #11]	MC	49	50			
15. Interpret a stem and leaf plot	MC			48	48	
16. Find change in y given x for linear equation	MC			47	48	
17. Identify true statement about graph of a line	MC				31	
18. Which of the following is an equation of a line that passes through the point (0, 5) and has a negative slope? [2011-M12 #7]	MC			26	29	31
19. Fill in the table below so that the points with coordinates (x, y) all lie on the same line [2013-M3 #15]	SCR			27	29	
20. Which of the following is the graph of the line with equation $y=-2x+1$? [2007-M11 #11]	MC	22	23	25		
21. In which of the following equations does the value of y increase by 6 units when x increases by 2 units? [2013-M3 #9]	MC			18	18	
22. The linear graph below describes Josh's car trip from his grandmother's home directly to his home [2011-M8 #15]	ECR				12	11
(a) Based on this graph, what is the distance from Josh's grandmother's home to his home?						90
(b) Based on this graph, how long did it take Josh to make the trip?						94
(c) What was Josh's average speed for the trip? Explain how you found your answer						54
(d) Explain why the graph ends at the x-axis						19
(a) through (d) all correct					12	11
Equations						
23. Mrs. Brown would like to pay off a loan in 180 months. According to the graph, what should be the approximate percent of the interest rate on her loan? [2011-M12 #10]	SCR			75	75	72
24. The admission price to a movie theater is $7.50 for each adult and $4.75 for each child. Which of the following equations can be used to determine the total admission price? [2011-M8 #3]	MC			70	70	70

#	Item	Type				
25.	The table below lists the coordinates of several points on a line. Which of the following is an equation of the line?	MC		55	60	65
26.	The length of a rectangle is 3 ft less than twice the width, w (in feet). What is the length of the rectangle in terms of w? [2009-M10 #8]	MC	48	50	51	
27.	The number of gallons of water, y, in a tank after x hours may be modeled by the linear equation $y = 800 - 50x$. Which of the following statements about the tank is true? [2011-M12 #15]	MC		46	48	48
Inequalities						
28.	On the number line shown, the arrow is pointing to a number that is closest to which of the following (all values are decimals)? [2013-M6 #8]	MC	71	73	74	
29.	Which of the following is the graph of the inequality given?	MC	43	44	45	
30.	Graph the solution set for $3 \leq x \leq 5$ on the number line below [2011-M9 #11]	SCR			24	22
Other						
31.	Archeologists use the formula $h = 73 + 2.5t$ to estimate the height of a dinosaur from the length of the tibia (t). If the tibia is 400 cm, what is the estimated height? [2013-M3 #14]	MC		38	40	
32.	The map above gives the distances, in miles, between various locations in a state park. Traveling the shortest possible total distance along the paths shown on the map, from the visitor center Teresa visits the cave, waterfall, and monument, but not necessarily in that order, and then returns to the visitor center. If she does not retrace her steps along any path and the total distance that Teresa travels is 14.7 miles, what is the distance between the cave and the monument? [2005-M3 #15]	MC	24	23		
33.	If an eruption of old faithful lasts t minutes, then the next eruption will occur approximately $12.5t$ minutes later. If the previous eruption lasted 6 min and ended at 1:23 p.m., when is the next eruption expected to occur? [2013-M7 #12]	SCR		21		

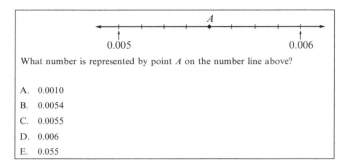

What number is represented by point A on the number line above?

A. 0.0010
B. 0.0054
C. 0.0055
D. 0.006
E. 0.055

Fig. 1 NAEP item 2005-M12 #12

focuses on the graph of a function and thus fits both in Table 1, which includes functions, and in Table 2, which includes algebraic representations. In each table, items were further sorted into subcategories, such as number line and graphs under algebraic representations (Table 2). The tables, along with patterns and trends in the data identified by the authors based on those tables, form the results section of this chapter. Note that while NAEP classifies items into one of five primary content strands, our coding scheme identified items in content strands other than algebra when those items required algebraic reasoning to complete. For example, Item 2005-M12 #12 (Item 4 in Table 2 and shown in Fig. 1) was identified as a number properties and operations item in the secure data set. Members of our project team, however, coded it as (a) number sense, (b) fractions, decimals, and percents (both under number properties and operations), and (c) algebraic representations (under algebra) because it involved aspects of each of these subtopics.

Results

Scale scores for overall NAEP and for the algebra strand for students in grade 8 between 1990 and 2013 are shown in Fig. 2. As can be seen in the figure, there was consistent growth at grade 8 although the rate of growth has varied somewhat across the years. Scores in 2013 were significantly higher ($p < .05$) than any previous year both in overall performance and in algebra.

Patterns, relations, and functions. Table 1 shows the 20 items used between 2003 and 2011 that required understanding and use of (a) patterns, (b) graphs of functions, or (c) other relation or function skills. The table includes either the actual wording or a description of each item and the percentage of students who responded correctly each year the item was administered. Within each subgroup in the table, items are ordered from highest to lowest with respect to proportion of students answering correctly.

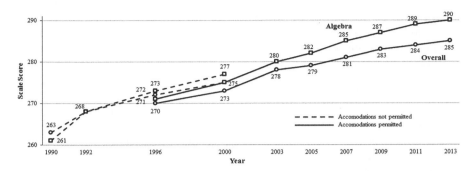

Fig. 2 NAEP overall scale scores and algebra scale scores. Data come from the *NAEP data explorer* (http://nces.ed.gov/nationsreportcard/naepdata/)

Item 1 in Table 1, which involved a pattern of shaded wedges within a circle, was the easiest algebraic reasoning item in the years analyzed. The item was also administered at grade 4 where 74 % of students answered correctly. The fact that the figure provided the first three and the fifth elements of the pattern along with the visual representation of the pattern likely contributed to the high percentage of students answering correctly. Item 2 in Table 1, which required students to find the next two numbers in the pattern 1, 6, 4, 9, 7, 12, 10, and write the rule used to find the numbers, was also given at grade 4 where 40 % answered correctly. The difference in performance between fourth and eighth grade is typical of trends found on Long-Term Trend NAEP, where items have been used at multiple grade levels more often than on Main NAEP (Kloosterman, 2014). It is interesting to note that these items, like many of those used at multiple levels on LTT NAEP, focus on content that is usually not taught after fourth grade yet older students did significantly better.

The three most difficult pattern questions for eighth graders were numbers 5 though 7. Items 6 and 7 required written explanation and students regularly have problems with items where they have to provide an explanation or justification (Arbaugh, Brown, Lynch, & McGraw, 2004). Item 5 tells students that there is a constant ratio between terms in the sequence 35, 280, 2,240 but even with that information, only about one third were able to calculate the next term. Given that calculators were not allowed for this item, it is hard to know whether students did poorly because they did not know what to do to find the next term or because they could not do the relatively complex paper-and-pencil calculations.

Few students could relate a linear equation to its graph (Item 9) or identify an equation where the value of y increased by 6 when the value of x increased by 2 (Item 20). On the other hand, almost all students could infer straightforward information from a graph (Items 10a and 10b) and over half could infer average speed from a graph of time and distance (Item 10c). The fact that only 19 % could explain why the line showing distance would end when the destination point was reached (Item 10d) is another example of how hard it is for most students to explain mathematical concepts in writing.

The items classified as "other" in Table 1 required a variety of skills with equations and relations. Item 19 was difficult (27 % answered correctly in 2007) likely because two pieces of information had to be added to a chart, and then a generalization had to be made for the item to be fully correct. On the positive side, 85 % of students were successful on at least one part of the item. Of the remaining items, performance was over 50 % only on Items 11 and 12.

Looking at Table 1 as a whole, one gets the sense that some, but certainly not all, eighth graders can understand and explain relationships between two variables in different settings and formats. The items in the table range from the purely visual (Item 1) to the connection between symbols and graphs (Item 9), to the purely graphical (Item 11). Taken together, they provide a good sense of what eighth-grade students know with respect to patterns and functions.

Algebraic representations. Table 2 shows performance on items involving algebraic representations. As was the case in Table 1, performance was stable or increasing modestly over time on most items. Items 1 and 2 were the only items that were also given at grade 4. Item 1 involved a number line where increments were tenths rather than whole numbers. Performance on this item was substantially better at grade 8 than at grade 4 (89 % vs. 44 % in 2005). A correct response for Item 2 involved writing a whole number (3) and fractional quantities (3½, 3¾) on a number line. Eighth graders did substantially better than fourth graders (77 % vs. 45 % in 2009). Items 7 and 9 required identifying the coordinates of a point and, taken together, suggest that 70 % or more of eighth graders have this skill.

Item 14 involved interpolating the x-intercept of a curve that crossed the x-axis nearly midway between 1 and 2. Half of students selected the correct answer (1.4) but 18 % selected −2, indicating that they had little sense of what the graph represented, and another 17 % selected 1.1, suggesting they knew the answer had to be more than 1 but had no sense of how much more than 1. Performance on Item 25 improved from 55 to 65 % between 2005 and 2009. This item focused on identifying the equation of a line when given a table with coordinates of 4 points on the line. The substantial improvement indicates that students are getting better at plugging x and y coordinates into an equation to see if they represent points on the line. Item 11 was similar to item 25 in that it required checking values but rather than the equation of the line, students had to find the points on a grid to see whether they were on the line. The 8 % gain on this item from 2005 to 2009, together with the gain on item 25 suggests that students are getting better at using graphical as well as numerical representations of linear functions. However, the percentage of students who could identify the graph of $y = -2x + 1$ (Item 20) was just above the chance level so even though many eighth graders can plot points, few are able to use slope and intercept or determine which points to plot to connect an equation and a graph.

Variables and expressions. Table 3 focuses on understanding and use of variables and expressions. Performance on these items was relatively stable except for items 2 and 12, where there was significant improvement. Item 2 required identification of an expression to represent 4 more than twice a given number indicating that more students are becoming proficient at this type of task. Item 12 required

Table 3 Performance on items involving understanding and use of variables and expressions

Item description	Item type	Percent correct				
		2003	2005	2007	2009	2011
Expressions						
1. If m represents the total number of months that Jill worked and p represents Jill's average monthly pay, which of the following expressions represents Jill's total pay for the months she worked? [2007-M11 #6]	MC	73	72	73		
2. A car can seat c adults. A van can seat 4 more than twice as many adults as the car can. In terms of c, how many adults can the van seat? [2013-M6 #2]	MC		63	67	68	
3. The expression $80n$ could represent which of the following? [2013-M7 #6]	MC				46	
4. Which of the following is equal to $6(x+6)$? [2005-M12 #3]	MC	41	44			
5. If n is any integer, which of the following expressions must be an odd integer? [2011-M8 #8]	MC				40	40
6. Simplify an algebraic expression	MC				38	
7. Consider each of the following expressions. In each case, does the expression equal $2x$ for all values of x? (a) 2 times x, (b) x plus x, (c) x times x [2007-M9 #2]	MC					
(a) 2 times x (Yes or No)						
(b) x plus x (Yes or No)						
(c) x times x (Yes or No)						
(d) a, b, and c are all correct						
8. The music place is having a sale. Write an expression that shows how to calculate the cost of buying n CD's at the sale [2011-M9 #13]	SCR	31	30	32	3	1
Equations						
9. If $x=2n+1$, what is the value of x when $n=10$? [2007-M7 #1]	MC		78	80		
Inequalities						
10. If $a>0$ and $b<0$, which of the following must be true? [2011-M12 #17]	MC			27	28	29
Other						
11. Solve for x in a problem involving similar triangles	MC				66	
12. Given a whole number square root less than 10, find the original number	MC	58	61	64		
13. $3+15\div3-4\times2=?$ [2003-M10 #12]	MC	52				

understanding of the term square root so the improved performance there indicates more middle school students understand square roots. The relatively strong performance on Item 1 (73 % in 2007) shows that by the time they are in eighth grade, many students can interpret expressions written with multiple variables ($m \times p$ in this case). The strong performance on Item 9 (80 % in 2007) shows that most eighth graders understand what it means to substitute a value into an expression. Less than half know that $6(x+6)$ is $6x+36$ (Item 4). Item 8 was by far the most challenging of any of the algebra items as only 1 % answered correctly in 2011. The item included a sign saying that the first CD is \$12 and additional CDs were \$6 (including tax) and students had to write an expression for the cost of buying n CDs. A common mistake was failing to account for the first CD in the expression and thus writing $12+6n$ rather than $12+6(n-1)$ or $6+6n$. Looking at Items 1, 2, 4, and 8 as a group indicates that many eighth graders can write simple variable expressions (Items 1 and 2) but far fewer know more than basic rules for combining variable expressions (Item 4) and very few can write linear expressions, where the constant is not obvious (Item 8).

Equations and inequalities. Table 4 includes items that involve equations and inequalities. Items 5 and 6 are related in that both involved problems that could be solved by writing and solving simple equations ($x+2x=18$ in Item 5, $x+3x=152$ in Item 6). The similarity in performance on the two items (52 and 47 % in 2007) may be misleading. One could argue that a higher proportion of students correctly answered Item 5 because they only had to identify the equation needed to solve the problem. However, because the numbers in Item 5 were small it is possible that if students had been asked to *solve* Item 5, use of guess and check would have made the percentage of students answering correctly higher. The numbers in Item 6 were large enough that, even though there were only 5 answer choices to test, guess and check was a relatively difficult strategy compared to writing and solving an equation. Sixteen percent of students selected 38 as the correct answer to Item 6, which suggests that these individuals found the number of hot dogs sold by solving an equation or by guess and check but failed to remember that they were being asked for the number of hot dogs that Carmen rather than Shawn sold. When those who incorrectly selected 38 are pooled with those who correctly solved the problem, it suggests that the number of individuals who can solve linear combination problems is higher than performance on either Item 5 or 6 indicates. The supposition that some students prefer informal methods of solving linear combination problems is supported by looking at Item 18, which is solvable using the equation $2.5x+1(5-x)=8$ or $1x+2.5(5-x)=8$. Performance on this constructed response item was quite good (76 % in 2007) and the need to use decimals when writing and solving equations for this item strongly suggests that many if not most successful students used a guess and check or other informal solution method. The fact that students were asked for two pieces of information (number of newly released movies and number of classics) probably improved performance because students had to think about what the number they calculated represented.

Table 4 Performance on items involving understanding and use of equations and inequalities

Item description	Item type	Percent correct				
		2003	2005	2007	2009	2011
Linear equations in one variable						
1. $\square-8=21$. What number should be put in the box to make the number sentence above true? [2009-M5 #6]	MC	87	87	87	87	
2. If $15+3x=42$, then $x=$? [2007-M9 #4]	MC	79	80	83		
3. Which of the following equations has the same solution as the equation $2x+6=32$? [2009-M10 #1]	MC		69	70	70	
4. Which of the following equations is NOT equivalent to the equation $n+18=23$? [2011-M8 #5]	MC				60	59
5. Robert has x books. Marie has twice as many books as Robert has. Together they have 18 books. Which of the following equations can be used to find the number of books that Robert has? [2011-M12 #5]	MC			52	52	53
6. At the school carnival, Carmen sold 3 times as many hot dogs as Shawn. The two of them sold 152 hot dogs altogether. How many hot dogs did Carmen sell? [2007-M11 #15]	MC	47	47	47		
Linear equations in two variables						
7. If $x=2n+1$, what is the value of x when $n=10$? [2007-M7 #1]	MC		78	80		
8. The admission price to a movie theater is $7.50 for each adult and $4.75 for each child. Which of the following equations can be used to determine T, the total admission price, in dollars, for x adults and y children? [2011-M8 #3]	MC				70	70
9. Angela makes and sells special-occasion greeting cards. The table above shows the relationship between the number of cards sold and her profit. Based on the data in the table, which of the following equations shows how the number of cards sold and profit (in dollars) are related? [2007-M7 #15]	MC		52	54		
10. Which of the following equations represents the relationship between x and y shown in the table above? [2005-M12 #17]	MC	51	54			
11. A rectangle has a width of m inches and a length of k inches. If the perimeter of the rectangle is 1,523 inches which of the following equations is true? [2011-M9 #10]	MC				48	49
12. The number of gallons of water, y, in a tank after x hours may be modeled by the linear equation $y=800-50x$. Which of the following statements about the tank is true? [2011-M12 #15]	MC			46	48	48

(continued)

Table 4 (continued)

Item description	Item type	Percent correct				
		2003	2005	2007	2009	2011
13. In the equation $y=4-x$, if the value of x is increased by 2, what is the effect on the value of y? [2005-M3 #10]	MC	33	34			
14. The point $(4, k)$ is a solution to the equation $3x+2y=12$. What is the value of k? [2011-M12 #12]	MC			30	32	33
Quadratic equations						
15. A reasonable prediction of the distance d in feet, that a car travels after the driver has applied the brakes can be found by using the formula $d=0.055t^2$. If Mario is driving at 60 miles per hour and applies the brakes, how far will Mario's car travel before it stops? [2011-M8 #12]	MC				52	50
16. The formula $d=16t^2$ gives the distance d, in feet, that an object has fallen t seconds after it is dropped from a bridge. A rock was dropped from the bridge and its fall to the water took 4 s. According to the formula, what is the distance from the bridge to the water? [2007-M7 #9]	MC		48	50		
Other						
17. Ravi has more tapes than magazines. He has fewer tapes than books. Which of the following lists these items from the greatest in number to the least in number? [2005-M3 #5]	MC	87	87			
18. At Jorge's local video store, "new release" video rentals cost $2.50 each and "movie classic" video rentals cost $1.00 each (including tax). Jorge rented five videos and spent a total of $8.00. How many of the five rentals were new releases and how many were movie classics? [2007-M11 #18]	SCR	77	76	76		
19. Which of the following values make the given inequality true?	MC	56	57	60		
20. The temperature in degrees Celsius can be found by subtracting 32 from the temperature in degrees Fahrenheit and multiplying the result by 5/9. If the temperature of a furnace is 393 °F, what is it in degrees in Celsius, to the nearest degree? [2007-M9 #14]	MC	35	35	37		
21. Which of the graphs below is the set of all *whole* numbers less than 5? [2005-M12 #10]	MC	36	37			
22. What are all values of n for which $-2n \geq n+6$? [2013-M7 #10]	MC				31	
23. Graph the solution set for $3 \leq x \leq 5$ on the number line below [2011-M9 #11]	SCR				24	22

NAEP as a Database of Student Understanding

The primary goal of the analyses reported here was to use performance on NAEP items to determine the strengths and weaknesses of American students when it comes to algebraic reasoning. A secondary goal was to identify items where performance changed substantially between 2003 and 2011. The large number of algebraic reasoning items used by NAEP since 2003, in addition to the large proportion of those that were released, makes NAEP an excellent source of student performance on a wide range of algebraic tasks. For example, the fact that only 25 % of students identified the equation of a line when shown a linear graph in 2007 (Item 9 in Table 1, a 5-option multiple-choice item) is strong evidence that by the spring of eighth grade, most students cannot connect simple linear equations and their graphs. Performance on Item 18 in Table 1 (31 % correct in 2011) shows they are not much better at identifying the equation of a line that passes through a given point and has a negative slope. The performance trend on Items 11 (48–56 %), and 25 (55–65 %) in Table 2 strongly suggests that students are getting better at plotting points to see the connection between coordinates, equations, and graphs. This is one aspect of grade 6 Common Core State Standard (CCSS) 6.NS.C.8 (see http://www.corestandards.org). While the improvement is encouraging, about 1/3 of grade 8 students had not mastered this part of a grade 6 standard. A major focus of the CCSS grade 8 Expressions and Equations domain is connecting lines with equations and using pairs of simultaneous linear equations to solve problems. Performance near the chance level on items that focus on relating equations and graphs (e.g., Items 20 and 21 in Table 2) suggests that it will take a while before grade 8 students are mastering grade 8 CCSS.

The fact that performance on most items has been relatively stable in recent years indicates that using items released several years ago as indicators of current student performance is justified. Even during the years 1990 through 2003 when performance in the area of algebra was increasing substantially, there was relatively modest gain on many NAEP algebra items (Chazan et al., 2007; Sowder et al., 2004); the minimal gains on items between 2003 and 2011 are, therefore, not that surprising. Given that the grade 8 NAEP items tend to focus on CCSS in grades 5 through 7, the low performance and modest rate of improvement on many items indicates that it may be many years before most grade 8 students are meeting the expectations of the CCSS.

Themes in the Algebra Data

In addition to documenting progress on various components of algebraic reasoning, the data reported here provide several insights into mathematics learning in general. The first of those is that many students learn skills long after the time at which those skills are introduced. This is consistent with other analyses of NAEP items used at more than one grade level (Blume, Galindo, & Walcott, 2007; Kastberg & Norton, 2007; Kloosterman, 2014; Warfield & Meier, 2007) and is also consistent with theories of learning that indicate that students are impacted by informal settings (Bransford et al., 2006).

Another major theme is that problem complexity makes a substantial difference in the proportion of students who answer an item correctly. This is consistent with analyses of NAEP items to see what, besides content knowledge, makes them difficult (D'Ambrosio et al., 2007). Test developers have long been aware that changes in context and wording affect problem difficulty (Ferrara, Svetina, Skucha, & Davidson, 2011) so this theme is not a surprise. NAEP data are useful because they include items with similar content but have substantial variation in the proportion of students answering correctly. Given that assessments of student knowledge of CCSS will be high stakes for students, teachers, and schools, it is essential that wording and context be considered in interpretation of performance on individual items.

A third theme, which has been identified in many smaller scale studies, is that students try to make sense of mathematics; but given the large number of details they are expected to master, many come to incorrect conclusions. Early research on incorrect conclusions included study of error patterns in mathematics (Ashlock, 2006) and much of the current psychology of mathematics learning is based on the idea that students attempt to make sense of mathematics (Cobb, 2007). With respect to algebra, Kieran (2007) notes that algebra students interpret problems and try to use algebraic procedures that make sense to them even though those interpretations and procedures are not applicable to the situation at hand. In this study, there were items such as Item 3 in Table 2 where the correct response looked like a thermometer laid on its side. Recall that performance on this item was considerably better than a similar item (Item 5), where there was no context on which to make an inference. Another example is the CD sale item (Item 8 in Table 3), where only 1 % provided an expression that accounted for the fact that the first CD bought was more expensive than additional CDs. The NAEP data set does not give a specific figure for the percentage of students who made this mistake but the examples of incorrect responses provided show that many students realized they were supposed to write an expression (i.e., they applied a procedure that made sense to them) yet failed to account for the price of the first CD. Because NAEP data are representative of the United States as a whole, performance on this item shows that eighth graders across the country have a very difficult time writing expressions that represent sequences where the first term is different from the others.

In brief, although released items are available online, this study went far beyond what is available online by (a) categorizing items with algebra content by subtopic, (b) documenting trends in performance over time on items, and (c) including data on secure items. This is important information for those trying to document strengths and weaknesses in algebra instruction. We now move to a discussion of issues related to research using large-scale databases.

Conceptual, Logistical, and Methodological Issues in the Use of NAEP Data

As can be seen from our analysis of algebraic reasoning of eighth graders, NAEP provides a powerful data set for documenting student performance on a wide variety of mathematical tasks. There are, however, a number of conceptual, logistical, and methodological issues that make research using NAEP data challenging. We now discuss some of those issues in relation to the results we have presented.

Analyses Are Limited by the Data Available

In most research studies, questions are identified from the literature or the field and then instruments and procedures are designed to collect data to answer those questions. When using NAEP, or any large-scale assessment data, researchers must define questions that can be answered, at least partially, by the data available. There are many answers to the question of what constitutes algebraic reasoning, but for the purposes of this study, it is simply performance on the algebraic reasoning items in Tables 1, 2, 3, and 4. This includes the ability to see geometric and visual patterns, to define and interpret variables, to construct and interpret graphs, and to solve word problems involving linear equations. As can be seen in the tables, there is substantial variation in performance depending on the exact nature of the tasks on these topics.

Many studies using NAEP data (e.g., Innes, 2012; Wei, 2012) assume that scores for mathematics in general, or in algebra, are valid. It is important to be as specific as possible about the NAEP items because verbatim reporting of items (in the case of released items) and clear item descriptions (in the case of non-released items) allows readers to see for themselves what is being measured and thus the extent to which NAEP items match their conceptions of what constitutes algebraic reasoning. We are also explicit about our categorization of items because others may have somewhat different classifications. For example, of the five number line items in Table 2, only Item 5 is in the NAEP algebra strand. Items 1–4 are in the number and operations strand but we included them because understanding of number lines and integers is essential background for coordinate graphing, which is normally considered part of the algebra curriculum.

It is interesting to note that Items 3 and 5 in Table 2 both require selecting a number line to meet specific criteria yet they come from different strands. Performance on Items 3 and 5 was substantially different (59 % on Item 3 vs. 37 % on Item 5), and we believe this is primarily due to two factors. The first is the fact that all the distractors for Item 3 were continuous lines running from −20 to +120 so students had to choose the line with dots filled in at −20 and +120 and a continuous segment between them. Item 5 was similar in that all the distractors were lines marked at −5 and +5. In Item 5, however, students were told to identify the number line that showed the *whole* numbers less than 5. None of the distractors had any values higher than 5 so the keys to getting a correct answer were (a) selecting a number line with dots at each whole number rather than a number line with a continuous segment, and (b) realizing that whole number meant only numbers greater than or equal to zero. The percentage of students selecting the distractor showing all values less than 5 was almost the same as the percentage answering correctly indicating that many did not know that negative numbers could not be whole numbers. The second factor explaining the difference in performance between Items 3 and 5 is context. Item 3 was about temperature range in a city while Item 5 simply asked for a set of whole numbers.

The correct answer for Item 3 included all the values from −20 to +120 and looked like a horizontal thermometer. These cues probably helped some students. Item 5 had no real world context so students had to rely exclusively on what they knew about number lines from their mathematics instruction.

Finally, we note that had we restricted our study of number lines in the context of algebraic reasoning to items that were in the NAEP algebra strand, it would appear that a relatively small percentage of eighth-grade students have the solid knowledge of number lines necessary to use and construct graphs in the $x-y$ plane. Including the number line items from the number and operations strand shows that many students have at least some knowledge of the principles behind coordinate graphs.

Access to Secure NAEP Data

User of NAEP data know that NCLB requires relatively quick release of fourth- and eighth-grade Main NAEP mathematics results, and thus when data are collected in the spring of a year, overall results are available in fall of the same year. The secure data set, however, is usually not released until much later. In the case of the LTT NAEP, the spring 2008 data set was not released until February of 2012. The only 2011 Main NAEP data used for this study came from the 2011 released items because the secure data set had not been released at the time this chapter was written—two and a half years after data collection was completed. Such delays mean, of course, that it is difficult to provide timely reports based on a secure data set.

In our experience, receiving and maintaining a secure site license has been relatively easy (see http://nces.ed.gov/pubsearch/licenses.asp) although there are restrictions on how the data are stored and used that need to be followed. One of these restrictions, for example, is that the data must be used on a computer that is not networked in a room with restricted access. For us, this meant that we had to buy special licenses for programs like SPSS and SAS because most licensed software is designed to periodically check the validity of the license by communicating with a server and that cannot happen unless a computer is networked. Another issue is that all users must sign non-disclosure agreements, and those agreements must be approved before users can access the data. Approval of new users has been relatively fast, but each license is currently restricted to seven users so we had to get a second license to accommodate our entire research group.

Another logistical issue for content specialists using the secure NAEP data is that while the data sets received from NCES contain all data collected for a given assessment, they only include very general descriptions of the items on which the data are based. Most researchers using the NAEP data do statistical analyses on the overall findings and for them, item content is not an issue—as noted previously they accept the items as valid measures of what students should know. For those interested in exactly what NAEP measures, there is a process that allows researchers to view non-released (secure) items by visiting the Institute for Education Sciences (IES) in

Washington, DC. The approval process has taken us months and notes taken at viewing sessions were restricted to information that was very general. With the large number of items to view, we have found it useful to go to viewing sessions with lists of items where we have seen unusual patterns in performance and then spend most of our viewing time analyzing those items. Otherwise, we cannot remember enough about the items to understand unusual patterns we see in the secure data.

Descriptions of secure items can be included in publications but only when they are described in general terms, as seen in the results tables provided in this chapter. We included secure algebra items in this chapter to provide readers with a sense of what can be said about such items although we normally only use a secure item in our item-by-item analyses when we can be specific enough about the item to show that it provides insight that released items do not provide. Given the substantial number of items that have been released every 2 years since 2003, it has become easier than it was a decade ago to rely on released items to draw conclusions about student knowledge on most topics. We use secure items more often when we build scales of items on a specific construct. As explained later in this chapter, we use fit analyses to be sure that the secure items are measuring the same latent variable as the released items in the scales.

Using Statistical Software with NAEP Data

In addition to the logistical issues of getting a license and authorizing users, there are issues in the use of a data set where the sampling design means that the majority of the data are missing for every respondent and that respondents in some demographic categories count less toward overall scores than others. For example, NAEP oversamples American Indian students in many states to have enough of these students to report results for American Indians at the state level. Thus, when state-level scores for all students are calculated, the responses of each American Indian student count less so that the overall score is representative of the student demography in the state.

Fortunately, there are tools available that take the large amount of missing data and the sampling weights into account. One option is *AM* (http://am.air.org). This software is available for download and does much more than the online *Data Explorer* (http://nces.ed.gov/nationsreportcard/naepdata) although it has significant limitations.

Along with the secure data set, NAEP provides open-source control statements for SPSS, SAS, and STATA so that NAEP data can be used with these software programs. Using these commands requires much more knowledge of the programs than is required for standard analyses but it does allow for relatively sophisticated analyses using NAEP data. In particular, multi-level modeling is not possible with the *Data Explorer* or *AM*, and thus any hierarchical analyses must be done using more robust software packages.

What Does It Mean to Say That a Certain Percentage of Students Answered an Item Correctly?

Although it is easy to tell from the secure data set whether a given student answered an item correctly, calculating the percentage of students who correctly answer a given item requires a set of decisions. Valid item response codes included in the secure data set include traditional item responses (correct and incorrect) and special codes (illegible, off-task, I don't know, nonrateable, omitted, not reached, and multiple responses). In official NAEP results, including those reported on the NAEP website, "not reached," "illegible," and "I don't know" responses are treated as incorrect while "omitted" and "multiple" responses are treated as incorrect for constructed response items and fractionally correct on multiple-choice items. All other responses are treated as missing data and not taken into account in the scaling process. When we calculated the percentage correct on the NAEP items used for this study, we took the view that only the codes representing correct responses by students account for their achievement. Thus, in this chapter the special responses except "I don't know" are treated as missing codes and not included in the analysis. This often means that the percent correct for released items reported in the online *Questions Tool* does not match the percent correct in our tables. At times it is useful to have analyses that are consistent with standard NAEP practice and, given that there is debate in the psychometric community about how to deal with special responses (Brown, Dai, & Svetina, 2014; Brown, Svetina, & Dai, 2014; Misley & Wu, 1996), we make the decision on how to deal with such responses based on the audience we are writing for.

Limitations on Analyses by Demographic Subgroup

Because of the number of students completing NAEP, there is a great deal of statistical power for analyses. With the large number of background and demographic variables available for NAEP, it is possible to run analyses on subgroups of the population—analyses that can be very helpful in specific settings. For example, it is relatively easy to see if the gap in performance between Hispanic students in Indiana who are eligible for free lunch and their counterparts who are not eligible is greater or less than the gap of their Black peers. However, when analysis is restricted to Hispanic and Black students within a given state, the standard errors for the populations are much larger than they are for national results for all students. When looking at differences in performance on individual items, statistical power decreases further because only those students who complete the item in question can be included in the analysis. This is not to say that analyses for subgroups are inappropriate, but larger differences are needed to make claims involving statistical significance. We have conducted item-level analyses similar to those reported in this chapter and found that when percentage correct of one group differs from another group by at least 2 %, that difference is usually statistically significant ($p < .05$) for the entire national sample but differences of 4 % may not be significant when analyses are restricted to specific states.

Looking Forward

Knowing how well students across the United States perform on specific aspects of mathematics will always be important to curriculum developers and, one would hope, to policy makers who impact what is being taught in schools. In particular, with the adoption of the Common Core State Standards (CCSS, http://www.corestandards.org), we project that there will be substantial interest in how well students perform on items that assess those standards. To the best of our knowledge, the National Assessment Governing Board has no immediate plans to shift the NAEP frameworks to be more in alignment with the CCSS. Even if such alignment eventually comes to pass, it will be several years before an NAEP assessment would have items based on the CCSS. Thus, for the time being, we assume there will be interest in identifying existing NAEP items that assess skills and concepts identified by CCSS. In our case, there are plans to report performance on both released and secure items that connect with CCSS. Looking at items reported in this chapter, the case can be made that some are good indicators of CCSS. For example, results for Item 9 in Table 3 (If $x = 2n + 1$, what is the value of x when $n = 10$?) provide information on the extent to which students have the background knowledge and skills to solve linear equations in one variable (CCSS Standard 8.EE.7). Other NAEP items, however, do not appear to be measures of any concepts addressed within the CCSS. Item 1 from Table 1, for example, involves a pattern of shading in a fraction circle. Although it might be possible to argue that this item represents patterns similar to what the CCSS outline for the elementary grades, it is hard to see how this item represents anything in the middle school CCSS.

Perhaps the biggest question at this point is whether NAEP's role in assessing student learning will change when students start completing the PARCC (http://www.parcconline.org/parcc-assessment) and Smarter Balanced (http://www.smarterbalanced.org/smarter-balanced-assessments/) assessments. Although these assessments will not provide data on students from all states, they take place at grades 3 to 8 and 11 and align with the CCSS. Moreover, they will be high-stakes tests for teachers and schools. Thus, it is possible that the NAEP governing board will consciously avoid any movement toward alignment with the CCSS in fear that NAEP results will be seen as an indicator of achievement on the CCSS when that was never the intent of the NAEP program.

Subscales for Specific Mathematics Skills

With the CCSS and the aforementioned categorization system in mind, we are doing more of the types of analyses reported here. We believe that more can be done to quantify performance on specific constructs by combining performance on clusters of items. For example, reasoning about similar triangles or solving linear equations in two variables can be viewed as distinct constructs, each of which has items that focus on those constructs. Assessment items associated with similar triangles or linear equations have a range of difficulties—some linear equations, for example, are harder to solve than others. However, within each construct, students and items are assumed to behave consistently with respect to their proficiencies and difficulties. If a student can answer a particular item, they probably can answer any item that is easier, and if a student cannot answer an item, they probably cannot answer any item that is more difficult. When a group of students and items behaves in this consistent manner, there is strong evidence that the items reflect a valid construct, and that the students possess a meaningful proficiency or cognitive skill that is necessary to respond successfully to those items. This method of analysis is called construct-referenced measurement (Brown & Wilson, 2011; Wilson, 2005). Identifying and describing constructs is of interest to the research community because they offer insight into the nature of specific mathematics skills. We see the identification of specific constructs, and the sets of NAEP items that measure those constructs, as a productive direction for further research. Our initial applications of construct-referenced measurement to the NAEP data are described below.

Psychometric Issues

Tables 1, 2, 3, and 4 in this chapter represent our perceptions of sets of items that should fit together to form constructs. We have begun testing the extent to which the items in the tables actually represent constructs using construct-referenced measurement. Specifically, we use the item response data from these items to build one-parameter item response models from the Rasch family. We use a combination of the basic Rasch model (Rasch, 1960/1980) for dichotomous data and the partial credit model (Masters, 1982) for polytomous data. We estimate student proficiencies and item difficulties using joint maximum likelihood (JML) estimation, using the psychometric software *ConQuest 3.0* (Adams, Wu, & Wilson, 2012).

The psychometric analysis produces a *fit statistic* for each item, a residual-based measure of how well the data associated with the item fits the assumption of a unidimensional construct. In other words, we examine the degree to which students respond to the item in a way that is consistent with their proficiencies and the item's difficulty. Although there are several types of fit statistics, we prefer to examine the weighted mean-square statistic, which acts as an effect size indicator that is not sensitive to outliers. If the item responses are consistent with the students' proficiencies and the items' difficulties, the weighted mean-square fit statistics will be approximately normally distributed with a mean of 1.00. Obvious deviations from this distribution represent a problem with the hypothesis that the items are associated with a single specific proficiency or skill.

The most common deviation from the expected distribution is when a fit statistic is a clear outlier. An outlying fit statistic indicates that an item is not representative of the hypothesized skill. This may be because the skill required to respond to that item is different than expected because the item requires one or more additional skills beyond the one expected, or because the item is confusing or tricky for some students. Importantly, an outlying fit statistic does *not* indicate that an item is necessarily easier or more difficult than the other items. When an outlying fit statistic identifies a misfitting item, content and pedagogical experts should be consulted to determine which of these possible explanations is most likely.

Another deviation from the expected distribution is when the fit statistics cluster into two groups, one less than 1.00 and one greater than 1.00. This generally indicates that these two sets of items are associated with different skills. One hypothesis we tested was whether different skills are required to respond to items involving equations in two variables when the equations are linear (Items 7 through 14 in Table 4) versus when they contain a quadratic term (Items 15 and 16 in Table 4). When we analyzed these items together as a subscale, the fit statistics for the items containing a quadratic term did not stand out as outliers, implying that despite being more difficult, they are associated with the same underlying skill. A second hypothesis we tested was whether different skills are required to respond to items involving inequalities in one variable (Items 19 and 22 in Table 4) versus equations in one variable (Items 1–6 in Table 4). When we analyzed these items together as a subscale, the fit statistics for the items containing inequalities clustered together and stood apart from the others, implying that solving inequalities does represent a distinct skill. Across the algebra items, there also appear to be distinct proficiencies associated with extending patterns, understanding and using variables and expressions, and understanding and using coordinate grids and graphs.

In closing, we reiterate the importance of NAEP (and other large common data sets) as a source of valuable information for mathematics curricula and teaching. Traditional item analyses that show how well students do on specific mathematics tasks are important because education professionals without training in statistics can understand them and because they provide a concrete picture of what students know or at least do under the constraints of the testing conditions. It is important, however, to look for new ways to build arguments about how performance in mathematics is *changing*. The construction of IRT scale scores for the 1990 NAEP allowed for comparison of overall mathematics performance across years for the first time. With the use of IRT and other statistical techniques we are showing that it is possible to quantify and track performance on skills that are more specific than the algebra scales NAEP now uses but also more general than what can be determined from individual items. This middle level of analysis is proving to be useful both in teasing out development of skills over time and in the identification of the different proficiencies that, together, contribute to this development.

Acknowledgement This chapter is based upon work supported by the National Science Foundation under the REESE Program, grant number 1008438. Opinions, findings, conclusions, and recommendations expressed in the chapter are those of the authors and do not necessarily reflect the views of the National Science Foundation.

References

Adams, R., Wu, M., & Wilson, M. (2012). *ACER ConQuest version 3.0.1: Generalised item response modeling software [Computer software and manual]*. Camberwell: ACER Press.

Arbaugh, F., Brown, C., Lynch, K., & McGraw, R. (2004). Students' ability to construct responses (1992–2000): Findings from short and extended-constructed response items. In P. Kloosterman & F. K. Lester Jr. (Eds.), *Results and interpretations of the 1990 through 2000 mathematics assessments of the National Assessment of Educational Progress* (pp. 337–362). Reston, VA: National Council of Teachers of Mathematics.

Ashlock, R. B. (2006). *Error patterns in computation: Using error patterns to improve instruction* (9th ed.). Upper Saddle River, NJ: Pearson.

Blume, G. W., Galindo, E., & Walcott, C. (2007). Performance in measurement and geometry from the viewpoint of the principles and standards for school mathematics. In P. Kloosterman & F. K. Lester Jr. (Eds.), *Results and interpretations of the 2003 mathematics assessment of the National Assessment of Educational Progress* (pp. 95–138). Reston, VA: National Council of Teachers of Mathematics.

Bransford, J., Stevens, S., Schwartz, D., Meltzoff, A., Pea, R., Roschelle, J., et al. (2006). Learning theories and education: Toward a decade of synergy. In P. A. Alexander & P. H. Winne (Eds.), *Handbook of educational psychology* (2nd ed., pp. 209–244). Mahwah, NJ: Lawrence Erlbaum.

Braswell, J. S., Lutkus, A. D., Grigg, W. S., Santapau, S. L., Tay-Lim, B., & Johnson, M. (2001). *The nation's report card: Mathematics 2000* (Report No. NCES 2001-517). Washington, DC: National Center for Education Statistics.

Brown, N. J. S., Dai, S., & Svetina, D. (2014, April). *Predictors of omitted responses on the 2009 National Assessment of Educational Progress (NAEP) mathematics assessment*. Paper presented at the annual meeting of the American Educational Research Association, Philadelphia.

Brown, N. J. S., Svetina, D., & Dai, S. (2014, April). *Impact of omitted responses scoring methods on achievement gaps*. Paper presented at the annual meeting of the National Council on Measurement in Education, Philadelphia.

Brown, N. J. S., & Wilson, M. (2011). A model of cognition: The missing cornerstone in assessment. *Educational Psychology Review, 23*, 221–234.

Chazan, D., Leavy, A. M., Birky, G., Clark, K., Lueke, M., McCoy, W., et al. (2007). What NAEP can (and cannot) tell us about performance in algebra. In P. Kloosterman & F. K. Lester Jr. (Eds.), *Results and interpretations of the 2003 mathematics assessment of the National Assessment of Educational Progress* (pp. 169–190). Reston, VA: National Council of Teachers of Mathematics.

Cobb, P. (2007). Putting philosophy to work: Coping with multiple theoretical perspectives. In F. K. Lester Jr. (Ed.), *Second handbook of research on mathematics teaching and learning* (pp. 3–38). Charlotte, NC: Information Age.

D'Ambrosio, B. S., Kastberg, S. E., & Lambdin, D. V. (2007). Designed to differentiate: What is NAEP measuring? In P. Kloosterman & F. K. Lester Jr. (Eds.), *Results and interpretations of the 2003 mathematics assessment of the national assessment of educational progress* (pp. 289–309). Reston, VA: National Council of Teachers of Mathematics.

Ferrara, S., Svetina, D., Skucha, S., & Davidson, S. (2011). Test development with performance standards and achievement growth in mind. *Educational Measurement: Issues and Practice, 30*(4), 3–15.

Innes, R. G. (2012). Wise and proper use of national assessment of educational progress (NAEP) data. *Journal of School Choice: Research, Theory, and Reform, 6*, 259–289.

Kastberg, S. E., & Norton, A. N., III. (2007). Building a system of rational numbers. In P. Kloosterman & F. K. Lester Jr. (Eds.), *Results and interpretations of the 2003 mathematics assessment of the national assessment of educational progress* (pp. 67–93). Reston, VA: National Council of Teachers of Mathematics.

Kieran, C. (2007). Learning and teaching algebra at the middle school through college levels: Building meaning for symbols and their manipulation. In F. K. Lester Jr. (Ed.), *Second handbook of research on mathematics teaching and learning* (pp. 707–762). Charlotte, NC: Information Age.

Kloosterman, P. (2010). Mathematics skills of 17-year-olds in the United States: 1978 to 2004. *Journal for Research in Mathematics Education, 41*, 20–51.

Kloosterman, P. (2011). Mathematics skills of 9-year-olds: 1978 to 2004. *The Elementary School Journal, 112*, 183–203.

Kloosterman, P. (2014). How much do mathematics skills improve with age? Evidence from LTT NAEP. *School Science and Mathematics, 114*, 19–29.

Kloosterman, P., & Lester, F. K., Jr. (Eds.). (2004). *Results and interpretations of the 1990 through 2000 mathematics assessments of the national assessment of educational progress.* Reston, VA: National Council of Teachers of Mathematics.

Kloosterman, P., & Lester, F. K., Jr. (Eds.). (2007). *Results and interpretations of the 2003 mathematics assessment of the national assessment of educational progress.* Reston, VA: National Council of Teachers of Mathematics.

Lobato, J., Ellis, A. B., & Munoz, R. (2003). How "focusing phenomena" in the instructional environment support individual student's generalizations. *Mathematical Thinking and Learning, 5*, 1–36.

MacGregor, M., & Stacey, K. (1997). Student's understanding of algebraic notion: 11–15. *Educational Studies in Mathematics, 33*, 1–19.

Masters, G. (1982). A Rasch model for partial credit scoring. *Psychometrika, 47*, 149–174.

Misley, R. J., & Wu, P. K. (1996). *Missing responses and IRT ability estimation: Omits, choice, time limits, and adaptive testing.* Retrieved from http://oai.dtic.mil/oai/oai?verb=getRecord&metadataPrefix=html&identifier=ADA313823

National Assessment Governing Board. (2010). *Mathematics framework for the 2011 national assessment of educational progress.* Washington, DC: U.S. Department of Education.

National Assessment Governing Board. (2012). *Mathematics framework for the 2013 national assessment of educational progress.* Washington, DC: U.S. Department of Education.

National Center for Education Statistics. (2012). *The nation's report card: Mathematics 2011* (Report No. NCES 2012-458). Washington, DC: Institute for Education Sciences, U.S. Department of Education.

No Child Left Behind (NCLB) Act of 2001. (2001). Public Law 107–110 Title I Part A, Section 1111.

Rasch, G. (1980). *Probabilistic models for some intelligence and attainment tests.* Chicago, IL: University of Chicago Press (Original work published 1960).

Sowder, J. T., Wearne, D., Martin, W. G., & Strutchens, M. (2004). What do 8th-grade students know about mathematics? In P. Kloosterman & F. K. Lester Jr. (Eds.), *Results and interpretations of the 1990 through 2000 mathematics assessments of the National Assessment of Educational Progress* (pp. 105–144). Reston, VA: National Council of Teachers of Mathematics.

Warfield, J., & Meier, S. L. (2007). Student performance in whole-number properties and operations. In P. Kloosterman & F. K. Lester Jr. (Eds.), *Results and interpretations of the 2003 mathematics assessment of the National Assessment of Educational Progress* (pp. 43–66). Reston, VA: National Council of Teachers of Mathematics.

Wei, X. (2012). Are more stringent NCLB state accountability systems associated with better outcomes? An analysis of NAEP results across states. *Educational Policy, 26*, 268–308.

Wilson, M. (2005). *Constructing measures: An item response modeling approach.* Mahwah, NJ: Lawrence Erlbaum.

Homework and Mathematics Learning: What Can We Learn from the TIMSS Series Studies in the Last Two Decades?

Yan Zhu

Homework Is an Important Issue Inside and Outside of Academia

Homework as a way to extend learning beyond the classroom is a generally recommended or common practice in many educational systems nowadays. However, the research community has not reached consensus about its importance, nor about the amount and type of homework students should be assigned. A brief review of the history of the homework debate in the USA may help to unravel the complexities of the issue. The debate can be traced back to the late 1800s and has gone in cycles ranging from more rote practice to more learner-centered tasks, and from more allotted time to less allotted time (Cooper, Robinson, & Patall, 2006; Gill & Schlossman, 2004).

According to Herrig (2012), although the late nineteenth century and early twentieth century marked a period of homework emphasis focusing on memorization and rote practice, it is actually the progressive education movement that began a public dialogue regarding homework. In concert with the public's concern about homework affecting the health of children, the movement led an attack on homework. At that time, some even declared that homework was nothing but "legalized criminality" (Nash, cited in Gill & Schlossman, 2004). In the 1950s and 1960s, with a decline in the progressive education movement and the USSR's successful launch of *Sputnik*, homework started to be viewed as a necessary condition guaranteeing learning (Canadian Council on Learning, 2009). However, the onset of the Vietnam War again led public opinion away from support for homework during the 1960s and 1970s as the public's attention was diverted from the academic excellence

Y. Zhu (✉)
Department of Curriculum and Instruction, School of Education Science, East China Normal University, 3663 Zhongshan Road (North), Shanghai 200062, China
e-mail: yzhu@kcx.ecnu.edu.cn

© Springer International Publishing Switzerland 2015   209
J.A. Middleton et al. (eds.), *Large-Scale Studies in Mathematics Education*,
Research in Mathematics Education, DOI 10.1007/978-3-319-07716-1_10

movement (also see Center for Public Education, 2007). Following the publication of *A Nation at Risk* in 1983, the public's focus and education's response was brought towards a new view of homework as providing both character-building and academic benefits throughout the 1980s and 1990s. Regarding these cyclical changes, Hallam (2004) suggested that they reflected the political and economic concerns of the time. In this sense, the pendulum may continue swinging between favor and disfavor of homework depending on the social and economic challenges of the day.

One focal point in the homework debate is whether doing homework benefits students' learning. Three basic types of research studies have emerged from efforts to examine the relationship between homework and students' achievement (Cooper, 1989; also see Zhu & Leung, 2012). One class of studies compares homework with no homework. Most of these studies revealed effects favoring homework (including 14 out of 20 studies (1962–1986) in Cooper's 1989 review and all six studies (1987–2003) in Cooper et al.'s 2006 synthesis). Most of these positive effects were further found to be grade-level related; that is, high school students benefited most followed by junior high students with the effect barely noticeable for elementary students. While Cooper and his colleagues synthesized the studies of the impact of having homework on students' learning in various school subjects, Austin's (1979) review focused on studies about mathematics homework between 1900 and 1977. This included 16 studies showing significant differences favoring students who were assigned mathematics homework and another 13 studies showing no difference. The second class of studies compares homework with in-class supervised learning (i.e., homework-like in-class activities). These also detected a strong grade-level effect. In particular, in-class supervised study was generally proved superior to homework at the primary level and a reversed pattern was observed at the junior high level and above. The third class of studies used statewide or national surveys to correlate the amount of time students spent on homework with their achievement scores. Cooper et al.'s (2006) work on 69 correlations revealed 50 in a positive direction, and his earlier work (Cooper, 1989) also showed that the majority (43 out of 50) favored homework. In these studies, a strong grade-level association again was prevalent.

Although many of the studies reviewed in Cooper and his colleagues' work reported a positive effect of homework on students' achievement, Cooper and many others (e.g., Cooper, 1989; Cooper et al., 2006; Cooper, Lindsay, Nye, & Greathouse, 1998; Kohn, 2006; Trautwein & Köller, 2003) claimed that the link between homework and achievement is far from clear. One essential reason is that most of these studies are simply correlational and make no claims about an explicit causal link (e.g., Blazer, 2009; Cooper, 2008). There are also researchers suggesting that the relationship between homework and achievement could be nonlinear (e.g., Keys, Harris, & Fernandes, 1997; NSW Department of Education and Communication, 2012). As a matter of fact, both proponents and opponents can easily find supporting evidence from research to sustain their respective views about the effects of homework. Many review works have been devoted to producing lists of such advantages and disadvantages from the perspective of students' learning (e.g., Center for Public Education, 2007; Queensland Department of Education & the Arts, 2004).

It is clear that homework has recently become an even more important issue not only inside academia but also outside of academia (Canadian Council on Learning, 2009). For instance, when the most recent PISA announced that Shanghai students again ranked first on all three assessed subjects, the number of hours these students spent on homework per week immediately aroused heated discussion among the public (13.8 h, the most among all the participating education systems). In the USA, Kralovec and Buells' book *The End of Homework* received massive media attention and spawned an ongoing debate between the anti-homework and pro-homework contingents (Vatterott, 2009). In the UK, a recent intense debate followed news about one British high school's announcement that it will go "homework free" but lengthen the school day.[1] In France, the president is even making the issue of homework a major part of his reelection campaign, promising to abolish homework if reelected.[2] In Australia, Rindlefleish and Alexander's media report *The War on Homework* published in *The Sunday Mail* (Queensland) has resulted in increased parental concerns about the amount of homework expected outside of school time (Queensland Department of Education & the Arts, 2004). In fact, the national umbrella organization of parents and citizens groups in Australia, the Australian Council of State School Organizations, argued that there is no evidence that students benefit from the practice and that it has become an overbearing invasion of family life.[3]

Effects of Homework Are Inclusive

Kohn (2006) points out that there is no conclusive evidence showing that homework provides any benefits, either academic or nonacademic, to students. A proper question that therefore might be posed is, "Under what conditions is homework beneficial for students' learning, and what kinds of homework are effective for various learning goals?"

Regarding criticism on the inconclusive and even incoherent effects of homework, Blazer (2009) may provide some insights. Homework is a difficult variable to study directly, uncontaminated by other variables. Researchers have found that the impact of homework on students is affected by many other factors (NSW Department of Education and Communication, 2012), such as students' understanding of the purpose of homework (e.g., Warton, 1997), their attitudes toward homework, learning and achievement (e.g., Cooper, Jackson, Nye, & Lindsay, 2001; Corno, 2000;

---

[1] See *The Homework Debate*, retrieved on September 6, 2013, from http://northshoremums.com.au/the-homework-debate/

[2] See *France's Hollande Promises Pupils "No More Homework"*, retrieved on October 11, 2012, from http://www.france24.com/en/20121010-hollande-promises-school-children-no-more-homework-education-reform-france/

[3] See *Parents in Australia Call for Ban on Homework*, retrieved on April 10, 2007, from http://www.foxnews.com/story/2007/04/10/parents-in-australia-call-for-ban-on-homework/

Epstein, Simon, & Salinas, 1997; Lange & Meaney, 2011; O'Rourke-Ferrara, 1998), and their feelings of empowerment in the design and allocation of homework tasks (e.g., Cooper & Valentine, 2001; Smith, 2000; Warton, 2001). However, Blazer (2009) maintained that homework itself is also influenced by more factors than any other instructional strategy, including student ability, motivation, grade level, variation in homework completion (e.g., when and how), and home environment (e.g., Cooper et al., 2006; McPherson, 2005).

Some other researchers attribute the "complex, fragmented, and contradictive" findings to methodological limitations in homework research (e.g., Corno, 1996; Inglis, 2005; Kralovec & Buell, 2001; Miller & Kelley, 1991; Paschal, Weinstein, & Walberg, 1984; Redmond, 2009; Trautwein, 2007; Trautwein, Köller, Schmitz, & Baumert, 2002; Trautwein, Lüdtke, & Pieper, 2007). Concerned about the lack of randomized procedures, for example, Cooper (2001) discussed the "extraordinary variability of results" (p. 28). Besides a similar critique about non-randomization, Vatterott (2009) further pointed out homework research was often conducted with small sample sizes such as a few students or a few classrooms. Trautwein and Köller (2003) challenged that much of homework research suffered from not taking the hierarchical nature of most homework data into account. According to them, a large percentage of variation in homework can be caused by classroom and school effects; in other words, a large portion of the effect of homework is related to the school a student attends and the specific class in which a student is enrolled (also see Maltese, Tai, & Fan, 2012). Consistent with this concern, Klangphahol, Traiwichitkhun, and Kanchanawasi (2010) advised that homework-related research studies should pay careful attention to the data with multilevel variance to avoid incorrect research conclusions and they claimed that the variables related to homework are, by nature, multilevel and hierarchical nested data. Moreover, Trautwein and Köller (2003) found that longitudinal data is sparse in homework research and they noted that the existing longitudinal studies mainly included only a small sample of students.

Bearing these cautions in mind, this study examines the role of homework played in students' school life through analyzing the homework-related data from the *Trends in International Mathematics and Science Study* (TIMSS), one of the most extensive large-scale international survey studies of students' schooling and achievement. With the analysis of the relevant data focusing on the subject of mathematics at the eighth grade level in all five rounds of the study series from 1995 to 2011, the study also aims to examine whether the role of homework has changed in the last two decades. It is believed that the changes in homework practice and its role can, to a certain degree, mirror the political and economic concerns of the time, as argued by Hallam (2004). Eight education systems were selected for the present study, including all five East Asian ones (Chinese Taipei, Hong Kong SAR, Japan, Korea, and Singapore) and three Western ones (i.e., Australia, England, and the USA). To get a fuller appreciation of the contexts in which students learn, TIMSS collected extensive background data from students, teachers, school principals/head-masters, and curriculum experts. Therefore, the homework issue can further be examined at different levels of curriculum (i.e., *intended curriculum* as expressed in policy rhetoric, *implemented curriculum* as practiced in real life and in school and classroom practices, and *attained curriculum* as manifested in learner experiences and outcomes). Furthermore, with such a stratified design, the hierarchical nature of homework data has been accounted for in TIMSS. Correspondingly, all the analyses in this study are using appropriate weighting (i.e., Students: TOTWGT; Mathematics Teachers: MATWGT; Schools: SCHWGT) to achieve more precise system-level data.

Changes in TIMSS Investigations About Homework from 1995 to 2011

As a worldwide comprehensive ongoing comparison of students' mathematics and science knowledge as well as their learning experience, TIMSS could not ignore homework as an important element in students' school life in its investigation. As a matter of fact, all five rounds of the surveys included specific items on homework practices in both the teacher and student questionnaires. In more recent rounds, specific sections entitled "homework" (2003T, 2007S, 2007T, and 2011S)[4] or "mathematics homework" (2011T) started to appear in the surveys with some additional items about homework in other sections.

In student questionnaires, frequency of assigning homework and time spent on doing homework are the two common themes in all rounds of the surveys. In contrast to TIMSS 2011, the other four TIMSS years consistently asked students about homework-related classroom activities. While the questionnaires in the 1990s suggested four such activities, only two were included in the 2000s. The settings of questions in the teacher questionnaires showed similar features to those in the student questionnaires across the years. Both frequency and amount of time again appeared in all the TIMSS years. Although all the teacher questionnaires included items about homework-related classroom activities, a greater variety can be seen in the 1990s (8 vs. 5). Furthermore, the teacher questionnaires in all but TIMSS 2011 included items about homework types with more types in the 1990s (10 vs. 3).

All these changes seem to imply that homework practice has received decreasing attention in the 2000s. However, two new items about parental involvement in students' homework practices first appeared in the TIIMSS 2011 student questionnaire. The earlier surveys (TIMSS 1999, TIMSS 2003, and TIMSS 2007) included only one relevant item asking school principals whether the school requested parents to ensure their children completed homework. This change may suggest that TIMSS researchers have started to pay more attention to parents' roles in students' learning experience. Regarding the topic of homework practices, the design of TIMSS 2011 is seemingly more comprehensive, as practices from all three important parties (i.e., students, teachers, and parents) have been investigated so that students' homework experiences can be depicted in a more holistic manner.

---

[4] T stands for teacher questionnaires, whereas S stands for student questionnaires.

Is There a System-Level Homework Policy Available?

When looking into homework policies in different education systems, the TIMSS studies assessed the issue from different perspectives in different TIMSS years; in other words, the data were collected from three parties, including school leaders, curriculum experts, and the TIMSS encyclopedia. Each of the parties were required to report different aspects of homework policy in their respective system. Correspondingly, the results about homework policy in this study are presented separately by these different information providers.

To school leaders, TIMSS 1995/1999 asked who had primary responsibility for establishing homework policies in schools. The possible options for the item include "not a school responsibility," "school's governing board," "principal," "department head," and "teachers." Figure 1 shows clearly that school teachers played a main role in Chinese Taipei, Japan, Korea, and the USA, whereas department heads took the primary responsibility in Hong Kong SAR and Singapore. Australia is the only system where no single party had primary responsibility; in other words, all four parties had some role in this work. Moreover, more responsibilities were handed over to school governing boards in Hong Kong SAR, department heads in Korea, and principals in the USA from TIMSS 1995 to TIMSS 1999. The within-system Chi-square tests[5] reveal

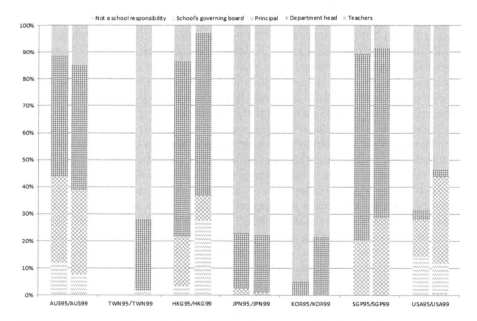

Fig. 1 Primary responsible parties for establishing homework policies in schools
Note. Chinese Taipei started to participate in the TIMSS series studies in 1999; no data on this item were available for England

---

[5] Effect size for a Chi-square test (on nominal data) is expressed by Cramer's φ (small: $\varphi = 0.10$, medium: $\varphi = 0.30$, large: $\varphi = 0.50$), when a significant χ^2 is detected.

that the changes between years were insignificant in Singapore, trivial in Australia, Hong Kong SAR, and Japan (i.e., $\varphi < 0.10$), and small in Korea and the USA (i.e., $\varphi < 0.30$). Moreover, the magnitude of cross-system differences grew from nearly large in TIMSS 1995 ($\varphi = 0.49$) to large in TIMSS 1999 ($\varphi = 0.57$).

The TIMSS 2007 curriculum questionnaire set one item directly asking curriculum experts whether there was a policy to assign mathematics homework at the eighth grade for schooling in their respective systems. Out of the eight systems, only Korea stated "yes." However, the experts from Korea further elaborated that the Korean policy was only applicable to probability and statistics, and homework in the other strands was assigned by teachers' discretion. Curriculum experts in both Australia and England highlighted that homework policy should be a school-based decision, whereas those in Chinese Taipei believed that teachers should be responsible. In comparison, the situation was most complex in the USA, where homework policies varied by state, district, school, and sometimes within districts and schools. Interestingly, curriculum experts in Singapore claimed that "There is no need for such a policy" and those in Hong Kong SAR commented that "Assignment guidelines are for teachers' reference only."

Besides the questionnaire-based data, TIMSS 2007/2011 provided rich information on homework policies across education systems in its encyclopedia (see Mullis et al., 2008, 2012). Although information on Australia, Chinese Taipei, Japan, and Korea was not included, the other four systems summarized how they dealt with homework at a policy level. While none of these systems established a system-wide official policy on homework, both Hong Kong SAR and England gave recommendations on the frequency and/or amount of homework for primary and secondary school students. In addition, Hong Kong SAR and Singapore highlighted the importance of providing feedback on homework.

How Often Do Students Receive Homework in Mathematics from Teachers?

In all the TIMSS years, mathematics teachers were asked to report the frequency that they assigned homework to their mathematics classes. In TIMSS 1995/1999/2011, the options[6] ranged from never (1) to every day (5), while TIMSS 2003/2007 asked for the relevant information via two separate questions, one checking whether the teachers assigned homework at all and the other asking for the frequency on a 3-point Likert scale ranging from every or almost every lesson (1) to some lessons (3).[7]

The data on the "never" option in TIMSS 1995/1999, the "I do not assign mathematics homework" option in TIMSS 2011, and the corresponding items in TIMSS 2003/2007 showed that in all eight education systems, more than 90 % of the

---

[6] A 5-point Likert scale was used in TIMSS 1995/1999/2011 (1: never, 2: less than once a week, 3: once or twice a week, 4: 3 or 4 times a week, 5: every day); the first option in TIMSS 2011 was stated as "I do not assign mathematics homework".

[7] A 3-point Likert scale was used in TIMSS 2003/2007 (1: every or almost every lesson, 2: about half the lessons, 3: some lessons).

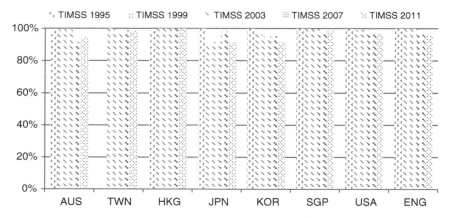

Fig. 2 Percentages of eighth grade mathematics teachers who reported that they assigned homework to their TIMSS classes

mathematics teachers assigned homework to their students (see Fig. 2). In particular, in all but TIMSS 1995, all the Hong Kong teachers reported that they assigned homework, while a higher percentage of mathematics teachers from Japan and Korea said they did not. Furthermore, all the systems except for Hong Kong SAR showed an overall downward tendency in frequency of assigning homework. In short, more teachers started to not assign mathematics homework to their students, though the changes were generally small (in particular, all φs are smaller than 0.20 and changes in Chinese Taipei and the USA were trivial, $\varphi < 0.10$).

Because the teacher questionnaires in TIMSS 1995/1999/2011 used nearly the same ascending 5-point Likert scale to measure teachers' homework assigning frequency, a comparison across the three TIMSS years was carried out. In addition, TIMSS 2003/2007/2011 asked students to report on a reversed Likert scale how often they received homework from their mathematics teachers.[8] Table 1 lists the average frequencies across the systems over the years, which showed that teachers from most systems tended to assign mathematics homework less frequently in later years from both the teachers' and the students' perspectives. However, a clear increasing tendency was observed in Chinese Taipei.

Though it appears that Japanese mathematics teachers gave students homework more frequently in TIMSS 2011 than TIMSS 1999, nearly returning to the TIMSS 1995 level, their homework assigning frequencies remain the lowest or next to the lowest among the eight investigated systems. From TIMSS 1995 to TIMSS 2011, both teachers and students from Japan reported that the average homework assigning frequencies were consistently below "3" on an ascending 5-point Likert scale; in short, mathematics homework was assigned less than "once or twice a week" on average in Japan. A similarly low frequency was also found in England. In contrast,

---

[8] Though TIMSS 1995/1999 student questionnaires also had one item on homework assigning frequency, they used a different 4-point Likert scale for the measurement (1: almost always, 2: pretty often, 3: once a while, 4: never). Due to the incomparability, no analysis was done on TIMSS 1995/1999 student data on this item.

Table 1 Mathematics homework assigning frequencies at the eighth grade level in eight education systems

	TIMSS 1995	TIMSS 1999	TIMSS 2003	TIMSS 2007	TIMSS 2011	
	TR	TR	ST	ST	TR	ST
AUS	3.78	3.78	3.65	3.47	3.52	3.40
TWN	NA	3.56	3.39	3.82	3.95	3.94
HKG	3.63	3.68	3.98	4.04	3.64	4.01
JPN	2.98	2.64	2.63	2.79	2.88	2.68
KOR	3.39	3.15	3.43	2.84	3.09	2.78
SGP	4.18	4.15	4.11	3.88	3.60	3.85
USA	4.11	4.20	4.39	4.35	4.02	4.21
ENG	3.01	2.99	2.89	2.81	2.81	2.87

Note: *TR* stands for teacher responses and *ST* stands for student responses (reversed). The homework assigning frequencies are measured on an ascending 5-point Likert scale (see footnote 7)

mathematics teachers from the USA consistently assigned homework at a high frequency. As a matter of fact, the USA is the only system among the eight where both teachers and students reported average homework assigning frequencies above "4" (i.e., 3 or 4 times a week).

Among the five rounds, only TIMSS 2011 used a similar Likert scale to measure the homework assigning frequency from both teachers' and students' perspectives. It is interesting to see that there was a high consistency between the two parties in some systems but low consistency in others. In particular, the Wilcoxon signed ranks tests[9] reveal the highest consistency in Chinese Taipei ($r=0.01$) and the lowest in Hong Kong SAR ($r=0.33$). Moreover, the comparison showed that the number of systems where teachers' self-reported frequencies were higher than those given by their students is the same as the number of systems having the reverse relationship (i.e., 4 vs. 4).

How Much Time Do Students Spend on Mathematics Homework?

In the TIMSS surveys, both teachers and students were asked to estimate how much time one student needs to spend on homework. The stems of the items[10] show that the estimation was not on a day/week/month-base but time per assignment.

---

[9] Effect size for a Wilcoxon signed rank test (on matched ordinal data between two groups) is expressed by r (small: $r=0.10$, medium: $r=0.30$, large: $r=0.50$), when a significant Z is detected.

[10] The item stem in the teacher questionnaire is "when you assign mathematics homework to the TIMSS class, about how many minutes do you usually assign?" and that in the student questionnaire is "when your teacher gives you mathematics homework, about how many minutes do you usually spend on your homework?"

Table 2 Estimated time spent on mathematics homework at eighth grade level in eight education systems

	1995	1999	2003		2007		2011	
	TR	TR	TR	ST	TR	ST	TR	ST
AUS	1.87	2.08	1.93	2.31	2.00	2.07	1.99	1.94
TWN	NA	2.52	2.42	2.30	2.48	2.41	2.54	2.31
HKG	2.27	2.46	2.30	2.44	2.36	2.43	2.31	2.20
JPN	2.05	2.05	2.03	2.18	2.02	2.09	1.94	1.99
KOR	2.45	2.24	2.04	1.96	2.16	1.90	2.22	1.84
SGP	2.75	2.82	2.65	2.69	2.48	2.65	2.57	2.49
USA	2.20	2.26	2.23	2.28	2.14	2.10	2.00	1.95
ENG	2.46	2.41	2.30	2.39	2.29	2.07	2.29	2.01

Note: TR stands for teacher responses and ST stands for student responses. The estimated time duration is measured on an ascending 5-point Likert scale (see footnote 12)

While teachers were advised to make the estimation based on an average student in their classes, students reported the time based on their own individual experiences. Different from the topic of homework assigning frequency, the TIMSS teacher surveys investigated the time-spent issue on a very similar 5-point Likert scale[11] across the years, making a cross-year comparison feasible. In addition, the measurement scales used in the student questionnaires[12] were very close to the ones in the teacher version in the 2000s.

The analysis reveals clearly that in all systems but Chinese Taipei, the time amount estimated by students declined, and the greatest declines were observed in Australia and England followed by the USA (see Table 2). In addition, the data showed that the big decline in Australia and England occurred from TIMSS 2003 to TIMSS 2007 but the rate of decline slowed recently. The reverse pattern was found in Hong Kong SAR and Singapore. The decline in Japan, Korea, and the USA was comparatively steady. Among the eight systems, students from Korea and Japan gave the shortest time duration estimations, while those from Singapore gave the longest.

Compared to their students' reports, the teachers' estimations were less consistent across the systems. Figure 3 reveals that the time amounts reported by the teachers from Japan, Singapore, and the USA had an overall downward tendency, while the teachers from some other systems appeared to start to have an increasing demand on the time necessary for doing homework, particularly between TIMSS

---

[11] A 5-point Likert scale was used on the items about time spent (1: fewer than 15 min (or "15 min or less" in TIMSS 2011), 2: 15–30 min (or "16–30 min" in TIMSS 2011), 3: 31–60 min, 4: 61–90 min, 5: more than 90 min). One additional option "I do not assign homework" in TIMSS 1995 questionnaire was merged with "less than 15 min" in the later analysis.

[12] In the two most recent TIMSS years, student questionnaires added one more option on the time spent item (TIMSS 2007: zero minutes; TIMSS 2011: my teacher never gives me homework in mathematics), which was combined with "1–15 min" to be recoded as "15 min or less" for the later analysis.

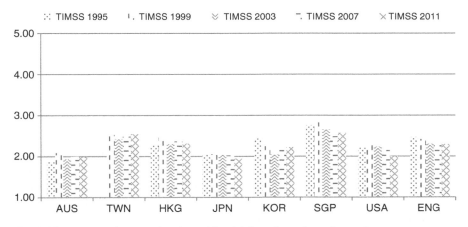

Fig. 3 Time spent on homework estimated by eighth grade mathematics teachers

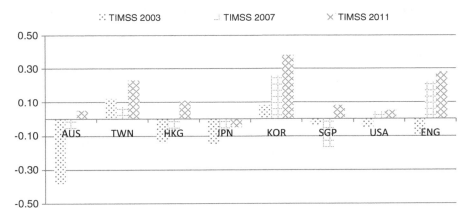

Fig. 4 Differences in estimations on time spent on homework between eighth grade mathematics teachers and their students in eight systems

Note. The differences are calculated via teachers' average estimations minus students' average estimations; the estimations from both parties were measured on a 5-point Likert scale

2003 and TIMSS 2011. Consistent with the findings from the students' estimations, homework assigned by teachers from Singapore required the longest period of time. However, from the teachers' perspectives, the least time-consuming homework was not found in Asia but Australia.

Due to the fact that mathematics teachers and their students gave the estimation of time spent on homework based on different perspectives, inconsistencies between the two parties are expected. Among the eight systems, the teachers and their students from the USA gave the closest estimations (see Fig. 4), while the largest gaps were observed in Korea. Interestingly, while TIMSS 2003 saw six systems in which students gave higher duration estimations than their teachers, only Japan maintained this pattern in TIMSS 2011 with the gap getting smaller. Teachers from Chinese Taipei and Korea continued to give higher duration estimations than their students and the gaps widened overall, with the teachers' estimation going up faster than their students.

What Types of Mathematics Homework Do Teachers Assign?

The issue of what types of tasks are included in homework is also an important aspect to be considered when assigning homework. It has been argued that the quality of homework assigned is likely to be more important than the quantity (e.g., Cai et al., 2012; Canadian Education, Ontario Institute for Studies in Education, & University of Toronto, 2010; Epstein & Van Voorhis, 2001; Van Voorhis, 2004; Walker, 2011). Although all but TIMSS 2011 included items to ask teachers about this aspect of homework, a greater variety of response choices for the types of homework items was shown in the 1990s. Compared to the practices on the most frequently used types of homework (i.e., "worksheets or workbook" and "problem/question sets in textbook"), greater differences were found in the usage of the least frequently used types of homework. Table 3 lists the top two and bottom two types of homework in terms of their utilization frequencies by eighth grade mathematics teachers in each system in TIMSS 1995 and TIMSS 1999.

Table 3 Top two and bottom two types of homework assigned by eighth grade mathematics teachers in terms of their utilization frequencies in TIMSS 1995 and TIMSS 1999

	TIMSS 1995				TIMSS 1999			
	(1)	(2)	(9)	(10)	(1)	(2)	(9)	(10)
AUS	PQ	WW	OR	KJ	PQ	WW	KJ	OR
TWN	NA	NA	NA	NA	WW	PQ	WI	KJ
HKG	PQ	WW	WG	KJ	PQ	WW	WI	KJ
JPN	PQ	WW	WI	WG	PQ	WW	WI	WG
KOR	PQ	WW	KJ	WI	PQ	RT	WI	KJ
SGP	PQ	WW	WI	KJ	PQ	WW	WR	KJ
USA	PQ	WW	KJ	OR	PQ	WW	KJ	OR
ENG	PQ	WW	OR	KJ	WW	PQ	WG	KJ

Note: (1) The numbers in parentheses are the ranks of each type of homework in terms of the utilization frequencies
(2) Ten types of homework were listed in TIMSS 1995/1999 surveys, including *WW* worksheets or workbook, *PQ* problem/question sets in textbook, *RT* reading in a textbook or supplementary materials, *WR* writing definitions or other short writing assignment, *SI* small investigation(s) or gathering data, *WI* working individually on long term projects or experiments, *WG* working as a small group on long term projects or experiments, *FC* finding one or more uses of the content covered, *OR* preparing oral reports either individually or as a small group, and *KJ* keeping a journal

It can be seen that tasks that heavily involved communication skills (i.e., OR or KJ) were assigned as homework by mathematics teachers least frequently in most systems (except for Japan). In fact, in the three western systems (i.e., AUS, ENG, and USA), both preparing oral reports and keeping journals were less assigned as homework than many other types of homework. The East Asian systems also shared some commonalities among themselves; that is, long term projects/experiments were one type of homework with a low assigning frequency.

The analysis reveals that there were big differences in the utilization frequencies[13] among different types of homework within a system as well as across systems in the 1990s. The most frequently used type of homework was assigned by mathematics teachers at least "sometimes" (i.e., $m > 3$) with the exception of Japan (TIMSS 1995/1999) and England (TIMSS 1999). Teachers from Hong Kong SAR (e.g., TIMSS 1995: 3.96) and Singapore (e.g., TIMSS 1999: 3.82) assigned "problem/ question sets in textbook" as homework nearly "always." In contrast, the least frequently used types of homework were assigned by mathematics teachers no more than "rarely" (i.e., $m < 2$). For instance, teachers from Japan used long term projects/ experiments as a group just slightly more than "never" (TIMSS 1995: 1.03; TIMSS 1999: 1.05). The within-system analyses revealed that the largest difference was observed in Hong Kong SAR between "problem/question sets in textbook" and "keeping a journal" (TIMSS 1995: $r = 0.92$; TIMSS 1999: $r = 0.90$). In fact, the corresponding differences in the other systems were about as large, with the smallest difference being in the USA between "problem/question sets in textbook" and "preparing oral reports either individually or as a small group" (TIMSS 1995: $r = 0.77$; TIMSS 1999: $r = 0.80$).

In contrast, there were only three types of homework investigated in TIMSS 2003/2007, and they were measured on a 3-point Likert scale.[14] A comparison between the types listed in TIMSS 1995/1999 and those in TIMSS 2003/2007 shows that the types stated in the later years are broader, whereas the earlier ones are more specific. In particular, both types WW and PQ in TIMSS 1995/1999 can be regarded as "doing problem/question sets" in TIMSS 2003/2007. Similarly, types SI, WI, and WG can be grouped into "gathering data and reporting" and type FC is close to "finding one or more applications of the content covered." Some types of homework in TIMSS 1995/1999 were not investigated in the later years, such as those involving communication skills.

It is clear that teachers from all eight systems used "doing problem/question sets" for students' homework most frequently (see Fig. 5). Except for Japan in TIMSS 2003 ($m = 2.40$), the average frequencies in both TIMSS 2003 and TIMSS

---

[13] The measurement is based on a 4-point Likert scale (1: never; 2: rarely; 3: sometimes; 4: always). A fifth option "I don't assign homework" was used in TIMSS 1995, which was recoded as missing data in the later analysis.

[14] TIMSS 2003/2007 used a 3-point Likert scale (1: always or almost always, 2: sometimes, 3: never or almost never), which was reversed into an ascending order in the later analysis so as to ease the interpretation.

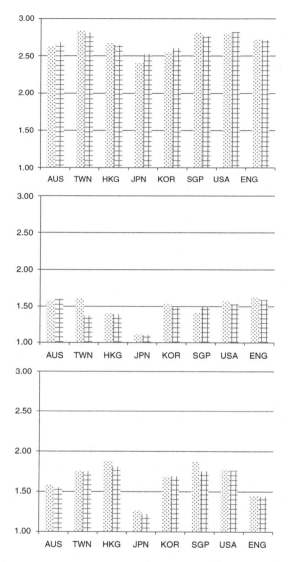

Fig. 5 Frequencies of three types of homework assigned by eighth grade mathematics teachers in eight systems in TIMSS 2003 and TIMSS 2007

2007 in all the systems were above 2.50 on a 3-point Likert scale. Mann–Whitney U tests[15] revealed insignificant or trivial between-year changes in frequency in all the systems but Japan ($r=0.12$). Homework about "gathering data and reporting" was assigned least frequently, which was particularly true in Japan (TIMSS 2003: 1.11, TIMSS 2007: 1.10). The greatest change in the use of this type of homework occurred in Chinese Taipei with a decrease of about 0.23 points in terms of utilization frequency ($r=0.22$). The changes in the other systems were either insignificant or trivial. On homework about "finding one or more applications of the content covered," Japanese teachers again reported a lower frequency than the other systems. These findings are consistent with those revealed in TIMSS 1995/1999. This suggests that in most cases, teachers from Japan tended to assign various types of homework at a low frequency. The between-year changes on application-related homework in all the systems were either insignificant or trivial.

In six systems, "applications of the content covered" used as homework was assigned more frequently than "gathering data and reporting" in both TIMSS 2003 and TIMSS 2007. However, the reverse was observed in Australia in TIMSS 2003 and England in both TIMSS years. The comparisons of the utilization frequencies between the two less used homework types found the smallest differences in Australia in both TIMSS years and the largest in Singapore in TIMSS 2003 ($r=0.52$) and Chinese Taipei in TIMSS 2007 ($r=0.50$). Furthermore, Friedman tests[16] showed that the differences in utilization frequencies across the three homework types in most systems were about trivial in both TIMSS years (i.e., $W>0.60$) but small in Australia (TIMSS 2003: $W=0.57$) and Hong Kong SAR (TIMSS 2003: $W=0.52$, TIMSS 2007: $W=0.56$).

What Mathematics Homework-Related Activities Were Carried Out in Classes?

While homework is mainly assigned for students' after-school studies, many teachers also often used it to organize in-class activities. TIMSS 1995/1999 teacher questionnaires suggested eight mathematics homework-related classroom activities and five were also investigated in TIMSS 2003/2007/2011. In the student questionnaires, four relevant classroom activities were found in TIMSS 1995/1999 and two in TIMSS 2003/2007. Table 4 lists these survey items across different TIMSS years.

Due to the fact that the measurement scales used in TIMSS varied across different versions of the surveys as well as among different years, some adjustments have been made to make the cross-version and cross-year comparisons feasible. As a result, all the scales on the relevant items were recoded into an ascending 3-point

---

[15] Effect size for a Mann-Whitney U test (on independent ordinal data between two groups) is expressed by r (small: $r=0.10$, medium: $r=0.30$, large: $r=0.50$), when a significant U is detected.
[16] Effect size for a Friedman test (on matched ordinal data across more than two groups) is expressed by Kendall's W (small: $W=0.60$, medium: $W=0.40$, large: $W=0.20$), when a significant χ^2 is detected.

Table 4 TIMSS survey items on mathematics homework-related classroom activities at eighth grade level

	1995	1999	2003	2007	2011
We (can) begin our homework in class (S)	✓	✓	✓	✓	
The teacher checks homework (S)	✓	✓			
We check each other's homework (S)	✓	✓			
Have students exchange assignments and correct them in class (T)	✓	✓			
We review homework (S)			✓	✓	
We discuss our completed homework (S)	✓	✓			
Use it as a basis for class discussion (T)	✓	✓	✓	✓	✓
Have students correct their own assignments in class (T)	✓	✓	✓	✓	✓
Record whether or not the homework was completed (T)	✓	✓			
Monitor whether or not the homework was completed (T)			✓	✓	✓
Collect, correct and keep assignments (T)	✓	✓			
Collect, correct assignments and then return to students (T)	✓	✓			
Give feedback on homework to whole class (T)	✓	✓			
Correct assignments and then give feedback to students (T)			✓	✓	✓
Use it to contribute towards students' grades or marks (T)	✓	✓	✓	✓	✓

Note: S stands for student questionnaire items and T stands for teacher questionnaire items. Items about similar activities but worded differently are listed together in the table

Likert scale, which is a reversed version of the scale used in the TIMSS 2003/2007/2011 teacher surveys.[17]

In all but TIMSS 2011, students were asked about how often they began their mathematics homework in class. The data showed that this occurred in classrooms in the USA most frequently, at least "sometimes" ($m > 2$). In contrast, students from Korea began their homework in class at the lowest frequency except for TIMSS 2007 (replaced by England). Regarding this practice, some researchers have compared the effectiveness of homework completed in school vs. homework completed at home. They generally found that the later practice has a greater effect on achievement, especially at the higher grade levels (e.g., Cooper, 1994; Keith & Diamond, 2004). Kirk and Ward (1999) suggested some possible explanations for the phenomenon, including that working on homework in school may take away instructional time and schools may not provide an environment allowing students to concentrate fully on their homework (also see Blazer, 2009).

Among the listed mathematics homework-related classroom activities, five were investigated in all the TIMSS years in teacher questionnaires, including homework-based discussion, students' self-correction of homework, monitoring of homework completion, giving feedback on homework, and homework contributing to students' grades/marks. The analyses showed that, of these five activities, mathematics teachers from all eight systems monitored students' homework completion most frequently except in three cases (i.e., AUS1999, TWN2011, and SGP1995; see Table 5). Moreover, teachers in the USA and England monitored more fre-

---

[17] TIMSS 2003/2007/2011 teacher surveys used a descending 3-point Likert scale (1: always or almost always, 2: sometimes, 3: never or almost never).

Table 5 Most-frequent (MF) and least-frequent (LF) mathematics homework-related classroom activities at eighth grade level in eight systems from TIMSS 1995 to TIMSS 2001

	1995		1999		2003		2007		2011	
	MF	LF	MF	LF	MF	LF	MF	LF	MF	LF
AUS	M	C	F	C	M	G	M	G	M	G
TWN	NA	NA	M	C	M	C	M	F	D	C
HKG	M	C	M	C	M	C	M	C	M	C
JPN	M	D	M	D	M	D	M	D	M	D
KOR	M	G	M	G	M	D	M	D	M	D
SGP	F	G	M	G	M	G	M	G	M	G
USA	M	C	M	C/D	M	F	M	F	M	F
ENG	M	C	M	C	M	C	M	C	M	C

Note: C students' self-correction of homework, D homework-based in-class discussion, F giving feedback on homework, G homework contributing to students' grades/marks, M monitoring on homework completion

quently than teachers in other systems, whereas teachers in Japan monitored least frequently expect in one case (TWN2007). Kruskal–Wallis tests[18] showed that the magnitudes of the utilization frequency differences in the monitoring practices across the systems was at a medium level, ranging from 0.10 (η^2 in TIMSS 2011) to 0.12 (η^2 in TIMSS 1995).

In contrast, on the least-frequent homework-related classroom activities, greater cross-system inconsistencies were observed. Furthermore, some cross-year changes in the least-frequent homework-related classroom activities were found in Australia and the USA, particularly between the 1990s and 2000s. For each type of homework-related classroom activity, a series of cross-system comparisons of their utilization frequencies were carried out for all five TIMSS surveys. Besides "monitoring of homework completion," "giving feedback on homework" and "students' self-correction of homework" were another two classroom activities for which the cross-system differences maintained a medium level in the 2000s (i.e., $\eta^2 < 0.14$). Though the cross-system differences for the other two classroom activities were at a large level in TIMSS 1995 (i.e., $\eta^2 > 0.14$), only on "homework-based in-class discussion" did the systems keep great discrepancies in all the years (highest η^2 in TIMSS 2011: 0.39). On "homework contributing to students' grads/marks," the magnitude of cross-system difference decreased to $\eta^2 = 0.17$ in TIMSS 2011. Finally, the eight systems had increasingly similar practices related to "giving feedback on homework" with the magnitude of the difference in the most recent TIMSS being $\eta^2 = 0.02$.

Besides the cross-system comparison, a series of cross-year comparisons within each system were carried out to examine how frequently each system used homework to organize different classroom activities across the TIMSS years. It was found that on the use of homework to organize in-class discussion, the cross-year

---

[18] Effect size for a Kruskal Wallis test (on independent ordinal data across more than two groups) is expressed by η^2 (small: $\eta^2 = 0.01$, medium: $\eta^2 = 0.06$, large: $\eta^2 = 0.14$), when a significant χ^2 is detected.

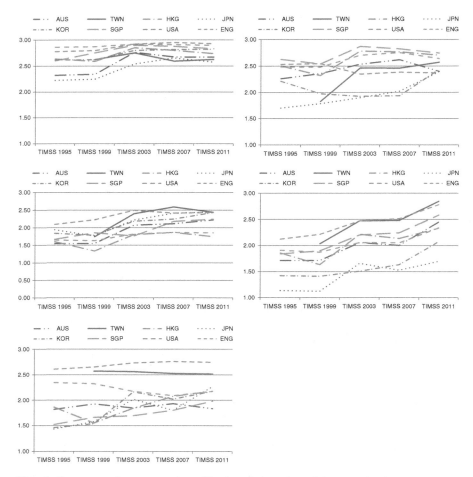

Fig. 6 Cross-year changes on the utilization frequencies of homework-related classroom activities in eight systems

differences in all eight systems were at least at a medium level, ranging from 0.09 (η^2 in England) to 0.19 (η^2 in Japan). The average utilization frequencies of this type of classroom activity in all the systems revealed an increasing tendency across TIMSS years with the highest frequency in Chinese Taipei (TIMSS 2011: $m = 2.84$). In comparison, the changing trends on the other four types of classroom activities across the systems showed greater variances, particularly on "homework contributing to students' grades/marks" (see Fig. 6).

In all the East Asian systems but Chinese Taipei, the magnitudes of the frequency changes in all the classroom activities except monitoring practices were about medium. In contrast, the magnitudes in the USA and England on all but homework-based in-class discussions were relatively small ($\eta^2 < 0.06$).

How Much Were Parents Involved in Students' Homework?

It is believed that parents are a valuable asset to promoting student success, especially during their children's after-school time (e.g., Balli, Wedman, & Demo, 1997; Chandler, Argyris, Barnes, Goodman, & Snow, 1986; Epstein, 1988; Horsley & Walker, 2008; Leone & Richards, 1989; Patall, Cooper, & Robinson, 2008; Van Voorhis, 2003; Xu & Corno, 1998). Beginning from TIMSS 1999, parental involvement in students' homework was investigated, though not specifically with respect to mathematics homework. In particular, TIMSS 1999/2003/2007 asked school principals/headmasters whether the schools expect/request parents to ensure their children completed homework. While this item was more at an expectation level, two new items appeared in the TIMSS 2011 student questionnaires asking for parents' actual monitoring behaviors on students' homework.

More than 90 % of the schools in all the six systems other than Japan and Korea expected/required students' parents to check their children's homework completion (see Fig. 7). Across the three TIMSS years, schools in Australia and Chinese Taipei did not have big changes on this "policy," while fewer schools from Hong Kong SAR, Singapore, and the USA seemed to have this expectation of parents in TIMSS 2007 as in TIMSS 1999. In contrast, the percentages of schools having this requirement in Japan increased steadily from 44.6 % in TIMSS 1999 to 84.6 % in TIMSS 2007. Korea maintained a comparatively low level of this expectation; in TIMSS 2007, there were only about 54.1 % of Korean schools having such a policy, which was far lower than the other seven systems.

From the students' perspective, TIMSS 2011 revealed that parents from Korea spent the least time on monitoring whether their children set aside time for

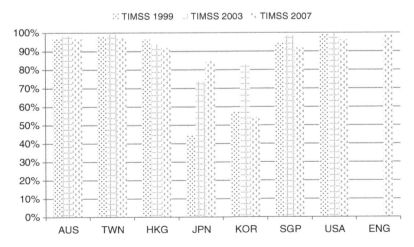

Fig. 7 Percentages of schools having expectations/requirements on parents to ensure their children complete homework at the eighth grade level in eight systems from TIMSS 1999 to TIMSS 2007
Note. No data were available in TIMSS 1999 and TIMSS 2003 in England

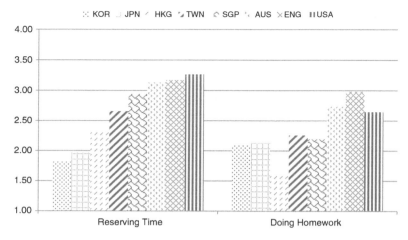

Fig. 8 Frequencies of two types of parental monitoring activities on homework at eighth grade level in eight systems in TIMSS 2011

homework.[19] Interestingly, parents from the three Western systems were found to be involved in such monitoring of time more frequently than those from East Asia (see Fig. 8). In particular, parents from Australia, England, and the USA did so at least "once or twice a week" (i.e., $m > 3$), whereas parents from Korea and Japan did so no more than "once or twice a month" (i.e., $m < 2$). Among the five East Asian systems, parents from Singapore and Hong Kong SAR seemed to have closer practices to those from Western systems.

Compared to monitoring whether students reserved time for homework, parents from Korea and Japan appeared to care more about whether their children did their homework. The reverse was observed in the other systems. However, once again, parents from the Western systems were found to check whether their children did their homework more frequently than those from the East Asian systems. The largest difference was revealed between Hong Kong SAR and England ($r = 0.34$); parents from Hong Kong SAR checked no more than "once or twice a month" ($m = 1.61$), while parents from England did so nearly "once or twice a week" ($m = 2.98$). In addition, within-system analyses found that parents from Hong Kong SAR ($\Delta = 0.73$, $r = 0.54$) and Singapore ($\Delta = 0.75$, $r = 0.55$) showed greater differences between the two homework monitoring behaviors than the other systems.

---

[19] A 4-point Likert scale was used on the two parent-related items in TIMSS 2011 (1: every day or almost every day, 2: once or twice a week, 3: once or twice a month, 4: never or almost never), which was reversed before the analysis to ease the interpretation.

Summary and Conclusions

Homework as "an important extension of in-school opportunities to learn" (Good & Brophy, 2003) has been around for decades. Over the years, considerable research has been conducted regarding the effects of homework on students' learning. As a result, the debate over homework has also raged on for years and continues to be a controversial topic today. Both supporters and adversaries of homework could produce a long list of consequences of homework to sustain their own standpoints. In fact, homework is a hot button issue not only inside academia but also outside. It appears that nobody is totally free from the homework debate, from the masses up to national leaders.

One important reason for the fragmented and often conflicting findings about homework is suggested to relate to the complex nature of homework. Not only is the impact of homework on students' learning affected by many other factors, but also homework itself is influenced by more factors than any other instructional strategies. From a methodological point of view, researchers have argued that many existing studies suffer from methodological flaws, such as small sample sizes and not taking the hierarchical nature of homework data into account. According to them, many findings about homework have been inflated.

Through analyzing the homework-related data from the *Trends in International Mathematics and Science Study* (TIMSS), this study aims to depict a comprehensive picture of the role homework plays in students' school lives. As one of the most extensive international survey studies, the TIMSS collected contextual data from students, teachers, school principals/headmasters, and curriculum experts every 4 years. In this sense, not only has the hierarchical element been accounted for, but also changes in homework practices over the years. Moreover, such multilevel data also allow each participating system to assess the alignment between the intended and implemented/attained curriculum from the perspective of homework.

At the *intended curriculum* level, though many systems reported that there was no system-wide policy about homework, assigning homework was recommended or common practice. There are also some systems providing recommendations on the frequency and amount of homework as well as guidance and feedback on homework. Many systems believed that schools and teachers are responsible for assigning homework mainly based on students' abilities and needs. The data from TIMSS 1995/1999 showed that the primary responsibility is mainly on teachers in Chinese Taipei, Japan, Korea, and the USA and on department heads in Hong Kong SAR and Singapore. Some changes were observed from TIMSS 1995 to TIMSS 1999; that is, school governing boards in Hong Kong SAR and principals in the USA started to take more responsibility for establishing homework policy.

At the *implemented/attained curriculum* level, more than 90 % of the eighth grade mathematics teachers from the eight investigated systems reported that they assigned homework to their students. However, the data also found that in all but Hong Kong SAR, over time more teachers started to not assign homework, though the changes were small. From both the teachers' and students' perspectives, teachers from most of the systems tended to assign mathematics homework less frequently than before. Among the eight systems, teachers from Japan assigned mathematics homework no more than "once or twice a week," while their colleagues in the USA did so at least "3 or 4 times a week." Furthermore, a comparison between teacher data and student data found that the two groups reported similar frequencies of assigning homework in Chinese Taipei, but students and teachers reported quite different frequencies in Hong Kong SAR and Korea.

Besides frequency, the amount of time spent on homework is another important factor when concerning students' workload. Across the, TIMSS years, based on their estimations for an average student, teachers from Japan, Singapore, and the USA appeared to make decreasing demands on the length of time necessary for doing homework. In contrast, teachers from the other systems tended to request more homework time in recent years. Among the systems, teachers from Singapore assigned mathematics homework requiring the longest period of time to complete and their Australian colleagues were found least demanding in this respect. However, the student data revealed somewhat different results. Across the, TIMSS years, according to the students, the amount of time they spent on homework generally decreased, except in Chinese Taipei. While Singapore again was the system where students needed to spend the longest period of time on homework, students from Korea and Japan worked on homework for the least amount of time. Furthermore, in all but Japan, teachers have recently tended to report longer homework time estimations than their students nowadays; the reverse was observed in earlier years.

As argued above, the quality of homework may be more influential in students' learning experiences than the quantity of homework. In all the systems, "worksheets or workbooks" and "problem/question sets in textbook" were the most commonly used homework types. Tasks heavily involving communication skills were rarely assigned as mathematics homework in most systems (except for Japan). Teachers assigned the most commonly used types of homework at least "sometimes" and in some systems, the frequency was even close to "always." In contrast, for the uncommonly used types, teachers in some systems almost "never" assigned them as students' mathematics homework. Great differences in utilization frequencies among various types of homework were observed in all the systems in all the TIMSS years. In addition, the data showed that teachers in Japan tended to use all types of tasks as homework at a lower frequency than their colleagues from other systems.

Besides its value in extending learning opportunities beyond school, homework is also valuable in building on students' learning at school. In fact, the TIMSS surveys have suggested various kinds of homework-related in-class activities. Of these activities, teachers from all eight systems almost always monitored their students' homework completion with the highest frequency across all the TIMSS years. However, teachers from different systems had greatly differing practices with respect to the least-frequent homework-related in-class activities. In particular, teachers in Japan and Korea rarely had homework-based discussions; teachers in Chinese Taipei and England rarely asked students to do self-correlation of homework; teachers in Singapore and Australia seldom used homework to contribute to students' grades/marks; and teachers in the USA did not give much feedback on homework. All these different practices could, to a certain degree, reflect teachers' different understandings about the function as well as the purpose of using homework for their students' learning.

The TIMSS surveys investigated homework practices at various curriculum levels. The results showed that the degree of alignment between these hierarchical levels varied across systems. In general, on popular practices, different systems showed a high commonality, while on those unpopular ones, more variation was revealed. Such information is particularly valuable for systems to reflect on their own practices in a global context. Moreover, given the cyclic design of the TIMSS, the data from the TIMSS allow each system to measure changes within itself over the years. This is useful for reviewing the coherence between the desired and actual state of the educational system, and is particularly informative for policy makers and curriculum developers.

Over the years, TIMSS has also made changes in its measurement designs, which imply some important messages about homework practices. First, homework-related items are now grouped under specific sections entitled "homework" or "mathematics homework" rather than being scattered as in the early TIMSS surveys. This may suggest that homework has become an indispensable component of students' school lives. In fact, when preparing the TIMSS encyclopedia, many systems included "homework policies" as one important section in their respective chapters. Though the analysis revealed that some teachers did not use homework in their mathematics teaching, the important role of homework in students' mathematics learning was generally recognized by the majority of the teachers around the world. In this aspect, teachers from Japan appeared to make the least use of homework, which might be related to the higher prevalence of cram schools in their system than in all the other systems.

Second, frequency and amount of homework are the two common themes investigated in all rounds of the TIMSS surveys. Though there are items on the types of tasks included in homework and homework-related in-class activities in almost all the TIMSS surveys, the variety of types goes down from the 1990s to 2000s. This may be related to the findings from the early surveys that certain types of tasks and in-class activities received more attention from teachers while others were hardly used. There are two possible reasons for the low level of variety. One is related to difficulty in implementation (particularly about designing and evaluating) and the other is related to teachers' beliefs about the value of homework in students' mathematics learning. More investigations will need to be carried out to uncover the underlying causes.

Third, both teachers' and students' perspectives about their homework practices were investigated in the TIMSS surveys. This two-level design takes into consideration the differences between the implemented and attained curriculum. In fact, this study did find many discrepancies between the two different perspectives in many systems and in most the TIMSS years. Moreover, the corresponding data are also more in line with the hierarchical nature of homework data. Therefore, they are more methodologically sound for further analysis on how homework practices by different parties influence students' learning.

Finally, starting from TIMSS 1999, parents' roles in students' homework received attention from the TIMSS researchers. In the most recent round, the relevant investigation on parental involvement has been moved from the "intended" level to "attained" level. That is, the focus shifted from what parents were expected/required to do to what parents have actually done with their children's homework practice at home. This change embodies one function of homework as a bridge between school and home. In other words, parents' involvement could play an important role in the process of students doing homework via their monitoring and assistance, and students' experiences with homework at home can be seen as a kind of extension of school learning. This change in the TIMSS design may reflect that students' learning is not only limited to the school context, but happens in a broader context.

References

Austin, J. D. (1979). Homework research in mathematics. *School Science and Mathematics, 79*, 115–121.

Balli, S. L., Wedman, J. F., & Demo, D. H. (1997). Family involvement with middle-grades homework: Effects of different prompting. *The Journal of Experimental Education, 66*, 31–48.

Blazer, C. (2009). *Literature review: Homework*. Miami, FL: Miami Dade County Public Schools. Retrieved December 10, 2012, from http://files.eric.ed.gov/fulltext/ED536245.pdf.

Cai, J., Nie, B., Moyer, J. C., Wang, N., Medhanie, A., & Hwang, S. (2012, April). *Learning mathematics using standards-based and traditional curricula: An analysis of homework problems*. Paper presented at the annual meeting of American Educational Research Association, Vancouver, British Columbia, Canada.

Canadian Council on Learning. (2009). *A systematic review of literature examining the impact of homework on academic achievement*. Ottawa: Author. Retrieved January 12, 2012, from http://www.ccl-cca.ca/pdfs/SystematicReviews/SystematicReview_HomeworkApril27-2009.pdf.

Canadian Education Association, & Ontario Institute for Studies in Education, University of Toronto. (2010). *The facts on education: How useful is homework?* Retrieved January 13, 2012, from http://www.cea-ace.ca/sites/cea-ace.ca/files/cea-2010-foe-homework.pdf

Center for Public Education. (2007). *Research review: What research says about the value of homework*. Retrieved May 10, 2009, from http://www.learndoearn.org/for-educators/Research-Review-on-Value-of-Homework.pdf

Chandler, J., Argyris, D., Barnes, W., Goodman, I., & Snow, C. (1986). Parents as teachers: Observations of low-income parents and children in a homework-like task. In B. Schieffelin & P. Gilmore (Eds.), *Ethnographic studies of literacy* (pp. 171–187). Norwood, NJ: Ablex.

Cooper, H. (1989). Synthesis of research in homework. *Educational Leadership, 47*(3), 85–91.

Cooper, H. (1994). *The battle over homework: An administrator's guide to setting sound and effective policies*. Thousand Oaks, CA: Corwin.

Cooper, H. (2001). *The battle over homework* (2nd ed.). Thousand Oaks, CA: Corwin.

Cooper, H. (2008). *Homework: What the research says*. Reston, VA: National Council of Teachers of Mathematics. Retrieved from http://www.nctm.org/news/content.aspx?id=13814.

Cooper, H., Jackson, K., Nye, B., & Lindsay, J. (2001). A model of homework's influence on the performance evaluations of elementary school students. *The Journal of Experimental Education, 69*(2), 143–154.

Cooper, H., Lindsay, J. J., Nye, B., & Greathouse, S. (1998). Relationships among attitudes about homework, amount of homework assigned and completed, and student achievement. *Journal of Educational Psychology, 90*(1), 70–83.

Cooper, H., Robinson, J., & Patall, E. (2006). Does homework improve academic achievement? A synthesis of research, 1987–2003. *Review of Educational Research, 76*, 1–62.

Cooper, H., & Valentine, J. C. (2001). Using research to answer practical questions about homework. *Educational Psychologist, 36*(3), 143–153.

Corno, L. (1996). Homework is a complicated thing. *Educational Researcher, 25*(8), 27–30.

Corno, L. (2000). Looking at homework difficulty. *The Elementary School Journal, 100*, 529–548.

Epstein, J. L. (1988). *Sample clinical summaries: Using surveys of teachers and parents to plan projects to improve parent involvement* (Parent Involvement Series, Report P-63). Baltimore: Johns Hopkins University, Center for Research on Elementary and Middle School.

Epstein, J., Simon, B. S., & Salinas, K. C. (1997). *Involving parents in homework in the middle grades* (Research Bulletin No. 18). Baltimore: Johns Hopkins University, Center for Evaluation, Department and Research.

Epstein, J. L., & Van Voorhis, F. L. (2001). More than minutes: Teachers' roles in designing homework. *Educational Psychologist, 36*(3), 181–193.

Gill, B. P., & Schlossman, S. L. (2004). Villain or savior? The American discourse on homework, 1850–2003. *Theory Into Practice, 43*(3), 174–181.

Good, T., & Brophy, J. (2003). *Looking in classrooms* (9th ed.). Boston: Allyn and Bacon.

Hallam, S. (2004). *Homework: The evidence*. London: Institute of Education.

Herrig, R. W. (2012, May 22). *Homework research gives insight to improve teaching practice. STEM—Science-technology-engineering-math*. Retrieved January 12, 2014, from https://www.mheonline.com/glencoemath/pdf/homework_research.pdf

Horsley, M., & Walker, R. (2008). Best practice in designing and managing after school homework support: A sociocultural interpretation of homework and affording learning through homework practices. In D. McInerney & A. Liem (Eds.), *Teaching and learning: International best practice* (Vol. 8, pp. 79–109). Greenwich, CT: Information Age.

Inglis, S. (2005). *A two-way street: Homework in a social context*. Centre for Education, Auckland College of Education. Retrieved October 29, 2007, from http://www.education.auckland.ac.nz/uoa/fms/default/education/docs/word/research/foed_paper/issue15/ACE_Paper_8_Issue_15.doc

Keith, T. Z., & Diamond, C. (2004). Longitudinal effects of in-school and out-of-school homework on high school grades. *School Psychology Quarterly, 19*(3), 187–211.

Keys, W., Harris, S., & Fernandes, C. (1997). *Third international mathematics and science study, second national report. Part 2: Patterns of mathematics and science teaching in upper primary schools in England and eight other countries*. Slough: NFER.

Kirk, P. J., Jr., & Ward, M. E. (1999). Homework in high school: Influence on learning. *Evaluation Brief, 1*(7), 1–4. Retrieved November 6, 2004, from http://www.ncpublicschools.org/docs/accountability/evaluation/evalbriefs/vol1n7-hwrk.pdf.

Klangphahol, K., Traiwichitkhun, D., & Kanchanawasi, S. (2010). Applying multilevel confirmatory factor analysis techniques to perceived homework quality. *Research in Higher Education Journal, 6*, 1–10. Retrieved January 3, 2010, from http://www.aabri.com/manuscripts/09373.pdf.

Kohn, A. (2006). Abusing research: The study of homework and other examples. *Phi Delta Kappan, 88*(1), 9–22.

Kralovec, E., & Buell, J. (2001). End homework now. *Educational Leadership, 58*, 39–42.

Lange, T., & Meaney, T. (2011). I actually started to scream: Emotional and mathematical trauma from doing school mathematics homework. *Educational Studies in Mathematics, 77*(1), 35–51.

Leone, C. M., & Richards, M. H. (1989). Classwork and homework in early adolescence: The ecology of achievement. *Journal of Youth and Adolescence, 18*, 531–548.

Maltese, A. V., Tai, R. H., & Fan, X. (2012). When is homework worth the time? Evaluating the association between homework and achievement in high school science and math. *The High School Journal, 96*(1), 52–72.

McPherson, F. (2005). Homework—Is it worth it? Retrieved August 25, 2008, from http://www.memory-key.com/improving/strategies/children/homework

Miller, D., & Kelley, M. (1991). Interventions for improving homework performance: A critical review. *School Psychology Quarterly, 6*(3), 174–185.

Mullis, I. V. S., Martin, M. O., Minnich, C. A., Stanco, G. M., Arora, A., Centurino, V. A. S., & Castle, C. E. (2012). *TIMSS 2011 encyclopedia: Education policy and curriculum in mathematics and science* (Vol. 1 & Vol. 2). Chestnut Hill, MA: TIMSS & PIRLS International Study Center, Boston College.

Mullis, I. V. S., Martin, M. O., Olson, J. F., Berger, D. R., Milne, D., & Stanco, G. M. (Eds.). (2008). *TIMSS 2007 encyclopedia: A guide to mathematics and science education around the world* (Vol. 1 & Vol. 2). Chestnut Hill, MA: TIMSS & PIRLS International Study Center, Boston College.

NSW Department of Education and Communication. (2012). *Homework policy: Research scan*. Retrieved from https://www.det.nsw.edu.au/policies/curriculum/schools/homework/Hwk_Res%20scan.pdf

O'Rourke-Ferrara, C. (1998). *Did you complete all your homework tonight, dear?* New York: Elementary and Early Childhood Education Clearinghouse. Unpublished manuscript. ERIC ED 425862. Retrieved from http://files.eric.ed.gov/fulltext/ED425862.pdf

Paschal, R. A., Weinstein, T., & Walberg, H. J. (1984). The effects of homework on learning: A quantitative synthesis. *The Journal of Educational Research, 78*, 97–104.

Patall, E., Cooper, H., & Robinson, J. (2008). Parental involvement in homework: A research synthesis. *Review of Educational Research, 78*(4), 1039–1101.

Queensland Department of Education & the Arts. (2004). *Homework literature review: Summary of key research findings*. Retrieved December 20, 2010, from http://education.qld.gov.au/review/pdfs/homework-text-for-web.pdf

Redmond, K. (2009). *Examining the effects of homework on achievement: A research synthesis*. Retrieved from http://facultyweb.cortland.edu/kennedym/courses/663/Literature%20Review%20Article/2009LRA/Kari%20FINAL%20copy.doc

Smith, R. (2000). Whose children? The politics of homework. *Children & Society, 14*(4), 316–326.

Trautwein, U. (2007). The homework-achievement relation reconsidered: Differentiating homework time, homework frequency, and homework effort. *Learning and Instruction, 17*, 372–388.

Trautwein, U., & Köller, O. (2003). The relationship between homework and achievement—Still much of a mystery. *Educational Psychology Review, 15*, 115–145.

Trautwein, U., Köller, O., Schmitz, B., & Baumert, J. (2002). Do homework assignments enhance achievement? A multilevel analysis in 7th-grade mathematics. *Contemporary Educational Psychology, 27*, 26–50.

Trautwein, U., Lüdtke, O., & Pieper, S. (2007). *Learning opportunities provided by homework*. Berlin, Germany: Max Planck Institute for Human Development. Retrieved January 12, 2012, from http://www.mpib-berlin.mpg.de/en/research/concluded-areas/educational-research/research-area-i/homework-halo.

Van Voorhis, F. L. (2003). Interactive homework in middle school: Effects on family involvement and science achievement. *The Journal of Educational Research, 96*(6), 323–338.

Van Voorhis, F. L. (2004). Reflecting on the homework ritual: Assignments and designs. *Theory Into Practice, 43*(3), 205–212.

Vatterott, C. (2009). *Rethinking homework: Best practices that support diverse needs*. Alexandria, VA: ASCD.

Walker, R. (2011). Should children do traditional homework? *The Sydney Morning Herald*. Retrieved January 13, 2012, from http://www.smh.com.au/federal-politics/the-question/should-children-do-traditional-homework-20110923-1kpaj.html

Warton, P. M. (1997). *Homework: Learning from practice*. London: The Stationery Office.

Warton, P. M. (2001). The forgotten voices in homework: Views of students. *Educational Psychologist, 36*(3), 155–165.

Xu, J., & Corno, L. (1998). Case studies of families doing third-grade homework. *Teachers College Record, 100*(2), 402–436.

Zhu, Y., & Leung, F. K. S. (2012). Homework and mathematics achievement in Hong Kong: Evidence from the TIMSS 2003. *International Journal of Science and Mathematics Education, 10*(4), 907–925.

Effect of an Intervention on Conceptual Change of Decimals in Chinese Elementary Students: A Problem-Based Learning Approach

Ru-De Liu, Yi Ding, Min Zong, and Dake Zhang

In this chapter, we described a study that compared a problem-based learning (PBL) approach to a traditional approach for teaching decimal concepts to 76 Chinese fifth graders. This chapter started with a review of literature regarding conceptual change, challenges in teaching decimals to elementary students, the PBL in relation to self-efficacy, and the rationales for conducting the present study. Then, we elaborated the PBL approach as an intervention approach in an independent sample of fifth graders. Finally, we discussed implication of PBL in educational settings.

Decimal fraction learning is considered one of the cornerstones of mathematics education internationally (Stacey et al., 2001). In the United States, formal instruction of decimal fractions begins in fourth grade and continues throughout all secondary grade levels (National Council of Teachers of Mathematics (NCTM), 2000). The NCTM Standards require third to fifth graders to be able to understand and convert fractions, decimals and percentages. And students older than sixth graders should flexibly solve problems involving fractions, decimals and percentages. In China, decimals and fractions are also introduced to students at the elementary level beginning in fourth grade (Zong, 2006).

R.-D. Liu, Ph.D. (✉)
School of Psychology, Beijing Normal University, Beijing 100875, P.R. China
e-mail: rdliu@bnu.edu.cn

Y. Ding, Ph.D. (✉)
Division of Psychological and Educational Services, School Psychology Program,
Graduate School of Education, Fordham University, 113 West 60th Street Room 1008,
New York, NY 10023, USA
e-mail: yding4@fordham.edu

M. Zong, M.S.
China Foreign Affairs University, Beijing, China

D. Zhang, Ph.D.
Rutgers University, New Brunswick, NJ, USA

© Springer International Publishing Switzerland 2015
J.A. Middleton et al. (eds.), *Large-Scale Studies in Mathematics Education*,
Research in Mathematics Education, DOI 10.1007/978-3-319-07716-1_11

A substantial number of studies have demonstrated that children have difficulties understanding decimals (Baturo, 1998; Hiebert & Wearne, 1983; Ni & Zhou, 2005; Resnick et al., 1989; Sackur-Grisvard & Leonard, 1985; Stacey & Steinle, 1998; Stafylidou & Vosniadou, 2004; Vamvakoussi & Vosniadou, 2004). Difficulty with fractions (including decimals and percent) has been identified as a pervasive problem and is a major obstacle preventing students from progressing in mathematics, including algebra (National Mathematics Advisory Panel, 2008). Even a considerable number of adults continue to hold such misconceptions (Putt, 1995; Silver, 1986; Stacey et al., 2001). Therefore, exploring how to help children develop their decimal knowledge is a priority for educational researchers.

In addition to the technical aspects of learning specific mathematics concepts, noncognitive variables play a role in student performance in mathematics. One such factor is students' self-efficacy. Bandura (1986) has argued that self-efficacy has a powerful impact on academic achievement. Research regarding mathematics self-efficacy has indicated that, in comparison to their counterparts with low self-efficacy, students with high self-efficacy demonstrate stronger persistence in difficult problem-solving situations and have better execution results in mathematics computation (Collins, 1982; Hoffman & Schraw, 2009). Thus, exploring self-efficacy in the context of mathematics learning has been of interest to educators and researchers.

A Conceptual Change Approach to Explain Children's Difficulties with Decimals

Rational numbers, including integers, terminating decimals, and repeating decimals, are numbers that can be expressed as a/b (both a and b are integers, and b can not be zero) (Vamvakoussi & Vosniadou, 2010). A single rational number, such as ½, can be represented in several ways (e.g., 5/10, 50/100, or 0.5), which all have the same value and are all alternative representations of the same rational number. In this chapter, we use the term "decimals" to refer to decimal representations of elements (i.e., a subset) of the set of rational numbers; we do not discuss decimals such as π that are not rational numbers.

The conceptual change approach has been recently used to explain students' persistent misconceptions regarding rational numbers (Vosniadou, 2007; Vosniadou & Verschaffel, 2004). Children's initial number frameworks are essentially natural numbers, which possess discreteness, whereas rational numbers have the feature of density, closely related to the concept of infinity (Hannula, Maijala, Pehkonen, & Soro, 2001; Malara, 2001; Merenluoto & Lehtinen, 2002). Natural numbers follow the successor principle (Vamvakoussi & Vosniadou, 2010) that all natural numbers are well ordered. Each natural number has a definite position in a sequence (e.g., 3 is the third number in the sequence of natural numbers), but rational numbers do not have this feature. When non-natural numbers, such as decimals and fractions, are introduced to students, their prior number frameworks based on natural numbers might hinder their understanding of the non-natural numbers.

Misconceptions. Previous literature has documented substantial information regarding the difficulties students usually encounter when they learn decimals. A common misconception is the notion that *longer is larger* (Moskal & Magone, 2000; Moss, 2005; Roche, 2005), in which students evaluate the value of a decimal number by comparing the number of its digits (e.g., 0.56 is larger than 0.8). A contrasting misconception is that *shorter is larger*, in which the students confuse decimals with fraction denominators (Steinle & Stacey, 2004). For example, in one study, children consistently judged that the larger number had fewer digits to the right of the decimal point; thus, 2.43 was larger than 2.897 (Sackur-Grisvard & Leonard, 1985). Another misconception, *multiplication makes bigger* (Fischbein, Deri, Nello, & Marino, 1985; Steffe, 1994), is true for natural numbers other than one, but is incorrect for decimal or fractional numbers less than one. Misconceptions can also arise in children's understanding of the density and infinity features of decimals. Finally, children often have difficulty with combining a string of digits into a single decimal quantity (Hannula et al., 2001; Malara, 2001; Merenluoto & Lehtinen, 2002; Resnick et al., 1989).

Children's misconceptions are often associated with over-generalization from their knowledge of natural numbers. For example, *longer is larger* and *multiplication makes bigger* may originate in children's experiences with comparing whole numbers. Children's difficulty with understanding the infinity feature (i.e., there are infinitely many decimal numbers between any two different decimals) of decimals can also be associated with their existing concept of whole numbers (Nunes & Bryant, 2007). In the domain of whole numbers, a number is a set of units of one, whereas in decimals, there is no minimum unit corresponding to ones. Instead, the minimum unit of decimals could be tenths, hundredths, thousandths, and so on. Children tend to intuitively generalize their mathematical reasoning skills regarding whole numbers to solving problems with decimals, which often leads to errors.

Existing Interventions for Teaching Decimals

The literature has documented programs that help children, sometimes identified as having had low achievement, to learn decimals. Two studies (Resnick, Bill, & Lesgold, 1992; Resnick, Bill, Lesgold, & Leer, 1991) emphasized the importance of helping low-SES African American parents to understand algebra and decimals and thus to provide their children with a better learning environment at home. Another study (Rao & Kane, 2009) helped children with intellectual disabilities to learn decimal calculation using a behavioral simultaneous prompting procedure in which the teacher delivered the target stimuli and the controlling prompt simultaneously; thus, the children did not have time to respond independently and therefore did not learn the task with errors.

Several existing interventions have taught students decimal concepts using representation techniques, such as using a number line and blocks to represent the place values. Hiebert and Wearne (Hiebert, 1988; Wearne, 1990; Wearne & Hiebert,

1988, 1989) conducted a group of studies using representation techniques. These studies emphasized using manipulatives (blocks) to promote conceptual under-standing of decimals. The researchers taught students using place-value blocks in which different shaped blocks were named to represent different place values. Specifically, large cubes represented a unit, a flat block represented a tenth of a unit, and a long block represented a hundredth of unit. Results showed that fourth through sixth graders made notable improvement based on the intervention. Similarly, Swan (1983) used representation techniques (a number line model) to help students under-stand the meaning of decimal notation and found that the students made consider-able progress. Woodward, Baxter, and Robinson (1999) also successfully used visual representations (e.g., wood block rectangles, squares, or cubes) to teach basic decimal concepts to children with learning disabilities.

One group of studies focused on the effects of exposing children to their miscon-ceptions regarding problem solving, where decimals were used. For example, Swan (1993) compared two classes whose teachers had adopted two different teaching styles. One class was taught with a "positive-only" teaching style. First, the teachers explained the concepts and methods of using a number line; then, the students prac-ticed using this method but were not asked to mark their work or diagnose errors. Another class was taught using a "conflict teaching style," in which the teacher initially gave students problems that exposed them to misconceptions and taught students a method using a number line; then, the teacher led a discussion of the students' errors and misconceptions. The "conflict teaching style" resulted in sig-nificantly more progress in children's achievement than the "positive-only style." The results suggested that exposing children to their misconceptions helped them to overcome their errors. Another study by Pierce, Steinle, Stacey, and Widjaja (2008) revealed the importance of identifying college students' difficulties with decimals. In this study, nursing students were given a decimal problem-solving test that identi-fied the students' misunderstandings of particular items. Next, the teacher used vari-ous models to illustrate the place value and base ten concepts, and students were encouraged to ask questions and provide responses. This study found significant improvement by the students on a delayed post-intervention test, and the research-ers concluded that it was necessary to expose students to their errors and plan for remediation of students' misconceptions before teaching procedure rules.

Similarly, Huang, Liu, and Shiu (2008) revealed the effectiveness of exposing students to incorrect examples to facilitate their conceptual understanding of deci-mals. Sixth graders were exposed to incorrect examples when learning the meaning of decimals (e.g., in 5.4, saying the .4 represents 4 ones instead of 4 tenths). After 4 weeks, these students performed better than students who were not presented with incorrect examples.

In summary, existing interventions have suggested the importance of exposing students to their mathematical misconceptions and errors. Researchers have reported that providing incorrect examples or examining students' own mistakes can pro-mote deeper reflection on correct concepts (VanLehn, 1999) and increase students' frequency of choosing correct strategies (Siegler, 2002). Based on these findings, it is plausible to assume that PBL would be effective for improving children's concep-tual understanding of decimals.

Problem-Based Learning and Self-Efficacy

PBL is a student-centered instructional strategy in which students learn through solving problems in groups and making reflections on their problem solving experiences. PBL is rooted in constructivist theories of learning that stress the importance of learners being engaged in constructing their own knowledge (Mayer, 2004; Palincsar, 1998). In PBL, students work in groups and are challenged with open ended and ill-defined problems. PBL is highly student-centered: Students are encouraged to explore the solutions and direct the problem solving process by themselves, and teachers only serve as facilitators (Hmelo & Guadial, 1996; Quntana, et al.). PBL is reported to be effective in enhancing content knowledge and fostering the development of communication, problem-solving, and metacognitive skills (Hmelo-Silver, Duncan, & Chinn, 2007). PBL has been shown to be effective in various empirical studies as described by Hmelo-Silver et al. For example, "there is an extensive body of research on scaffolding learning in inquiry- and problem-based environments (Collins, Brown, & Newman, 1989; Davis & Linn, 2000; Golan, Kyza, Reiser, & Edelson, 2002; Guzdial, 1994; Jackson, Stratford, Krajcik, & Soloway, 1996; Reiser, 2004; Toth, Suthers, & Lesgold, 2002" (p. 100, Hmelo-Silver, Duncan, & Chinn, 2007). Theory based and empirically validated strategies for effectively scaffolding students during PBL have been developed by many researchers (Hmelo-Silver, 2006; Hmelo-Silver, Duncan & Chinn 2007; Reiser et al., 2001). PBL is often used to assist learning of complex tasks. Complex tasks often require scaffolding to help students engage in sense making, self-management of their problem-solving processes, and facilitate students to articulate their thinking and reflect on their learning experiences (Quintana et al., 2004). Scaffolding helps to reorganize complex tasks and reduce cognitive load by structuring a task in a way that allows the learners to focus on relevant aspects of the task (Hmelo-Silver, 2006).

Many challenging tasks require both adequate skills and self-efficacy, which is about one's beliefs about whether or not one can successfully complete a task (Bandura, 1986). The relationship between PBL and self-efficacy has gained increasing attention. For example, self-efficacy was a significant predictor of science achievement in middle school students in a computer-enhanced PBL environment (Liu, Hsieh, Cho, & Schallert, 2006). For adult learners, specific instructional strategies (i.e., authentic problems of practice, collaboration, and reflection) used in PBL were reported to improve levels of self-efficacy in undergraduate computer science students (Dunlap, 2005). For educators, those with higher scores of self-efficacy demonstrated a significantly higher use of a PBL approach, direction instruction, and communication skills in mathematics teaching (Ordonez-Feliciano, 2010).

However, there are no studies dealing with the application of PBL to the instruction of decimal fractions in Chinese elementary students with consideration of students' self-efficacy. We were particularly interested in the PBL approach in Chinese students partially due to the fact that traditional Chinese mathematics instruction often follows a curriculum-centered approach with relatively large student–teacher ratios, making few opportunities available for students to be exposed to a PBL environment. Given that previous studies have found that exposing students to their misconceptions or errors and to challenging problems was effective in enhancing their conceptual understanding of decimals, the purpose of the present study was to investigate the following questions: (a) Does a PBL approach outperform a traditional instructional approach to enhance conceptual change in decimal computation? (b) Does a PBL approach outperform a traditional instructional approach to promote metacognition, measured by explicit interpretation of strategy use? and (c) Does a PBL approach lead to a higher level of self-efficacy and academic interest than a traditional instructional approach?

Method

Design

This study utilized a quasi-experimental design to compare pretest and posttest measures. The independent variable was the instructional method, consisting of a PBL approach in the experimental group and a traditional instructional approach in the control group.

Participants and Setting

The instructors were two experienced mathematics education teachers. One investigator majoring in educational psychology was on site for training, progress monitoring, and data collection. Each classroom had one experienced teacher as the lead teacher. The two classes of students had mathematics classes at different time periods on each day, so the investigator was able to observe classroom activities and collected data both in the control group and the experimental group for similar amounts of time to avoid the Hawthorne Effect. The participants were 76 fifth graders at an elementary school in the urban area of Beijing in Mainland China. The students were in two parallel classes, which were chosen because they were equivalent in terms of the students' performance in mathematics. Both classes followed the same mathematics curriculum, had a similar pace (teaching unit by unit according to textbook), used the same curriculum-based exams (designed by curriculum committees at the school), and the two teachers had comparable teaching experiences (i.e., years of teaching mathematics, teaching similar students at similar schools). Both teachers were new to the two groups because data collection started in the beginning of the school year. One class ($n=38$) received experimental PBL instruction that emphasized problem-based scenarios for teaching and students' own computation errors and prior experiences for discussion and problem analysis. The other class ($n=38$) received traditional instruction that emphasized curriculum-centered lecture and use of demonstration examples from the textbooks. All students were with normal intellectual abilities and were enrolled in regular classroom settings. Students' prior whole number and decimal number knowledge was similar due to the highly uniform school instruction implemented in Chinese schools.

Dependent Measures

Decimal computation test. To quantitatively measure students' conceptual change in decimals, we developed two sets of decimal computation tests. The measurement instruments used in this study involved a pretest (eight items in total) and a posttest (ten items in total). The pretest and posttest involved computation of both decimal and whole number problems. The pretest had three items that involved whole number computation only. Of the other five items relevant to decimals in the pretest, three of them involved decimal computation only and two of them involved mixed computation. The posttest had four items that involved whole number computation only. Of the other six items relevant to decimals in the posttest, four of them involved decimal computation only and two of them involved mixed computation. Each test included pairs of corresponding decimal and whole number items, as explained in Table 1. The computations included addition, subtraction, multiplication, and division. These items were chosen from *Beijing Compulsory Education Curriculum Reform Experimental Materials of Mathematics in Elementary School-Volume IX* (Lu & Yang, 2005). The pretest and posttest items were not identical due to consideration of the curriculum taught during the 22 classes. The pretest functioned as a placement test to examine whether the experimental group had similar prior knowledge as the control group. The posttest functioned as a summative test to measure whether students had mastered designated computation skills after receiving 22 classes of formal instruction. The difficulty levels of the pretest and posttest were consistent with curriculum content. Cronbach's alpha was .63 for the pretest and 0.72 for the posttest. Because of the limited testing time, we only designed eight items for the pretest and ten items for the posttest. The relatively low reliabilities might be attributable to the number of testing items we had. Sample items are listed in Table 1.

Qualitative measure of students' conceptual understanding of decimal division. Students' conceptual change was qualitatively measured by an open-ended question on the posttest to examine students' conceptual understanding of decimal division. The open-ended question asked, "Currently, there is a student who does not understand decimal division. Please elaborate your procedures of problem solving. For example, tell this student what to do first, what to do as a second step, and then what else."

Self-efficacy survey. A self-report questionnaire was developed based on Qin (2003) and Zhang (2005) to explore (a) social self-efficacy, (b) academic self-efficacy, and (c) academic interest. The questionnaire utilized a 6-point Likert scale ranging from *completely unlike me* to *completely like me*. The subtest of social self-efficacy included six items, with 36 points as the highest score. The subtest of academic self-efficacy included seven items, with 42 points as the highest score. The subtest of academic interest included seven items, with 42 points as the highest score.

Table 1 Sample problems in probes

Measures	Sample items	
Computation tests (All items included)		
Pretest computation	$10.1 \div 0.2$ (decimal)	$0.9 + 2.32$ (decimal)
	15×0.8 (mixed)	5.85×0.60 (decimal)
	$101 \div 2$ (whole number)	$9 + 232$ (whole number)
	585×60 (whole number)	$100 - 2.56 \times 5 + 32.5 \div 10$ (mixed)
Posttest computation	$1.21 \div 0.2$ (decimal)	$0.9 + 3.25$ (decimal)
	$5.58 - 0.9$ (decimal)	7.8×0.60 (decimal)
	$120 \div 20$ (whole number)	$9 + 325$ (whole number)
	78×60 (whole number)	$2.56 \times 5 + 32.4 \div 10 - 4.85$ (mixed)
	$2.5 \times 18 - 0.67 + 0.5 \div 5$ (mixed)	$558 - 9$ (whole)
Sample items of self-efficacy and interest survey		
Students are asked to rate on a Likert scale (i.e., 1–6, 1 stands for completely disagree and 6 stands for completely agree) according to each item		
Social self-efficacy questionnaire	Sample A: I can successfully interpret my thoughts to my classmates	
	Sample B: When other students talk with me, I do not know what I should talk about with them	
Academic self-efficacy questionnaire	Sample A: If I have sufficient time, I can learn mathematics well	
	Sample B: I can learn math even if some contents are very difficult	
Academic interest questionnaire	Sample A: I like math class more than I do other subjects	
	Sample B: The problems discussed in math class are very interesting	

Note: In the pretest computation, $10.1 \div 0.2$ (decimal) corresponds with $101 \div 2$ (whole number); $0.9 + 2.32$ (decimal) corresponds with $9 + 232$ (whole number); 5.85×0.60 (decimal) corresponds with 585×60 (whole number). In the posttest computation, $1.21 \div 0.2$ (decimal) corresponds with $120 \div 20$ (whole number); 7.8×0.60 (decimal) corresponds with 78×60 (whole number); and $0.9 + 3.25$ (decimal) corresponds with $9 + 325$ (whole number). Between pretest and posttest computation, $10.1 \div 0.2$ (decimal division) corresponds with $1.21 \div 0.2$; $101 \div 2$ (whole number division) corresponds with $120 \div 20$; $0.9 + 2.32$ (decimal addition) corresponds with $0.9 + 3.25$; and 5.85×0.60 (decimal multiplication) corresponds with 7.8×0.60; $9 + 232$ (whole number addition) corresponds with $9 + 325$. There are two mixed (whole number and decimal number) computation problems in pretest and posttest, respectively

The internal consistency coefficients were 0.71, 0.92, and 0.92 for the three subtests, respectively. The survey was administered to students during both the pretest and the posttest. Sample items are listed in Table 1.

Coding and Scoring

Decimal computation test. First, a graduate assistant who was unaware of the purpose of the study scored the decimal computation test using an answer key. Specifically, items on the test were scored as correct or incorrect, with one point awarded if the correct answer was given. In the pretest, there were five items involving decimal computation. In the posttest, there were six items involving decimal computation. We calculated students' total scores on the pretest and posttest and also calculated their decimal computation scores in pretest and posttest, respectively.

Second, as there were paired whole number and decimal items in each of the two tests, the students' answers were classified into four categories: (1) whole number computation is correct, and decimal computation is also correct; (2) whole number computation is correct, but decimal computation is incorrect; (3) whole number computation is incorrect, and decimal computation is also incorrect; and (4) whole number computation is incorrect, but decimal computation is correct. The second category of responses indicated that the students were unable to correctly apply whole number computation rules to decimal computation.

Third, we coded for errors to examine the mistakes students made during their problem solving. The investigators used a coding system to categorize seven types of computation errors in both groups: aligning place value, carrying, displacement of decimal point, carelessness, mnemonics, computation order, and missing values (see Table 2). Most Chinese textbooks have a student version and an instructor version, and the instructor version provides details such as exercise items, solutions, and common types of errors. The classification of computation errors was designed based on the types of errors suggested by the instructor version of the mathematics textbook utilized in the school.

Qualitative measure of students' procedural understanding of decimal division. There was one open-ended question in the posttest to qualitatively examine students' procedural understanding of decimal division. Students' levels of awareness of the strategies they used were categorized into three types, including missing or inaccurate (i.e., incorrect answers), nonessential (i.e., answers regarding general computation rules that applied to whole numbers but did not apply to decimal numbers), and essential answers (i.e., answers that were essential for decimal computations or answers showing correct examples or decimal computation rules) (see Table 2).

Self-efficacy survey. Scores on negatively worded items were reverse coded. A higher score indicated a higher level of self-efficacy. Students were rated as "0" when they chose *completely unlike me* and were rated as "6" when they chose *completely like me.*

Inter-rater agreement. Another graduate assistant rescored 30 % of the tests. Inter-rater agreement was computed as the percentage of the number of agreements divided by the total number of rated items. Inter-rater agreement was 95 %.

Procedures

Following the pretest assessment, one intact class became the PBL group and the other intact class became the control group. Because the two participating classes had identical curriculum, similar class schedules and similar instructional approaches, the selection of the PBL group and control group was totally random. Students in the PBL group received the intervention during five classes per week for

Table 2 Coding scheme for computation errors and awareness of strategy use

	Descriptors	Examples
Coding for different types of computation errors		
Aligning	Lining up the decimal points as below	
	0.9	0.9
	+3.25	+3.25
	4.15	3.34
Carrying	Students made mistakes during carrying numbers between different unit positions	0.9
		+3.25
		3.15
Displacement	Students placed the decimal point at wrong place after calculation	$7.8 \times 0.60 = 0.468$
Carelessness	Due to carelessness, students made mistakes like miscopying of numbers, omission, or skipping of calculation steps	"I accidentally put 0.12 as 0.18"
Mnemonics	Students retrieved incorrect multiplication facts	"3 times 7 is 22"
Computation order	Students did not calculate according to computation order, such as (1) calculating from left to right, (2) calculation in parenthesis should be done first, (3) exponents or radicals should be done next, (4) multiplication and division should be done in the order in which it occurs, and (5) addition and subtraction should be done in the order in which it occurs	In the example of "$2.56 \times 5 + 32.4 \div 10 - 4.85 = ?$," the student did not calculate multiplication and division before they calculated addition and subtraction
Missing	During pretest measures, students had not yet learned decimal division, thus students chose to give up on some of the items	"Can I skip this problem?"
Coding for conceptual understanding of decimal computation		
Missing or inaccurate	Students provided inaccurate answers or did not provide any answers	"Well, I am not supposed to explain that"
		"Let me think about it"
		"That is good"
Nonessential	Students provided only general computation rules that applied to whole numbers but did not apply to decimal numbers	"Decimal division is pretty much like division of whole numbers"
		"You compute it like division of whole numbers, then add a decimal point afterwards"
Essential	Students provided answers that were essential for decimal computations, used correct examples, or mentioned important decimal computation rules	"To divide by a decimal, multiply that decimal by a power of 10 great enough to obtain a whole number"
		"When we multiply the divisor, we also need to multiply the dividend"

4½ weeks during the school day, for a total of 22 classes. Each session of class lasted approximately 40–45 min. The control group continued to have their regular mathematics classes (i.e., five classes per week for 4½ weeks), for a total of 22 sessions. The major teaching content for both classes was decimal multiplication and division, which covered two units of the textbook. The complete content for that semester included seven units.

Teacher training. The instructor for the experimental group had utilized a PBL approach for more than 5 years and was very familiar with PBL. She received 1 week of additional training on the PBL approach before the intervention started. The purpose of the training was to help the teacher to conduct the intervention in the designated manner and to train the teacher to be proactive. The teacher relearned the PBL approach, had opportunities to practice how to teach students using the PBL approach, and received feedback from the investigator during the training. The investigator developed the teaching scripts (see Appendices 1 and 2), which were studied by the instructor of the experimental group to prepare for teaching the lessons. For the control group, the investigator observed classroom activities and collected data. For the experimental group, the investigator was on site for observation, progress monitoring (i.e., making sure the teacher was following the teaching scripts), and data collection. The investigator spent a similar amount of time in each classroom for observation and data collection.

Assessment conditions. Assessment conditions refer to the pretest assessment prior to the intervention and the post-intervention assessment. Pre- and post-intervention tests were administered using paper and pencil for all students. Experimenters did not provide any prompting or feedback regarding the accuracy of students' solutions. Students were provided with sufficient time to complete the test and the survey.

Experimental group. The experimental group adopted a PBL approach. The students began with specific problem scenarios and the teacher provided them with opportunities to reveal their computation errors and prior experiences. The teacher encouraged the students not only to explain the patterns of computation errors, but also to analyze the causes of computation errors. The teacher in the PBL group encouraged an open learning atmosphere and supported the students' reliance on prior learning experiences to guide their learning behaviors.

The instructional materials included projectors, experiment record sheets, and reminder cards. The reminder cards provided hints to the students when the problems were presented; for example, after the computation, the reminder cards helped the students to self-check the computational procedure, such as "I have checked the placement of the decimal point." Therefore, reminder cards were considered an effective method to monitor students' metacognition (Tong & Zhang, 2004).

The teacher gradually faded out the use of reminder cards as the instruction progressed. Specifically, at the beginning of the instruction, the teacher provided the students with complete reminder cards. The teacher determined the instructional framework and distributed the reminder cards to every student in the classroom. After discussion, the students summarized the types of computation errors made by all of the students in the class and noted the computation errors on the reminder cards according to a scaffolding framework. As the instruction progressed to approximately halfway through the intervention, the teacher provided the students with partially completed reminder cards. Each student analyzed only computation errors that he or she made and then noted the types of errors on the reminder cards. Each student could individualize his or her reminder card.

For both new lecture and review classes, the PBL curriculum followed similar procedures, including class preparation, instruction, and PBL. The focus was to analyze the students' prior experiences and design a PBL environment to motivate the students to think through problems and work out solutions. During the instructional procedure, the focus was to facilitate group discussion, analyze problems, and guide students to come up with solutions to solve problems. The initial PBL sessions helped students to identify errors and analyze prior experiences. The later PBL sessions emphasized exercises tapping into metacognition, such as analyzing types of computation errors, discussing the rationales for errors, and self-revising computation errors. Appendices 1 and 2 present examples for a new lecture and a review class. Appendix 3 presents a flowchart of the PBL approach guiding our intervention.

Control group. The teacher in the control group closely followed the instructional guidelines used for the regular curriculum. Traditional Chinese mathematics instruction focuses on a curriculum-based teaching approach. Due to a relatively large student–teacher ratio (e.g., 40 or 50:1), lecture that closely follows the curriculum is often the main teaching method. Due to the mandatory teaching content specified by the Ministry of Education of the People's Republic of China, teachers often closely follow guidelines in the curriculum as a typical practice. Although the teacher of the control group had opportunities to question students, few opportunities were available for small-group discussion, close interaction between the teacher and students, and students' reflection on their own errors and prior experiences. The 4½-week instructional activities included new lectures and review classes for decimal multiplication and decimal division. For the new lectures, the teacher introduced new concepts based on the textbook, started with demo exercises, explained rules of computation, asked the students to complete exercises, and provided students with opportunities to ask questions. During the review classes, the teacher primarily relied on demo items in the textbook to explain computation errors, and the discussion of patterns of computation errors was based on teaching experience rather than on actual computation errors that occurred during the students' exercises. Thus, the discussion of computation errors was not specific, and the teacher did not provide the students with opportunities to reflect on their own computation errors. The most frequently used method was to discuss classical computation errors addressed by the textbook as examples. Although there were opportunities for teacher–student interaction, most of the demo items had fixed answers, which were not likely to challenge students' higher levels of reasoning. The teacher typically gave students general praise but did not provide specific feedback.

Treatment Fidelity

In addition to the first investigator who was on site for data collection and progress monitoring, a second investigator independently observed ten treatment sessions in the experimental group to assess fidelity or quality of implementation of specific performance indicators. Treatment fidelity checklists are provided in Appendix 4. Half of the sessions were new lectures and half of the sessions were review classes. The observation sessions were equally distributed throughout the intervention period. The teacher used a teaching script to guide the teaching strategy during each class session. In addition, for each session, the first and third author used a checklist, which listed the key instructional components, to evaluate teachers' adherence to the assigned instructional condition type. The second investigator judged the adherence of the instructor's teaching based on the presence or absence of the features listed on the fidelity checklist. The overall treatment fidelity was .92 for the sessions observed.

Results

Pretreatment Group Equivalency

We used ANOVA tests to examine pretreatment group equivalency on the decimal computation test, self-efficacy questionnaire, and academic interest survey. Results indicated no significant difference between the two groups on the total computation test, $F(1, 74) = 0.044$, $p = 0.835$; the decimal computation test (i.e., decimal computation in computations involved decimals only and mixed numbers), $F(1, 74) = 0.633$, $p = 0.429$; the social self-efficacy questionnaire, $F(1, 73) = 0.048$, $p = 0.828$; the academic self-efficacy questionnaire, $F(1, 73) = 3.783$, $p = 0.056$; or the academic interest survey, $F(1, 73) = 3.633$, $p = 0.061$ (see Table 3).

We also compared computation errors made by the two groups of students during the pretest. Both groups appeared to make the most frequent computation errors in aligning place value, carrying, and displacement of the decimal point. In both groups, a large number of students chose to skip the questions because they had not learned decimal division prior to the intervention. The chi-square test indicated nonsignificant differences in the distribution of the seven computation errors with the exception of displacement of the decimal point, $\chi^2 = 5.775$, $p = 0.038$ (see Table 6), with better performance in the control group.

Quantitative Measure of Students' Conceptual Change in Decimals

Total computation. We performed an ANOVA test (with the pretest difference as a covariate) on the posttest scores to assess the effects of instruction on students' total computation performance. Results indicated a significant difference between groups at posttest, $F(1, 74) = 10.063$, $p = 0.002$ (see Table 3), favoring the PBL group.

Table 3 Means, standard deviations, paired samples T-test, and ANOVA by treatment conditions

Test	M PBL	M Control	SD PBL	SD Control	N PBL	N Control	ES	Paired samples T-test PBL	Paired samples T-test Control	ANOVA F/Sig.
Pretest-total	4.32	4.24	1.69	1.60	38	38	0.05	NA	NA	.044/.835
Posttest-total	8.32	7.08	1.53	1.87	38	38	0.73			10.063/.002
Pre-De	2.39	2.21	0.99	1.03	38	38	0.18	NA	NA	.633/.429
Post-De	4.42	3.45	1.46	1.33	38	38	0.69			9.215/.003
Pre-SSE	27.24	27.49	4.71	5.18	38	37	−0.05			.048/.828
Post-SSE	31.68	27.84	4.13	5.76	38	37	0.77	−6.141/.000	−.455/.652	35.723/.000
Pre-ASE	30.24	32.54	4.88	5.37	38	37	−0.45			3.783/.056
Post-ASE	32.61	33.03	4.95	4.51	38	37	−0.09	−5.070/.000	−.709/.483	2.30/.134
Pre-AC	27.76	30.59	6.29	6.57	38	37	−0.44			3.633/.061
Post-AC	34.55	30.57	6.85	5.69	38	37	0.63	−6.203/.000	.031/.976	18.950/.000

Note: Pretest-Total, Posttest-Total pretest, posttest test total scores on all computation items, *Pre-De* all decimal computation pretest (including decimal computation in mixed computation), *Post-De* all decimal computation posttest (including decimal computation in mixed computation), *Pre-SSE* social self-efficacy pre-survey, *Post-SSE* social self-efficacy post-survey, *Pre-ASE* academic self-efficacy pre-survey, *Post-ASE* academic self-efficacy post-survey, *Pre-AC* academic interest pre-survey, *Post-AC* academic interest post-survey, *ES* effect size by Cohen's *d*, which is calculated as the two conditions' mean differences divided by the pooled standard deviation (Hedges & Olkin, 1985)

Table 4 Univariate analysis of variance of posttest decimal computation in two groups

Source	Type III sum of squares	df	Mean square	F	Sig
Corrected model	23.375[a]	2	11.687	6.125	.003
Intercept	132.046	1	132.046	69.200	.000
Pre-De	5.361	1	5.361	2.810	.098
Group	16.104	1	16.104	8.439	.005
Error	139.296	73	1.908		
Total	1,339.000	76			
Corrected total	162.671	75			

Note: [a]R squared = .144 (adjusted R squared = .120); *Pre-De* all decimal computation pretest (including decimal computation in mixed computation), *Post-De* Dependent variable, all decimal computation posttest (including decimal computation in mixed computation)

Decimal computation. An ANOVA test on the posttest scores revealed that the experimental group outperformed the control group on decimal computation (i.e., the decimal computation and mixed items), $F(1, 74) = 9.215, p = 0.003$ (see Table 3). Due to the fact that the pretest (which served as a placement test) and posttest (which served as summative evaluation) did not have the same question items, within-group comparison of pretest and posttest scores for each group could not be conducted. We conducted univariate analysis of variance to further control for differences in pretest decimal computation performance (termed Pre-De in Table 3). The analysis results in Table 4 indicated a significant main effect caused by group difference (i.e., experimental group vs. control group) and a nonsignificant main effect of pretest decimal computation performance.

Students' Self-Efficacy and Academic Interest

Self-efficacy. An ANOVA test (with the pretest difference as a covariate) on the post-survey of self-efficacy revealed significantly higher social self-efficacy in the experimental group, $F(1, 73) = 35.723, p = 0.000$. Although the control group reported relatively higher academic self-efficacy, the test did not indicate a significantly higher score than the score of the experimental group, $F(1, 73) = 02.30, p = 0.134$ (see Table 3). In terms of within-group comparison, we conducted a paired samples t test. The control group did not show significant improvement on either social self-efficacy or academic self-efficacy after 22 sessions of classes. The experimental group showed significant improvement on both social self-efficacy ($p = .000$) and academic self-efficacy ($p = .000$) after receiving the entire intervention.

Academic interest. An ANOVA test (with the pretest difference as a covariate) on the post-survey of academic interest indicated significantly higher academic interest in the experimental group over the control group, $F(1, 73) = 18.950, p = 0.000$ (see Table 3). We also conducted a paired samples t test to examine within-group

Table 5 Qualitative analysis of students' conceptual understanding of decimal computation

Type of answers	Control	MA	Total	χ^2/Sig.
Missing	26	14	40	15.857/.000
Nonessential	11	10	21	
Essential	1	14	15	
Total	38	38	76	

improvement. The control group did not show significant improvement on academic interest from pretest to posttest measures, whereas the experimental group demonstrated significant improvement.

Qualitative Measure of Students' Conceptual Change in Decimals

The posttest included an open-ended item that asked, "Currently, there is a student who does not understand decimal division. Please elaborate your procedures of problem solving. For example, tell this student what to do first, what to do as a second step, and then what else." Approximately two thirds of the students in the control group chose to give up, and another one third of the students provided answers showing no conceptual understanding of decimal division (e.g., using whole number rules for decimal computation). Only one student in the control group provided an answer showing a conceptual understanding of decimal division. In contrast, 14 students in the experimental group explained essential features associated with computation of decimal division. We used the coding scheme listed in Table 2 to classify the narrative responses provided by the students, including missing or inaccurate, nonessential, and essential answers. The Monte Carlo chi-square test revealed significant differences in the distribution of the three types of answers in the two groups, $\chi^2 = 15.857$, $p = 0.000$ (see Table 5). Students in the experimental group were more likely to explicitly describe their computation procedures and demonstrated understanding of unique features of decimal computation that differ from whole number computation (see Table 5).

Students' Computation Errors

Given that the pretest and posttest instruments did not consist of the same number of testing items, a comparison of absolute numbers of computation errors on the pretest and posttest measures was not conducted. There were no significant differences among the computation errors between the experimental group and the control group during the pretest measures, with one exception (more errors occurred on displacement of the decimal point for the experimental group). In other words, prior to the treatment, students in the experimental group had similar or slightly worse

Table 6 Analysis of pre- and-posttest computation errors

Error type	Pretest		Chi-square	Posttest		Chi-square
	Control	MA	χ^2/Sig.	Control	MA	χ^2/Sig.
Aligning	15(12)	27(19)	3.619/.282	23(18)	11(6)	7.664/.006
Carrying	15(12)	15(13)	.252/1.000	28(20)	13(9)	2.427/.119
Displacement	14(12)	24(22)	5.775/.038	22(19)	12(10)	5.573/.018
Carelessness	4(4)	7(6)	1.151/.734	13(12)	8(6)	1.891/.169
Mnemonics	5(4)	9(8)	2.023/.523	20(12)	5(5)	7.649/.006
Computation order	5(5)	8(8)	.835/.361	4(4)	3(3)	.642/.423
Missing	86(37)	66(32)	7.060/.173	NA	NA	NA/ NA

Note: Numbers within the parentheses indicate the number of students who made the errors

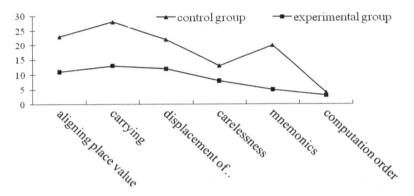

Fig. 1 Comparison of number of posttest computation errors in each group

computation skills than did those in the control group. Students in the experimental group did not have a better computation foundation before the treatment. For every single type of computation error made during the posttest, those in the control group made more errors than did those in the experimental group. The students in the control group had relatively more computation errors on aligning, carrying, displacement of the decimal point, and mnemonics. Computation errors made by the students in experimental group were primarily errors on aligning, carrying, and displacement of the decimal point. The experimental group had fewer students who made computation errors and as a group made fewer total computation errors (see Table 6).

The data in Figs. 1 and 2 present the total number of items with computation errors and the total number of students who made computation errors in the two groups. The trends in the two figures consistently indicate that students in the experimental group made fewer computation errors and had fewer students who made errors. For each type of computation error, a chi-square test was performed with a 2 (pretest, posttest) × 2 (control group, experimental group) contingency table on the number of errors the students made of that type. The results showed significant

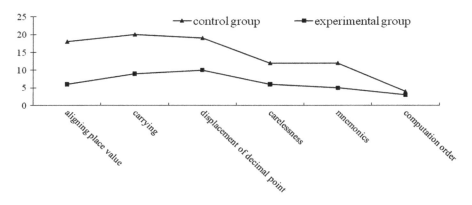

Fig. 2 Comparison of number of students in each group who made errors on posttest

differences in the distribution of aligning, displacement of the decimal point, and mnemonics, $\chi^2=7.664$, $p=0.006$; $\chi^2=5.573$, $p=0.018$; $\chi^2=7.649$, $p=0.006$ (see Table 6), for which students in the experimental group had significantly fewer errors. In terms of carrying, carelessness, and computation order, there were no significant differences between the two groups, $p=0.119$, $p=0.169$, $p=0.423$ (see Table 6).

Analysis of Relations Between Whole Number and Decimal Computation

Some computation rules for whole numbers are similar to those for decimals; however, other computation rules are different. The students' answers to the paired whole number and decimal test items were classified into four categories: (1) whole number computation is correct, and decimal computation is also correct; (2) whole number computation is correct, but decimal computation is incorrect; (3) whole number computation is incorrect, and decimal computation is also incorrect; and (4) whole number computation is incorrect, but decimal computation is correct. The second category of responses indicated that the students were unable to correctly apply rules of whole number computation to decimal computation. Thus, the second type of error tapped into our research interest regarding the relations between whole number computation and decimal number computation. If students demonstrated that their whole number computation was correct but decimal computation was incorrect, we assumed that the students did not achieve conceptual change. A chi-square test was conducted to compare differences between the two groups in this type of error pattern for addition, subtraction, multiplication, and division, respectively (see Table 7). It appeared that the two groups of students significantly differed in this type of error on multiplication. Students in the control group tended to make more

Table 7 Posttest mistakes in applying rules of whole number to decimal computation for each operation

Second category of response	Control	MA	χ^2/Sig.
Division	12	8	1.09/.297
Multiplication	17	6	7.54/.006
Addition	5	4	0.13/.723
Subtraction	4	8	1.58/.208

such errors on multiplication than did the experimental group. For division, addition, and subtraction, the two groups of students did not show significant differences in this type of error, $p = 0.297$, $p = 0.723$, $p = 0.208$, respectively (see Table 7).

Discussion

The purpose of this study was to evaluate and compare the effectiveness of a PBL instructional approach and a traditional instructional approach for teaching decimal multiplication and division to Chinese fifth-grade elementary students. We examined students' conceptual change in decimal computation both quantitatively and qualitatively. The results showed that the students in the experimental group had a higher accuracy rate on computation and were more likely to explain their computation procedures and principles of computation strategically.

PBL and Improvement in Computation Skills

The findings revealed a significant intervention effect for computation skills in the experimental group when compared to the control group. In other words, PBL outperformed a traditional instructional approach in enhancing students' computation skills involving both whole numbers and decimal numbers.

Effects on Enhancing Students' Self-Efficacy and Academic Interest

This study also examined the intervention effects on students' self-efficacy and academic interest. The PBL approach primarily improved the students' social self-efficacy, whereas it had little impact on their academic self-efficacy compared to the traditional approach. One interpretation might be that the experimental group had ample opportunity for teacher–student and student–student interactions. The mathematics class was no longer a competitive environment in which the students needed to compete to answer the questions. If there was a disagreement, the students had opportunities to share differences and express their ideas and suggestions, which may have resulted in a higher level of willingness to collaborate among these students

(Hmelo, Gotterer, & Bransford, 1997). This might explain the higher level of social self-efficacy in the experimental group. The limited impact on academic self-efficacy might be due to the intervention duration of only 1 month on one instructional unit. During an intervention with a relatively short duration, it might be difficult to change students' overall impressions and attitudes toward mathematics learning. In addition, the learning of decimals was a relatively challenging unit, and the students might have experienced some level of anxiety. Thus, it might be unrealistic to expect a rapid change in students' academic self-efficacy after a 1-month intervention.

In terms of academic interest, the students in the experimental group reported a higher level of academic interest than did those in the control group. The PBL approach emphasized student-centered instruction during the choice of problem situations, collaboration and discussion in class, and reflections on solutions. The teachers were facilitators of learning. The focus of the class was to maintain the students' interest and provide more opportunities for self-exploration. The instruction included interesting and challenging problem situations (examples provided in Appendix 5), and the students were able to freely express their opinions and experience a sense of accomplishment after they solved the problems. As a result, they reported a higher level of academic interest. Although similar problem situations provided by the textbooks might be available to students in the control group, no efforts were made to give students opportunities for self-exploration and self-reflection.

Effects on Enhancing Students' Metacognition

Vosniadou (1999) emphasized the importance of metacognition during children's mathematical problem-solving processes. In the present study, the measure of students' awareness of their strategy use was to examine students' metacognition. The results indicated that students in the experimental group were more likely to explicitly describe their own computation procedures and were more likely to discover essential features of the computation procedures. It appeared that the PBL approach not only guided the students to explore the rationales for computation, but also provided opportunities for the development of students' metacognition. The error clinic was designed for review classes. The teachers provided reminder cards to guide the students to externalize the metacognitive procedures, such as analyzing and exploring the rationales of computation errors, and self-revising computation procedures (Alan & Hennie, 1990; Tong & Zhang, 2004).

Effects on Conceptual Change in Decimals

When presented with the open-ended prompt regarding decimal division, 14 students in the experimental group were able to explicitly explain essential features of decimal division. Only one student in the control group provided an answer showing a conceptual understanding of decimal division. This could be due to the fact that students in the control group were given few opportunities in class to self-reflect on computational procedures and errors, and thus they might have a lower level of metacognition when asked to verbalize procedures and essential features of decimal computation. Chinese students are traditionally trained to execute computations or work out problems, but are not provided with ample opportunities to verbally elaborate their understanding of specific concepts or computations. The PBL approach appeared to help students develop their metacognition.

Moreover, in addition to being able to elaborate on decimal computation, it is also important to examine whether students can actually execute the computation procedure correctly. If a student can only explain how to do decimal computation, but fails in actual computation, then the student may not have achieved conceptual understanding. The posttest revealed that the experimental group had significantly fewer students making computation errors and that the group as a whole made significantly fewer computation errors than did the control group, which suggests that PBL had a positive impact on conceptual change.

The analysis of types of pretest computation errors revealed that students in both the experimental group and the control group frequently made mistakes of aligning place value, carrying, and displacement of the decimal point. During the posttest, the experimental group had fewer students making mistakes and as a group made fewer mistakes on the three types of computation errors. The three types of computation errors revealed essential differences between whole number computation and decimal computation. In whole number computation, the last place is the units place; thus, students automatically aligned place value based on the last digit (units place), instead of the same place value (e.g., units place to units place, tens place to tens place). When the students were conducting decimal computation, they also mechanically aligned the place value based on the last digit of the decimal numbers. The decimal point is unique for decimal numbers. Sometimes students arbitrarily placed the decimal point and randomly deleted "0" after the decimal point. The errors related to aligning place value indicated the need for students to understand the computation rules of both whole numbers and decimal numbers. Although strategies of borrowing and carrying in whole numbers and decimal numbers are not considerably different from each other, the introduction of decimal point concepts results in increased cognitive workload and increased use of working memory. The use of reminder cards in the experimental group helped the students to divide complex computations into smaller steps, which might have decreased their computation errors. For computation errors that were unique to decimal numbers, the intervention showed a positive impact. In contrast, for general computation errors that did not differ between whole numbers and decimal numbers, such as carelessness and computation order, the intervention did not show as much impact because the students could directly transfer computation knowledge and skills from whole numbers to decimal numbers.

In the posttest, the researchers designed some whole number computations and decimal computations with identical digits, with the only difference being the decimal point placed in decimal computation (e.g., 7.8×0.60 vs. 78×60). Some students correctly completed the whole number computation, but made errors on the corresponding decimal computations. Because of the introduction of the decimal point, some students made errors by directly transferring whole number computation rules to decimal computation. In the multiplication computation in particular, some students disregarded the differences between the computation rules for whole numbers and those for decimal numbers, such as deleting the redundant "0" after a decimal point or arbitrarily applying whole number computation rules to decimal computation (Markovits & Even, 1999). The control group made significantly more second category of responses (i.e., whole number computation is correct, but decimal computation is incorrect) on multiplication than did the experimental group. For division, addition, and subtraction, the two groups of students did not show significant differences with respect to this type of error. This suggests that the decimal point is a challenging concept and that decimal computation, particularly in decimal multiplication, is difficult to master. Prior knowledge of whole number computation might interfere with decimal computation, and so the students' computation errors varied.

In short, the analysis of computation errors and how students explicitly explained decimal division indicated that students tended to rely on prior knowledge and computation rules to work out decimal computation. PBL helped the students to deal explicitly with rationales of computation rules and to differentiate between whole number and decimal computations. To some degree, this approach promoted conceptual change regarding some erroneous conceptions of computation rules.

Limitations and Conclusions

This study has implications for educational practitioners and future researchers. However, there are a number of limitations of the study that suggest caution in generalizing the results. First, the students were not randomly assigned to two groups, although they shared many commonalities and showed similar performance on most measures during our pretests. Second, the number of problems in the decimal computation tests was relatively small, which might explain the relatively low internal consistencies for the pretest and posttest. There was not the same number and type of items in the pretest as in the posttest, although the difficulty level of items in the pretest and posttest was similar, according to the textbook we referred to. Third, some variables could not be controlled, such as students' prior beliefs about learning mathematics and about decimal and whole number computation, teachers' beliefs about mathematics learning and decimal computation, and teachers' knowledge about students' misconceptions. Fourth, ideally, learning behavior is better assessed by using a variety of methods (e.g., qualitative and quantitative methods) to provide a relatively comprehensive view of an individual's learning behavior. Because of limited resources, we were unable to videotape teacher–student interactions and were unable to provide a systematic qualitative analysis of changes in learning behaviors. Fifth, the traditional PBL approach is often utilized in small group settings. Due to the reality of the Chinese school system, it was impossible to have a very small student–teacher ratio to conduct the PBL. Thus, our study primarily relied on group discussions and activities that could take place simultaneously with all students. Although this was not an ideal way to implement the PBL, it provided insights for future Chinese teachers who might implement a similar approach in large classroom settings. Finally, the intervention was implemented intensively over a 1-month period, and there was no longitudinal follow-up to examine the persistence of the treatment effect.

This study revealed the importance of exposing students to their mathematics errors. Computation errors helped students to discover computation problems and provided opportunities for conceptual change. The PBL approach is driven by challenging, open-ended, ill-defined, and ill-structured PBL (as in the examples provided in Appendix 5). Our conclusions are in line with prior research findings where students were exposed to challenges and were guided to reflect on their misconceptions.

This study also has implications for teachers' roles during instruction. During PBL instruction, teachers serve only as facilitators. In contrast, during traditional instruction, teachers often emphasize the teaching of computation rules instead of the conceptual understanding of decimals. Discussion of computation errors is not encouraged in traditional instruction, which might result in students' resistance to disclosing their computation errors, and some students might hide exercise books to avoid sharing them with other students. When new learning content is introduced, it is important to allow students the opportunities to reflect on errors and causes, to enhance metacognition, and to promote the construction of new knowledge.

Appendix 1: Teaching Scripts for Teaching New Decimal Division

1. Introduction to the problems and divide students into five groups to solve the problems.
   "We have successfully overcome the decimal in multiplication. Now, it occurs in computation of division. We have new challenges now. Does anybody have any ideas to solve the problem? Now, let us divide the class into five groups and we will work on five division problems, including whole number divided by decimal, decimal smaller than 1, and decimal larger than 1. We want to see which group can come up with more solutions. When we explain the solutions, you need to tell us the procedures to reach the solution. What types of principles do you use to solve the problem?"

2. Encourage students to solve the problems with their own problem-solving methods. Based on previous experience, students are asked to create hypotheses, such as dealing with decimal division like division for whole numbers, ignoring the decimal point, and following rules of division for whole numbers first and then placing the decimal point.

3. Guided practice
   Students are asked to report their problem-solving methods to the class, and the teacher guides the students to reflect on these methods. The teachers guide the students to differentiate the differences between these questions. They encourage the students to reflect on the principles of multiplication and division that they have learned in their previous classes. The teachers ask the students to use multiplication to verify the results for division. The students are asked to raise questions and summarize the principles for decimal division.

4. Students modify their results and report their results to the class.
5. Transfer and application.
   Provide additional problems for students to practice, including two-digit decimals.

Appendix 2: Teaching Scripts for Reviewing Previous Contents (Decimal Division Error Clinic)

1. Introduction to the problems.
   The students form five groups to examine errors in the worksheet. Say: "Thank you for joining error clinic. Today, we will focus on examining errors in decimal division. It is hoped that we will all be able to solve different challenges in decimal division." Then, the teacher distributes the worksheets.
2. Encourage students to solve the problems and explain the errors that they found, such as errors due to misunderstanding of principles or careless errors.
3. Guided practice.
   The teacher guides students to draw conclusions about their problem-solving methods, help students make their reminder cards, and encourage students to reflect on their problem-solving experiences. Ask the students to summarize the causes of mistakes, such as assuming decimal division is similar to division of whole numbers.
4. Students modify their results, establish their own reminder cards, and report their results to the class.
5. Transfer and application.
   Ask the students to give an example of mistakes they made in decimal computation during the previous week.

Appendix 3: PBL Procedure

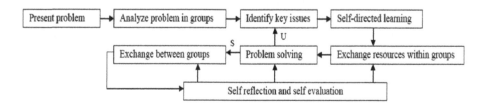

A flowchart of the PBL procedure. S=successful; U=unsuccessful.

Appendix 4: Treatment Fidelity Checklists

New lectures	Yes/No
1. Provide problem-based learning situations at the beginning of classes	
2. Small group discussion	
3. Students form hypotheses and try to work out solutions	
4. Students report discussion results group by group	
5. Teacher provides feedback and analyzes differences and relations between different problems	
6. Students break into small groups for further discussion	
7. Students report further discussion results	
8. Application and transfer	

Review classes	Yes/No
1. Provide problem-based learning situations and welcome students to error clinic	
2. Small group discussion to find own mistakes	
3. Teacher provides guidance	
4. Teacher provides reminder cards, uses scaffolding, and coaches students to be able to fill out reminder cards	
5. Students report their discussion report and report what is on reminder card	
6. Application and transfer, students create problems for others	

Appendix 5: Sample Problems from the Curriculum

Samples for ill-defined and ill-structured problems

Directions: Please check the following computation procedures and see whether they are right. Please correct computations that were executed incorrectly.

7.下面各题计算的对吗? 把不对的改正过来。

$$50.4 \times 1.9 - 1.8 \qquad 3.76 \times 0.25 + 25.8$$
$$= 50.4 \times 0.1 \qquad = 0.094 + 25.8$$
$$= 5.04 \qquad = 25.894$$

Samples for ill-defined and ill-structured problems

Directions: Are the following computation procedures right? If no, please check them.

下面的计算对吗? 如果不对,错在哪里?

$$24 \div 15 = 16 \qquad 1.26 \div 18 = 0.7$$

```
        1 6                        0.7
  1 5 ) 2 4               1 8 ) 1.2 6
        1 5                        1 2 6
        9 0                              0
        9 0
        ─────
          0
```

Samples for challenging problem

Directions: (1) In a parking lot, the parking rate is 2.50 Yuan/h if it is within 1 h. (2) After the first hour, the parking rate is 2.50 Yuan/0.5 h. Uncle Li paid 12.5 Yuan, then how many hours did he park in the parking lot?

(Sources: Lu & Yang, 2005)

Acknowledgement This project was supported by Beijing Educational Research Eleventh-Five Project (ABA07014) [北京市教育科学"十一五"规划重点课题] funding to Ru-De Liu. The project was partially supported by an internal grant from Fordham University to Yi Ding. Thanks to the parents, teachers, and students in participating schools. Thanks to Agnes DeRaad for editorial support.

References

Alan, J. C., & Hennie, P. (1990). Training in metacognition: An application to industry. In K. J. Gilhooly, M. T. G. Keane, R. H. Logie, & G. Erdos (Eds.), *Lines of thinking* (Vol. 2, pp. 313–328). New York: Wiley.

Bandura, A. (1986). *Social foundations of thought and action: A social cognitive theory.* Englewood Cliffs, NJ: Prentice-Hall.

Baturo, A. R. (1998). Year 5 students' available and accessible knowledge of decimal-number numeration. In C. Kanes, M. Goos, & E. Warren (Eds.), Teaching mathematics in new times. *Proceedings of the 21st annual conference of the Mathematics Education Research Group of Australasia* (pp. 90–97). Brisbane, Queensland, Australia: MERGA.

Collins, J. L. (1982, March). *Self-efficacy and ability in achievement behavior.* Paper presented at the meeting of the American Educational Research Association, New York.

Collins, A., Brown, J. S., & Newman, S. E. (1989). Cognitive apprenticeship: Teaching the crafts of reading, writing, and mathematics. In L. B. Resnick (Ed.), *Knowing, learning, and instruction: Essays in honor of Robert Glaser* (pp. 453–494). Hillsdale, NJ: Erlbaum.

Davis, E. A., & Linn, M. C. (2000). Scaffolding students' knowledge integration: Prompts for reflection in KIE. *International Journal of Science Education, 22,* 819–837.

Dunlap, J. C. (2005). Problem-based learning and self-efficacy: How a Capstone course prepares students for a profession. *Educational Technology Research & Development, 53*(1), 65–85.

Fischbein, E., Deri, M., Nello, M. S., & Marino, M. S. (1985). The role of implicit models in solving verbal problems in multiplication and division. *The Journal for Research in Mathematics Education, 16*(1), 3–17.

Golan, R., Kyza, E. A., Reiser, B. J., & Edelson, D. C. (2002, April). *Scaffolding the task of analyzing animal behavior with the Animal Landlord software.* Paper presented at the annual meeting of the American Educational Research Association, New Orleans, LA.

Guzdial, M. (1994). Software-realized scaffolding to facilitate programming for science learning. *Interactive Learning Environments, 4*, 1–44.

Hannula, M., Maijala, H., Pehkonen, E., & Soro, R. (2001, June). *Taking a step to infinity: Students' confidence with infinity tasks in school mathematics.* Paper presented at the Third European Symposium on Conceptual Change, Turku, Finland.

Hedges, L. V., & Olkin, I. (1985). *Statistical methods for meta-analysis.* Orlando, FL: Academic Press.

Hiebert, J. (1988). A theory of developing competence with written mathematical symbols. *Educational Studies in Mathematics, 19*, 333–355.

Hiebert, J., & Wearne, D. C. (1983, April). *Students' conceptions of decimal numbers.* Paper presented at the annual meeting of the American Educational Research Association, Montreal, Québec, Canada.

Hmelo, C. E., Gotterer, G. S., & Bransford, J. D. (1997). A theory-driven approach to assessing the cognitive effects of PBL. *Instructional Science, 25*(6), 387–408.

Hmelo, C. E., & Guzdial, M. (1996). Of black and glass boxes: Scaffolding for learning and doing. In D. C. Edelson & E. A. Domeshek (Eds.), *Proceedings of ICLS 96* (pp. 128–134). Charlottesville, VA: AACE.

Hmelo-Silver, C. E. (2006). Design principles for scaffolding technology based inquiry. In A. M. O'Donnell, C. E. Hmelo-Silver, & G. Erkens (Eds.), *Collaborative reasoning, learning and technology* (pp. 147–170). Mahwah, NJ: Erlbaum.

Hmelo-Silver, C. E., Duncan, R. G., & Chinn, C. A. (2007). Scaffolding and achievement in problem-based and inquiry learning: A response to Kirschner, Sweller, and Clark (2006). *Educational Psychologist, 42*(2), 99–107.

Hoffman, B., & Schraw, G. (2009). The influence of self-efficacy and working memory capacity on problem-solving efficiency. *Learning and Individual Differences, 19*, 91–100.

Huang, T.-H., Liu, Y.-C., & Shiu, C.-Y. (2008). Construction of an online learning system for decimal numbers through the use of cognitive conflict strategy. *Computers & Education, 50*, 61–76.

Jackson, S., Stratford, S. J., Krajcik, J. S., & Soloway, E. (1996). Making system dynamics modeling accessible to pre-college science students. *Interactive Learning Environments, 4*, 233–257.

Liu, M., Hsieh, P., Cho, Y., & Schallert, D. L. (2006). Middle school students' self-efficacy, attitudes, and achievement in a computer-enhanced problem-based learning environment. *Journal of Interactive Learning Research, 17*(3), 225–242.

Lu, J., & Yang, G. (2005). *Beijing compulsory education curriculum reform experimental materials of mathematics in elementary school* (Vol. IX). Beijing, P. R. China: Beijing Education Press.

Malara, N. (2001). From fractions to rational numbers in their structure: Outlines for an innovative didactical strategy and the question of density. In J. Novotná (Ed.), *Proceedings of the 2nd Conference of the European Society for Research Mathematics Education II* (pp. 35–46). Praga, Czech Republic: Univerzita Karlova v Praze, Pedagogická Faculta.

Markovits, Z., & Even, R. (1999). The decimal point situation: A close look at the use of mathematics-classroom-situations in teacher education. *Teaching and Teacher Education, 15*, 653–665.

Mayer, R. E. (2004). Should there be a three-strikes rule against pure discovery learning? *American Psychologist, 59*, 14–19.

Merenluoto, K., & Lehtinen, E. (2002). Conceptual change in mathematics: Understanding the real numbers. In M. Limon & L. Mason (Eds.), *Reconsidering conceptual change: Issues in theory and practice* (pp. 233–258). Dordrecht, The Netherlands: Kluwer.

Moskal, B. M., & Magone, M. E. (2000). Making sense of what students know: Examining the referents, relationships and modes students displayed in response to a decimal task. *Educational Studies in Mathematics, 43*, 313–335.

Moss, J. (2005). Pipes, tubes, and beakers: New approaches to teaching the rational-number system. In M. S. Donovan & J. D. Bransford (Eds.), *How students learn: Mathematics in the classroom* (pp. 121–162). Washington, DC: National Academic Press.

National Council of Teachers of Mathematics (NCTM). (2000). *Principles and standards for school mathematics*. Reston, VA: Author.

National Mathematics Advisory Panel. (2008). *Foundations for success: The final report of the national mathematics advisory panel*. Washington, DC: U.S. Department of Education.

Ni, Y., & Zhou, Y. (2005). Teaching and learning fraction and rational numbers: The origins and implications of whole number bias. *Educational Psychologist, 40*, 27–52.

Nunes, T. & Bryant, P. (2007). Paper 3: Understanding rational numbers and intensive quantities. In Key understandings in mathematics learning: A review commissioned by the Nuffield Foundation. Retrieved from http://www.nuffieldfoundation.org/sites/default/files/P3_amended_FB2.pdf.

Ordonez-Feliciano, J. (2010). Self-efficacy and instruction in mathematics. *Dissertation Abstracts International Section A: Humanities and Social Sciences, 70*(12-A), 4615.

Palincsar, A. S. (1998). Social constructivist perspectives on teaching and learning. *Annual Review of Psychology, 45*, 345–375.

Pierce, R. U., Steinle, V. A., Stacey, K. C., & Widjaja, W. (2008). Understanding decimal numbers: A foundation for correct calculations. *International Journal of Nursing Education Scholarship, 5*(1), 1–15.

Putt, I. L. (1995). Pre-service teachers' ordering of decimal numbers: When more is smaller and less is larger! *Focus on Learning Problems in Mathematics, 17*(3), 1–15.

Qin, X. (2003). *The influence of self-efficacy, goal orientation and curriculum perception on the help-seeking behaviors of mathematics in primary students*. Unpublished doctoral dissertation, Beijing Normal University, Beijing, China.

Quintana, C., Reiser, B. J., Davis, E. A., Krajcik, J., Fretz, E., Duncan, R. G., et al. (2004). A scaffolding design framework for software to support science inquiry. *Journal of the Learning Sciences, 13*, 337–386.

Rao, S., & Kane, M. T. (2009). Teaching students with cognitive impairment a chained mathematical task of decimal subtraction using simultaneous prompting. *Education and Training in Developmental Disabilities, 44*(2), 244–256.

Reiser, B. J. (2004). Scaffolding complex learning: The mechanisms of structuring and problematizing student work. *Journal of the Learning Sciences, 13*, 273–304.

Reiser, B. J., Tabak, I., Sandoval, W. A., Smith, B. K., Steinmuller, F., & Leone, A. J. (2001). BGuILE: Strategic and conceptual scaffolds for scientific inquiry in biology classrooms. In S. M. Carver & D. Klahr (Eds.), *Cognition and instruction: Twenty-five years of progress* (pp. 263–305). Mahwah, NJ: Erlbaum.

Resnick, L. B., Bill, V. L., Lesgold, S. B., & Leer, M. N. (1991). Thinking in arithmetic class. In B. Means, C. Chelemer, & M. S. Knapp (Eds.), *Teaching advanced skills to at-risk students: Views from research and practice* (pp. 27–67). San Francisco: Jossey-Bass.

Resnick, L. B., Bill, V., & Lesgold, S. (1992). Developing thinking abilities in arithmetic class. In A. Demetriou, M. Shayer, & A. Efklides (Eds.), *Neo-Piagetian theories of cognitive development: Implications and applications for education* (pp. 210–230). London: Routledge.

Resnick, L. B., Nesher, P., Leonard, F., Magone, M., Omanson, S., & Peled, I. (1989). Conceptual bases of arithmetic errors: The case of decimal fractions. *Journal of Research in Mathematics Education, 20*, 8–27.

Roche, A. (2005). Longer is larger? Or is it? *Australian Primary Mathematics Classroom, 10*(3), 11–16.

Sackur-Grisvard, C., & Leonard, F. (1985). Intermediate cognitive organizations in the process of learning a mathematical concept: The order of positive decimal numbers. *Cognition and Instruction, 2*(2), 157–174.

Siegler, R. S. (2002). Microgenetic studies of self-explanation. In N. Garnott & J. Parziale (Eds.), *Microdevelopment: A process-oriented perspective for studying development and learning* (pp. 31–58). Cambridge, MA: Cambridge University Press.

Silver, E. A. (1986). Using conceptual and procedural knowledge: A focus on relationships. In J. Hiebert (Ed.), *Conceptual and procedural knowledge: The case of mathematics* (pp. 181–197). Hillsdale, NJ: Lawrence Erlbaum.

Stacey, K., Helme, S., Steinle, V., Baturo, A., Irwin, K., & Bana, J. (2001). Preservice teachers' knowledge of difficulties in decimal numeration. *Journal of Mathematics Teacher Education, 4*(3), 205–225.

Stacey, K., & Steinle, V. (1998). Refining the classification of students' interpretations of decimal notation. *Hiroshima Journal of Mathematics Education, 6*, 49–59.

Stafylidou, S., & Vosniadou, S. (2004). The development of student's understanding of the numerical value of fractions. *Learning and Instruction, 14*, 503–518.

Steffe, L. P. (1994). Children's multiplying schemes. In G. Harel & J. Confrey (Eds.), *The development of multiplicative reasoning in the learning of mathematics*. Albany, NY: State University of New York Press.

Steinle, V., & Stacey, K. (2004). Persistence of decimal misconceptions and readiness to move to expertise. In M. J. Hoines & A. B. Fuglestad (Eds.), *Proceedings of the 28th Conference of the International Group for the Psychology of Mathematics Education* (Vol. 4, pp. 225–232). Bergen, Norway: PME.

Swan, M. (1983). *Teaching decimal place value: A comparative study of conflict and positive only approaches*. Nottingham, England: Shell Centre for Mathematical Education.

Swan, M. B. (1993). Becoming numerate: Developing conceptual structures. In S. Willis (Ed.), *Being numerate: What counts?* (pp. 44–71). Hawthorne, Victoria: Australian Council for Educational Research.

Tong, S., & Zhang, Q. (2004). The effect of metacognition training on the improvement of secondary students' problem solving ability. *Psychological Development and Education, 2*, 62–68.

Toth, E. E., Suthers, D. D., & Lesgold, A. M. (2002). "Mapping to know": The effects of representational guidance and reflective assessment on scientific inquiry. *Science Education, 86*, 244–263.

Vamvakoussi, X., & Vosniadou, S. (2004). Understanding the structure of the set of rational numbers: A conceptual change approach. *Learning and Instruction, 14*, 453–467.

Vamvakoussi, X., & Vosniadou, S. (2010). How many decimals are there between two fractions? Aspects of secondary school students' understanding of rational numbers and their notation. *Cognition and Instruction, 28*(2), 181–209.

VanLehn, K. (1999). Rule learning events in the acquisition of a complex skill: An evaluation of Cascade. *Journal of the Learning Sciences, 8*(2), 179–221.

Vosniadou, S. (1999). Conceptual change research: State of the art and future directions. In W. Schnotz, S. Vosniadou, & M. Carretero (Eds.), *New perspectives on conceptual change* (pp. 1–14). Amsterdam: Pergamon.

Vosniadou, S. (2007). The conceptual change approach and its reframing. In S. Vosniadou, A. Baltas, & X. Vamvakoussi (Eds.), *Reframing the conceptual change approach in learning and instruction* (pp. 1–15). Oxford, England: Elsevier.

Vosniadou, S., & Verschaffel, L. (2004). Extending the conceptual change approach to mathematics learning and teaching. *Learning and Instruction, 14*, 445–451.

Wearne, D. (1990). Acquiring meaning for decimal fraction symbols: A one year follow-up. *Educational Studies in Mathematics, 21*, 545–564.

Wearne, D., & Hiebert, J. (1988). Constructing and using meaning for mathematical symbols: The case of decimal fractions. In J. Hiebert & M. Behr (Eds.), *Research agenda for mathematics education: Number concepts and operations in the middle grades* (pp. 220–235). Reston, VA: Erlbaum/National Council of Teachers of Mathematics.

Wearne, D., & Hiebert, J. (1989). Cognitive changes during conceptually based instruction on decimal fractions. *Journal of Educational Psychology, 81*(4), 507–513.

Woodward, J., Baxter, J., & Robinson, R. (1999). Rules and reasons: Decimal instruction for academically low achieving students. *Learning Disabilities Research & Practice, 14*(1), 15–21.

Zhang, A. (2005). *The use of concept map in fostering and evaluating concept change of science in primary students*. Unpublished doctoral dissertation, Beijing Normal University, Beijing, China.

Zong, M. (2006). *An investigation of concept change in elementary mathematics: A problem learning perspective*. Unpublished master's thesis, Beijing Normal University, Beijing, China.

A Longitudinal Study of the Development of Rational Number Concepts and Strategies in the Middle Grades

James A. Middleton, Brandon Helding, Colleen Megowan-Romanowicz, Yanyun Yang, Bahadir Yanik, Ahyoung Kim, and Cumali Oksuz

Introduction

Research in the area of rational number knowledge and proportional reasoning has produced many important findings on how students think about and operate with rational numbers (Behr, Harel, Post, & Lesh, 1992; Behr, Lesh, Post, & Silver, 1983; Empson, Junk, Dominguez, & Turner, 2006; Kieren, 1976). The complex nature of this research has yet to discover a clear picture or model of how rational number knowledge develops over time. Some conjectures have been made concerning rational number development from cross-sectional studies, but without longitudinal evidence such trajectories are difficult to confirm. Defining a framework for interpreting students' understanding along a

This material is based upon work supported by the National Science Foundation under Grant No. 0337795. Any opinions, findings, and conclusions or recommendations expressed in this material are those of the authors and do not necessarily reflect the views of the National Science Foundation.

J.A. Middleton (✉)
Arizona State University, Tempe, AZ, USA
e-mail: jimbo@asu.edu

B. Helding
Boulder Language Technologies, Boulder, CO, USA

C. Megowan-Romanowicz
Arizona State University, Tucson, AZ, USA

Y. Yang
Florida State University, Tallahassee, FL, USA

B. Yanik
Anadolu University, Eskişehir, Turkey

A. Kim
Ewha Women's University, Seoul, South Korea

C. Oksuz
Adnan Menderes University, Aydın, Turkey

© Springer International Publishing Switzerland 2015   265
J.A. Middleton et al. (eds.), *Large-Scale Studies in Mathematics Education*,
Research in Mathematics Education, DOI 10.1007/978-3-319-07716-1_12

developmental path, if one exists, is a desired goal. Without such a framework, the research base will remain fragmented and primarily focused on further examination of understandings of a particular subconstruct's origin and phenomenology instead of the essential transitions among contexts and conceptions that should mark a more mature rational number understanding among constructs (Streefland, 1993).

Investigations into the way that young children are introduced to whole number operations have revealed certain barriers to rational number learning due to the inconsistencies between the mathematics of whole numbers and the mathematics of fractions (Bransford, Brown, & Cocking, 1999; Mack, 1993; Middleton, van den Heuvel-Panhuizen, & Shew, 1998). For example, the rules of thumb *multiplying makes larger* and *dividing makes smaller* when working with whole numbers become problematic when students must consider cases involving multiplication or division by proper fractions (Kieren, 1993). Recent research has led to a belief that the common part–whole introduction of fractions is not as effective in removing early-knowledge barriers to the mathematics of fractions as an approach emphasizing the ideas of partitioning and unit which are more closely related to thinking about fractions as quotients (Empson, 1999; Lamon, 2006; Mack, 1993; Streefland, 1993). But how might this partitioning approach assist students in gaining conceptual knowledge in other subconstructs such as measurement? Understanding the transitional paths from one subconstruct to another across the field of rational number concepts is paramount to building a meaningful model of rational number learning. It is clear, moreover, that this sought-after developmental trajectory is complex and will not follow a simple one-dimensional path, moving in an orderly, linear fashion from one subconstruct to another. Rather, it depends upon content and representations emphasized in instruction as well as contextual referents that give rise to initial conceptions of multiplicative quantities (Lamberg & Middleton, 2002, 2009). In fact, middle-school children who traverse this complex path will no doubt face barriers and perhaps take detours that divert, prolong, or even stall their progress toward rational number understanding in the course of compulsory instruction.

Longitudinal Analysis

Rational number understanding has been termed a "watershed concept" (Kieren, 1976). Fractions, ratios, and proportional reasoning are key underpinnings of algebra, calculus, statistics, and other higher mathematics that are becoming more and more critical for the development of workplace skills (Oksuz & Middleton, 2005). Cross-sectional studies of students at different ages are the norms for the field in examining students' reasoning and development (see for example, the work of Empson et al., 2006). This body of work has aided in the development of new curricular tasks and sequences aimed at providing a more theoretically defensible and psychologically connected approach to the teaching and learning of rational number (Carpenter, Fennema, & Romberg, 2012; Lamberg & Middleton, 2009; Lesh, Post, & Behr, 1988; Streefland, 1993; Toluk & Middleton, 2004).

However, due to their cross-sectional design, this body of work does not provide a coherent developmental picture of rational number knowledge as students move across several grade levels (Carraher, 1996). More recent studies, however, give us a glimpse of how this knowledge might develop, beginning with the ideas of unit and equivalence, then gradually developing the five interconnected interpretations or "subconstructs" that predominate the language of the field: Part–whole, measure, operator, quotient, and ratio (Lamon, 2006).[1] The work reported here, supported through a grant from the National Science Foundation, has allowed us to trace these changes in understanding related to learning rational number concepts as they developed over the middle-school years where this content is most heavily stressed. The results of the study are intended to contribute theoretically to the understanding of numbers and operations and pragmatically to the further design of curriculum materials and pedagogical strategies that will positively impact students' ability to think, represent, and communicate their understanding of rational number concepts and procedures over time.

The importance of knowing how rational number knowledge and proportional reasoning develop through the middle grade levels is prompted in part by the fact that such knowledge forms the foundation for the study of higher mathematics. This need is further evidenced by the fact that students in the United States have demonstrated weaknesses in these topic areas in comparative studies with other international student populations such as the Trends in International Mathematics and Science Study (Kelly, Mullis, & Martin, 2000; Mullis, Martin, Gonzalez, & Chrostowski, 2004). Some investigators have shown that even postsecondary students have difficulty representing fraction magnitudes (Bonato, Fabbri, Umiltà, & Zorzi, 2007).

Besides these reasons that pertain specifically to academic progress and global competitiveness, fundamental understanding of rational number is necessary for a well-informed citizenry which includes but is not limited to interpreting graphs and other data displays, projecting trends and forecasts, comparing quantities multiplicatively, and basic consumer and home skills.

Issues in Mapping Students' Growing Knowledge

In this study, we traced individual students' development of each of the rational number subconstructs through a constructivist lens. On the individual level, we utilized individual interviews, following a target sample of students from the sixth grade through the eighth grade to assess their growth individually. Yet we also recognize that the development of rational number knowledge in a classroom is distributed across members of student groups or the class, coordinated between internal and external structures, and across time where results of earlier tasks and events transform

---

[1] It must be noted that these five subconstructs are not the only way to parse student reasoning or mathematical manifestations of these concepts. Confrey, Maloney, Nguyen, Mojica, and Myers (2009), for example, provide a rich alternative framework.

the nature of later events (Hollan, Hutchins, & Kirsh, 2000; Roth & McGinn, 1998). We therefore observed students' mathematics classes twice per week, coordinating our understanding of their individual growth with their classroom experiences.

The inscriptions or representational tools recorded and analyzed in student interviews and in class observations provided a way to describe the propagation of rational number knowledge across classroom participants and within a single student's mind over time (e.g., Lamberg & Middleton, 2002). Examining student inscriptions was essential in our study due as they documented the form of knowledge at the moment of instruction and developmental sequence in which the knowledge arose. Inscriptions also served as the object of collective negotiations of meaning between the student and class, student and teacher, and student and researcher, and were appropriated (transported from one person to another) allowing us to trace the diffusion of knowledge across the 3 years of the study, when they appeared spontaneously in interview sessions.

In summary, this study is aimed at understanding the intellectual resources individual children bring to bear in developing rational number understanding and the classroom norms and practices that constrain and enable individual development *longitudinally*. Specifically, the scope of work is intended to advance the field of rational number learning by:

1. Uncovering patterns and mechanisms of individual development in students' understanding of rational numbers and proportional reasoning
2. Integrating the current piecemeal body of research on rational number into a coherent developmental model by examining how understanding of rational number subconstructs evolve concurrently and interactively.
3. Developing insight into the ways in which classroom instruction, especially the use of and talk around inscriptions impact students' ability to think about, represent, and communicate their understanding of rational number concepts and operations as it develops over time.
4. Generating transportable models of rational number development that can be factored into teacher pre- and in-service staff development to promote quality instructional practices in the future.

Method

Setting and Participants

This study analyzes data collected over a 20-month period in a longitudinal study conducted in an urban K-8 school located in the southwestern United States. The approximately 850 students enrolled in the school were predominately from a Hispanic lower-middle-class background. Over 90 % of the students received free or reduced lunch. Sixth-, seventh- and eighth-grade students participated in the study. Their classrooms were equipped with whiteboards on two walls, lined with low bookshelves and were furnished with round and rectangular tables at which students were typically seated in groups of four to six students. Students attended mathematics class daily. Each class lasted for 70 min except Wednesdays, when classes were shortened to 55 min to accommodate after-school teachers' meetings. The District-selected mathematics curriculum consisted of the NSF-sponsored, *Mathematics in Context* (2003) series supplemented with Arizona Instrument to Measure Standards (AIMS) test preparation materials, which the teachers used on an alternating basis. Some teachers favored drill and practice more than others, and these sessions lasted from 10 to 45 min in a typical 70-min class period. A significant number of the students in all three classes were English Language Learners (ELL). As a school norm, teachers tried to seat the ELL students with classmates whose English was sufficient to assist them as needed. Participating teachers often used overhead projectors during instruction.

Although the exact enrollment in each class varied over the 3 years of the study, the average ratio of teacher to students in the sixth-, seventh-, and eighth-grade classes was 1–30. The sixth-grade class was self-contained, where a single teacher conducted instruction in all subjects. The seventh- and eighth-grade classes followed a middle-school format where students traveled to different classrooms for subject instruction. Additionally, some seventh- and eighth-grade students were given the opportunity to attend a resource class for extended mathematics instruction. In this special resource class (held twice per week), students worked in small groups on challenging problems outside of the regular mathematics curriculum. As a part of the classroom norms in the resource class, students were expected to work together and present group solutions to the whole class.

Data Collection Procedures: Interviews and Classroom Observations

Interviews

To make comparisons across students possible, we designed parallel interview protocols to assess rational number knowledge across all five subconstructs (Behr et al., 1992; Lamon, 2006). These protocols were administered to all students enrolled in the study in the first and last two interview cycles of the school year, regardless of the grade level. The tasks in the first pair of these parallel protocols were the same in terms of context and level of difficulty, and they covered the subconstructs of operator, quotient, and part–whole. The other parallel pair involved the subconstructs of measurement and ratio. Both pairs of parallel protocols were administered in the fall and spring semesters to assess individual growth over time, which included both ability to correctly solve problems, and also, changes in preferred strategies for solving problems.

In addition to these parallel protocols administered to all interviewees, we captured the impact of curricular tasks and instruction using class-specific individual interview protocols with prompts adapted from tasks in the district-adopted *Mathematics in Context* (2003) curriculum. Like the parallel protocols described above, these additional tasks focused on one or more of the five subconstructs of rational number, but utilized the inscriptions and language that we observed being developed in students' classes.

Interview Procedures and Coding

All interviews (common, parallel protocols, and grade-specific protocols) were videotaped. Special attention was given to recording the students' written inscriptions and their verbal "think aloud" responses. Interviewers attempted to capture students' intuitive, procedural, and conceptual knowledge of rational numbers and track their change over time. Interviewers were trained to listen closely and carefully prompt students for additional thinking without commenting on the appropriateness of any of their solution strategies. Students who spoke little English were interviewed by interviewers fluent in both English and Spanish.

Each protocol was coded across five dimensions: (1) Problem subconstruct (the anticipated conception of rational number we hypothesized the problem would elicit); (2) Students' solution strategies (Convert to common fractions; Use of equivalent ratios; Measurement division; Multiply by a scale factor (operator); Part/Whole; Proportional Reasoning; Relating to a similar problem; or No Strategy observed/Strategy not code-abled); (3) Whether the strategy utilized was developed ad hoc, or if it had been previously observed in the student's class; (4) Whether the problem was solved correctly; and (5) Whether the problem strategy led to a sensible answer mathematically even if the answer was technically incorrect.

Analyses traced the proportion of strategies utilized across each of the interviews as students moved from early sixth grade, through the seventh grade, and finally, as they prepared to finish the eighth grade, comparing differences in strategy use for each of the four other variables.

Table 1 displays the number of students in cohorts who were individually interviewed by grade and by year. Arrows represent student groups followed up through successive grade levels. During the first year of the study, 53 sixth graders and 11 seventh graders participated. Eleven new sixth graders, four new seventh graders, and seven eighth graders entered the interview process in the second year, while 38 of the previous sixth graders and 8 of the previous seventh graders continued into the seventh grade and eighth grade, respectively.

Table 1 Participants involved in individual interviews

	6th grade	7th grade	8th grade	Total
1st year	53	11	--	64
2nd year	11	38+4	8+7	68
3rd year	12	9	32+4+4	61
Total	76	62	55	102

In the third year, 12 new sixth graders and 4 eighth graders entered the interview process, while 9 of the previous seventh graders and 38 of the previous eighth graders remained in the study. Among these 38 students, 32 were retained from the sixth grade across the 3 years. As a result, a total of 102 students took part in individual interviews during the 3-year study, and the 32 students who were followed over 3 years became our focus group in this paper.

Classroom Observations

In addition to individual interviews, the mathematics classes of students participating in the study were videotaped twice weekly. These 70-min observations were conducted to provide a contextual reference within which we embedded individual interviews and analyses. Interviewers were able to see their student interviewees engaging in mathematical activities within a social setting, to see what inscriptions occurred in the classroom, which were favored, and to look for clues to the origins of the problem solving and reasoning strategies students used in interview settings. While targeted students interacted in groups or whole class situations, our cameras recorded their development of mathematical notations and representations within the sociolinguistic structure of the classroom.

Assessment of Students' Rational Number Performance

There were two major purposes for collecting performance data: (1) to compare performance of our sample to a national/international sample; and (2) to describe student growth over time quantitatively. Quantitative assessment data were gathered at four time points: at the end of the fall semester in year 1, the beginning and end of year 2, and the beginning and end of year 3. Questions were drawn from released items from national/international mathematics assessments, the Trends in International Mathematics and Science Study (Martin & Kelly, 1998; International Association for the Evaluation of Educational Achievement, 2001; International Association for the Evaluation of Educational Achievement, 2005) and the National Assessment Educational Progress (NAEP). Utilizing questions from TIMSS and NAEP tests also allowed for comparisons of these students with students of similar age throughout the country and around the world.

To determine the rational number constructs the test items represented, the original form was piloted using a separate sample to ensure appropriate content and discrimination across the three grades. Three items were excluded due to the students' extremely low percentage of correct responses. As a result, the assessment consisted of 27 items, assessing 11 categories of rational number including: Ordering fractions, part–whole, ratio, relationships between fraction and decimal, proportion, linear measurement, rates, percent, equivalent fraction, operator, and decimal notation. Among the 27 items, 4 were free-response (item 1, 9, 18, and 21), while the remaining items were multiple-choice (see Table 2).

Table 2 Test item information

Item #	Source	Type	Item content
1	TIMSS 95	Ordering fractions	Write a fraction that is larger than 2/7
2	TIMSS 99	Part–whole	Which shows 2/3 of the square shaded?
3	NAEP 98	Part–whole	What fraction of the rectangle ABCD is shaded?
4	TIMSS 99	Ordering fractions	Given two common fractions. Which of these fractions is smallest?
5	NAEP 03	Ratio	Given two ratios. Which of the following ratios is equivalent to the ratio of 6:4?
6	NAEP 92	Relation between fraction and decimal	Given a common fraction, which is closest in value to 0.52?
7	TIMSS 99	Part–whole	Given a picture. What fraction of the circle is shaded?
8	TIMSS 99	Part–whole	Given a picture. Robin and Jim took X cherries from a basket. What fraction of the cherries remained in the basket?
9	TIMSS 99	Proportion	John and Mark sold X magazines. Knowing the total amount of money, how much money did Mark receive?
10	TIMSS 99	Part–whole	Penny had a bag of marbles. How many marbles were in the bag to start with?
11	TIMSS 95	Ratio	Given a picture with numbers, what is the ratio of red paint to the total amount of paint?
12	NAEP 03	Linear measure	Given a picture, the distance from Bay City to Exton is 60 miles, what is the distance from Bay City to Yardville?
13	TIMSS 99	Rates	A runner ran 3,000 m in exactly 8 min. What was his average speed in meters per second?
14	TIMSS 95	Percent	From 60 cents to 75 cents, what is the percent increase in the price?
15	TIMSS 03	Ordering fractions	Given two common fractions. In which of these pairs of # is 2.25 larger than the first number but smaller than the second number?
16	TIMSS 99	Proportion	If there are 300 calories in 100 g, how many calories are there in a 30 g portion of this food?
17	TIMSS 95	Ratio	3/5 of the students are girls. Add 5 girls and 5 boys, which statement is true of the class?
18	NAEP 03	Linear measure	Given a picture, a dot shows where 1/2 is. Use another dot to show where 3/4 is
19	TIMSS 99	Equivalent fraction	In which list of fractions is all of the fractions equivalent?
20	NAEP 03	Linear measure	3/4 of a yard of string is divided into pieces; they are 1/8 yard long each. How many pieces?

(continued)

Table 2 (continued)

Item #	Source	Type	Item content
21	TIMSS 95	Operator	Luis runs 5 km each day, the course is 1/4 km long. How many times through the course does he run each day?
22	TIMSS 99	Decimal	Which of these is the smallest number?
23	TIMSS 95	Ordering fractions	Which list shows the numbers from smallest to largest?
24	TIMSS 95	Ratio	The ratio of girls to boys is 4:3. How many girls are in the class
25	TIMSS 99	Ratio	The tables show some values of x and y, what are the values of P and Q?
26	TIMSS 03	Decimal	Divide a number by 100. By mistake multiplying it by 100, obtained an answer of 450. What was the right answer?
27	TIMSS 03	Decimal	45 L of fuel; consumer 8.5 L per 100 km. After traveling 350 km, how much remained?

Each test had two forms, A and B, which differed only in item order. Assessments were given to the entire sixth, seventh, and eighth grades, and students sitting next to each other received different test forms to prevent cheating. Since students included in the study were predominately Hispanic, a Spanish version of the test was created, translated by faculty and graduate assistants fluent in Spanish. Students were asked for their test language preference, and although most students were of Hispanic decent, only a few students preferred to take the Spanish version. Data were collected based on the students' original responses to test items, and were coded according to their correct (1) or incorrect (0) answers to items. Summing the number of correct responses formed a student's total score. Scores were also computed for items within each rational number subconstruct represented in the NAEP and TIMSS items.

Results

Comparison of Performance of Sample to a National/International Sample

Table 3 displays the number of students and gender distribution in each grade tested. Numbers in parentheses represent the number of classes involved in the testing at each grade level.

To benchmark our students against (inter)national norms, we compared mean performance and proportion correct for each of the 24 comparable items on the performance assessment. In terms of overall performance, students in our sample students scored at or just below the level of middle schoolers around the nation (for NAEP items) and the world (for TIMSS items). Only 10 of the 24 comparable items

Table 3 Number of students participating in each test administration broken out by gender

		Sixth grade N (n)	Seventh grade N (n)	Eighth grade N (n)	Total N
Test 1		74 (3)	27 (1)	–	101
	Female/male	33/41	14/13	–	47/54
Test 2		22 (1)	84 (3)	62 (2)	168
	Female/male	12/16	35/49	29/33	76/92
Test 3		27 (1)	74 (3)	51 (2)	152
	Female/male	16/11	32/42	26/25	74/78
Test 4		28 (1)	65 (3)	80 (3)	173
	Female/male	16/12	35/30	34/46	85/88
Test 5		26 (1)	61 (3)	85 (3)	172
	Female/male	15/11	32/29	43/42	90/172

Table 4 Mean and standard deviation of student test scores

Test	Sixth grade	Seventh grade	Eighth grade
1 (Fall, year 1)	**9.39 (3.16)**	8.56 (3.33)	–
2 (Fall, year 2)	6.73 (3.15)	**9.71 (3.70)**	12.81 (5.52)
3 (Spr, year 2)	8.48 (2.62)	**10.46 (4.05)**	14.73 (6.46)
4 (Fall, year 3)	7.89 (2.62)	8.88 (3.36)	**10.88 (4.45)**
5 (Spr, year 3)	8.96 (3.23)	9.51 (4.52)	**13.22 (4.96)**

Note: The bold items show the trajectory of sixth graders in year 1 as they matriculated through seventh and eighth grade

showed statistically significant differences in percent, correct. These differences centered around the predominant focus on Part/Whole fraction instruction in our observed classes. We propose that this instructional bias, which is typical of fraction instruction in the United States, resulted in a predominance of the use of Part/Whole strategies to the exclusion of other learned strategies—strategies which ultimately are more efficient, conceptually meaningful, and that are useful for more sophisticated ratio and proportional reasoning problems.

Comparison of Performance at Different Grade Levels

Table 4 presents students' average score and standard deviation by grade level for each administration. One way Analysis of Variance was performed on percent correct using grade as an independent variable. Post hoc Scheffe tests show that, eighth-grade students outperformed sixth and seventh graders for all administrations ($p < 0.05$). Seventh graders outperformed sixth graders on administration 2 only. Sixth graders scored on average, higher than seventh graders on the first administration, but the difference is not statistically significant ($p > 0.05$). Students grew significantly over time, with greatest gains appearing, not surprisingly during the academic years, with very little, but some growth occurring over the summer periods.

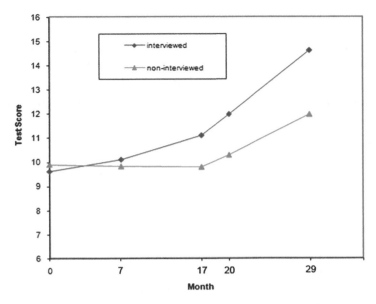

Fig. 1 Growth in rational number performance for students interviewed in the study versus non-interviewed students

Describing Students' Mathematics Achievement over Time

As we studied the results of individual interviews (see below), it became apparent that the students we were interviewing displayed more capability for solving ratio and proportion problems than the larger sample of students in the school that did not receive interviewing. To determine if a Hawthorne effect explained this difference in student abilities, average test scores for students who were interviewed in the study and peers who were never interviewed were separately computed and plotted in Fig. 1. The number of students in the former group was 33 and latter group was 56. Figure 1 presents a mean plot for these two groups of students across five test points.

It is obvious from the figure that interviewed students' mathematics performance increased steadily, and even accelerated over time. While non-interviewed students' mathematics performance did not increase from the beginning of study till the time when the third test was given, they linearly increased starting at about the third test (after 15 months of school time had elapsed). Although separate hierarchical linear/nonlinear models could be specified for each of these two groups to examine and compare student's growth on mathematics achievement over time, we decided to apply a two-level linear model to only the interviewed group, with the following justification:

1. Thirty-three students were target students in this study, and we had a large body of qualitative data for each of these 33 students. This made it possible to combine both qualitative and quantitative data outcomes to describe students' learning trajectories.

2. Among the 33 students in the interviewed group, only 4 data points were missing.
3. Among 56 students in the non-interviewed group, 50 % of data points were missing, and only 9 students completed all 5 tests. It would not have satisfactory power to apply linear model to this group.

The following two-level linear model was specified to interviewed group as following:

Level 1: $\text{Total}_{ti} = \beta_{0i} + \beta_{1i}\left(\text{month}_{ti}\right) + r_{ti}$

Level 2: $\begin{aligned}\beta_{0i} &= \beta_{00} + \gamma_{0i}\\ \beta_{1i} &= \beta_{10} + \gamma_{1i}\end{aligned}$

where:

Total_{ti}: the observed math achievement score of individual i at month level t

β_{0i}: the estimated status when month$=0$

β_{1i}: the estimated growth rate for individual i per month$_{ti}$ where month is a time-related variable

r_{ti}: the residual of individual i at month level t, which was assumed to have a mean of zero and equal variance of σ^2 across grades

β_{00}: the average true status when month$=0$

β_{10}: the average slope for the population

γ_{0i}: the difference between the individual intercept and the average true status when month$=0$

γ_{1i}: the difference between individual slope and average slope

γ_{0i} and γ_{1i} are assumed to have MVN with a mean of zero and equal variance

We estimated fixed effects: β_{00}, β_{10} and random effects: $e_{ti}, \gamma_{0i}, \gamma_{1i}$

In this model, predictors in level 1 (i.e., β_{0i} and β_{1i}) became criterion variables in Level 2, allowing students to have different starting points and growth rates. This model assumed that a straight line adequately represented each person's true change over time and that any deviations from linearity observed in the sample data resulted from random measurement error r_{ti}. The model was examined by using HLM 6.0 software. Table 5 presents the results.

Table 5 Linear model of growth in math achievement (unconditional model)

Fixed effect	Coefficient	SE	t ratio	p value
Mean status at month$=0$, β_{00}	9.03	0.57	15.52	0.000
Mean slope, β_{10}	0.16	0.02	7.93	0.000
Random effect	Variance component	df	χ^2	p value
Status at month$=0$, γ_{0i}	8.21	32	120.99	0.000
Slope, γ_{1i}	0.0055	32	50.94	0.018
Level-1 error, e_{ti}	4.63			
Correlation between γ_{0i} and γ_{1i}	0.74			

The estimated mean intercept, $\hat{\beta}_{00}$, and mean growth rate, $\hat{\beta}_{10}$, for the math achievement data was 9.03 and 0.16, respectively. This means that the average math achievement score at month=0 was estimated to be 9.03 and students were gaining an average 0.16 of a score per month. Both the mean intercept and growth rate have large t statistics indicating that both parameters are necessary for describing the mean growth trajectory of math achievement.

The estimates for the variances of individual growth parameters β_{0i} and β_{1i} were 8.21 and 0.0055, respectively. The χ^2 statistics for γ_{0i} was 120.99 (df = 32, $p < .05$), leading us to reject the null hypotheses and conclude that students vary significantly in month=0. The χ^2 statistics for γ_{1i} was 0.0055 (df = 32, $p < .05$), leading us to reject the null hypotheses and conclude that there is also significant variation in students' math achievement growth rates. The variance of $\gamma_{1i} = 0.0055$ implied an estimated standard deviation of 0.074. Thus, a student whose growth was one standard deviation above average was expected to grow at the rate of $0.16 + 0.074 = 0.234$ scores per month. The correlation between mean and slope was 0.74, suggesting that students with a higher score at the starting point tended to learn faster.

In other words, interviewed students showed slightly, but significantly lower initial performance than non-interviewed students, but over time, they learned more, and at a faster rate, resulting in a set of learners with markedly different capabilities than the uninterviewed students in the school. Recall that there were nonsignificant differences overall in the performance of our sample with the (inter)national norms. Some kind of Hawthorne effect, therefore, must have occurred as a function of the student interview process. The reasons for this will be discussed following the rest of the results.

Interview Results

We present two cases to illustrate key transitional points in students' development for two of the subconstructs distinguishing our sample's performance from that of the (inter)national sample: Part–whole and ratio. These cases do not capture all students' developmental details, not even all of the details for the two students chosen. However, they illustrate common cognitive challenges and dilemmas students faced, and they show common realizations that moved students towards a deeper and more useful understanding in the two primary rational number subconstructs where sample students differed from their (inter)national peers. As such they can be thought of as representative of the larger sample of student growth patterns in these two areas for sampled students, but illustrative of the differences in international curriculum and learning. We are developing a full account of students' individual trajectories in a follow-up paper.

Level 2 (discrete units with multiple part-whole relationships possible)

A) The following week Jorge's sister is having a birthday party and there are 20
guests. Jorge's mom asks him to go to the Food City and buy Pepsi for all of the
guests at the party. He notices that the Food City only sells six packs of Pepsi. If
Jorge buys 4 six packs of Pepsi and gives one can to each guest, what part of the
whole (4 six packs of Pepsis) does Jorge have left over?

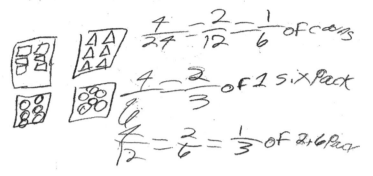

Fig. 2 Elias' flexible unitization

Elias: Part–Whole and Unitization

The case of Elias presents an example of how a student can extend a well-developed
understanding of part–whole concepts and unitization to navigate through other
rational number subconstructs, employing this knowledge to guess and check solutions in less familiar contexts (Fig. 2). Elias, like most of the interviewed samples,
reflected a well-developed notion of fractions as part–whole concepts. When asked
to express part–whole responses to contextualized questions, he responded with
fraction notation, languages, and labels indicating an understanding of units and
what each portion or unit represented. His was flexible, moving among suggested
units, appropriately representing new, equivalent part–whole ratios correctly.

In describing the different units, for example, in a case of 4 cans out of a case of
24 cans of soda, Elias' was able to flexibly change the unit from 24 cans to one
6-pack and then to two 6-packs. With each new given unit, Elias correctly calculated
the correct fraction and labeled his answer in terms of the appropriate unit. Thus the
4 cans became one-sixth of the 24 cans, two-thirds of a 6-pack, or one-third of two
6-packs.

Within the other rational number subconstructs, Elias' intuitive knowledge
appeared to lack the depth necessary to transition smoothly into formal. For example, although he had an implicit understanding of ratio and could correctly solve
simple ratio problems, he was not able to use this implicit understanding to explain
his reasoning and computation in ratio terms (e.g., a to b, a per b, a for b, etc.). The
following vignette illustrates his difficulty when he had to alter a recipe that called
for 2 cups of flour and 1 cup of sugar because the cook only had ½ cup flour. In this
particular context, the relationship between flour and sugar is a fairly simple

part–part ratio (two parts flour to one part sugar). Elias immediately identified the correct numerical answer, but his explanation emphasized the partitioning of two cups of flour into four ½ cups. He then described a process of partitioning one cup of sugar until he finally revealed a method of taking away three ¼ cups, leaving one ¼ cup as the amount of sugar needed.

Elias: Hm…sugar… you would need ¼.

Interviewer: How did you get that?

Elias: Cause if you cut 1 into half, wait…if you cut 2 into half it would equal 1, and if you cut 1 into half you cut…I am getting myself confused. I'm gonna do it another way. If you take 2 minus ¼ it would be ¼ . .. ½ I mean would equal 1 ½, take away ½ again, and it would equal 1 and it would equal ¼ of a cut, so it would be 1, 2, 3, 4, so it would be 4.

Interviewer: Draw a picture if you need to.

Elias: Oh yeah, if you have ¼ + ¼ + ¼ and how much sugar would you need there are 3 of these and take away to get ¼. This is my strategy, but you won't get it.

Elias was able to solve this problem quickly, without visible calculations, yet, as we have seen, when encouraged to reveal his thinking process, he expressed frustration in making himself clear to the interviewer and never explicitly described the proportional relationship between the flour and sugar quantities given in the original recipe. If Elias' understanding of the ratio subconstruct was developed beyond familiar part–whole relationships to part–part or part–part–whole, we would expect him to better attend to and express the multiplicative relationship involved in changing the quantities of flour and sugar (i.e., the amount of sugar is ½ the amount of flour). What we see here is a reliance on Part–whole reasoning, with a fallback on a Measure conception as evidenced by Elias's iteration of a ¼ unit. Like the majority of our sample, Elias used these two conceptions approximately 60 % of the time in his interviews. Rarely did he utilize equivalent ratios, proportional reasoning, or multiplication by a scale factor (operator conception) to solve rate and proportion problems.

Inez: Ratio Subconstruct

One of the most dramatic examples of growth in the ratio subconstruct was seen in the test scores and protocol work of Inez. During her mid-sixth grade year Inez was only able to correctly answer 4 out of the 13 ratio problems on the common test drawn from TIMMS and NAEP questions. By the fall semester of her seventh-grade year, she was able to answer 10 of the 13 correctly, dipping slightly to 8 at the end of that year, but coming back strongly in her eighth-grade year to a score of 11. What made this development interesting was her admitted lack of familiarity with ratio vocabulary and instructor-initiated inscriptions. In several conversations with her interviewer, Inez expressed a limited knowledge of the word "ratio" and with the ratio table method, which was used extensively by her seventh-grade teacher. Comments by Inez such as "What is the ratio?" and "I heard about ratio table but I

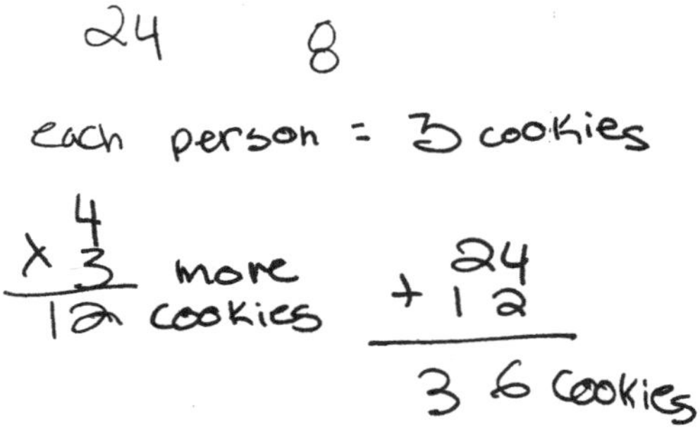

Fig. 3 Inez's use of a unit ratio

$$24 \qquad 8$$

each person = 3 cookies

$$\times \frac{4}{3} \text{ more} \atop \overline{12 \text{ cookies}}} \qquad +\frac{24}{12}$$

$$3\ 6 \text{ cookies}$$

don't know about ratio." seemed to indicate that she wasn't aware of the formal language or notations typically associated with ratio problems.

In one protocol session during her sixth-grade year, Inez was given a problem to find the amount of calories in 30 g of ice cream given the fact that 450 calories were in 100 g. She tried to divide 450 by 30 and stated "We already have 30 g, so we can ignore the 100." It wasn't until the interviewer prompted her to determine the amount of calories in 1 g of ice cream, that she seemed to recognize the relationship between the original quantities of calories and grams and was able to find a specific ratio, a unit rate. By mid-year Inez started to show her own usage of a unit rate. Figure 3 shows her work in determining how many more cookies must be added to a given amount of cookies to maintain the initial ratio of cookies to guests.

Interviewer: You are shopping for a party and you buy 24 cookies for 8 people. Your cell phone rings and you are told that four more people are coming to the party. How many more cookies will you have to buy to keep the ratio the same? How many total cookies will you need?

Inez: We have 24 cookies and only 8 people. So each person will get three cookies. So when 4 more people are coming, we have to multiply by 4, so 12 more cookies and all together we need 36 cookies.

Despite her lack of familiarity with formal ratio symbols and operations, Inez, like many of our sample students, was able to solve a variety of contextual ratio problems by *using her own personal notation* for assigning correspondence between ratio quantities. From the fall semester of her sixth-grade year to protocols through-out her seventh-grade and eighth-grade year, Inez used an "=" to pair ratio quantities and then worked efficiently with this pairing to build up or down to a desired solution. In Fig. 4, for example, she established a relationship between 90 lions in the zoo with 1,800 kgs of food. Once she wrote this "equality" on her paper, she then divided or multiplied both sides as needed to create other equivalent ratios, often also adding corresponding parts of these pairs to solve given problems. Eventually she found the unit rate of 20 kg for one lion and then demonstrated her knowledge of how to use this rate to determine the number of kilograms for any given number of lions.

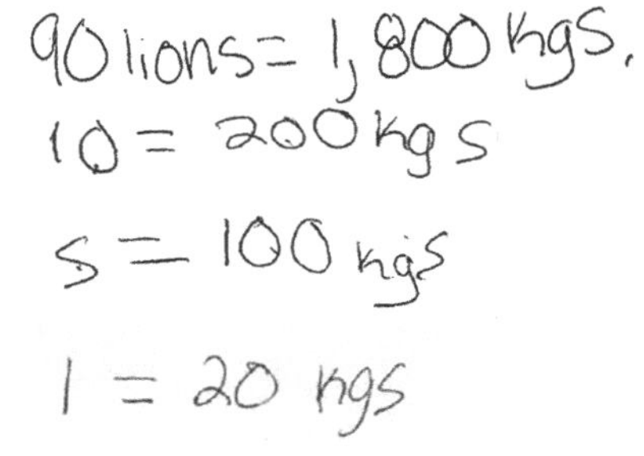

Fig. 4 Inez's equals sign
used as "colon" ratio symbol

$$90 \text{ lions} = 1,800 \text{ kgs},$$
$$10 = 200 \text{ kgs}$$
$$5 = 100 \text{ kgs}$$
$$1 = 20 \text{ kgs}$$

During the last year of the study, Inez was confronted by the interviewer about her use of the "equal sign" inscription for a ratio problem. Inez was quick to say that she knew that the two numbers were not really "equal," but that this was her own way of organizing the information in the problem. It was clear from her work that this method of organization provided a structure within which she could move easily to create equal ratios as needed.

Inez is also indicative of our sample students in that, informal, in-the-moment notations were used extensively, along with very few teacher-sanctioned inscriptions (like the ratio table, for example). These ad hoc inscriptions had meaning for each individual student, but were not capitalized on by the teachers in an attempt to systematically make them more formal and precise.

Summary of Interview Data

Interview protocols were coded based on the type of problem presented (Part–Whole, Measure, Quotient, Ratio, or Operator), strategies employed to solve the problem (including the use of heuristics, super-strategies, and taught procedures), and the sensibility of students' strategies and the correctness of their answers. Emulating the wonderful interpretive method of Carpenter and Moser (1984) for young children's arithmetic strategy development, we represent the development of children's strategies as graphs showing the proportion of each coded strategy over time. The following four figures show demonstrably that students *enter* into rational number instruction with a *variety* of strategies, both informal and formal for solving a wide variety of problem types. These strategies echo the general research on rational number development in that the predominant way of approaching problems appears to be conceptualizing them as Part–Whole, with smaller proportions of strategies focusing on using benchmark fractions, common denominator strategies, and even some proportional reasoning, though this was very rare (<5 % of total strategies). Through instruction, certain strategies became preferred in the participating students' classrooms. In particular, the use of fraction bars and ratio tables favored the development of benchmark fractions, measure strategies, and some use of equivalent ratios and proportional reasoning.

Fig. 5 Strategy type leading
to sensible answers across
four parallel administrations
of interview protocols
2005–2007

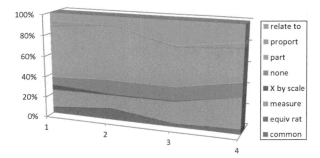

Fig. 6 Strategy type leading
to technically correct answers
across four parallel
administrations

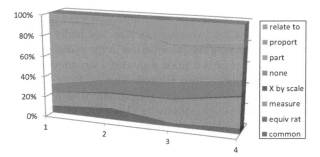

Findings show that sensible approaches to problem solution tended to lead to
generation of correct answers (see Figs. 5 and 6). The proportions of strategies
coded as sensible were nearly identical to the proportions of correct answers. The
trend for the development of sensible and correct strategies shows that the proportion of problems solved using Part/Whole reasoning decreased over time, being
supplanted by measure (quotitive division) strategies, relating problem quantities to
benchmark fractions, and also by a variety of ad hoc strategies made up on the spot
to solve the problem. Part–Whole, however, remained the dominant strategy preferred by students, even by the end of the eighth grade.

Examining errors, we show that, ad hoc strategies, were the most prominent
strategies chosen when the strategies did not make sense for the problem situation
(see Fig. 7). This indicates that for a large number of interview problems, students
neither learned nor were able to generate, a meaningful method of solution, and
instead relied on trial-and-error and other means–end solution methods. Fully 40 %
of students' responses were idiosyncratic, and this trend remained relatively constant over the entire course of the longitudinal study. Part–Whole methods were the
second-most prominent strategies used in ways that did not make sense for the problem context. Moreover, Part–Whole strategies were used *most* in cases where students' strategies yielded incorrect solutions. This echoes our comparisons with the
(inter)national sample.

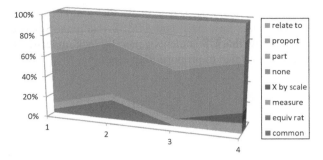

Fig. 7 Strategy type leading to non-sensible answers across four parallel administrations of interview protocols 2005–2007

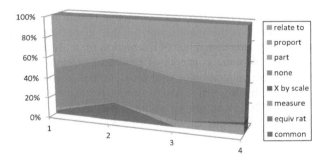

Fig. 8 Strategy type leading to technically incorrect answers across four parallel administrations of interview protocols 2005–2007

The most disappointing trend in our data suggests that potentially powerful methods of solution to rational number problems, such as proportional reasoning, the use of equivalent ratios, and common denominator strategies were neither stressed in students' classes (until well into the eighth-grade year), nor evident in their acquisition of strategies across the middle grades. Instead, the teachers focused class time on the use of robust-but-inefficient conceptual strategies such as the use of the fraction bar and ratio table. These strategies were heavily used in the students' textbooks, but were not exclusively emphasized there, indicating considerable teacher preference in the kinds of strategies legitimized in their instructional practices (Fig. 8).

Because our data are drawn from primarily poor, urban, largely Latino schools, we want to be careful generalizing the exact developmental trajectory of students' strategies to the rest of the United States. However, inasmuch as other middle-school curricula continue to overemphasize the use of part–whole conceptualizations of fractions, underemphasize the notion of fractions as indicated division, and underemphasize methods of computation that build strong understanding of factors and multiples, units, and partitions (e.g., Lamon, 2002; Ni & Zhou, 2005; Sophian,

2007; Thompson & Saldanha, 2003), it is likely that similar trends are occurring across many schools. Data on NAEP and TIMSS (National Science Foundation, 2002; NCES, 1999) show that US students generally lack understanding and skills in these and other areas related to proportional reasoning.

Discussion

A major objective of this longitudinal study was to contribute significantly to the research base on rational number learning by exposing patterns and mechanisms of development in students' understanding of rational numbers and by reorganizing the current fragmented body of research into a more coherent developmental model; illustrating how rational number subconstructs evolve concurrently and interactively along the road to a more profound knowledge of rational number concepts. The former goal is realized in this manuscript, but the latter is still a major challenge for the field. In particular, the coherence of instruction for teaching rational number, and especially the problematic concepts of ratio and proportion is still lacking, resulting in continued fragmentation of knowledge in the US children, favoring less sophisticated conceptualizations in this subject matter, than evidenced by students in the TIMSS 1999 and 2003 samples. Curriculum and teaching appear to be the key levers here (e.g., Saxe, Diakow, & Gearhart, 2012; Saxe, Gearhart, & Seltzer, 1999) as we were able to show that instructional strategies that favored Part–Whole conceptions predominated in our sample classrooms, leading to an overreliance on Part–Whole conceptualizations by students, yielding performance deficits in comparison to the international norm on more powerful concepts of rate, ratio, and proportional reasoning.

Despite the narrow demographics of our studied samples, we see our results as transportable to the US educational system in general. In the United States, research clearly shows that instruction in rational number tends to favor Part–Whole interpretations far more than other interpretations of fractions (Ni & Zhou, 2005; Sophian, 2007; Thompson & Saldanha, 2003). Use of measure, quotient, and ratio subconstructs are much less evident. The unfortunate point of this is that a Part–Whole understanding of fractions does not allow the student to deal with units other than one without tremendous difficulty. As a result, improper fractions become confusing (Mack, 1993). Fraction division, in particular becomes conceptually impossible. In countries like Japan and China, who traditionally perform better on international assessments of fractions and algebra, rational numbers are explained in terms of measurement models like the number line or area models, as the result of any division problem, and as a multiplicative comparison of dividend and divisor, numerator, and denominator (Moseley, Okamoto, & Ishida, 2007).

Our study reinforces earlier work that suggests that in rational number development, students tend to utilize a small number of robust-but-inefficient strategies which are applicable across a variety of situations. In the reported project, for example, we found that students who did not have ready access to procedures for determining factors and multiples of whole numbers were greatly hampered in their capacity to solve complex problems involving fractions, particularly fractions in proportional relationships. Conversely, students who DO have ready access to efficient procedures are able to solve problem subgoals in real time and progress towards successful problem resolution much more readily (Kim et al., 2007).

Moreover, in our data, overall, we have seen children using powerful iterative methods, such as the repeated halving strategy, far beyond their proficiency with other methods of computing fractions (e.g., finding common denominators, dividing numerator and denominator by a common factor). Students persisted in the use of these iterative strategies even though they had earlier demonstrated the ability to conceptualize fractions as indicated division, knowledge of and the ability to use factors and multiples, and the ability to solve complex problems using a division procedure.

Siegler, Thompson, and Schneider (2011) show that sixth graders show great variability of strategy use. They found that selection of strategies depended upon students' familiarity with solving problems with some arithmetic operations but not others. They found, however, that strategy use was highly variable within arithmetic operations. The sixth grade in the United States appears to be a key transitional grade, where students struggle to consolidate learned strategies for whole number arithmetic, and reconcile these with new rational number strategies they are currently learning.

Empson, Levi, and Carpenter (2011) show "there is a broad class of children's strategies for fraction problems motivated by the same mathematical relationships that are essential to understanding high-school algebra and that these relationships cannot be presented to children as discrete skills or learned as isolated rules. The authors refer to the thinking that guides such strategies as Relational thinking." What we found in our current study is that our studied children came into the sixth grade armed with a number of fine strategies for thinking about fraction problems. What failed to happen for many of our studied students is that over the course of their 3 years in middle school, they were not able to develop relational thinking for fractions much beyond Part–Whole and Measure conceptions.

Conclusions

In conclusion, we found that:

1. Children come to the middle grades with many useful ways of thinking about and solving rational number problems.
2. Children leave middle school with only a slightly expanded set of skills. They tend to rely on ad hoc, means–end reasoning and reliance on simple Part/Whole conceptions of fractions as opposed to developing more efficient and powerful methods of computation.

3. Teaching of fractions overemphasizes conceptual strategies using inscriptions like the fraction bar and ratio table, leaving little time to develop proportional reasoning, common denominator, and other equivalent fraction methods of solution.

4. The very act of interviewing children, only once every 3 weeks, is an intervention that leads them to learn more and achieve better than their matched counterparts. Even though teaching the children was not a goal of this study, interviewed children demonstrated significantly higher gains on TIMSS and NAEP items than their peers.

Commentary on the Issue of Scale in Intensive Interview and Observational Methods

At first glance, the scale of our study, comprising 204 students—of which all 204 were administered quantitative assessments of their rational number knowledge, 102 were interviewed, and 32 remained in the study for the full 3 years of the project—would generally not be considered *large* when compared to the samples reported in other studies in this book. However, as pointed out in the introductory chapter of this volume, *scale* depends on a variety of factors, not just the size of the sample. In our case, the scale is determined by two factors: (1) methods utilized; and (2) characteristics of the measurement.

Individual constructivist teaching experiments of approximately 45 min took roughly 2¼ hours to transcribe. Analysis of each interview took an additional 1.5 h on average. Multiplying these factors by 102 students, interviewed 9 times per year for 3 years, we get a total experimenter time of roughly 12,500 h for our qualitative work. For the quantitative assessments, administration of tests to all 204 students pre- and post- each year, coding responses and analysis of the quantitative data took roughly 300 additional experimenter hours. A bit more can be added to account for cleaning up TIMSS and NAEP data to bring our rough estimate close to 13,000 h of work (we do not count reading, writing, meetings, and other preparatory/reflective work in these estimates, nor do we count the 2 h per week of classroom observations, plus transcription and analysis). Clearly, the qualitative methods employed to uncover students' thinking constituted the vast majority of our researcher time. Every additional student added to our sample added an additional week (40.5 researcher hours equivalent) of effort. Given restrictions of funding, relative to the sensitivity of measurement we needed to track students' development of strategies over time, 102 students as ongoing informants was at the upper limits of scale possible.

So *scale,* as a construct in mathematics education must be thought of in terms of the complexity of the questions asked, the intensity of the data collection process, and the density of the data record. We benefited from this understanding of scale in that our interview protocols and performance assessments shared a common scheme by which problems could be coded. By utilizing and combining sensitive idiographic techniques such as interviews and observations, with (inter)nationally validated tasks on the performance assessment, we were able to identify a key weakness in the instruction of our sample, and tie this weakness to inadequate development of proportional reasoning.

References

Behr, M., Harel, G., Post, T., & Lesh, R. (1992). Rational number, ratio and proportion. In D. A. Grouws (Ed.), *Handbook of research on mathematics teaching and learning* (pp. 296–333). New York: Macmillan.

Behr, M., Lesh, R., Post, T., & Silver, E. (1983). Rational number concepts. In R. Lesh & M. Landau (Eds.), *Acquisition of mathematics concepts and processes* (pp. 91–125). New York: Academic Press.

Bonato, M., Fabbri, S., Umiltà, C., & Zorzi, M. (2007). The mental representation of numerical fractions: Real or integer? *Journal of Experimental Psychology: Human Perception and Performance, 33*(6), 1410.

Bransford, J. D., Brown, A., & Cocking, R. (1999). *How people learn: Mind, brain, experience, and school*. Washington, DC: National Research Council.

Carpenter, T. P., Fennema, E., & Romberg, T. A. (Eds.). (2012). *Rational numbers: An integration of research*. New York: Routledge.

Carpenter, T. P., & Moser, J. M. (1984). The acquisition of addition and subtraction concepts in grades one through three. *Journal for Research in Mathematics Education, 15*, 179–202.

Carraher, D. W. (1996). Learning about fractions. In L. P. Steffe, P. Nesher, G. Goldin, P. Cobb, & B. Greer (Eds.), *Theories of mathematical learning* (pp. 241–266). Mahwah, NJ: Lawrence Erlbaum.

Confrey, J., Maloney, A., Nguyen, K., Mojica, G., & Myers, M. (2009). Equipartitioning/splitting as a foundation of rational number reasoning using learning trajectories. In *33rd Conference of the International Group for the Psychology of Mathematics Education*. Thessaloniki, Greece.

Empson, S. B. (1999). Equal sharing and shared meaning: The development of fraction concepts in a first-grade classroom. *Cognition and Instruction, 17*, 283–343.

Empson, S. B., Junk, D., Dominguez, H., & Turner, E. (2006). Fractions as the coordination of multiplicatively related quantities: A cross-sectional study of children's thinking. *Educational Studies in Mathematics, 63*(1), 1–28.

Empson, S. B., Levi, L., & Carpenter, T. P. (2011). The algebraic nature of fractions: Developing relational thinking in elementary school. In J. Cai & E. Knuth (Eds.), *Early algebraization* (pp. 409–428). Berlin, Germany: Springer.

Hollan, J., Hutchins, E., & Kirsh, D. (2000). Distributed cognition: Toward a new foundation for human-computer interaction research. *ACM Transactions on Computer-Human Interaction (TOCHI), 7*(2), 174–196.

International Association for the Evaluation of Educational Achievement (IEA). (2001). *TIMSS 1999 mathematics items*. Boston: International Association for the Evaluation of Educational Achievement. Retrieved from http://timssandpirls.bc.edu/timss1999i/study.html.

International Association for the Evaluation of Educational Achievement (IEA). (2005). *TIMSS 2003 mathematics items: Released set, 8th grade*. Boston: International Association for the Evaluation of Educational Achievement. http://timss.bc.edu/timss2003i/released.html.

Kelly, D. L., Mullis, I. V., & Martin, M. O. (2000). *Profiles of student achievement in mathematics at the TIMSS international benchmarks: US performance and standards in an international context*. TIMSS International Study Center, Lynch School of Education, Boston College.

Kieren, T. (1993). Rational and fractional numbers: From quotient fields to recursive understanding. In T. P. Carpenter, E. Fennema, & T. A. Romberg (Eds.), *Rational numbers: An integration of research* (pp. 49–84). Hillsdale, NJ: Erlbaum.

Kieren, T. E. (1976). On the mathematical, cognitive, and instructional foundations of rational numbers. In R. Lesh (Ed.), *Number and measurement: Papers from a research workshop.* Columbus, OH: ERIC/SMEAC.

Kim, A., Hernandez, L. S., Megowan-Romanowicz, M. C., Middleton, J. A., Kim, Y., & Ellis, K. (2007). *The impact of unbalanced development between conceptual understanding and procedural skills.* Paper presented at the annual meeting of the American Educational Research Association, Chicago, IL.

Lamberg, T., & Middleton, J. A. (2009). Design research perspectives on transitioning from individual microgenetic interviews to a whole-class teaching experiment. *Educational Researcher, 38*(4), 233–245.

Lamberg, T., & Middleton, J. A. (2002). The role of inscriptional practices in the development of mathematical ideas in a fifth grade classroom. In A. Cockburn (Ed.), *Proceedings of the 26th annual meeting of the International Group for the Psychology of Mathematics Education.* Norwich, England: PME.

Lamon, S. J. (2002). Part-whole comparisons with unitizing. In B. Litwiller & G. Bright (Eds.), *Making sense of fractions, ratios, & proportions: 64th yearbook* (pp. 71–73). Reston, VA: National Council of Teachers of Mathematics.

Lamon, S. J. (2006). Teaching fractions and rations for understanding. Essential instructional strategies for teachers. Mahwah, NJ: Lawrence Erlbaum.

Lesh, R., Post, T., & Behr, M. (1988). Proportional reasoning. In J. Hiebert & M. Behr (Eds.), *Number concepts and operations in the middle grades* (pp. 93–118). Reston, VA: National Council of Teachers of Mathematics.

Mack, N. K. (1993). Learning rational numbers with understanding: The case of informal knowledge. In T. P. Carpenter, E. Fennema, & T. A. Romberg (Eds.), *Rational numbers: An integration of research* (pp. 85–105). Hillsdale, NJ: Erlbaum.

Martin, M. O., & Kelly, D. L. (1998, Eds.). Third International Mathematics and Science Study Technical Report Volume II: Implementation and Analysis-Priimary and Middle School Years. Boston, MA: International Association for the Evaluation of Educational Achievement.

Middleton, J. A., van den Heuvel-Panhuizen, M., & Shew, J. A. (1998). Using bar representations as a model for connecting concepts of rational number. *Mathematics Teaching in the Middle School, 3*(4), 302–312.

Moseley, B. J., Okamoto, Y., & Ishida, J. (2007). Comparing US and Japanese Elementary School Teachers' Facility for linking rational number representations. *International Journal of Science and Mathematics Education, 5*(1), 165–185.

Mullis, I. V. S., Martin, M. O., Gonzalez, E. J., & Chrostowski, S. J. (2004). *TIMSS 2003 international mathematics report: Findings from IEA's trends in international mathematics and science study at the fourth and eighth grades.* Chestnut Hill, MA: Boston College.

National Center for Education Statistics. (1999). *Highlights from TIMSS: The third international mathematics and science study.* Washington, DC: National Center for Education Statistics.

National Science Foundation. (2002). *Science and engineering indicators, 2002.* Arlington, VA: National Science Foundation.

Ni, Y., & Zhou, Y. D. (2005). Teaching and learning fraction and rational numbers: The origins and implications of whole number bias. *Educational Psychologist, 40*(1), 27–52.

Oksuz, C., & Middleton, J. A. (2005). *Children's understanding of algebraic fractions as quotients.* Paper presented at the annual meeting of the American Educational Research Association, Montreal, Canada.

Roth, W. M., & McGinn, M. K. (1998). Inscriptions: Toward a theory of representing as social practice. *Review of Educational Research, 68*(1), 35–59.

Saxe, G. B., Diakow, R., & Gearhart, M. (2012). Towards curricular coherence in integers and fractions: A study of the efficacy of a lesson sequence that uses the number line as the principal representational context. *ZDM, 45*, 343–364.

A Longitudinal Study of the Development of Rational Number Concepts... 289

Saxe, G. B., Gearhart, M., & Seltzer, M. (1999). Relations between classroom practices and student learning in the domain of fractions. *Cognition and Instruction, 17*(1), 1–24.

Siegler, R. S., Thompson, C. A., & Schneider, M. (2011). An integrated theory of whole number and fractions development. *Cognitive Psychology, 62*(4), 273–296.

Sophian, C. (2007). *The origins of mathematical knowledge in childhood.* Mahwah, NJ: Lawrence Erlbaum.

Streefland, L. (1993). Fractions: A realistic approach. In T. P. Carpenter, E. Fennema, & T. A. Romberg (Eds.), *Rational numbers: An integration of research* (pp. 289–325). Hillsdale, NJ: Erlbaum.

Thompson, P., & Saldanha, L. (2003). Fractions and multiplicative reasoning. In J. Kilpatrick, G. Martin, & D. Schifter (Eds.), *Research companion to the principles and standards for school mathematics* (pp. 95–114). Reston, VA: National Council of Teachers of Mathematics.

Toluk, Z., & Middleton, J. A. (2004). The development of children's understanding of quotient: A teaching experiment. *International Journal of Mathematics Teaching and Learning.5*(10). Article available online http://www.ex.ac.uk/cimt/ijmtl/ijmenu.htm

Part IV
Methodology

Measuring Change in Mathematics Learning with Longitudinal Studies: Conceptualization and Methodological Issues

Jinfa Cai, Yujing Ni, and Stephen Hwang

Learning is about growth and change. Learning is often demonstrated by changes in student achievement from one point in time to another. Therefore, researchers and educators are interested in academic growth as a means to understand the process of student learning. In mathematics education, there has been a growing interest in using longitudinal designs to examine and understand student learning over time. Researchers face a number of issues of measuring change using such designs. In this chapter, we draw on our experience gained from two longitudinal studies of mathematics learning to discuss various issues of measuring change in student learning. We start with a brief introduction of the two studies. Then we discuss the conceptualization and measures of change in mathematics learning. Third, we discuss issues of analyzing and reporting change. Finally, we discuss how to interpret changes in mathematics achievement in longitudinal studies appropriately.

Two Longitudinal Studies Examining Curricular Effect on Student Learning

This chapter draws on two longitudinal projects that studied the effects of curriculum on student learning. The first project was conducted in China and addressed the question, "Has curriculum reform made a difference?" by looking for changes in classroom practice and consequently in student learning. This project (hereafter called the China project) compared the effect of a new, reform-oriented elementary mathematics curriculum to that of the conventional curriculum on classroom

J. Cai (✉) • S. Hwang
University of Delaware, Ewing Hall 523, Newark, DE 19716, USA
e-mail: jcai@udel.edu; hwangste@udel.edu

Y. Ni
The Chinese University of Hong Kong, Hong Kong, China

© Springer International Publishing Switzerland 2015
J.A. Middleton et al. (eds.), *Large-Scale Studies in Mathematics Education*,
Research in Mathematics Education, DOI 10.1007/978-3-319-07716-1_13

293

practice and student learning outcomes. The second project—the LieCal project (Longitudinal Investigation of the Effect of Curriculum on Algebra Learning)—was conducted in the USA. This project was designed to investigate both the ways under which a reform curriculum did or did not have an impact on student learning in algebra, and the characteristics of the curricula that led to student achievement gains. Both projects looked into changes in classroom practice by examining the nature of classroom instruction, analyzing cognitive features of the instructional tasks implemented in different classrooms, the characteristics of classroom interactions, and changes in student learning outcomes.

The China project and the LieCal project shared similarities in their designs and data analyses. In particular, both projects addressed a set of common and critical questions about teaching and learning using reform-oriented curricula, including: (1) Does the use of the reform-oriented curriculum affect the quality and nature of classroom teaching; (2) Do students improve at solving problems, as the developers of the reform-oriented curricula claim; (3) Do students sacrifice basic mathematical skills with the reform-oriented curriculum; and (4) To what extent does the use of the reform-oriented curriculum improve learning for all students?

Conceptualizing and Measuring Change in Student Learning

Student learning takes place in various domains; two major domains are cognitive and affective (Krathwohl, 2002), each with multiple factors influencing what is learned, how it is learned, and how it is remembered and used. Here, we will focus on the cognitive domain, and in particular on mathematical thinking, to illustrate the issues of how to conceptualize and measure change in student learning. We will briefly touch on the affective domain afterwards.

Although there is no consensus on what mathematical thinking is, it is widely accepted that there are many aspects of mathematical thinking that warrant examination (Cai, 1995; Ginsburg, 1983; Schoenfeld, 1997; Sternberg & Ben-Zeev, 1996). Studies of mathematics learning over the years have included a focus on identifying those ways that students demonstrate a propensity to "think mathematically" in their actions. For example, Polya found that capable problem solvers employ heuristic reasoning strategies to solve problems (Polya, 1945). Being able to self-generate useful analogies while solving a problem is an example of a heuristic that capable solvers demonstrate as they solve problems. In addition, Krutetskii (1976) found that able students are more likely than less able students to use generalizations in their mathematical problem solving. Other researchers have described and explained mathematical thinking as distinct from the body of mathematical knowledge, focusing on processes such as specializing, conjecturing, generalizing, and convincing (Burton, 1984). More recently, mathematical thinking has been characterized in terms of the learner being able to develop strong understandings in mathematical situations (Kieran & Pirie, 1991) and making connections among concepts and procedures (Hiebert & Carpenter, 1992).

These studies suggest that we need to use multiple measures to assess the mathematical thinking of students. For example, although we know that it is important for students to have algorithmic knowledge to solve many kinds of problems, this does not ensure that they have the conceptual knowledge to solve nonroutine or novel problems (Cai, 1995; Hatano, 1988; Steen, 1999; Sternberg, 1999). Hence, it is crucial that studies of mathematical thinking include tasks that measure students' high-level thinking skills as well as their routine problem-solving skills that involve procedural knowledge. Indeed, as the heart of measuring mathematical performance is the set of tasks on which achievement is to be assessed, it is desirable to use various types of tasks to measure the different facets of students' mathematical thinking and gauge student growth in mathematics learning (Betebenner, 2008; Mislevy, 1995; National Research Council (NRC), 2001).

Recognizing the need to assess mathematical thinking broadly, both the China project and the LieCal project used multiple measures of student achievement. Most of the assessment tasks used in both projects came from Cai's earlier work (1995, 2000), in which he investigated Chinese and US students' mathematical thinking. The design of the achievement measures in each project was guided by the following considerations: (1) a combination of multiple-choice and open-ended assessment tasks should be used to measure students' performance; (2) different cognitive components, specifically, the four components of Mayer's (1987) cognitive model (translation, integration, planning, and computation), should be attended to in the multiple choice tasks; and (3) in responding to open-ended tasks, students should show their solution processes and provide justifications for their answers.

Because of their potential for broad content coverage and objective scoring, their highly reliable format, and their low cost, multiple-choice questions were used to assess whether students had learned basic knowledge and skills in mathematics. However, it is relatively difficult to infer students' cognitive processes from their responses to multiple-choice items; such questions are more appropriate for measuring procedural knowledge and basic skills than conceptual understanding. Thus, open-ended tasks were also included to assess student achievement in both projects. The open-ended tasks provided a better window into the thinking and reasoning processes involved in students' problem solving (Cai, 1997). The use of various types of assessment tasks provided the information to address questions such as, "Does the curricular emphasis on conceptual understanding come at the expense of fluency with basic mathematical skills?" For example, the China project showed that both students who received the reform-oriented curriculum and those who did not receive the curriculum had significant improvement in performance on computation and on routine and open-ended problem solving over time. However, the non-reform group showed a faster rate of improvement on the measure of computation. The LieCal project demonstrated that students receiving the reform-oriented CMP curriculum (Connected Mathematics Program, a *Standards*-based curriculum) showed a faster rate of improvement than the students receiving non-CMP curricula on the measures of solving open-ended tasks. However, the two groups did not differ in growth rate on the measure of computation and equation solving.

Research has also shown that changes in learning experiences can lead to changes in feelings towards mathematics, perception of mathematics, and consequently commitment to think mathematically. For example, Schoenfeld (1992) demonstrated how students' beliefs about mathematics could be changed with the experience of being engaged in solving authentic mathematical problems. Reform-oriented mathematics curricula aim not only to help students think mathematically but also to nurture their positive beliefs and attitudes toward learning mathematics. Therefore, the China project administered multiple measures of affective outcomes (interest in learning mathematics, classroom participation and views of what mathematics is about) several times. It was found that, although the students showed significant gains in the three measures of cognitive achievement, their interest in learning mathematics declined from the start of fifth grade to the end of sixth grade for both the reform and non-reform group, with a steeper decline for the non-reform group. This highlights the importance of considering change in students' mathematical learning broadly so that changes can be understood in a broader context of learning. In particular, it highlights the importance of longitudinal analyses so that growth rates can be estimated for key learning variables.

Analyzing and Reporting Change

The major purpose of a longitudinal study is to examine change and the correlates or causes of change over time. Because learning is fundamentally about growth and change, analyzing and reporting change in students' academic achievement is a significant endeavor for the study of learning. However, change is often difficult to document well, given the myriad variables and factors that may influence changes in students' learning. It is even more challenging to identify the causes of a change when change is detected. A sound analysis of longitudinal data relies on a sound study design that includes the use of multiple measures of the same variables over time to help enhance the internal validity of the study (Fisher & Foreit, 2002; Linn, 2007). Given the multifaceted nature of the mathematical thinking that the LieCal and China projects were studying, both projects used three cognitive measures of mathematics achievement (computation, routine problem solving, and complex problem solving) to gain a detailed picture of student growth in mathematics achievement and a possible curricular correlate to the growth.

Within the confines and constraints of non-randomized experimental design, the primary question about change in student achievement that our studies were designed to answer was whether or not there was any meaningful difference in growth rate in mathematics achievement among groups of students using different curricula (Cai, Wang, Moyer, Wang, & Nie, 2011; Ni, Li, Li, & Zhang, 2011).

At the same time, the projects were also designed to address other factors that might affect the students' mathematics achievement growth rate. For example, the LieCal project considered how the conceptual or procedural emphasis of classroom instruction might moderate the curricular influence on the growth rate of students' mathematics achievement. To measure these classroom variables, as the students progressed from sixth through eighth grade, we conducted over 500 lesson observations of over 50 mathematics teachers participating in the project. Each LieCal class was observed four times, during two consecutive lessons in the fall and two in the spring. Trained observers recorded extensive minute-by-minute information about each lesson using a detailed, 28-page observation instrument. The data from these observations were used to characterize key aspects of each lesson, including the degree of conceptual and procedural emphasis of instruction in the CMP and non-CMP classrooms (Moyer, Cai, Wang, & Nie, 2011).

In the China project, each of 60 participating teachers and their classrooms was observed for three lessons on three consecutive days. The videotaped lessons were analyzed in terms of cognitive features of implemented instructional tasks and patterns of classroom discourse. The project found significant differences in instruction between the reform and non-reform classrooms (Li & Ni, 2011). With the measured aspects of classroom instruction, it became possible to examine the relations between curriculum, classroom instruction, and student learning.

In addition, both the LieCal project and the China project attended to elements of the students' sociocultural backgrounds that might influence change in student achievement. Classrooms in the USA have become increasingly ethnically diverse, and there have been persistent concerns about disparities in the mathematics achievement of different ethnic groups. This is particularly true with respect to areas such as algebra and geometry, where success has been shown to help narrow disparities in post-secondary opportunities (Loveless, 2008). Given that middle school mathematics experiences can lay the foundation for students' development of algebraic thinking, the LieCal project explored potential differential effects of reform and traditional curricula on the mathematics performance of students from different ethnic groups (Cai, Wang et al., 2011; Hwang et al., 2015).

The China project took into consideration socioeconomic status (SES) as well. This variable was measured because one purpose of the project was to examine whether achievement gaps between higher SES students and low SES students would decrease or increase in the different aspects of mathematics achievement over time in relation to the different mathematics curricula.

Analyzing and Reporting Change Quantitatively

With these purposes in mind, both studies employed a panel design in which a cohort is followed for a period of time and a common set of instruments is administered repeatedly over that period (Ma, 2010). The studies produced data with a hierarchical structure of individual students nested within classes, classes nested within schools, etc. For this type of hierarchically structured data, the technique of hierarchical linear modeling (HLM), and in particular multilevel growth modeling, is appropriate and effective for examining change at both the individual and the group level. This is because this method is able to account for the correlated observations of the different levels due to the clustering effects and thus relax the assumption of independence of observations for the traditional regression analysis (Raudenbush & Bryk, 2002). Therefore, both projects used HLM models to answer their research questions. The HLM analyses revealed that, in the China project, the students showed a faster growth rate in computation and solving routine problems than in solving open-ended problems, and that this trend was more pronounced for the students receiving a conventional curriculum than those receiving a reform curriculum. The LieCal project used four two-level HLM models (one for each outcome measure) with the mean of conceptual emphasis or procedural emphasis across 3 years as a teaching variable together with student ethnicity and curriculum type nested in schools (Cai, Wang et al., 2011). The results of the HLM analysis showed that students who used CMP had a significantly higher growth rate than non-CMP students on open-ended problem-solving and translation tasks while maintaining similar growth rates on computation and equation-solving tasks. Thus, the relatively greater conceptual gains associated with the use of the CMP curriculum did not come at the cost of basic skills.

In addition, to gain a finer-grained picture of the curricular impact and also as a validation of the results of the HLM analyses, Cai, Wang et al. (2011) compared the percentage of students receiving the CMP curriculum who obtained positive gain scores to the percentage of students receiving non-CMP curricula who obtained positive gain scores. These calculations showed the relative sizes of the groups of students whose performance increased on each of the outcome measures whereas the results of the HLM analyses estimated an overall difference in the means of the gain scores between the two groups of students. For example, we found that 89 % of CMP students had positive gains in open-ended problem-solving tasks over the course of the middle grades. This was a statistically significantly larger percentage than for the non-CMP students, of whom 83 % showed gains in open-ended problem solving. With respect to computation, despite the fact that the mean gains were not significantly different between the CMP and non-CMP students, we found that a larger percentage of non-CMP students than of CMP students showed positive gains (78 % vs. 60 %). With respect to equation-solving, however, the two groups were not significantly different either in mean gains or in percentage of students with positive gains (e.g., 50 % of student group A receiving non-CMP curricula obtaining positive gain scores and 70 % of student group B receiving the CMP curriculum doing so) (Cai, Wang et al., 2011).

Using a broad set of measures over time within a study also allows for the collection of information on what trade-offs may be faced with different curricula and about what can be realistically expected in typical classrooms (Brophy & Good, 1986). The China project showed that the non-reform group demonstrated faster growth in proficiency in computation skills from the fifth grade to the sixth grade, and they outperformed the reform group students in the final assessment. Also, the reform group students kept their initial advantage in solving open-ended problems, as they performed better than the non-reform group on the first assessment and the growth rates for the two groups were similar. Nevertheless, given the nature of the design, it could not be concluded that the reform group's better performance on complex problem solving was merely due to the curriculum or to their better initial status. However, the reform group appeared to have achieved a relatively more balanced development in the three measures of mathematics achievement, computation, routine problem solving, and complex problem solving.

The China project was also concerned with whether or not the different curricula would help reduce achievement gaps between students from different family backgrounds. The project found that the achievement gaps in computation skills between students of high SES backgrounds and those of low SES were narrowed significantly from their fifth grade to sixth grade, but there was no narrowing of the gap in solving open-ended mathematics questions. This was the case for both groups using either a reform curriculum or conventional curriculum. The closing achievement gap in computation but not in solving open-ended mathematics questions suggested that instructional conditions that facilitate mathematical explaining, questioning, exchanging, and problem solving are most valuable for students from low SES families because low SES families are less likely to be able to afford the conditions to facilitate high-order thinking (Ni et al., 2011).

Analyzing and Reporting Change Qualitatively

To deepen analyses of curricular effect on change in student learning it is necessary to look beyond measuring performance differences in terms of mean scores on various types of tasks between groups of students receiving different types of curricula. As useful as such comparisons may be, they do not provide a complete profile of what students who use different curricula can and cannot do. Two students may receive the same score on a task but use very different solution strategies or make very different types of errors. To inform these comparisons of performance on individual tasks, some additional exploration of the thinking and methods that led students to their answers is required.

The use of open-ended assessment tasks makes it possible not only to measure students' higher-order thinking skills and conceptual understanding, but also to analyze students' solution strategies, representations, and mathematical justifications (Cai, 1997). The strategies that students employ and the ways that they represent their solutions can provide insight into their mathematical ideas and thinking processes. For example, in the LieCal project, we supplemented our analysis of the correctness of answers with a longitudinal analysis of the changes in students' strategies over time (Cai, Moyer, Wang, & Nie, 2011). Figure 1 shows the doorbell problem, an open-ended task used in the LieCal assessments. In this problem, students were asked to generalize from the given pattern of doorbell rings.

Student performance on this task were analyzed longitudinally over the course of 3 years and found that, in general, both CMP and non-CMP students increased their generalization abilities over the middle school years and that CMP students developed, on average, greater generalization abilities than non-CMP students. More specifically, the success rate for each question improved over time for both CMP and

Making Generalizations

Sally is having a party.

The first time the doorbell rings, 1 guest enters.

The second time the doorbell rings, 3 guests enter.

The third time the doorbell rings, 5 guests enter.

The fourth time the doorbell rings, 7 guests enter.

Keep going in the same way.   On the next ring a group enters that has 2 more persons than the group that entered on the previous ring.

A. How many guests will enter on the 10th ring? Explain or show how you found your answer.

B. How many guests will enter on the 100th ring? Explain or show how you found your answer.

C. 299 guests entered on one of the rings.   What ring was it? Explain or show how you found your answer.

D. Write a rule or describe in words how to find the number of guests that entered on each ring.

Fig. 1 The doorbell problem used in the LieCal open-ended assessment

non-CMP students, but the CMP students' success rate increased significantly more than that of the non-CMP students on questions A and C in the doorbell problem over the course of the middle grades (Cai, Moyer et al., 2011).

By examining the students' solution strategies on this open-ended task, we obtained further data to inform and confirm this finding. We coded the solution strategies for each of these questions into two categories: abstract and concrete. Students who chose an abstract strategy generally formulated an algebraic representation of the relationship between the ring number and the number of guests entering at that ring (e.g., the number of guests who enter on a particular ring of the doorbell equals two times that ring number minus one). These students then were able to use their generalized rule (e.g., to determine the ring number at which 299 guests entered). In contrast, those who used a concrete strategy made a table or a list or noticed that each time the doorbell rang two more guests entered than on the previous ring and so added 2's sequentially to find an answer.

Looking at the changes over time in the solution strategies students employed to solve the doorbell problem, we found that both CMP and non-CMP students increased their use of abstract strategies over the middle grades. Indeed, in the fall of 2005, only one CMP student and none of the non-CMP students used an abstract strategy to correctly answer question A, but in the spring of 2008, nearly 9 % of the CMP students and 9 % of the non-CMP students used abstract strategies to correctly answer question A. Similarly, nearly 20 % of the CMP students and 19 % of non-CMP students used an abstract strategy to correctly answer question B by the spring of 2008. Although only a small proportion of the CMP and non-CMP students used abstract strategies to correctly answer question C in the spring of 2008, the rate of increase for the CMP students who used abstract strategies from the fall of 2005 to the spring of 2008 was significantly greater than that for non-CMP students ($z=2.58$, $p<.01$).

Thus, these results provided additional detail that informed our conclusion that both CMP and non-CMP students increased their generalization abilities over the middle school years, but that on average, the CMP students developed their generalization ability more fully than did non-CMP students.

The China project did a similar qualitative analysis of the solution strategies that students employed to solve open-ended mathematics questions. A similar observation was obtained that the students receiving the new curriculum were more likely to use a more generalized strategy (e.g., algebraic or arithmetic representation) to solve open-ended questions such as the doorbell problem than the students receiving the conventional curriculum (Ni, Li, Cai, & Hau, 2009). The advantage of using the more generalized strategy became evident in students' solutions to the part of the doorbell problem where 299 guests enter.

Analyzing and Reporting Change Beyond the Grade Band

Generally speaking, mathematics curricula are designed to address the needs of students within a particular grade band, whether it be the elementary, middle, or secondary grades. Analyses of curricular effect, however, should not be limited to the grades in which students encounter the curriculum. Indeed, students' experiences with mathematics curricula can set them up for success or failure in their future mathematics classes. Thus, it is important for longitudinal curriculum analyses to follow students beyond the grade band in which they experience a curriculum to gauge the long-term effects of the curriculum.

The LieCal project initially measured curricular effect on students' learning of algebra while they were still in middle school. The middle school results suggested a potential parallel with findings from studies of Problem-Based Learning (PBL) in medical education (Hmelo-Silver, 2004; Vernon & Blake, 1993). Specifically, medical students who were trained using PBL approaches performed better than non-PBL (e.g., lecturing) students on clinical components in which conceptual understanding and problem solving ability were assessed, but performed as well as non-PBL students on measure of factual knowledge. When the medical students were assessed again 6 months to a few years later, the PBL students were found to perform better than their counterparts on clinical components and measures of factual knowledge (Vernon & Blake, 1993).

Thus, the LieCal project subsequently followed 1,000 of the CMP and non-CMP students into high school to investigate the hypothesis that the superior conceptual understanding and problem solving abilities gained by CMP students in middle school might result in better performance on delayed assessments of procedural skill, conceptual understanding, and problem solving. We used measures of open-ended problem solving in the ninth grade, basic mathematical skills (on the state test) in the tenth grade, and problem solving and posing in the 11th grade to probe the long-term effects of the CMP and non-CMP curricula that the students had used in middle school. On all three measures, we found that the use of the CMP curriculum in middle school had positive effects, not only on students' middle school performance, but also on their high school performance (Cai, Moyer, & Wang, 2013).

More specifically, we found that, controlling for middle school achievement, the ninth grade, former CMP students performed as well as or significantly better than the non-CMP students on open-ended mathematics problems. On the tenth grade state standardized test of basic mathematical skills, we found that the CMP students had a significantly higher scaled mean score than the non-CMP students (Cai, Moyer, & Wang, 2013). This result held for a series of analyses of covariance controlling for the students' sixth grade baseline scores on LieCal multiple choice and open-ended tasks as well as for their sixth, seventh, and eighth grade state standardized mathematics test scores. Similarly, on problem-posing tasks administered in the 11th grade, we examined the performance of groups of CMP and non-CMP students who had performed similarly on their sixth grade baseline examinations (Cai, Moyer, Wang, Hwang, et al., 2013). We found that the CMP students were more likely to pose problems that correctly reflected the mathematical conditions of the given problem situation than the comparable non-CMP students. Moreover, a detailed analysis of the students' problem-solving performance and strategy use showed that the CMP students appeared to have greater success algebraically abstracting the relationship in the problem-solving task (Cai, Silber, Hwang, Nie, Moyer, & Wang, 2014). Together, these results point to the longer-term effects of curriculum and thus highlight the importance of analyzing and reporting change beyond the immediate grade band in which a curriculum is implemented.

Interpreting Change in Mathematics Achievement

Interpreting change in mathematics achievement means identifying the causes that may be responsible for the observed change. This is an extremely important task for advancing knowledge of how educational inputs are related to educational outputs and thus to inform educational practice. It is also an extremely difficult task to accomplish. Below we describe our approach to interpreting change in our longitudinal studies and the lessons we have learned in the process (Cai, Ni, & Lester, 2011). In particular, we focus on the importance of establishing equivalent groups of students in comparative curricular studies and on the need for a conceptual model that informs an initial hypothesis.

Equivalence of Student Sample Groups

Both the LieCal and China studies were designed to investigate curricular influence on change in student learning outcomes by comparing two curricula. To infer any causal links between a curriculum and observed change in student learning outcomes in this type of comparative study, it is of paramount importance to set up equivalent groups of students to receive the curricula (NRC, 2004). However, it is often challenging to implement random assignment of students to one or the other curriculum because of administrative and ethical constraints. When this is not possible, it is wise to collect as much information as possible about the student sample and consider how any observed change in student achievement may be associated with characteristics of the student sample in addition to the curriculum factor. The LieCal project randomly selected reform curriculum schools, and was able to obtain information on the prior achievement of the students to create statistically comparable groups by selecting comparable non-reform schools. However, this was not possible in the Chinese project. The researchers could not equate the groups statistically because they lacked prior achievement data. This resulted in a high degree of uncertainty about the observed changes in student achievement being due to the different curricula the students had received. The problem might have been mitigated if the Chinese project had, for example, administered an intelligence test and used it as a control variable in the analyses. However, a problem would still have remained because intelligence test scores are only moderately correlated with school achievement. This underscores the importance of obtaining adequate information about student populations prior to the beginning of a comparative study.

Initial Conceptual Model

One must have a theory or hypothesis, regardless how rudimentary it may be at first, to design a curriculum study that can test how curricular influence is related to classroom instruction and, in turn, to students' mathematics achievement (Christie & Fierro, 2010; NRC, 2004; Weiss, 1998). In the LieCal project, we used the conceptual model shown in Fig. 2 of the relations among curriculum, teaching and learning to frame our investigation of the factors or processes that likely caused the observed changes in students' mathematics achievement (e.g., Cai & Moyer, 2006). We considered that curriculum materials including curriculum standards, textbooks, and teacher manuals would affect the kinds of learning tasks that the teachers selected and implemented and the types of classroom discourse that the teachers engaged in with their students. The nature of the learning tasks and classroom discourse implemented in the classroom would in turn affect learning processes and learning outcomes for students.

It would be ideal to test the entire set of relations described in Fig. 2 simultaneously and conclusively. However, this is almost impossible to implement technically. Among other issues, one major obstacle is that a measurement model involving so many variables would produce a covariance matrix so complicated that it would be impossible to make a sensible estimation of the parameters concerned (Ni, Li, Cai, & Hau, in press; Raudenbush & Bryk, 2002). This complication is made even more acute by the difficulty in reliably measuring the variables.

Facing this challenge in our projects, we used the problem-solving heuristic of "divide-and-conquer" to address our research questions. After having observed the changes in students' mathematics achievement and their association with the type of

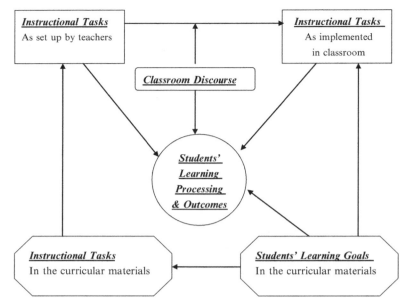

Fig. 2 Framework used in the two projects (Cai, 2007; Cai & Moyer, 2006; Ni et al., in press)

curriculum being implemented, the LieCal project used HLM analyses to investigate whether the conceptual or procedural emphasis of classroom instruction moderated the curricular influence on the achievement gains of the students. However, these variables did not show any meaningful influence. We then looked into the effect of the cognitive demand of instructional tasks. Using the classification scheme of Stein and Lane (1996), the instructional tasks actually used in the CMP and non-CMP classrooms were classified into four increasingly demanding categories of cognition: memorization, procedures without connections, procedures with connections, and doing mathematics. We found that the distributions of types of instructional tasks in the CMP and non-CMP classrooms were significantly different, with CMP teachers implementing a higher percentage of cognitively demanding tasks (procedures with connections and doing mathematics) than non-CMP teachers (Cai, 2014). In contrast, non-CMP teachers implemented a significantly higher percentage of tasks with low cognitive demand (memorization or procedures without connections). Moreover, we found that this variable was a significant predictor of achievement gains in the students receiving either curriculum.

Similarly, following the conceptual framework in Fig. 2, the China project examined the relationships of the cognitive features of instructional tasks (high cognitive demand, multiple representations, and multiple solution-strategies) to teacher–student classroom discourse on the one hand (Ni, Zhou, Li, & Li, 2014) and to students' mathematics achievement gains on the other hand in the Chinese mathematics classrooms (Ni, Zhou, Li, & Li, 2012). The results showed that high cognitive demand tasks were associated with teachers' high-order questions, which in turn led to students' highly participating responses. However, teachers tended to be more authoritative in evaluating student responses when they used high cognitive demand tasks or high-order questions. It was unexpected that teachers tended to ask low-order Yes or No questions when they elicited multiple solution methods from students for an instructional task. It appeared that the teachers just wanted students to talk more but did not press students to be accountable for their answers when pursuing multiple solution methods. Concerning the effects of the cognitive features of instructional tasks on student learning, the China project found that the cognitive features did not predict achievement gain on any of the cognitive learning outcomes (computation, routine problem solving, and complex problem solving). However, high cognitive demand of instructional tasks was shown to positively predict affective outcomes including students' expressed interest in learning mathematics, classroom participation, and a more dynamic view about mathematics. In turn, the indicators of students' positive attitude towards learning mathematics were significantly associated with their cognitive learning outcomes. These results illustrated the richness, complexity, and uncertainty of the links from the written curriculum to the implemented curriculum in classrooms and then to the achieved curriculum as shown in changes in student learning.

Our experience with the two projects indicates that a conceptual framework, such as the one in Fig. 2, is a necessary tool for planning and executing a quality longitudinal study of students' mathematics learning in relation to curricula and classroom instruction.

Conclusion

The LieCal project and the China project provide opportunities for us to consider the challenges in conducting high-quality longitudinal research into student learning. It is clear that the constructs we are interested in measuring are broad, requiring both careful definitions and well-chosen measures to address properly. If we wish to measure growth and change in students' academic achievement, it is necessary to use a variety of measures that address multiple facets of that growth and change. To characterize the effects of curriculum on student learning, diverse measures of conceptual understanding, procedural skill, problem-solving and problem-posing abilities, and interest and attitude toward learning mathematics are all useful tools.

In addition, the contexts and structures within which students learn guarantee that the data we collect will be complex. The methods of analyses we choose must therefore be suitable for the structure of the data and be sufficiently robust to take into consideration the many influences on student learning. Social and socioeconomic factors, the nature of classroom instruction, and many other factors can influence student learning, and thus the design of studies that include these factors must be carefully considered. Of course, no study design, however solid it may be, can address all of the potential influences. As we have done in planning the LieCal and China projects, researchers must use their conceptual models and hypotheses strategically to choose what to address and how, given the constraints of experimental design and ethical considerations.

As we consider the results from these two projects, we look forward to continued longitudinal research that seeks to conceptualize, measure, analyze, and interpret change in student learning. We conclude with a final note on the role of experimental studies and our expectations for them. It is important to note that the analyses done by both the LieCal project and the China project about the relations between classroom processes and gains in student mathematics achievement were descriptive in nature. Therefore, experimental studies are yet required to test and prove a causal link of the classroom processes to student learning outcomes. However, these correlational findings were derived from naturalistic situations in which the classrooms differed with respect to factors such as teachers' allocation of time to academic activities, classroom organization, and student backgrounds. The patterns of association observed in these situations do provide meaningful results that can guide further experimental studies and classroom practice (Brophy & Good, 1986).

Of course, not every experimental study using random assignment will produce causal links between a set of assumed factors and the observed outcomes. Conversely, it is always questionable for a non-randomized study to draw such causal links. Indeed, caution is always appropriate when interpreting the results of any single study. Consequently, consistency and replication of findings is the key to the generalization of any finding. A good example of this is the evaluation of the federally funded early childhood programs in the USA (Heckman, Doyle, Harmon, & Tremblay, 2009; Reynolds, 2000). On the one hand, the implementation of early childhood education varied in different states and communities. This made generalization of any particular finding about its effectiveness difficult. On the other hand, the assemblage of evaluations of programs that were carried out in diverse situations provided an excellent opportunity to examine whether or not a given finding about the effects of the programs could be observed across different circumstances. Converging evidence was obtained that indicated that the cognitive advantages for the children participating in the programs tended to disappear approximately 3 years after leaving the programs. However, those children who participated did benefit in terms of increased likelihood of retention in grade school, high school graduation, college education, and employment. The conclusions that arose from the convergence of consistent findings and the replication of those findings across diverse contexts have subsequently contributed to well-informed educational policy and practice for early childhood education. Similar concerted efforts are required to examine the robustness of findings about the influences of curricular and classroom variables on gains in student mathematics achievement in different circumstances and with different methods.

Acknowledgments Preparation of this article was supported by grants from the National Science Foundation (ESI-0454739 and DRL-1008536) and Research Grant Council of HKSAR, China (CERG-462405; CERG-449807) and the National Center for School Curriculum and Textbook Development, Ministry of Education of People's Republic of China. Any opinions expressed herein are those of the authors and do not necessarily represent the views of the funding agencies.

References

Betebenner, D. W. (2008). Toward a normative understanding of student growth. In K. E. Ryan & L. A. Shepard (Eds.), *The future of test-based educational accountability* (pp. 155–170). New York: Taylor & Francis.

Brophy, J., & Good, T. L. (1986). Teacher behavior and student achievement. In M. Wittrock (Ed.), *Handbook of research on teaching* (pp. 328–366). New York: Macmillan.

Burton, L. (1984). Mathematical thinking: The struggle for meaning. *Journal for Research in Mathematics Education, 15*, 35–49.

Cai, J. (1995). A cognitive analysis of U.S. and Chinese students' mathematical performance on tasks involving computation, simple problem solving, and complex problem solving. *Journal for Research in Mathematics Education monograph series 7*. Reston, VA: National Council of Teachers of Mathematics.

Cai, J. (1997). Beyond computation and correctness: Contributions of open-ended tasks in examining students' mathematical performance. *Educational Measurement Issues and Practice, 16*(1), 5–11.

Cai, J. (2000). Mathematical thinking involved in U.S. and Chinese students' solving process-constrained and process-open problems. *Mathematical Thinking and Learning, 2*, 309–340.

Cai, J. (2007). *Empirical investigations of U.S. and Chinese students' learning of mathematics: Insights and recommendations*. Beijing, China: Educational Sciences Publishing House.

Cai, J. (2014). Searching for evidence of curricular effect on the teaching and learning of mathematics: Some insights from the LieCal project. *Mathematics Education Research Journal*. doi:10.1007/s13394-014-0122-y.

Cai, J., & Moyer, J. C. (2006). *A conceptual framework for studying curricular effects on students' learning: Conceptualization and design in the LieCal Project*. Paper presented at the annual meeting of the International Group of Psychology of Mathematics Education, Prague, Czech Republic: Charles University in Prague.

Cai, J., Moyer, J. C., Wang, N., & Nie, B. (2011). Examining students' algebraic thinking in a curricular context: A longitudinal study. In J. Cai & E. Knuth (Eds.), *Early algebraization: A global dialogue from multiple perspectives* (pp. 161–186). New York: Springer.

Cai, J., Ni, Y. J., & Lester, F. (2011). Curricular effect on the teaching and learning of mathematics: Findings from two longitudinal studies in China and the United States. *International Journal of Educational Research, 50*(2), 63–143.

Cai, J., Wang, N., Moyer, J. C., Wang, C., & Nie, B. (2011). Longitudinal investigation of the curricular effect: An analysis of student learning outcomes from the LieCal project in the United States. *International Journal of Educational Research, 50*, 117–136.

Cai, J., Moyer, J. C., & Wang, N. (2013a). Longitudinal investigation of the effect of middle school curriculum on learning in high school. In A. M. Lindmeier & A. Heinze (Eds.), *Proceedings of the 37th Conference of the International Group for the Psychology of Mathematics Education* (pp. 137–144). Kiel, Germany: PME.

Cai, J., Moyer, J. C., Wang, N., Hwang, S., Nie, B., & Garber, T. (2013b). Mathematical problem posing as a measure of curricular effect on students' learning. *Educational Studies in Mathematics, 83*, 57–69.

Cai, J., Silber, S., Hwang, S., Nie, B., Moyer, J. C., & Wang, N. (2014). Problem-solving strategies as a measure of longitudinal curricular effects on student learning. In P. Liljedahl, C. Nicol, S. Oesterle, & D. Allan (Eds.), *Proceedings of the 38th Conference of the International Group for the Psychology of Mathematics Education and the 36th Conference of the North American Chapter of the Psychology of Mathematics Education* (Vol. 2, pp. 233–240). Vancouver, BC, Canada: PME.

Christie, C. A., & Fierro, L. A. (2010). Program evaluation. In P. Peterson, E. Baker, & B. McGaw (Eds.), *International encyclopedia of education* (Vol. 3, pp. 706–712). Oxford: Elsevier.

Fisher, A., & Foreit, J. (2002). *Designing HIV/AIDS intervention studies: an operations research handbook*. Washington, DC: Population Council.

Ginsburg, H. P. (Ed.). (1983). *The development of mathematical thinking*. New York: Academic.

Hatano, G. (1988). Social and motivational bases for mathematical understanding. In G. B. Saxe & M. Gearhart (Eds.), *Children's mathematics* (pp. 55–70). San Francisco: Jossey Bass.

Heckman, J., Doyle, O., Harmon, C., & Tremblay, R. (2009). Investing in early human development: Timing and economic efficiency. *Economics and Human Biology, 7*, 1–6.

Hiebert, J., & Carpenter, T. P. (1992). Learning and teaching with understanding. In D. A. Grouws (Ed.), *Handbook of research on mathematics teaching and learning* (pp. 65–97). New York: Macmillan.

Hmelo-Silver, C. E. (2004). Problem-based learning: What and how do students learn? *Educational Psychology Review, 16*, 235–266.

Hwang, S., Cai, J., Shih, J., Moyer, J. C., Wang, N., & Nie, B. (2015). Longitudinally investigating the impact of curricula and classroom emphases on the algebra learning of students of different ethnicities. In J. A. Middleton, J. Cai, & S. Hwang (Eds.), *Large-scale studies in mathematics education*. New York: Springer.

Kieran, T., & Pirie, S. B. (1991). Recursion and the mathematical experience. In L. P. Steffe (Ed.), *Epistemological foundations of mathematical experience* (pp. 78–101). New York: Springer.

Krathwohl, D. R. (2002). A revision of Bloom's taxonomy: An overview. *Theory Into Practice, 41*, 212–218.

Krutetskii, V. A. (1976). *The psychology of mathematical abilities in schoolchildren*. Chicago: University of Chicago Press.

Li, Q., & Ni, Y. J. (2011). Impact of curriculum reform: Evidence of change in classroom practice in the mainland China. *International Journal of Educational Research, 50*, 71–86.

Linn, R. L. (2007). Performance standards: What is proficient performance? In C. E. Sleeter (Ed.), *Facing accountability in education: Democracy and equity at risk* (pp. 112–131). New York: Teachers College Press.

Loveless, T. (2008). The misplaced math student. *The 2008 Brown Center Report on American Education: How well are American students learning?* Washington, DC: Brookings.

Ma, X. (2010). Longitudinal evaluation design. In P. Peterson, E. Baker, & B. McGaw (Eds.), *International encyclopedia of education* (Vol. 3, pp. 756–764). Oxford, England: Elsevier.

Mayer, R. E. (1987). *Educational psychology: A cognitive approach*. Boston: Little & Brown.

Mislevy, R. J. (1995). What can we learn from international assessments? *Educational Evaluation and Policy Analysis, 17*(4), 419–437.

Moyer, J. C., Cai, J., Wang, N., & Nie, B. (2011). Impact of curriculum reform: Evidence of change in classroom practice in the United States. *International Journal of Educational Research, 50*, 87–99.

National Research Council. (2001). *Knowing what students know: The science and design of educational assessment*. Washington, DC: National Academy Press.

National Research Council. (2004). *On evaluating curriculum effectiveness: Judging the quality of K-12 mathematics evaluations*. Washington, DC: The National Academies Press.

Ni, Y. J., Li, Q., Cai, J., & Hau, K. T. (2009). *Has curriculum reform made a difference? Looking for change in classroom practice*. Hong Kong, China: The Chinese University of Hong Kong.

Ni, Y. J., Li, Q., Cai, J., & Hau, K. T. (in press). Has curriculum reform made a difference in classroom? An evaluation of the new mathematics curriculum in the Mainland China. In B. Sriraman, J. Cai, K-H. Lee, F. Fan, Y. Shimuzu, C. S. Lim, K. Subramanium (Eds.). *The first sourcebook on Asian research in mathematics education: China, Korea, Singapore, Japan, Malaysia and India*. Charlotte, NC: Information Age.

Ni, Y., Li, Q., Li, X., & Zhang, Z.-H. (2011). Influence of curriculum reform: an analysis of student mathematics achievement in mainland China. *International Journal of Educational Research, 50*, 100–116.

Ni, Y. J., Zhou, D., Li, Q., & Li, X. (2012). *To feel it to better learn it: Effect of instructional tasks on mathematics learning outcomes in Chinese primary students*. Paper presented at the third meeting of the EARLI SIG 18 Educational Effectiveness, Zurich, Switzerland, 29–31 August 2012.

Ni, Y. J., Zhou, D. H., Li, X., & Li, Q. (2014). Relations of instructional tasks to teacher-student discourse in mathematics classrooms of Chinese primary schools. *Cognition and Instruction, 32*, 2–43.

Polya, G. (1945). *How to solve it.* Princeton, NJ: Princeton University Press.

Raudenbush, S. W., & Bryk, A. S. (2002). *Hierarchical linear models.* Newbury Park, CA: Sage.

Reynolds, A. J. (2000). *Success in early intervention: The Chicago Child-Parent Centers.* Lincoln, NE: University of Nebraska Press.

Schoenfeld, A. H. (1992). Learning to think mathematically: Problem solving, metacognition, and sense making in mathematics. In D. A. Grouws (Ed.), *Handbook of research on mathematics teaching and learning* (pp. 334–371). New York: Macmillan.

Schoenfeld, A. H. (Ed.). (1997). *Mathematical thinking and problem solving.* Mahwah, NJ: Erlbaum.

Steen, L. A. (1999). Twenty questions about mathematical reasoning. In L. V. Stiff & F. R. Curcio (Eds.), *Mathematical reasoning in grades K-12 (1999 Yearbook of the National Council of Teachers of Mathematics).* NCTM: Reston, VA.

Stein, M. K., & Lane, S. (1996). Instructional tasks and the development of student capacity to think and reason: An analysis of the relationship between teaching and learning in a reform mathematics project. *Educational Research and Evaluation, 2,* 50–80.

Sternberg, R. J. (1999). The nature of mathematical reasoning. In L. V. Stiff & F. R. Curcio (Eds.), *Developing mathematical reasoning in grades K-12* (pp. 37–44). Reston, VA: National Council of Teachers of Mathematics.

Sternberg, R. J., & Ben-Zeev, T. (Eds.). (1996). *The nature of mathematical thinking.* Hillsdale, NJ: Erlbaum.

Vernon, D. T., & Blake, R. L. (1993). Does problem-based learning work? A meta-analysis of evaluative research. *Academic Medicine, 68,* 550–563.

Weiss, C. H. (1998). *Evaluation: Methods for studying programs and policies* (2nd ed.). Upper Saddle River, NJ: Prentice Hall.

A Review of Three Large-Scale Datasets Critiquing Item Design, Data Collection, and the Usefulness of Claims

Darryl Orletsky, James A. Middleton, and Finbarr Sloane

Introduction

The number of government-coordinated longitudinal educational studies currently available to mathematics educators is somewhat staggering. There exist international studies such as the Trends in International Mathematics and Science Study (TIMSS) and the Program for International Student Assessment (PISA). Additionally, national-level studies exist such as the National Assessment of Educational Progress (NAEP), the National Education Longitudinal Studies (NELS) program, High School and Beyond (HS&B), Educational Longitudinal Study of 2002 (ELS:2002), and the High School Longitudinal Study of 2009 (HSLS:09) among others. This is just a partial listing of all of the studies completed or currently underway. If the number of longitudinal studies is staggering, then the number of research papers and reports based on the longitudinal studies is overwhelming. ELS:2002 has several hundred reports and papers associated with it on the NCES website alone; these reports do not include externally published reports from universities and private research firms which adds more layers to the information available.

In this chapter, we offer some background information on the design of large-scale studies and then focus on three studies as important cases illustrating critical issues for secondary data analysts interested in issues of mathematics teaching, learning, policy, and practice. The three were chosen because of their distinctly different missions. We then discuss item design and data collection after mentioning some

D. Orletsky (✉) • J.A. Middleton
Arizona State University, Tempe, AZ, USA
e-mail: darryl.orletsky@gmail.com; jimbo@asu.edu

F. Sloane
National Science Foundation, Arlington, VA, USA
e-mail: fsloane@nsf.gov

© Springer International Publishing Switzerland 2015
J.A. Middleton et al. (eds.), *Large-Scale Studies in Mathematics Education*,
Research in Mathematics Education, DOI 10.1007/978-3-319-07716-1_14

salient details of each study. This is followed by an examination of the issues raised by cluster sampling, a technique common to the three studies and, increasingly, to others. We conclude with a discussion of the validity and usefulness of the claims made based on these studies.

The three studies we chose to examine in detail are the Education Longitudinal Study of 2002 (ELS:2002), TIMSS, and the NAEP. ELS:2002 was chosen because of its longitudinal design and its focus on the social, economic, and other environmental factors affecting a student. TIMSS was chosen because it is an international assessment, and NAEP was chosen because of its focus on academic change in the USA.

Education Longitudinal Study of 2002

ELS:2002 is a nationally representative, longitudinal study of 15,000 tenth graders and their parents, also surveying students' mathematics and English teachers, and school administrators. Data collection began in 2002, when students were in tenth grade. Mathematics achievement was assessed in the tenth grade, and again in the 12th grade. Students were followed up through their postsecondary years in 2005, 2006, 2012, and most recently in 2013. High school transcripts were also recorded. The intent of the study was to assess students' trajectories from the beginning of high school into postsecondary education, the workforce, and beyond, hopefully highlighting key trends in course-taking, achievement, support systems, and patterns of college access and persistence that occur in the years following high school completion.

In all, over 6,000 data points were recorded for each student. The data included questions for parents, teachers, and schools (see nces.ed.gov/surveys/els2002/policy.asp for a comprehensive list). The mathematics test consisted of 81 questions covering five conceptualized levels: (1) simple arithmetical operations with whole numbers; (2) simple operations with decimals, fractions, powers, and roots; (3) simple problem solving, requiring the understanding of low-level mathematical concepts; (4) understanding of intermediate-level mathematical concepts and/or multistep solutions to word problems; and (5) complex multistep word problems and/or advanced mathematics material (Bozick & Lauff, 2007).

Mathematics items are developed using the following method: (1) Items are written with a specific framework in mind, sampling from different mathematical content domains (e.g., number, algebra, geometry, and statistics), with numerous items from each domain written in several formats such as multiple choice, open response, and so on; (2) Items in the pool are then typically reviewed by experts and teachers and subjected to extensive editing and refinement; (3) Items are then field tested with students; (4) Items are then calibrated using Item Response Theory (IRT) modeling; and then (5) a subset of items are chosen that exhibit utility for discrimination among student responses. This process may be iterated several times before any given year's item pool is selected. Additionally each year items are "retired" from the pool for two reasons: (1) Some items do not display good discrimination among student responses and (2) representative items are used to inform the public about the general content of the test.

A special note about IRT methods is appropriate here, because their use is so ubiquitous among large-scale assessments. IRT uses patterns of answers to determine an estimate of students' achievement across different test forms. Each item on the test is rated for difficulty, discriminating ability, and guessing factor. Unfortunately, ELS, like TIMSS and NAEP, does not provide information regarding the specific process for item development. Additionally, it is impossible, without traveling to the Department of Education headquarters in Washington, DC, to view the actual items on the assessment. This is true of all three studies examined. Beyond describing utilizing an iterative process using committees of stakeholders, the process of item development remains obscure. The TIMSS and NAEP technical manuals provided substantially more background information and theoretical justification than the ELS:2002.

National Assessment of Educational Progress

NAEP is the largest ongoing, nationally representative and continuing assessment of American students. 10,000–26,000 students are tested each year, providing the nation with a "Report Card" of sorts by which to judge state, regional, and national-level trends. Two forms of NAEP exist: (1) the long-term trend assessment (LTTNAEP) measures subject-matter achievement in mathematics and reading every 4 years, whereas (2) the main NAEP measures subject-matter achievement in mathematics, reading science, writing, the arts, civics, economics, geography, US history, and technology and engineering literacy (TEL) every 2 years. Both forms of NAEP gather data on instructional experiences, and examine school-level environments, albeit LTTNAEP only surveys students while the main NAEP surveys students, teachers, and gathers data concerning school facilities. Both forms gather data in a manner to facilitate reporting the results for various populations (grades) of students and also for groups within those populations based on characteristics such as sex and ethnicity (http://nces.ed.gov/nationsreportcard). Since NAEP reports on populations and groups within populations, individual school or student scores are not available. State-level reports are available along with some large urban district reports.

Important features of NAEP include that it is essentially the same assessment from year to year, allowing comparison of cohorts on roughly the same metric. Due to its rather complex sampling procedure, data does not represent students, but grades, states, regions, demographics, and higher levels of social organization. Representative samples of students are gathered at grades 4, 8, and 12 for the main NAEP, and samples of students at ages 9, 13, or 17 years are assessed for LTTNAEP (nces.ed.gov/nationsreportcard/).

Items for mathematical achievement questions for NAEP are based on a framework developed by the National Assessment Governing Board. The framework provides four points of consideration: (a) the theoretical basis for the assessment; (b) the types of items that should be included in the assessment; (c) how the items should be designed; and (d) how the items should be scored (nces.ed.gov/nationsreportcard). The mathematics framework conceptualizes five content areas and requires that each question measure one of them. The five areas include number properties and operations, measurement, geometry, algebra and data analysis with statistics and probability.

Both LTTNAEP and the main NAEP include items from all five areas and both measure basic skills along with the recall of definitions; however, each form of the NAEP has a different focus. In particular the main NAEP emphasizes problem solving and reasoning. The framework also requires a balance among content, complexity, format, and context. Both NAEP forms allow a calculator on some items and allow manipulatives for other items which is unique among the three studies being examined.

Data collection for NAEP varies depending on grade level, content, and assessment level, e.g., national, state, or district. At the state level, assessments for mathematics average 100 public schools and 2,500 students with 30 students per grade per subject being selected randomly in each school. A method of stratified random sampling within categories of schools is used for schools with similar characteristics. At the national level samples of schools are selected to represent the diverse population of students in the USA based on the Common Core of Data (CCD), which is a comprehensive list of operating public schools in the USA. The number of schools and students selected is sufficient to separately analyze the four NAEP regions of the country. Data on sex, race, degree of urbanization of school location, parent education, and participation in the National School Lunch Program (NSLP) is also collected. Students may also be classified as students with disabilities or as English language learners.

A separate sampling of private schools is taken to produce data concerning their performance at the regional and national level. To avoid nonresponse bias, in 2003 the NAEP changed their required reporting rate for schools in the sample from 70 to 85 %.

Trends in International Mathematics and Science Study

TIMSS is an international study, conducted every 4 years. The latest iteration (2011, the next administration is slated for 2015) included over 500,000 students in 60 countries. Over 20,000 students from more than 1,000 schools in the USA participated (http://nces.ed.gov/timss/).

Data on content knowledge and cognitive abilities is collected along with background information about the student, teacher, school resources, curriculum, and instruction. Since its inception (TIMSS 1999, but see also its precursors, the First and Second International Mathematics Studies, 1966–1972 and early 1980s, respectively), TIMMS has assessed the mathematics and science achievement of US fourth- and eighth-grade students. The program began testing 12th-grade students on the third TIMSS in 2007. The framework for TIMSS assessments generally, and for item design specifically, is different from ELS:2002 and NAEP because of its international aspect. The item design and data collection process is "an extensive collaborative process involving many individuals and expert groups from around the world..." (Mullis, Martin, Ruddock, O'Sullivan, & Preuschoff, 2009).[1]

Item design for TIMSS is based on two dimensions: content and cognitive. The content dimension specifies the domains to be tested. Domains are broad mathematical strands such as algebra, geometry, and number. The domains vary based on whether it is the grade 4 or grade 8 assessment. The cognitive dimension specifies the thinking process that is to be assessed: knowing, applying, or reasoning. Unlike the content domains, the cognitive domains remain the same across grades. The *TIMSS 2011 Assessment Frameworks* handbook gives detailed justifications and explanations for each aspect of the dimensions and domains. Like NAEP, TIMSS uses a matrix clustering process to develop eight test books of equal difficulty, content, and cognitive coverage. "Equal difficulty" is determined using probabilities of answering correctly for items in field trials.

The design framework specifically requires simplicity of language to decrease the reading load for both questions and answers, thus taking into account language differences across participating nations. Additionally, multiple choice answers and distractors must be "written to be plausible, but not deceptive." Constructed-response questions require the development of scoring rubrics that capture gradients of correct answers and also wrong answers since "[D]iagnosis of common learning difficulties in mathematics and science as evidenced by misconceptions and errors is an important aim of the study" (Mullis et al., 2009, p. 129).

Sampling Issues

Analyzing mathematical achievement on these three case assessments requires understanding the nature of the sampling process utilized for both *data units* (students and/or organizations) and items. For data units, because they utilize cluster sampling the NAEP and TIMSS studies require that three considerations be made regarding measurement and parameter estimation. First, multiple *plausible* values for each student must be used for each analysis. Second, the sampling design is not simple random sampling within blocks; therefore, sampling weights for key units of school organization, geographic region, and demographics must be used to generate unbiased estimates of population parameters. Third, cluster sampling changes the calculation of standard errors so special procedures must be used to adjust for the design when computing confidence intervals.

With regards to item sampling, the design of ELS:2002 and TIMSS nests items within booklets; not every mathematics item is in every booklet, and therefore levels of items are not crossed with levels of students. Instead, equivalent clusters of items are assigned to each test booklet. This process increases measurement error in the

---

[1] Bracey (1997) reported that Albert Beaton, while Director of TIMSS, remarked on several occasions that there are two things one doesn't want to watch being made (sausage and legislation), and this should include a third: international test items.

scores for each student. To adjust for this TIMSS recommends that five plausible values taken at random from each student's estimated distribution of achievement scores be used as an index of student proficiency.

Regardless of the sampling issue confronted, if the researcher is interested in generalizing findings to sub-populations, understanding and utilizing sampling weights is critical for effective parameter estimation. Fortunately, each study provides sampling weights to be used during analysis in the dataset. According to the TIMSS technical report (2011), statistical software programs like SAS and SPSS calculate error estimates based on simple random samples. TIMSS and ELS:2002 recommend the adoption of the jackknife repeated replication (JRR) technique which splits single samples into multiple subsamples (Efron & Stein, 1981). The technique then uses fluctuations among the subsamples to estimate the overall variability in sampling. TIMSS further suggests using the SPSS version of the *International Database Analyzer* (IDB) software.

ELS:2002 conceptualized data analysis at four levels (ELS:2002 in NCES 2006-344, 2005): (1) Cross-sectional profiles of the nation's high school sophomores and seniors (as well as dropouts after spring of the sophomore year); (2) Longitudinal analysis (including examination of life-course changes); (3) Intercohort comparisons with American high school students of earlier decades; and (4) International comparisons: US 15-year-olds to 15-year-olds in other nations. Each of these levels requires a different set of weights, since the sub-population to which student responses are generalized involves imputing scores across different booklets, different student populations each subsequent year of the longitudinal study, and different demographics (e.g., age, gender, ethnicity, SES).

Validity

It is necessary to address two issues before critiquing the validity of these three studies and by extension other large-scale studies. First, to avoid confusion or a definitional debate, we utilize definitions from the landmark work of Shadish, Cook, and Campbell (2002), hereafter abbreviated as SCC for readability. Statistical and internal validity are found in the second chapter while construct and external validity are located in Chap. 3.

Second, before analyzing the validity of a study, it is important to understand the stated goal of the study. If, for example, a study makes no causal claims, then to criticize the external validity of the study misapplies the concept of external validity. We note, however:

> …in the case of generalizations to persons, settings, treatments, and outcomes that were not studied, no statistical test of (threats to external validity) is possible. But this does not stop researchers from generating plausible hypotheses about likely interactions, sometimes based on professional experience and sometimes on related studies, with which to criticize the generalizability of experimental results and around which to design new studies. (SCC, p. 86)

This means that using study results beyond the author's intended usage is permissible if the purpose is to generate plausible hypothesis and design new studies that *can* test these hypotheses rigorously; nonetheless, a first level critique must be based on whether the study met its goals in a valid manner and this is only possible if the goals of a study are apparent.

After an extensive search, we conclude that ELS:2002 did not have a formally stated mission or goal. It did have two stated focus questions that provide some help in determining the unstated goals of the program:

- What are students' trajectories from the beginning of high school into postsecondary education, the workforce, and beyond?
- What are the different patterns of college access and persistence that occur in the years following high school completion? (http://nces.ed.gov/surveys/els2002)

Additionally, the NCES website breaks the ELS:2002 into six "policy and research issues": (1) transitions, (2) equity, (3) cognitive growth, (4) course taking, (5) school effectiveness, and (6) social capital. Each of these issues has multiple (10–25) listed research questions. The cognitive growth questions pertinent to this critique are:

- What are the trajectories of cognitive growth over time for different groups of students?
- How do course taking patterns relate to different rates of cognitive growth?
- At what levels of proficiency do students score in reading and mathematics?
- Which background factors are associated with higher achievement?
- What is the role of student involvement/academic and extracurricular engagement in predicting cognitive growth?
- What is the influence of language proficiency and use on tested achievement?
- Which background factors and family variables are associated with higher achievement?
- What is the role of parental involvement in student achievement and other outcomes?
- How is teacher quality related to achievement gains over time?
- How are school-level characteristics related to cognitive growth?
- How well are high schools preparing students for higher education and the world of work?
- What school and teacher factors are associated with different levels of cognitive growth?

These research questions each carry multiple issues of validity and usefulness. Many of these issues will be examined in the following sections.

The primary stated mission of the NAEP is to assess what America's students know and can do in various subject areas. However, there are now two faces to the NAEP: the main NAEP and the LTTNAEP study. The NCES website claims that these two parts to the NAEP "...makes it possible to meet two important objectives: to measure student progress over time, and as educational priorities change, to develop new assessment instruments that reflect current educational content and assessment methodology."

To measure student progress over time, the LTTNAEP assessment stays essentially the same from year to year, with only carefully documented changes. This permits NAEP to provide a clear picture of student academic progress over time. The main NAEP is changing towards computer-based assessments matching the stated objective of developing new assessment instruments. It is unclear how the types of items, and the sampling procedure for NAEP will change, if at all, to take advantage of computer-based administration (http://nces.ed.gov/nationsreportcard).

The International Association for the Evaluation of Educational Achievement (IEA), the parent organization for TIMSS, states that its mission is to "conduct comparative studies of student achievement in school subjects to inform educational policies and practices around the world" (TIMSS 2011 Executive Summary, p. 1). By comparing education systems in terms of their organization, curricula, instructional practices, and corresponding student achievement, TIMSS situates its goals clearly within policy analysis. It also dedicates itself to the improvement of teaching and learning in science and mathematics, and its curricular and instructional components align with this goal.

Our validity critiques are organized into four separate types: statistical, internal, construct, and external. Each type of validity has a brief introductory explanation followed by a breakdown into subtypes with specific example threats related to the three studies. The example threats we cite include both published results and our own musings. Some of these example threats are connected to multiple subtypes, e.g., homeschooling. In these cases a detailed explanation is offered when the specific example first appears.

Throughout the critique, words such as treatment, curriculum, ethnicity, funding, and teacher qualifications are conceptualized as an independent variable, whereas mathematical achievement is considered a dependent variable. We also consider the total "treatment" duration to be from the start of formal schooling as opposed to just the particular year in school, i.e., a student in fourth grade has 4–5 years of treatment (depending on kindergarten attendance), not just one—the fourth grade.

Statistical Validity

Statistical validity considers whether a presumed cause and effect are linked, or whether two hypothetically related factors covary. If they do covary, then the strength of the covariance must be considered to determine possible causal relationships. Nine threats to statistical validity are listed by SCC. We offer example threats for eight of them.

Violated Assumptions of Statistical Tests. The assumption of independent distributed errors is fundamental to most statistical tests. Violating it introduces bias into the estimation of standard errors. Nested designs account for much of this issue; however, large international studies may find it difficult to account for the diversity of the nested samples. For example, a small country with a relatively homogeneous culture such as Norway will have a simpler nested structure than the USA

Fishing and Error Rate Problem. It has long been known that, if hypothesis testing is repeatedly used on the same data set then the probability of making a Type I error grows substantially with each successive test (exceeding 60 % with 20 tests and 90 % with 50 tests) unless corrections such as a sequential Bonferroni technique are applied (Benjamini & Hochberg, 1995). All the large-scale studies reported here are at their core *data acquisition studies*; none exist to test a specific hypothesis. Given the nature of the studies and the large data sets gathered, multiple teams of researchers separated by time and distance looking at the same data set constitute a kind of distributed and extended fishing expedition. However, few studies utilizing these data apply appropriate correction factors for Type I error rate, nor is the cumulative effect of fishing across research groups estimated.

Unreliability of Measures. It seems obvious that if either the dependent or independent variable is measured poorly, then any findings of covariance may be spurious. All three studies reported checking the reliability of the mathematics portion of their assessments; however, upon reading multiple technical manuals for all three studies, we found no references to the reliability of the survey questions (e.g., Garden & Orpwood, 1996). For example, ELS:2002 surveys mathematics teachers concerning whether a student has fallen behind and further queries why the student has fallen behind. The responses available to the teacher were student health problems, disciplinary problems, lack of effort, lack of English language proficiency, and other. No evidence is available that a check was conducted on the accuracy of teachers' responses, their interpretation of the prompts, and students' actual reasons for falling behind.

Restriction of Range. Restrictions on the dependent variable (mathematical achievement) introduce possible floor and ceiling effects. This may cause a decrease in variance and a lack of power. When the independent variable is overly restricted possible effects may be eliminated completely. Considering TIMSS and the dependent variable, it is possible that certain countries have a substantial number of students that are capable of performing well above their grade level in mathematics; however, the exceptional ability of these students is lost because the item pool does not include questions beyond grade level. IRT models may then smooth over any subsequent skew.

However, examination of item response patterns at least insures a normal distribution of IRT scores, making rankings of student performance across the midspread of the data reasonably accurate. Coupled with the relatively small standard errors afforded by such samples, restriction of range is a relatively minor problem, if at all, for these data sets.

Unreliability of Treatment Implementation. This refers to the uneven or nonstandard implementation of independent variables. Examples include measuring or changing:

- School funding but not how the funding is spent
- Teacher professional development but not the type or rigor of the program (Garet, Porter, Desimone, Birman, & Yoon, 2001)

- Teacher education in credit hours but ignoring the type and rigor of course work
- National or intended curriculum, but ignoring the implemented and achieved curriculum

We found evidence of these types of issues in all three studies. For example, ELS:2002 surveyed teachers asking them how many hours of professional development they have had in the last year on how to teach special education students. There were no further questions asked concerning rigor, content, or implementation. Teachers were also asked on the ELS:2002 how many mathematics classes (not hours or credits) they took as part of their undergraduate studies. The survey made no distinction between a 2-day conference; a non-credit remediation course taken at a community college; and a five credit-hour accelerated freshman calculus course. The TIMSS teacher survey did not ask the number of undergraduate hours in mathematics; it simply asked if the teacher had a BA and if education, mathematics, or mathematics education was the primary area of study.

Extraneous Variance in Experimental Settings. Covariance may be incorrectly determined if errors are inflated due to environmental conditions. This may manifest itself as an uncomfortable temperature or background noise in the classroom on the day of the assessment. It may include extreme weather, power outages, or a recent unusual event at the school. The literature review found minimal references to these sorts of issues beyond minimal cautions in the directions for the test administrator. It can be assumed that such variables are relatively random for sampled units; schools are unlikely to change the temperature of their classroom systematically for the testing situation, and therefore the only source of error is likely random error.

Heterogeneity of Units. This may reduce the probability of detecting covariance between the independent and dependent variables. This is a significant threat in diverse countries such as the USA (Bracey, 2000). SCC suggest sampling students who are homogeneous in regards to characteristics correlated with major outcomes (p. 51). For example, all three studies record classroom characteristics such as the teacher–student ratio. Policy decisions regarding classroom size, the number of teachers needed, and funding, may be made based on achievement scores and this ratio in an attempt to boost mathematics scores. However, it may be that smaller classroom sizes will have no effect on students with high socioeconomic index (SEI) scores whereas students with lower SEI scores will benefit. This result cannot be parsed out of the data if the classrooms are heterogeneous in regards to SEI. If a limited number of classrooms are SEI homogeneous, then a significant result may be found, but the external validity (generalization) of the result would be weak.

Internal Validity

The internal validity of a study concerns the systematic design and conduct of the study so that the only variables influencing the results are the ones being studied. In brief, internal validity concerns design of instruments, assignment of units to experimental conditions, and to equivalence of measures and conditions over administrations in longitudinal designs. Internal validity is critical to the establishment of claims from data in survey research in that it enables the researcher to estimate the extent to which spurious effects, extraneous variation, and equivalence of experimental units across administrations contribute to the variability in the response variable being modeled. For example, each of the reviewed studies took extreme care in defining the sampling frame for selecting students across political and demographic strata. Because randomization into experimental units is not possible for survey methods, random selection of large-sized samples within demographic blocks enables researchers to create randomly assigned, functionally equivalent subgroups that may differ on one salient independent variable (say, number of homework hours per week; e.g., Perie, Moran, & Lutkus, 2005). While not as powerful as a true experiment, such statistical controls, when repeated for numerous random subsamples (using bootstrapping or permutation methods, for example), can isolate the overall effects of an independent factor on the response variable. Then, disaggregation into important subgroups as a whole can help estimate the differential effect of the independent factor across those groups. In such a manner, policy can be critiqued for its equity and impact.

Of the nine different threats to internal validity identified by SCC, we focus on five that particularly impact large-scale studies, and particularly among those, studies that attempt to collect coherent longitudinal data: Selection, History, Regression, Attrition, and Instrumentation.

Selection. If particular differences in the average characteristics of experimental groups exist prior to treatment then any differences measured after treatment may not be due to the treatment. Randomization is the theoretical method of avoiding selection bias; however, the issue is the representativeness of samples. True randomization and selected representative samples introduce issues raised under the statistical validity subtype of *heterogeneity of units*. True randomization may introduce extreme heterogeneity and attempting to select representative samples introduces tautological issues. For demographics, both ELS:2002 and NAEP oversample key demographics (e.g., ethnicity) to ensure appropriate sample size and heterogeneity of units.

For TIMSS, it is important to keep in mind that some participating nations did not strictly follow sampling protocol. Only six of the 21 countries that participated in the advanced math test met the standards for sampling on the third TIMSS. On the standard math and science tests, eight countries met the sampling criteria (Bracey, 2000). In particular, countries took samples from varying percentages of their total in-school student population. These rates varied from 97 to 77 % (Bracey, 1998). Questions about which groups of students were not selected from, what were their common characteristics if any, and why were they not selected from remain largely unanswered (Bracey, 1998). Holliday and Holliday (2003) suggested that some education ministries intentionally ignored sampling protocols.

Additionally, the average age differences in tested student populations for the "end of secondary school" advanced mathematics and physics tests varied by nearly 3 years. It depended on each country's definition of "end of secondary school." For some countries it is 12th grade, and for others it is the 14th grade, including 2 years of specialized mathematics and science instruction (Bracey, 1998, 2000).

Another issue was raised by Hong (2012) who pointed out that international studies generally only sample countries with sufficient resources and commitment to conduct the study; however, the IES did report that it is increasing financial assistance on the next TIMSS to convince more countries to participate (Reddy, 2006).

All three analyzed assessments undersample students who leave school early or drop out. TIMSS collected data on the size of the school-leaving cohort and acknowledged that these students were not sampled. Loveless (2008) wrote, "This percentage varied greatly, from more than 50 % in South Africa to about 35 % in the United States to 12–15 % in Norway and France" (p. 29).

History. This threat refers to all relevant events that occur between the start of treatment and posttest that might have produced the effect even without the treatment.

As an example, consider a large urban school district that after doing poorly on the mathematics section of the NAEP decides to improve their mathematics classrooms by remodeling them with the latest technology. The next time the district takes the NAEP their scores improve and credit is given to the upgraded mathematics facilities. However, a few researchers note that because the district fared poorly last time on the NAEP,

- Several new mathematics teachers have been hired in the district.
- Mathematics teachers modified their assessments to match the layout (feel) of the NAEP.
- Mathematics teachers modified their lessons to spend more time on topics covered by the NAEP; of course, this means that they spent less time on other topics.
- Several practice tests were given in the weeks before the NAEP.

All of these are plausible reasons for the district's improvement on the NAEP; hence, any study or claim that concluded that the classroom improvements led to the increased scores could be threatened with claims of poor internal validity.

Multi-year age differences of students taking the same test (discussed earlier under *Selection*) can also be considered a historical threat to internal validity. A student will not get better at algebra by simply living longer; however, life experiences may make story problems more meaningful by allowing the student to better understand their context (Holliday & Holliday, 2001).

TIMSS sometimes seems inconsistent concerning the "length of treatment" for which the independent variable was applied. TIMSS clearly understands that ethnicity affects the entire time span of students' lives. However, the manner in which it decides which items to include on the fourth-grade assessment indicates that it considers the length of the fourth grade to be the treatment time since it ignores earlier intended curriculum. TIMSS decides items by examining the intended national curriculum for all participating nations at the fourth-grade level and then suggesting items which participating countries score on a scale running from acceptable to unacceptable (TIMSS Technical Report, Volume I, 1995, p. 2–5). An item appearing on TIMSS must appear in at least 70 % of the national curriculums and not be unacceptable to more than 30 % of the countries. Cogan and Schmidt (2002) found that several countries had a limited national curriculum meaning that a large number of questions which appeared on the assessment were not covered in the fourth grade of these countries. This seemingly should have led to poor scores; however, several of these countries did well. An analysis of their entire national curriculum showed that the material had been covered in earlier years. Studies which examine the fourth-grade inputs (independent variables) of countries that did well on TIMSS but that ignore the entire history of the students could be threatened with claims of poor internal validity.

Regression. SCC pointed out that treatment selections are sometimes made because a respondent received a very low score on some measure (p. 57); however, this low measure may be a one-time event for the respondent who typically would have scored much higher. In the case of this critique, a country or district may receive funding for some intervention (e.g., teacher professional development) because of low mathematics scores; however, the low scores may have been the result of an undetermined one-time factor, such as bad weather in the days prior to the assessment which interrupted the regular review and preparations of the students.

We did not find any convincing instances of this threat *external* (e.g., bad weather) to the reviewed studies This is not surprising because the large number of districts, schools, and students sampled makes this threat less likely (not to mention nations in TIMSS). However, Cogan and Schmidt (2002) concluded that particular sets of items disadvantaged particular countries and that different sets of questions would have likely changed the rankings of some countries. This means that a poor showing on TIMSS may be rectified on the next TIMSS simply by choosing a different set of questions—a regression to the mean for that country. Any study of an intervention administered between the assessments which appeared to have positive results could have its internal validity challenged.

Attrition. The internal validity issue of attrition is that if a particular subgroup of people tend to drop out of the study, then the study may show treatment results even though none actually exist. All studies that use longitudinal data collection are threatened by attrition and corresponding incomplete or missing data sets.

Bozick and Lauff (2007) reported the following rates for ELS:2002. The initial number of selected sophomores was 17,600. Approximately 15,400 completed a base-year questionnaire, for a weighted response rate of 87 %. When the sophomores had become seniors a follow-up interview took place. The follow-up interview looked at 16,500 students, of whom 15,000 participated, for a weighted response rate of 89 %. One year after the sample members should have graduated from high school, transcripts were requested. At least one transcript was collected for 14,900 of the 16,400 eligible students, for a weighted response rate of 91 %. These numbers show that the data collected by ELS:2002 generally met the de facto guidelines that a study must maintain an 85 % response rate to avoid validity threats due to attrition—a remarkable accomplishment! However, it is also necessary to examine the response rates of any subgroup for which claims are to be made. Several authors (e.g., Strayhorn, 2009) caution and give techniques for adjusting large databases especially if a particular group seems to be more prone to dropping out of the study.

Instrumentation. A change in a test may mimic a treatment effect, causing spurious claims about the effects of policies or practices established in the interim period between administrations. Since the assessments reviewed in this paper all use cluster sampling of items, not every student gets the same items on each administration. This puts a premium on establishing equivalencies across test booklets. IRT controls for this issue fairly well, selecting items that have similar performance characteristics and response probabilities.

Construct Validity

Two major issues exist when examining construct validity: clearly specifying the construct and then measuring it. Specifying the construct requires identifying prototypical units and this is often difficult because researchers do not agree on the features of the prototype. See Shadish et al. (2002), Lakoff (1985), and Mervis and Rosch (1981) for a discussion of categorization and prototype difficulties. Even if the construct can be agreed upon, measuring it is an imprecise science due to both imprecise construct definition, and poor operationalization. SCC list 14 threats to construct validity. We found applicable examples for eight of these threats that pertain especially to large-scale survey methods: Inadequate explication of constructs; Construct confounding; Mono-Operation bias; Mono-method bias; Confounding constructs with levels of constructs; Reactive self-report changes; Experimenter expectancies; and Resentful demoralization. Threats we do not identify here, such as *Novelty and Disruption Effects*, and *Treatment Diffusion*, are certainly present in all studies with such complicated designs, but are not easily assessable from secondary documentation and therefore are not analyzed in this chapter. Threats such as *Treatment Sensitive Factorial Structure* are also inherent to the highly complex teaching and learning systems being studied: As the number of potential subgroups, independent factors, and potential group × factor interactions increases, the ways in which those permutations impact the variation in assessment subscales also increases.

Inadequate Explication of Constructs. SCC suggested that there are four common errors in explicating constructs (p. 74): (1) the construct is too general; (2) the construct is too specific; (3) the construct used is wrong; and (4) two or more constructs are unknowingly being measured in one study operation. In all three studies the explication of constructs was their weakest part.

Each of the assessment programs contains a component that administers an assessment focused on mathematics; however, the overriding construct of the mathematics assessment is never explained by any of the studies. Though it is never explained, perhaps it is "mathematical achievement." Beaton, Linn, and Bohrnstedt writing for the NAEP Validity Studies (NVS) panel, wrote in 2012 that the NAEP had always measured achievement in the past but that it was now time to consider measuring predictive achievement with an emphasis on preparedness "to qualify for entry-level college courses leading to a degree or job training without the need for remediation" (p. 1). Changing the construct definition forces test designers to change the operational definition, making comparison across administrations of NAEP (should this new definition be applied) difficult at best.

Studying Hong (2012), it appeared that teacher effectiveness was being measured by TIMSS, though their measurements were operationalized through multiple choice mathematics questions. If this appearance is correct, then both internal and construct validity are difficult to defend: internal validity is threatened by causal direction confusion, and construct validity is threatened by an error of multiple constructs being measured in a single operation.

Wang (2001) cautioned that the instrument and items being used by TIMSS to measure mathematical achievement will de facto become mathematics. He argued that reforms in mathematics and science teaching in the USA and Japan "...stress higher order thinking, complex scientific and mathematical reasoning and hands-on experience...and call for problem solving and first-hand, original investigations" (p. 19). However, Lange (1997) found that the TIMSS test "...measures mostly lower learning outcomes by means of predominantly multiple choice format..." (p. 3). Further, Lange pointed out that there were 429 multiple choice but only 29 extended-response items on the assessment. Wang continued, "Because students were tested on subsets of questions from this item pool, the TIMSS scores cannot reflect the kinds of outcomes that are emphasized in reform initiatives that focus on higher order thinking and hands-on experiences" (p. 19). Cogan and Schmidt (2002) mentioned that countries that did well on TIMSS are considered to have a "world class" curriculum (p. 2); hence, if your students do well on lower order thinking problems, then you have a world class curriculum.

Ndlovu and Mji (2012) suggested that, if South Africa wants to improve its results, then it should align the curriculum to teach items on the test. One-third of the items on TIMSS are released after each assessment, so curricular alignment is possible. Construct validity is threatened because the construct has morphed into the operation; however, internal validity is also threatened because it would not be treatments per se that improved results; rather, it would be the test itself that changed the results to the test.

TIMSS has the additional problem of having to seek consensus for each item on the assessment (see Internal Validity: *History* for details). This interferes with selecting the best item to operationalize a construct. A similar problem exists for all assessments which select from a pool of potential items. The problem is that selected items require a level of variance for reliability and to avoid floor and ceiling effects. This means that the best items (in the sense of prototypical) may not be selected to operationalize the construct, but to optimize response variability.

As mentioned at the beginning of this section, the construct of mathematics achievement is never adequately explained by any of the studies; however, all three of the studies operationalized the construct as being able to correctly answer multiple choice questions (though NAEP and TIMSS also use alternate formats for some items). There are at least two issues with this operational concept: (1) multiple choice questions can be successfully solved in a variety of ways which do not measure the construct (e.g., utilizing test-taking strategies) and (2) few of the items resemble real-world mathematics usage.

In the first case, facilities that prepare students to take large-scale tests recommend first eliminating impossible answers and then back calculating the remaining answers into the question rather than solving the problem directly to determine the answer. This solution technique is especially simple and quick if calculators are allowed. An analysis by Tarr, Uekawa, Mittag, and Lennex (2000) on eighth-grade calculator usage as part of TIMSS missed these inappropriate uses of a calculator. We were not able to find a direct reference to such uses with respect to any of the three studies included in this paper, but the general idea is illustrated in this discussion of the SAT test.

> When dealing with algebra, some equations and word problems can get nasty. However, you've got choices, right? Plug them in and see which one works. Who needs to do any real calculations when you can beat the system like that? When you're given general variables in specific ranges, plug in numbers in that range to see what works. Remember that the test [SAT] is designed to be possible even without a calculator. If you're getting weird, unfriendly answers where you have to use a calculator, like a quadratic equation that can't be factored nicely for a solution, you're probably off the right track. (http://sat.learnhub. com/lesson/1713)

Something is being tested, but construct validity claims are definitely threatened by the use of such techniques by students. TIMSS mitigated this threat somewhat with a small percentage of open-ended questions, but that still leaves potential problems related to the second case.

In the second case, every multiple test question lacks external validity in terms of format: mathematics problems in the real world do not provide multiple answers to choose from. The answers must be constructed by the learner. In addition, most released problems from these assessments are not situated within realistic application contexts. Contexts are not strictly necessary for external validity as defined by SCC, but are necessary for external validity of the type suggested by Beaton, Linn, and Bohrnstedt (2012).

Construct Confounding. It is difficult to devise an experimental operation that measures only one construct. For example, a test operation in the form of a question concerning geometry is attempting to measure students' understanding of the right triangle construct. The question must either use abstract notation or actual measures (in degrees for example). The students' abilities to understand abstract notation is not the construct being tested, but it is tested nonetheless. This is also true if degrees are used; students' understanding of measurement is tested even though it is not the construct of interest. Related to this is the definitional fact that all mathematics word problems require reading and mathematics skills to successfully solve them.

Holliday and Holliday (2003) had several such concerns with TIMSS. The concerns were centered on the issue of language (see Griffo, 2011 for similar concerns on the NAEP). On a particular item, English speaking students were asked to calculate using monetary units of "zeds." No doubt, this was an attempt by the item writers to not feature any one country's currency; however, "zed" is a common term as the last letter of the alphabet in all English speaking countries except the USA. A second issue that W. G. Holliday and B. W. Holliday mention is the exclusive use of the metric system. They acknowledge that US students are taught the system, but that the students do not use it outside of the classroom. Both of these issues are an example of an operation measuring more than one construct (i.e., the item is measuring mathematics *and* metric system knowledge).

Rather than the language, Wang (1998) analyzed the technical content of questions on TIMSS and found several example items that high-knowledge students might have missed because of their deep thinking. For example, Wang noted (p. 38) that a particular mathematics item asked students

> A chemist mixes 3.75 mL of solution A with 5.625 mL of solution B to form a new solution. How many milliliters does this new solution contain? (Item #K2, http://www.csteep.bc.edu/timssl/Items.html)

The correct answer according to TIMSS is the simple sum of 9.375 mL. However, high-knowledge students would potentially answer incorrectly for at least two reasons: (1) they already have knowledge of significant figures and (2) they know that depending on the solutions being mixed (e.g., an alcohol-based and a water-based solution) the total is not additive and will be less than expected. This item is particularly interesting since the reason for students answering it "incorrectly" could be different across students. That is, the second construct being tested is unknown and may be different across students.

Mono-Operation and Mono-Method Biases. SCC clearly state this threat: "Because single operations both underrepresent constructs and may contain irrelevancies, construct validity will be lower in single-operation research than in research in which each construct is multiply operationalized" (p. 75). This threat can be avoided by including multiple items on a test for each construct being studied; however, Wang (2001) reminded readers

> Because students tested in TIMSS had been enrolled in school for several years, measuring the cumulative achievement in a short testing period represented a considerable challenge. As with any large-scale assessment, a short test may not sufficiently cover what students have learned so far. As it usually demands more time and effort from the test takers, a lengthy test can cause a low response rate (p. 18).

Wang's reminder applies to any large-scale assessment including the NAEP and ELS:2002. Further complicating this are issues of language, culture, and necessary item variance. Language and culture both introduce irrelevancies (see the subtype *Construct Confounding*) which must be mitigated with additional items to preserve construct validity.

In addition, if all constructs are operationalized or measured in the same manner then the method of answering may introduce unintended treatment effects. Referring back to Wang's (2001) warning (Construct Validity: *Inadequate Explication of Constructs*) that the assessment is de facto becoming mathematics, if the constructs of large-scale tests were measured in varying ways then both Wang's issue and the mono-method bias would be mitigated. Possible variations include oral-response items; multiple choice questions with 12 answers available (and time constraints); manipulatives required to answer correctly; and reference materials available that are not test specific. Attempts to mitigate this threat are being implemented, particularly for newer iterations of NAEP, which has added computer administration and extended response items to its assessment repertoire.

Confounding Constructs with Levels of Constructs. A construct, such as homework, may be confused with the level of the construct, such as homework for 30 min or more. The construct validity of a study is threatened if it does not differentiate between binomial, polynomial, and various scaled measures.

The ELS:2002 collected data on student involvement in extracurricular activities. It collected frequency of involvement and the type of involvement: academic, athletic, service, or social. Publication NCES 2005-338 presented several statistics concerning students' success rates and their involvement in extracurricular activities. Some of the data was analyzed and presented based on membership alone; that is, without regard to the time a student actually spent on the activity or the achievement of the student (e.g., intramural, varsity, or captain). This part of the report should be carefully examined concerning the construct validity of any claims (positive or negative) concerning student achievement based on or influenced by extracurricular activities.

All three studies collected information on the construct of mathematics homework from students. Questions posed to the students included the amount of time spent, number of pages read, number of problems solved, and whether a computer or calculator was used. The studies ignored or were unable to measure the level of efficiency or intensity with which students approached their homework. Any study conclusion concerning the amount of homework time spent should be carefully analyzed for threats to construct validity from the lack of considering levels of efficiency or intensity—to draw an analogy, it is not the time spent in a gym, but rather what is done in the gym that counts.

Reactive Self-Report Changes. This threat to construct validity is really a question of honesty. Do teachers who desire more technology in their mathematics classrooms answer questions concerning the availability of it honestly? Do students who desire less homework each evening tend to underreport or overreport the amount of time they are spending? Did curriculum experts actually know what was being taught in a nation's classrooms and did they act honorably when accepting and rejecting items for TIMSS? Bracey (1998) pointed out that curriculum matching was not performed by the US delegation for the third TIMMS: "U.S. participants in TIMSS decided to accept uncritically all items at all grades—just to see how the students would cope" (p. 35). On the advanced mathematics part of the test, this led to US pre-calculus students receiving tests which had 18 % of the questions covering calculus topics.

Experimenter Expectancies. This threat is sometimes referred to as the Pygmalion, Rosenthal, or golem effect. In essence, the students will try a little harder or a little less depending on the expectations of their teacher or other authority figure.

Cultural differences play a large role in expected student outcomes (Bracey, 1998; Cogan & Schmidt, 2002; Holliday & Holliday, 2003), and increased expectancies had varying effects on different groups of students (Wei, 2012). Some countries view TIMSS as an international competition and attempt to drum up patriotic fervor to win TIMSS. In contrast, some US seniors saw TIMSS as just another government-mandated test administered just before graduation (May) and which would not affect them in any way (Bracey, 1998). Clearly, such biases impact comparisons across nations. The extent to which such cultural differences are "natural" to an assessment event, versus introduced for a specific assessment may differ by country, district, or school, making direct comparison of change in mathematics performance across administrations difficult to ascribe to curricular or instructional influences.

Resentful Demoralization. When a group knows they are not receiving a treatment, then they may change their behavior because they are demoralized by not being included in the treatment group. Demoralization could cause a change in the control group's dependent scores which would tend to magnify the actual treatment effects for the group receiving treatment. Ndlovu and Mji (2012) wrote that by some measures South Africa had previously done poorly on TIMSS. Thus the test received a reputation of being difficult, which demoralized future TIMSS test takers.

External Validity

External validity concerns generalization of the research findings beyond the situation in which the data was collected. Here, *generalizability* means the degree to which patterns in the data can be faithfully applied across populations, across contexts and situations, and especially for longitudinal data, over time. SCC identify four "threats" to external validity: Interaction of the causal relationship with units; Interaction of the causal relationship over treatment variations; Interaction of the causal relationship with outcomes; and Interaction of the causal relationship with settings.

For the reviewed studies, that do not employ experimental methods, these "threats" are typically the foci of scholarship. Kloosterman et al. (2015), for example, utilize longitudinal NAEP data to determine potential causal relationships between different mathematics content strands extracted from examination of items, and student performance across demographic subgroups over time. Settings (states, regions), treatment variations (school characteristics and state policies), units (demographics), and outcomes (performance), and time are all assumed to be related in a web of causality. The importance of *longitudinality* of the data set allowed the authors to make claims about the ways in which causal implications among these variables change and mutually impact each other over time. So, for the most part, we do not treat these factors as threats to validity per se, but as important potential outcomes of the research.

The one threat to validity that has been shown to be particularly problematic is interaction of the causal relationship with outcomes. We treat this issue briefly before moving on to general conclusions regarding the validity of our reviewed studies.

Interaction of the Causal Relationship with Outcomes. Claiming that a cause–effect relationship exists in an educational setting relies strongly on definitions. An educational treatment such as enhanced opportunities for tutoring may be seen as successful or failing depending on definitions. One stakeholder may judge the success of the program by the decrease in the number of students failing the class, while another stakeholder may measure the program's success by the number of students receiving an above average grade.

Ndlovu and Mji (2012) considered their TIMSS results for South Africa as misrepresenting the tremendous strides that the country had recently made in their education system. They considered successes in improving access and structure as equally or even more important than achievement scores. Similarly, the NAEP in the USA has gathered extensive data concerning the achievement of various ethnic groups. Examination of the data leads to claims and counterclaims that treatments have been successfully generalized (e.g., Hemphill & Vanneman, 2011; Howard, 2010).

Conclusions on Validity

The authors and researchers associated with all three studies were open to receiving criticisms concerning validity. In some cases they defended their decisions (e.g., Forgione, 1998) and in others (e.g., Beaton et al., 2012) they were open to modification of their claims or changed the assessment acknowledging the veracity of the critique.

Many issues that threaten validity do so for multiple types (e.g., external or internal) and subtypes. SCC posited that sampling bias was the most pervasive threat to validity. After critically examining these three long-term studies, we have come to the same conclusion.

Given the size and influence of these studies we expected to find more research attempting to quantify the threats to validity. SCC suggested that the authors themselves conduct research to quantify the threats as part of the study rather than waiting for others to do post hoc critiques (p. 41). For the most part, threats to validity for large-scale studies, like the rest of mathematics education research, is underreported.

Discussion: Usefulness

The perceived importance of large-scale studies leapt forward in 1983 with the publication of *A Nation at Risk* (e.g., Beaton & Robitaille, 2002; Gardner, 1983). More recently, *The Manufactured Crisis* (Berliner & Biddle, 1995) and Stedman's (1996) rebuttal have ignited controversy regarding what help results from large-scale national and international studies can really provide to the educator and practitioner; particularly problematic is the question of how useful any of the large-scale studies have been or even can be in improving student achievement.

Published results from all three of our reviewed datasets have given us great insight into the inequities of opportunity and achievement among students in the USA and approximately 60 other countries (Beaton et al., 2012; Hemphill & Vanneman, 2011; Howard, 2010; Lee, 1996). Eliminating these inequities requires tremendous social change (e.g., Marks, Cresswell, & Ainley, 2006). It also requires us to further understand how to teach mathematics to reach differing groups of students; this research theme was prevalent across all three data sets. In addition, all three studies posited that students need more mathematical knowledge and skill to compete in a world that is undergoing an ever-increasing rate of technological expansion. Hong (2012), for example, argued that a nation's future GDP was significantly correlated with TIMSS scores.

However, none of the three studies gathered data to determine if anything being done in relation to educational reform is actually being carried forward from early childhood into adulthood. Girls are now taking more STEM classes, but we do not know if their learning, retaining, and feelings about their experiences are the same or different than boys'; furthermore, if they are different, then in what ways? (Willms & Jacobsen, 1990). Similar questions abound and could be asked about any group of students. The point is, few studies from large-scale assessments shed light onto pedagogical strategies, nuances of curricular organization or the interaction of the two in impacting student learning and abilities.

In terms of learning, one of the key purposes of mathematics education research should be to positively influence the development of mathematical *abilities*. Part of the problem we have found critiquing the utility of these large-scale studies concerns the fact that such studies measure elementary and secondary school mathematics *achievement*—yet we are not even sure what is meant by *achievement* when curricular reform, changing needs of business and higher education, and international competition are thrown into the mix (e.g., Kupermintz & Snow, 1997).

When we examine the impact of large-scale studies on curriculum, we find more substantive utility. Of the three studies, TIMSS was the clearest in expressing the goal of determining relationships between the intended, implemented, and attained curriculum. Several researchers have shown that comparing various nations' intended curricula has led to questioning the status quo and subsequent adjustment and improvement (e.g., Macnab, 2000; Valverde, 2002). However, Cogan and Schmidt (2002) (see also Wiseman, 2010) posited that if a country scored poorly on TIMSS, then the national authority tended to focus on whether changes to the intended and/or implemented curriculum should be made that would fix the score. Of course, the questions on the next TIMSS will generally be different and the national authority may find itself chasing the past by making improvements to fix scores on an assessment that is no longer useful for the current educational need. If Cogan and Schmidt are correct, then this is not a fault of the assessment per se, but of the user; nonetheless, chasing past test results does not seem to be useful. So how can large-scale test results be used fruitfully for curricular innovation? It seems fruitful to conduct careful cross validation of test items with curricular frameworks to establish the degree to which the assessments' measure key features of the frameworks, in the manner in which the framework designers intended (for a lovely example, see Kloosterman et al., 2015).

According to Bracey (1998), in 1991 Tjeerd Plomp, an IEA official, referring to TIMSS sighed, "We can only hope that the tests are equally unfair to everyone." The usefulness of directly comparing the mathematical achievement of various countries, given that the tests are admittedly unfair for the reasons cited above, seems dubious. TIMSS clearly stated that it is not the intent of the study to rank countries in numerical order, yet it publishes the results in a numbered list.

Indeed, the rank ordering of superordinate units in large-scale assessments obscures the fact that often raw score differences are negligible or nearly so from a practical standpoint (Berliner & Biddle, 1996). TIMSS suggests understanding the results as three bands of achievement. Supporting the position of TIMSS, Beaton (1998) found that changing items only had a slight effect on nation's positions—but some did change positions. Wang (2001) calls the bands misleading and meaningless because even minor changes in scores due to a slightly different mix of questions or the use of a different score from the five statistically probable scores would change some rankings. If Wang was correct and the rankings were too broad, then knowing which countries should be used as exemplars is problematic. There is little disagreement, however, about the distinction between the top and bottom tiers. This is possibly useful to the bottom tier countries if they want to consider changing their systems since they would know the characteristics of a high-achieving system. However, it is unclear how such comparative information is of equal utility for high-achieving countries.

In summary, the reviewed assessment programs are, each of them, technological marvels of our age. The careful design of items, scales and sampling frames has enabled researchers to understand broadly the achievement characteristics of students in the US and abroad, and they have enabled the field to identify inequities in opportunity and achievement and work to find effective remedies. The longitudinal studies have provided critical evidence regarding trends in student achievement and on the effectiveness (or not) of curricular change on student outcomes. TIMSS has intrigued scholars with subtle differences in instructional practices and curricular organization and emphases, to the extent that many natural and contrived experiments in professional development and pedagogy have been instituted with much success (e.g., Hiebert & Stigler, 2000).

Large-scale studies will continue to be conducted, threats to validity of the kinds reviewed in this paper notwithstanding. We see their utility limited at the level of individual student variability, but at the level of curriculum and policy, they have clearly provided value to the field of mathematics education. Their usefulness and validity will continue to be questioned—as they should be. Eventually, perhaps with the innovations being proposed for computer administration and scoring, a base of knowledge and technical expertise will emerge that makes large-scale studies of mathematics learning useful at the individual level.

References

Beaton, A. E. (1998). Comparing cross-national student performance on TIMSS using different test items. *International Journal of Educational Research, 29*(6), 529–542.

Beaton, A. E., Linn, R. L., & Bohrnstedt, G. W. (2012). *Alternative approaches to setting performance standards for the national assessment of educational progress (NAEP)*. Washington, DC: American Institutes for Research.

Beaton, A. E., & Robitaille, D. F. (2002). A look back at TIMSS: What have we learned about international studies? In D. F. Robitaille & A. E. Beaton (Eds.), *Secondary analysis of the TIMSS data* (pp. 409–417). Dordrecht, The Netherlands: Springer.

Benjamini, Y., & Hochberg, Y. (1995). Controlling the false discovery rate: A practical and powerful approach to multiple testing. *Journal of the Royal Statistical Society: Series B Methodological, 57*(1), 289–300.

Berliner, D. C., & Biddle, B. J. (1995). *The manufactured crisis: Myth, fraud, and the attack on America's public schools.* Cambridge, MA: Longman.

Berliner, D. C., & Biddle, B. J. (1996). Making molehills out of molehills: Reply to Lawrence C. Stedman's review of "the manufactured crisis". *Education Policy Analysis Archives, 4*(3), 1–13.

Bozick, R., & Lauff, E. (2007). *Education longitudinal study of 2002 (ELS:2002): A first look at the initial postsecondary experiences of the sophomore class of 2002. NCES 2008-308.* Washington, DC: National Center for Education Statistics, Institute of Education Sciences, U.S. Department of Education.

Bracey, G. W. (1997). The seventh Bracey report on the condition of public education. *Phi Delta Kappan, 79*(2), 120–136.

Bracey, G. W. (1998). Tinkering with TIMSS. *Phi Delta Kappan, 80*(1), 32–36.

Bracey, G. W. (2000). The TIMSS "final year" study and report: A critique. *Educational Researcher, 29*(4), 4–10.

Cogan, L. S., & Schmidt, W. H. (2002). "Culture Shock"—Eighth-grade mathematics from an international perspective. *Educational Research and Evaluation, 8*(1), 13–39.

Efron, B., & Stein, C. (1981). The jackknife estimate of variance. *The Annals of Statistics, 9*(3), 586–596.

Forgione, P. (1998). Responses to frequently asked questions about 12th-grade TIMSS. Phi Delta Kappan, 79(10), 769–772. These remarks were originally distributed at "Heads in the Sand," a Brookings In- stitution Panel Discussion, Washington, DC, April 7, 1998.

Garden, R. A., & Orpwood, G. (1996). Development of the TIMSS achievement tests. In: *Third International Mathematics and Science Study. Technical Report, 1.*

Gardner, D. P. (1983). *A nation at risk.* Washington, DC: The National Commission on Excellence in Education, US Department of Education.

Garet, M. S., Porter, A. C., Desimone, L., Birman, B. F., & Yoon, K. S. (2001). What makes professional development effective? Results from a national sample of teachers. *American Educational Research Journal, 38*(4), 915–945.

Griffo, V. (2011). *Examining NAEP: The effect of item format on struggling 4th graders' reading comprehension.* PhD Dissertations, University of California, Berkeley.

Hemphill, F. C., & Vanneman, A. (2011). *Achievement gaps: How hispanic and white students in public schools perform in mathematics and reading on the national assessment of educational progress. statistical analysis report.* NCES 2011-459. Washington, DC: National Center for Education Statistics.

Hiebert, J., & Stigler, J. W. (2000). A proposal for improving classroom teaching: Lessons from the TIMSS video study. *The Elementary School Journal, 101*(1), 3–20.

Holliday, W. G., & Holliday, B. W. (2003). Why using international comparative math and science achievement data from TIMSS is not helpful. *The Educational Forum, 67*(3), 250–257.

Hong, H. K. (2012). Trends in mathematics and science performance in 18 countries: Multiple regression analysis of the cohort effects of TIMSS 1995-2007. *Education Policy Analysis Archives, 20*(33), 1.

Howard, T. C. (2010). *Why race and culture matter in schools: Closing the achievement gap in America's classrooms. Multicultural education series ERIC.* New York: Teachers College Press.

Kloosterman, P., Walcott, C., Brown, N. J. S., Mohr, D., Perez, A., Dai, S., et al. (2015). Using NAEP to analyze 8th-grade students' ability to reason algebraically. In J. A. Middleton, J. Cai, & S. Hwang (Eds.), *Large-scale studies in mathematics education.* New York: Springer.

Kupermintz, H., & Snow, R. E. (1997). Enhancing the validity and usefulness of large-scale educational assessments: III. NELS:88 mathematics achievement to 12th grade. *American Educational Research Journal, 34*(1), 124–150.

Lakoff, G. (1985). *Women, fire, and dangerous things*. Chicago: University of Chicago Press.

Lange, J. D. (1997). *Looking through the TIMSS mirror from a teaching angle*. Retrieved from http://www.enc.org/topics/timss/additional/documents/0, 1341, CDS-000158-cd158, 00

Lee, V. E. (1996). *Understanding high school restructuring effects on the equitable distribution of learning in mathematics and science* (Rev. ed.). Madison, WI: Center on Organization and Restructuring of Schools.

Loveless, T. (2008). *Lessons learned: What international assessments tell us about math achievement*. Washington, DC: Brookings Institution Press.

Macnab, D. (2000). Raising standards in mathematics education: Values, vision, and TIMSS. *Educational Studies in Mathematics, 42*(1), 61–80.

Marks, G. N., Cresswell, J., & Ainley, J. (2006). Explaining socioeconomic inequalities in student achievement: The role of home and school factors. *Educational Research and Evaluation, 12*(02), 105–128.

Mervis, C. B., & Rosch, E. (1981). Categorization of natural objects. *Annual Review of Psychology, 32*(1), 89–115.

Mullis, I. V., Martin, M. O., Ruddock, G. J., O'Sullivan, C. Y., & Preuschoff, C. (2009). *TIMSS 2011 assessment frameworks*. Chestnut Hill, MA: TIMSS & PIRLS International Study Center Lynch School of Education, Boston College.

Ndlovu, M., & Mji, A. (2012). Alignment between South African mathematics assessment standards and the TIMSS assessment frameworks: Original research. *Pythagoras, 33*(3), 1–9.

Perie, M., Moran, R., & Lutkus, A. D. (2005). *NAEP 2004 trends in academic progress: Three decades of student performance in reading and mathematics*. Washington, DC: National Center for Education Statistics, US Department of Education, Institute of Education Sciences.

Reddy, B. V. (2006). *Mathematics and science achievement at South African schools in TIMSS 2003*. Cape Town, South Africa: HSRC Press.

Shadish, W., Cook, T., & Campbell, D. (2002). *Experimental and quasi-experimental designs for generalized causal inference*. Boston: Houghton Mifflin.

Stedman, L. C. (1996). Respecting the evidence. *Education Policy Analysis Archives, 4*, 7.

Strayhorn, T. L. (2009). Accessing and analyzing national databases. In T. Kowalski & T. Lasley (Eds.), *Handbook of data-based decision making in education* (pp. 105–122). New York: Routledge.

Tarr, J. E., Uekawa, K., Mittag, K. C., & Lennex, L. (2000). A comparison of calculator use in eighth-grade mathematics classrooms in the United States, Japan, and Portugal: Results from the third international mathematics and science study. *School Science and Mathematics, 100*(3), 139–150.

Valverde, G. A. (2002). *According to the book: Using TIMSS to investigate the translation of policy into practice through the world of textbooks*. Dordrecht, The Netherlands: Kluwer.

Wang, J. (1998). A content examination of the TIMSS items. *The Phi Delta Kappan, 80*(1), 36–38.

Wang, J. (2001). TIMSS primary and middle school data: Some technical concerns. *Educational Researcher, 30*(6), 17–21.

Wei, X. (2012). Are more stringent NCLB state accountability systems associated with better student outcomes? An analysis of NAEP results across states. *Educational Policy, 26*(2), 268–308.

Willms, J. D., & Jacobsen, S. (1990). Growth in mathematics skills during the intermediate years: Sex differences and school effects. *International Journal of Educational Research, 14*, 157–174.

Wiseman, A. W. (2010). Introduction: The advantages and disadvantages of national education policymaking informed by international achievement studies. *International Perspectives on Education and Society, 13*, xi–xxii.

Methodological Issues in Mathematics Education Research When Exploring Issues Around Participation and Engagement

Tamjid Mujtaba, Michael J. Reiss, Melissa Rodd, and Shirley Simon

Background

In common with many other countries, the government in England is committed to increasing the number of STEM (science, technology, engineering and mathematics) professionals as it sees this as crucial for England to be able to compete in an increasingly competitive global economy (Department for Business, Innovation and Skills, 2009). While the number of students choosing to study mathematics after the age at which it is no longer compulsory (16 in England) has been rising in recent years, there is still a problem with the relatively low proportion of English students, compared with other countries, who continue with mathematics in post-compulsory education (Royal Society, 2011). There have been a number of pieces of research that have concluded that this is at least in part due to the high levels of disaffection of many students taking secondary mathematics courses (e.g. Brown, Brown, & Bibby, 2008; Nardi & Steward, 2003). A shortage in the number of students undertaking post-compulsory mathematics has implications for the number who can go on to do careers that require mathematics, including specialist mathematics teacher training courses, which impacts the availability of good quality mathematics teaching for school students.

Existing research has demonstrated the importance of gender, as well as prior attainment, socio-economic status and ethnicity, on whether students continue with post-compulsory mathematics (e.g. Noyes, 2009). Feminist-inspired work has looked at why girls too often conclude that mathematics is not for them. In a qualitative study of young people in schools and colleges Mendick (2006) drew on theorisations of masculinities by Connell (1995), amongst others, and concluded that to understand gender difference we need to start from social context, processes and actions and see gender difference as relational. More recently, the importance of mathematical relationships in education has been stressed by Black, Mendick,

T. Mujtaba (✉) • M.J. Reiss • M. Rodd • S. Simon
Institute of Education, University of London, 20 Bedford Way, London WC1H 0AL, UK
e-mail: T.Mujtaba@ioe.ac.uk

© Springer International Publishing Switzerland 2015
J.A. Middleton et al. (eds.), *Large-Scale Studies in Mathematics Education*,
Research in Mathematics Education, DOI 10.1007/978-3-319-07716-1_15

and Solomon (2009) and others. For instance, the importance of pedagogy was investigated by Palmer (2009) who found that female teacher-education students became much more positive about mathematics after they undertook a course that adopted a feminist post-structural approach based on critical pedagogy and deconstructive theory.

A number of the factors that influence engagement with mathematics are to do with schooling, which is susceptible to a range of influences: changes in schools, changes in subject teachers, introduction of new learning plans. The family, though, is an important influence that is resistant to changes that take place in the school environment. Noyes (2003) qualitative study found that students' family backgrounds played a key role in how students identified with mathematics. Those students for whom the family habitus resonated with the culture of the school benefited more from school than did students for whom mathematics and the learning culture at home did not so resonate. This identification is partly the result of such cultural forces and an individual's relationship with their school, but it is the individual's affective response, both conscious and unconscious, that ultimately attracts, or fails to attract, each person to the subject (cf. Boaler, 2009; Middleton & Jansen, 2011).

The Context of This Study

This chapter aims to identify the factors that relate to students' intended choices with respect to mathematics in schools in England, using a mixed methods longitudinal approach. The data are drawn from the Understanding Participation rates in post-16 Mathematics And Physics (UPMAP) project which was conducted from 2008 to 2011. The quantitative element of the study, part of which we draw on here, surveyed students aged 12–13 (year 8) and 14–15 (year 10). Throughout we highlight the methodological issues that surfaced in our study; indeed, we have structured the paper as a sort of narrative of the various stages of the analysis, recounting the different methodological decisions we made at each stage.

Whilst there is a considerable literature in mathematics education pertaining to extrinsic factors affecting choices and achievement, comparatively little has been reported on the relationship between intrinsic factors, such as personality, attitudes to mathematics and achievement in mathematics, and their relationships to subject choice, achievement and post-16 participation. Accordingly, we designed student questionnaires to include items derived from established psychological constructs. Given that the focus of the study was to find factors that influence post-16 participation in mathematics and/or physics it was a deliberate part of the sampling to over-represent in our sample schools which were above average in either or both of mathematics and physics attainment and post-16 participation. In addition, given our research agenda, we targeted classes that contained students who were said by the teachers to be of above average or average attainment in mathematics and physics/science. This focus was intentional because, although all barriers to participation are important, we are particularly interested in factors that affect the choices of those students who have the opportunity, including fulfilment of attainment criteria, to study mathematics (or physics) post-16.

Student questionnaires were designed following a review of the literature (Reiss et al., 2011) that considered factors that may influence post-compulsory participation rates. Alongside questions related to intentions to continue to study mathematics post-16, the survey included mathematics-specific items to determine attitudes to the subject, attitudes to lessons, self-concept, perceptions of teachers, support for learning, intrinsic and extrinsic motivation for learning, personality and mathematical understanding. A factor analysis using principal components affirmed some of the constructs though also led to minor changes in others. Cronbach's alphas were used to assess the internal consistency of all constructs, which were found to have fair to high reliability (.6–.9). All of the items within each construct were scored so that a high score represents strong agreement.

This chapter also draws on qualitative data to provide further insights into the statistical findings and suggest new analyses. We use extracts from semi-structured interviews undertaken with three 15-year-old girls. Each interview was conducted by one of the authors and was around 30 min in length.

Introduction to Findings

For all of our surveys and for each year group (year 8 and year 10) we used factor analysis to determine the underlying dimensions of the constructs. For the mathematics surveys we found there were three mathematics-specific constructs related to motivation and values: intrinsic value, extrinsic social gain motivation and extrinsic material gain motivation (these constructs are explained below). In addition, there were seven mathematics-specific constructs which were related to perceptions of learning and students' mathematics education: home support for achievement in mathematics; perceptions of mathematics teachers; emotional response to mathematics lessons; perceptions of mathematics lessons; mathematics self-concept; advice-pressure to study mathematics and social support in mathematics learning. Items were on a six-point Likert scale with scores above three representing agreement/more favourable answers. Our year 8 survey also obtained data on four underlying personality dimensions: competitiveness (a measure of how competitive in life students are), self-direction (whether students report they can change what is going to happen to them), emotional stability (whether students report they are generally happy or upset) and extroversion. The surveys can be downloaded from www.ioe.ac.uk/UPMAP and information about the specifics of instrument design is available in Reiss et al. (2011).

We used a six-point Likert item that asked students whether they were intending to continue with mathematics post-16; this item was used as the dependent variable within our multi-level analysis. For the sake of brevity, 'intention to participate' refers to expressed intentions to continue with a mathematics course at post-compulsory education (i.e. after the age of 16). A high score (4, 5 or 6) represents a stated intention to continue with mathematics post-16 with 6 being 'strongly agree'; the other end of the scale (1, 2 or 3) represents disinclination to continue, with 1 being 'strongly disagree'. Table 1 indicates the overall mean response (4.36) for the year 8 students with statistically significant differences between boys and girls ($t = 5.508$, $p < .001$) in favour of boys with an effect size of .155 ($p < .001$).

Table 1 Year 8 students' perceptions of mathematics

Overall perceptions of maths — Item	All students[a]				Boys			Girls			Comparison (boys vs. girls)		
	N	Corr.	M	SD	N	M	SD	N	M	SD	t	df	Effect size Cohen's d
I intend to continue to study maths after my GCSEs (key variable of interest)	5,154	1	4.36	1.42	2,294	4.49	1.42	2,825	4.27	1.42	5.508	4,904.349	0.155***
Self-concept (original construct)	5,342	0.444***	4.09	1.01	2,369	4.30	0.99	2,938	3.91	0.99	14.489	5,076.146	0.400***
Intrinsic perceived value of mathematics (original construct)	5,270	0.516***	4.13	0.90	2,324	4.27	0.86	2,913	4.03	0.92	9.600	4,830.896	0.269***
Extrinsic material gain motivation (original construct)	5,334	0.572***	4.84	0.80	2,364	4.93	0.78	2,935	4.78	0.83	6.643	4,905.685	0.185***
Extrinsic social gain motivation (original construct)	5,168	0.363***	3.08	1.13	2,273	3.17	1.07	2,863	3.01	1.19	4.760	4,606.280	0.133***
I am good at maths	5,169	0.451***	4.52	1.15	2,298	4.77	1.16	2,836	4.31	1.09	14.487	5,016.390	0.404***
I do not need help with maths	5,196	0.204***	3.83	1.48	2,291	4.08	1.42	2,872	3.63	1.50	11.006	4,786.352	0.310***
To be good at maths, you need to be creative	4,850	0.267***	2.87	1.36	2,151	3.08	1.27	2,670	2.71	1.44	9.336	4,313.640	0.274***
I think maths will help me in the job I want to do in the future	5,151	0.506***	4.60	1.42	2,283	4.77	1.43	2,833	4.45	1.39	8.094	4,949.433	0.227***
Being good at maths makes you popular	4,659	0.235***	2.19	1.22	2,045	2.34	1.13	2,584	2.08	1.30	7.132	4,070.883	0.215***
Maths is important in making new discoveries	4,805	0.391***	4.20	1.33	2,152	4.35	1.29	2,623	4.07	1.36	7.253	4,494.904	0.212***
People who are good at maths get well-paid jobs	4,797	0.326***	4.65	1.06	2,140	4.75	1.06	2,627	4.56	1.06	6.321	4,566.066	0.184***
To be good at maths, you need to work hard	5,127	0.251***	5.01	1.08	2,265	5.12	1.09	2,830	4.92	1.05	6.553	4,927.604	0.184***

Those who are good at maths are clever	4,928	0.165***	4.27	1.36	2,179	4.40	1.35	2,718	4.17	1.37	5.840	4,685.001	0.168***
I think maths is an interesting subject	5,182	0.568***	3.95	1.42	2,299	4.08	1.44	2,850	3.85	1.40	5.897	4,857.858	0.166***
Maths is interesting	5,119	0.517***	3.76	1.53	2,252	3.90	1.57	2,834	3.65	1.49	5.589	4,723.161	0.159***
Maths teaches you to think logically	4,844	0.390***	4.79	1.06	2,150	4.88	1.08	2,664	4.72	1.04	5.034	4,529.416	0.146***
In maths, it is interesting to find out about the laws that explain different phenomena	4,994	0.434***	4.07	1.45	2,209	4.18	1.46	2,756	3.97	1.43	5.071	4,683.308	0.145***
Maths improves your social skills	4,776	0.301***	3.15	1.46	2,127	3.23	1.52	2,619	3.08	1.40	3.411	4,380.065	0.100***
These days, everybody needs to know some maths	5,129	0.279***	5.09	1.01	2,262	5.14	1.01	2,835	5.05	1.00	3.248	4,846.272	0.092**
I think maths is a useful subject	5,245	0.515***	5.19	0.97	2,324	5.24	1.01	2,887	5.15	0.94	3.171	4,815.377	0.089**
Maths helps you in solving everyday problems	5,111	0.335***	4.75	1.19	2,260	4.80	1.21	2,818	4.71	1.18	2.416	4,775.948	0.068*
Being good at maths impresses people	4,780	0.335***	3.85	1.35	2,105	3.88	1.40	2,644	3.84	1.32	1.042	4,379.670	0.031
There is only one right way to solve any maths problem	4,823	−0.067***	4.69	1.41	2,138	4.70	1.49	2,656	4.68	1.34	0.413	4,338.214	0.012

Notes: *N* number; *M* mean; *SD* standard deviation; comparisons between girls and boys

[a]Shaded area is a separate set of analyses (correlations (Corr.)) between items that explore all students' perceptions of mathematics with students' intentions to continue to study maths post-16; unshaded area explores gender issues amongst items; ***significant at .001; **significant at .01; *significant at .05

Multi-level Findings: Intention to Participate in Mathematics Post-16 Amongst Year 8 Students

To ascertain which factors were the most important in explaining intended mathematics participation, multi-level modelling (MLM) procedures were used to establish which combinations of factors were best able to explain the variation in year 8 students' intentions to study mathematics post-16. These findings represent our initial key results which helped create new avenues of research and lines of enquiry. At this stage we felt that reliance on MLM procedures was appropriate for the sort of data we were analysing, given that MLM enabled us to recognise the nature of student responses by including students as one of the levels within our nested multi-level model. Students' intentions to continue with mathematics post-16 are likely to be influenced by factors operating at a number of levels and for the data we collected we were able to explore influences at the individual student level and at the school level. The variance in MLM procedures is therefore partitioned out between the student and school levels. The standard errors are smaller than those obtained using traditional regression techniques and so MLM procedures are less likely to have type 1 errors.

We began our analysis of data from year 8 students' questionnaires by fitting a variance components model for the outcome measure 'intention to study mathematics post-16'; the intra-school correlation demonstrated that around 7 % of the variation in students' intention to study mathematics post-16 is attributable to differences between secondary schools with the rest of the variation reflecting differences between students. Given this low intra-school correlation, our analytical approach focused more on exploring student level factors; we began with more basic models that evolved as we explored the importance of various influences on intended participation. The chi-square likelihood ratio test and the deviance statistic were used to establish whether the addition of new (statistically significant) variables provided better model fit than earlier models.

Our Initial Construct-Based Multi-level Analysis

The final model presented in Table 2 went through a number of stages; we will refer to findings from earlier models to create a more complete picture about what we tested and which student level variables were removed in the final model. To maximise sample sizes, scores on constructs from the survey were divided into quartiles; this allowed us to maximise the number of students within the models as it enabled us to retain students for whom we had scores for some but not all items within a construct.

Controlling for Only Background Characteristics

Student background characteristics were the first variables we controlled for, primarily because of the known influence of prior attainment, gender, ethnicity and socio-economic status on actual participation. We did find an independent influence of the

Table 2 Estimates of fixed effects on year 8 England students' intentions to study mathematics post-16

Parameter	Estimate	Std error	df	t	Sig.	Effect size
Intercept	5.766	0.103	1,161.470	56.244	0.001	
Gender	0.064	0.053	1,243.693	1.217	0.224	0.062
Maths self-concept (comparison group: top quartile)						
(Bottom quartile)	−0.038	0.072	1,703.892	−0.527	0.598	−0.036
(Lower middle quartile)	−0.241	0.077	1,686.183	−3.146	0.002	−0.231
(Upper middle quartile)	−0.588	0.084	1,685.733	−7.010	0.001	−0.564
Emotional response to maths lessons (comparison group: top quartile)						
(Bottom quartile)	−0.052	0.078	1,703.868	−0.665	0.506	−0.049
(Lower middle quartile)	−0.183	0.079	1,703.452	−2.317	0.021	−0.176
(Upper middle quartile)	−0.316	0.088	1,700.024	−3.591	0.001	−0.303
Advice-pressure to study maths (comparison group: top quartile)						
(Bottom quartile)	−0.170	0.076	1,702.531	−2.225	0.026	−0.163
(Lower middle quartile)	−0.351	0.084	1,703.920	−4.151	0.001	−0.336
(Upper middle quartile)	−0.964	0.081	1,698.910	−11.913	0.001	−0.925
Intrinsic value of maths (comparison group: top quartile)						
(Bottom quartile)	0.011	0.078	1,703.367	0.136	0.892	0.010
(Lower middle quartile)	−0.163	0.082	1,703.986	−1.989	0.047	−0.157
(Upper middle quartile)	−0.463	0.094	1,701.561	−4.946	0.001	−0.444
Extrinsic prospects (comparison group: top quartile)						
(Bottom quartile)	−0.313	0.081	1,703.610	−3.880	0.001	−0.300
(Lower middle quartile)	−0.530	0.082	1,701.668	−6.501	0.001	−0.509
(Upper middle quartile)	−1.044	0.096	1,703.619	−10.875	0.001	−1.002
Random-effects parameters						
Variance (Level 2)	0.008	0.007				
Variance (Level 1)	1.087	0.037				
Deviance (−2 × log restricted-likelihood)	5,093.092					

first three background characteristics although they are not reported in the final model in Table 2 because all of these lost significance when we began to control for students' attitudes and perceptions of their mathematics education. However, and somewhat surprisingly, the analysis indicated that even at this initial stage there was no influence of free school meal status (a measure of social deprivation—about one in six school children are entitled to receive meals at school without paying for them because of low household income). In line with other existing research we found that girls were less likely than boys to express intentions to continue with mathematics post-16. Students of Asian heritage were more likely than those of other ethnicities to express intentions to continue with mathematics post-16. However, once we accounted for more of the survey measures no ethnicity effects were statistically significant and so this measure was removed from the final model. Gender effects also lost statistical significance in later models but we retained gender as a control primarily because, as our analysis will indicate, when these same students were in year 10, gender differences were significant even after accounting for a range of survey measures. This finding, that as students progress through secondary school the gap in future mathematics aspirations widens between boys and girls, is important.

Controlling for Personality Traits

Within the UPMAP project we were interested in the association between intended participation and the psychological traits of students. The four core constructs for which we collected data were competitiveness (indicates that students have a tendency to self-enhancement); emotional stability (measures a state of composure and calmness); extroversion (measures a tendency to gain gratification through social interactions with others) and locus of control (which measures the extent to which students feel they have an influence over issues that impact them). Some of our earlier work found that girls with high intentions to study mathematics had statistically significantly higher competitive personalities than (a) girls with low intentions and (b) boys, whether boys had high or low intentions to study mathematics (Mujtaba & Reiss, under revision), with similar findings when exploring such trends in post-compulsory physics intentions (Mujtaba & Reiss, 2013c). Within our multi-level analysis we expected to find the association between competitiveness and intended mathematics participation (which we did initially); however, once we included mathematics-specific measures of motivation (see below), the influence of any general underlying personality trait, including competitiveness, was not significant.

Measures of Motivation and Support for Learning (Non-mathematics Specific)

We tested for the importance of our construct 'general motivations and aspirations towards learning' and 'home support for achievement in general' with mathematics aspirations. Initial findings indicated that both were positively associated with mathematics aspirations. This is hardly surprising (cf. Eccles, 2009; Schunk, Pintrich, & Meece, 2010). However, once we included (in later models, see Table 2) mathematics-specific measures of motivation and support for learning, these more general measures lost statistical significance. These findings demonstrate that without the inclusion of mathematics-specific measures we might have concluded that a general emphasis on learning in the home and at school will boost mathematics aspirations.

Inclusion of Measures That Explore Student Perceptions of Their Mathematics Education

To substantiate the impact of students' perceptions of their mathematics education (mathematics self-concept, perception of teachers, lessons and emotional response to lessons) on future mathematics aspirations, such measures were included in the model prior to the inclusion of constructs that tapped into students' attitudes towards mathematics-specific issues (e.g. extrinsic material gain motivation, intrinsic value, self-concept) and encouragement in mathematics learning and choice (e.g. advice-pressure to study mathematics, social support in mathematics learning, home support for achievement in mathematics). There was a statistically significant independent influence of 'perceptions of mathematics teachers' although this lost significance in later models once we controlled for other measures of students' mathematics education. The only constructs that measure perceptions of mathematics education that continued to have an influence in the final model were 'emotional response to mathematics lessons' and 'mathematics self-concept'.

Mathematics-Specific Measures of Motivation and Support for Learning

The constructs 'social support in mathematics learning' and 'extrinsic social gain motivation' (a measure which explores students' desire to continue with mathematics for social gain) were associated with intended participation. We also found there was an association between 'home support for achievement in mathematics' (a construct which measures support that students derived from the family in raising mathematics attainment). However, once we introduced 'extrinsic material gain motivation' (which identifies students wanting to continue with mathematics for some tangible reward, such as future career prospects) and 'advice-pressure to study mathematics' (a construct which measures the encouragement students receive from a range of people about continuing with mathematics) the constructs 'social support in mathematics learning', 'extrinsic social gain motivation', and 'home support for achievement in mathematics' did not have an independent influence in demonstrating their association with future mathematics aspirations; these measures were subsequently removed.

The Final Model

In the final model the following five constructs were found to be significantly associated with the post-16 mathematics intentions of year 8 students: 'mathematics self-concept', 'emotional response to mathematics lessons', 'advice-pressure to study mathematics post-16', the 'intrinsic value' students accord to mathematics and 'extrinsic material gain motivation'. The largest effect size (ES) out of all of the measures we tested within our models (and whilst controlling for the influence of other measures) was for 'extrinsic material gain motivation' (ES = 1.002), followed by 'advice-pressure to study mathematics' (ES = .925), 'mathematics self-concept' (ES = .564), 'intrinsic value of mathematics' (ES = .444) and positive 'emotional response to mathematics lessons' (ES = .303). As expected, boys were more likely to express intentions to participate in mathematics than girls, though at year 8, once we controlled for other survey responses, the influence of gender lost significance. However, we decided to retain gender in the final model (Table 2) to help illuminate issues around gender with further analysis (as discussed below).

Other Considerations

Our modelling explored the influence of a range of predictors and it was apparent that the non-inclusion of 'extrinsic material gain motivation' would have led to a conclusion that 'other factors' were important in explaining intended participation. Our initial modelling found that students' perceptions of their 'home support for achievement in mathematics' was a better predictor than the construct 'home support for achievement in general', the latter being a measure of support from the family for all types of learning, which lost significance once mathematics-specific measures were introduced. In addition, general measures of students' motivation for learning (our constructs 'general motivations' and 'aspirations towards learning') or personality-based measures of general motivation in life, such as 'competitiveness', were not significant predictors of intended participation in post-16 mathematics once we used mathematics-specific measure of motivation in our model. We found 'extrinsic material gain motivation' to be more precisely related to intention to study mathematics post-16 than any of our other measures of motivation in education (or life in general, as in the competitiveness measure). Without having mathematics-specific measures, we would not have been able to come to such conclusions and could easily have suggested that support for learning from the home and students' own motivation towards learning were not important in intended mathematics choice.

The Importance of Looking Beyond What the Immediate Findings Suggest

It is also worth emphasising the importance of the order in which variables are introduced in the steps of multi-level modelling (MLM). The results from the construct-based MLM analysis indicated that 'perception of mathematics lessons' and 'perception of mathematics teachers' lost statistically significant association with year 8 students' intentions to continue with mathematics post-16 once the constructs 'extrinsic material gain motivation' and 'advice-pressure to study mathematics' were introduced.

Does this mean that in school there should be less of a focus on the teacher–student relationship and on how students perceive their mathematics lessons and more of an emphasis on creating an awareness about the material gain of a post-16 mathematics qualification? We do not think that this would be an appropriate conclusion. For one thing, at least part of the influence of mathematics teachers and lessons may be absorbed by such constructs as 'self-concept' and 'extrinsic material gain motivation'. It is rare for any attitude to exist in isolation from another. Although the constructs that measure the influence of teachers and lessons were not as strong/ effective predictors of intended mathematics participation as other measures in our final construct-based multi-level analysis, we wondered whether there might be individual items within these constructs that have a strong effect on intended participation in mathematics. We reasoned there was a possibility that the importance of specific items might have been lost, once these were combined with other items within an overall construct. Such thinking was further influenced by the fact we did find within our qualitative work, as discussed below, that perceptions of mathematics teachers and mathematics lessons were very important in the decision making of some students. This led us to go back to some of our original constructs and analyse at the item level to help bridge the findings between the qualitative and quantitative work. In this next section we discuss some of the findings from the qualitative work which helped us to re-think how we ought to approach our survey analysis and the conclusions we were drawing before we return to deconstructing our constructs via an item-level analysis.

The Emergence of the Importance of Teachers via Qualitative Work

The qualitative element of this chapter focuses on interviews with 15 year olds (year 10 students). Although this section supports some of the key quantitative findings reported in Table 2 it also brings new insights, namely the importance of teachers in student choice. There is now a considerable body of evidence to suggest that the quality of teaching is a major determinant of student engagement and feelings of success in all school subjects. However, subject choices are not made solely on the quality of teaching. A substantial amount of research on subject choice has established that students are more likely to study subjects that they see as interesting and useful and ones in which they expect to do well in (Eccles, 1994; Mujtaba & Reiss, 2013a, 2013b), factors that may correlate with teaching quality but are not entirely contained within this. Students' feelings of success at mathematics (mathematics self-concept) can also contribute to their perceptions of mathematics and to intended subject choice.

Our qualitative work indicated that for some students a close, supportive relationship with mathematics teachers was important in future mathematics intention. The extracts below support the quantitative work by drawing out the role of self-concept, extrinsic material gain motivation and the intrinsic value of mathematics whilst also indicating that teachers' encouragement to some extent may have underpinned mathematics self-concept and students' intended choices. The analysis of three student interviews exemplifies the importance of teachers in students' decision-making processes; the students were specifically chosen from the larger pool of interviewees to portray three very different ways that teachers can have an influence on students' feelings about mathematics. In the first case, the teacher serves a role in connecting the student with mathematics; the student had a very weak relationship with mathematics prior to this teacher's long-term support and this encouragement eventually led to a choice to take an academic course in post-16 mathematics (A-Level mathematics). In the second case, the student already had a strong attachment to and self-concept in mathematics with an intention to continue with mathematics post-16; the teacher served simply to encourage and reinforce the student's mathematics choice. In the final example, the student had an attachment with mathematics which developed from the home; the teacher and class environment unfortunately served to break that mathematics attachment. All three students are female and were interviewed approximately at the same time.

Alice in Yellow-Wood School

Alice attended a semi-rural low socioeconomic status school. Her parents were both employed, her mother as an accountant and father as a landscaper/builder. Alice was one of two non-identical twin girls and said she had learning difficulties and low expectations until she reached secondary school:

> I have always struggled in previous years because I had a learning difficulty when I was younger. I couldn't read properly and I was always really slow at processing things in my mind and when I was a child my parents were told that I would never be able to learn.

Prior to year 9 Alice was not particularly fond of mathematics and struggled with it. However, her twin sister stated that mathematics was one of her easiest subjects and she intended to continue with it. In year 10, Alice's relationship with mathematics became linked to her relationship with her mathematics teacher (Mrs. S) who was also the Head of Mathematics at Alice's school. In a separate interview we undertook with Mrs. S, she acknowledged the importance of student performance in mathematics, more so than in other subjects. She also stated that though she felt she and her colleagues were under pressure to maximise student attainment in mathematics, the department strove to develop a culture where having rounded students who learnt to value mathematics as an end in itself, rather than simply increasing attainment, was seen as the objective. Perhaps this explains why a number of students within this school, a higher proportion than in other schools, said that they enjoyed mathematics and wanted to work hard at it, without necessarily intending to continue with it post-16.

Prior to year 10 Alice was a student of below-average attainment in mathematics who had not intended to continue with the subject. By the time of her year 10 interview she was considering doing mathematics post-16, which coincided, thanks to Mrs. S, with an increased confidence in her mathematical ability. The extracts below were chosen because they signify the importance of Mrs. S to Alice and mathematics, and how such encouragement translated into an increased self-concept and a more positive relationship Alice had with mathematics:

> Because having Miss S I've actually developed a load of skills in maths. I know a lot more than I thought I would know before and my grades have actually increased than what they were before. I went from down from a C grade to … [meaning an increase from a C grade to a B grade] and I've found it fun as well because my teacher isn't boring and I've managed to get on with the homework and I am pushing myself in maths because I come to see Miss S if I'm struggling and that … It's really organised and so you're never sort of stuck with what you're doing and she really goes through it really clearly so it's sort of a step-by-step guide but not in a patronising way. Like, if you get stuck she will definitely come and help you—she doesn't ignore you—she comes straight over. It's just a lot she does really, it's

really helpful. Previous teachers I had are quite good, but I've never really got on with the style that they worked with. Like, they have taught it well but I've never felt confident enough to go and see them if I was stuck on something. Whereas with Miss S you have that confidence to say 'I don't get this, can I have the help, please'.

In many ways Alice's interview demonstrated how she felt she had found someone who believed in her educational capabilities. Alice held onto that attachment to support her through her schooling:

I feel I am doing and achieving the most when it comes to maths ... it wasn't the case before. But after these 2 years with Miss S I have improved.

The extract below indicates how her teacher's encouragement helped her overcome what had appeared to be an on-going problem with an aspect of mathematics (percentages); overcoming this problem clearly had a role in increasing her mathematics self-concept. Alice was asked what her most memorable mathematics lesson was:

I've never grasped doing percentages—no matter how hard I tried—but Miss S just explained it in the way she does and I finally got it and I think that's just been probably the best time at maths because when you finally know something, after not knowing it for so long, it is so much better isn't it? And it just made me feel really good.

Nearly all other interviewees either gave a bland answer to the same question ('What is your most memorable mathematics lesson?') such as 'There's nothing I find particularly memorable for maths lessons' (a male student from her school) or talking about something unrelated to mathematics or a lesson which was different from the normal mathematics lessons such as 'In year 8 we went into the Tom Smith Hall and played all different maths games and Splat and everything like that'.

Given the encouragement Alice received, she chose to study mathematics at year 12, though she subsequently dropped the subject after finding the lessons difficult. In her interview what came across was her intrinsic liking of mathematics and how that relationship with mathematics developed through a teacher. There was no evidence from her interview that she was intending to choose mathematics because of the extrinsic material gain of the subject.

Sandy in Yellow-Wood School

Sandy was in the same school as Alice. Her mother was an administrator and her father a surveyor; both graduates. Sandy was also taught by Mrs. S, and also chose mathematics at year 12. Although the extracts below lend some support to the quantitative findings in Table 2, they also highlight how important individual relationships with teachers and perceptions of teachers are in subject choice.

In her year 10 interview, Sandy talked about the importance of her mathematics teacher's encouragement in her intention to continue with mathematics: 'We had parents evening and my maths teacher said I could be perfectly capable studying maths, that I'll be a good student, I was encouraged by that'. However, it was also evident that she was aware of the material gain of having a post-16 qualification in mathematics, a sentiment expressed by the great majority of students who continued with mathematics at year 12: 'Maths and physics are quite hard to take, but I just want the best available options, keep the door open for later in life.' Although Sandy did not especially express how important encouragement from her mathematics teacher was, she did indicate how important it was to 'like' and 'be liked' by teachers when deciding what subjects to continue with. When discussing influences, she noted that hers included:

> Probably relationships with the teachers and how the school works because like if there's a subject that you've been put off from the lower years you're not going to want to continue with it … because sort of year 7 and 8 I was really good at art and I took it in year 9 but the teachers were just awful teachers, I didn't like them at all they didn't like me and so then I didn't bother because there was no way that we were going to get on with at GCSE [the examinations sat by the great majority of school students in England at age 16 in year 11].

This issue of personal relationships and their importance to choice is an issue we examined when we decided to explore whether certain individual items within the 'perceptions of teachers' construct were more important in explaining future mathematics aspirations and gender differences in perceptions (Table 3). Supporting the key findings of the quantitative multi-level analysis (see Table 1), Sandy was also very aware of the material gain of having a mathematics qualification and indicated that she was probably going to continue with mathematics after compulsory education (as she indeed did):

Because I like maths and I like physics and I believe they will give me the greatest gateway for work after I go to university and I'm just generally interested in them … I suppose because I've always been quite good at it [mathematics] and again it's logical as well apart from when I thought I don't like it anymore there were some proofs that weren't very good but now I just generally enjoy it.

Elira in Cherry Blossom School

Just as teachers were important in encouraging students to continue with mathematics or build their self-concept and relationship with mathematics, teachers could also damage the relationship students had with mathematics. A prime example of this was Elira who attended a high-attaining Church of England school that had a high proportion of minority ethnic students. She was a second generation Muslim from Kosovo[1] who came to England at the age of three with refugee status. In Kosovo her mother was a doctor although it took quite a few years until she managed to do further training and find work as a gynaecologist in England. Her father graduated in physics or geology (Elira could not recall), though the only jobs open to him in England entailed unskilled work. By the time of Elira's interview he had managed to create a business in buying and renting out homes in Albania and Bulgaria as well as owning restaurants in England. Her parents worked very hard to

---

[1] Her background is raised because she has raised it, which was distinct from other interviewees who largely did not indicate their cultural or religious heritage.

Table 3 Year 8 students' perceptions of their mathematics teachers

Overall perceptions of maths — Item	All students[a]		Boys			Girls			Comparison (boys vs. girls)		Effect size Cohen's d
	N	Corr.	M	SD	N	N	M	SD	t	df	
Students' perception of teacher (original construct)	5,293	0.277***	4.62	0.95	2,342	2,917	4.61	0.97	0.211	4,937.741	0.006
My maths teacher is good at explaining maths	5,072	0.226***	4.63	1.37	2,220	2,820	4.55	1.34	4.559	4,835.753	0.129***
My maths teacher has high expectations of what the students can learn	4,737	0.191***	5.08	0.95	2,108	2,597	5.03	0.94	4.132	4,525.251	0.121***
My teacher thinks that I should continue with maths beyond my GCSEs	2,640	0.425***	4.89	1.25	1,220	1,400	4.83	1.20	2.998	2,603.995	0.117**
My maths teacher wants us to really understand maths	5,008	0.177***	5.29	0.89	2,190	2,788	5.26	0.87	3.239	4,784.903	0.092**
My maths teacher does not only care about students who get good marks in maths	4,731	0.138***	4.72	1.46	2,075	2,626	4.78	1.52	−3.043	4,266.580	0.090**
My maths teacher believes that all students can learn maths	4,902	0.164***	5.27	0.88	2,156	2,714	5.24	0.89	2.434	4,605.209	0.070*
I like my maths teacher	5,193	0.234***	4.25	1.57	2,295	2,864	4.29	1.61	−2.441	4,824.092	0.069*
My maths teacher seems to like all the students	4,684	0.185***	4.12	1.56	2,067	2,588	4.16	1.60	−2.027	4,319.770	0.060*
My maths teacher is interested in what the students think	4,756	0.206***	4.59	1.33	2,088	2,636	4.57	1.33	1.199	4,476.392	0.035
My maths teacher does not let us get away with not doing our homework	4,996	0.027	5.06	1.27	2,191	2,775	5.05	1.30	1.033	4,583.120	0.030
My maths teacher believes that mistakes are OK as long as we are learning	4,957	0.178***	5.01	1.13	2,173	2,753	4.99	1.16	−0.606	4,562.587	0.017
My maths teacher marks and returns homework quickly	4,869	0.158***	4.44	1.43	2,133	2,705	4.43	1.42	0.589	4,623.098	0.017
My maths teacher treats all students the same regardless of their maths ability	4,860	0.163***	4.64	1.44	2,146	2,683	4.63	1.45	0.426	4,549.322	0.012
My maths teacher sets us homework	5,139	0.071***	5.40	0.94	2,252	2,855	5.39	0.94	0.300	4,821.925	0.008
My maths teacher is interested in me as a person	3,927	0.225***	3.60	1.53	1,743	2,159	3.60	1.58	0.154	3,636.737	0.005

Notes: *N* **number;** *M* **mean;** *SD* **standard deviation; comparisons between girls and boys**

[a]Shaded area is a separate set of analyses (correlations (Corr.)) between items that explore all students' perceptions of their teachers with students' intentions to continue to study maths post-16; unshaded area explores gender issues amongst items; ***significant at .001; **significant at .01; *significant at .05

ensure that the family were able to rebuild their lives in England and instilled the same emphasis on hard work within Elira. Her mother was the deciding force behind Elira's year 9 subject choices and the pervasiveness of that influence is apparent when she talked about her future subject choices:

> And it's kind of—I want to have something in common with her, in a way … she thinks I'm her in a way, she thinks I'm more academic … my mum wants me to do sciences like physics, chemistry, coz she's a doctor … so that is quite a big influence in my life and she kind of encourages me … but, at the same time, I personally like and enjoy my subjects like maths, physics, chemistry, biology, I enjoy them.

Her interview suggested that her relationship with most subjects was through relative performance. However, her interview also suggested that she was considering mathematics because of its material gain:

> And I was thinking of taking maths because it's like a really important subject most jobs look for that … You need maths. It's like there and it looks good on your CV if you got an A or something.

Furthermore, choice in mathematics was also tied in with Elira's mathematics self-concept (not dissimilar to other students) and also her parents' expectations:

> My personal achievement will be to get an A or an A*. If I get a B I would probably be upset but I will still continue it. I don't think I'll continue if I got a C, I would just think I was kind of not good at it … my parents don't accept anything under an A; they'll be like "What are you doing?". And so they're strict on education … they're like "You get anything underneath an A you know you're not gonna go out …."

Elira's 'mathematics identity' stemmed from her earlier life experiences when her parents tested her mathematics knowledge to help strengthen her mathematics competencies. Her interview also indicated that such testing left her feeling quite anxious about mathematics as a child and she recalled thinking 'Oh God, don't make me get it wrong'. Nevertheless, as a 15-year-old she was able to identify positively with mathematics. However, in year 10 Elira's relationship with mathematics began to crumble. In the following extract she contrasts her mathematics lesson and teacher with that of physics:

> I like my teacher as well [in reference to physics], he's quite—it makes it interesting—and then the class actually reacts well to the lesson, and in maths, for example, our class is usually noisy, no-one concentrating, it's kind of hard to control them even though we're supposed to be one of the top sets—second top—it's still kind of—it kind of distracts the whole class.

Although Elira stated 'I think I'm quite good at maths', she also notes that:

> I was kind of not concentrating at all and everything; just talking and kind of being noisy … my mocks I got a D. It's kind of hard to find it fun in our class coz our class is really bad—even our Head of Year had to come and shout at us coz the grades we were getting weren't acceptable for the standard we're all supposed to be working on, and our ability. And it's like no one cares about it and it kind of influences everybody else.
> Interviewer: Why do you think nobody cares about it?
> Elira: Because no one does the work—Sir tries to explain, everybody's talking, no-one listens, it's kind of hectic in the room. We're always noisy. Even in exams we talk. And it's kind of hard to control the class. Some people in the class are kind of rude to Mr. W as well. And they go 'Oh, Sir, you're being unfair, we don't know this, we don't know this' but if they listened then obviously they would.

In a later interview in year 11, Elira gave a detailed account of how in year 10 she felt the maths lessons were so awful that she was unable to learn anything. Disruptive students continued to make the working life of the teacher difficult and, according to Elira, the entire class got left behind in mathematics since the norm became discussing anything other than mathematics. Her anxiety with mathematics became more pronounced as she was encouraged by her parents to be good at it and to continue with it. However, as she felt she did not do well at GCSE she did not continue with mathematics at A-Level.

Deconstructing What Our Original Constructs Actually Measured: Perceptions of Mathematics Teachers, Mathematics and Mathematics Lessons

Students' Perceptions of Their Mathematics Teachers

The MLM analysis (Table 2) indicated that the 'perceptions of teachers' construct did not have an independent statistically significant influence in explaining year 8 students' intentions to continue with mathematics post-16 after controlling for a range of other student level factors. Such findings are inconsistent with our qualitative research where encouragement and support from teachers were important in enhancing or severing students' relationship with mathematics. To see if findings from the two separate strands of our project could be aligned, we decided quantitatively to deconstruct what we meant by 'perception of teachers' and therefore conducted a series of item-level analyses.

Our perception of teachers construct explored two key dimensions: encouragement in learning and personal relationships. Students (as a group) reported positive perceptions of their teachers as indicated by their scores on the individual items; the mean for the actual construct 'perceptions of teachers' was also fairly high (4.62). Preliminary work suggested that particular items within constructs might be of especial significance. We decided to include an item which was a part of our original 'advice-pressure to study mathematics' construct, namely 'my teacher thinks that I should continue with maths post-16', on the grounds that teacher advice seems likely to be of importance, and analyse this item along with the remaining items that created the construct 'perception of teachers'. The means in Table 3 indicate that students were most positive about their teachers setting them homework (mean of 5.40); this was followed by their teachers really wanting them to understand maths (5.29) and teachers believing that all students can learn maths (mean of 5.27). These findings somewhat mirror results we found with year 10 physics students (see Mujtaba & Reiss, 2013b); two of these items were the top two most positive responses: teachers really wanting them to understand physics (mean of 4.93), teachers believing that all students can learn physics (mean of 4.90). Year 8 students were least positive about their mathematics teachers being interested in them as people (mean of 3.60) and liking all students (mean of 4.12); these findings also mirror those we found with physics (see Mujtaba & Reiss, 2013b) where we found means 3.33 and 3.80, respectively.

Table 3 also shows how boys and girls responded to each item and whether gender differences were statistically significant. In total, eight of the 15 items showed statistically significant differences between the responses of boys and girls, as well as some of the items having stronger associations with intended participation than others. However, the overall construct 'perceptions of teachers' indicated that there was no statistically significant difference between girls and boys. This is rather worrying given that we used the 'perceptions of teachers' construct to explore associations with year 8 students' intended post-16 participation in mathematics and could have concluded that this construct was not important in explaining intended participation or gender differences in participation. There is a possibility that there are particular items within this overall construct that are individually better able to explain intended participation and that their effect(s) are masked by being immersed in an overall construct.

If we continued simply to use this construct to explore gender differences in students' perceptions of their mathematics teachers without looking at individual-level items our findings would have also missed issues that can help explain gender differences in perceptions of mathematics teachers. Of the eight statistically significant items, the largest effect size in gender differences was for 'my teacher is good at explaining maths' (ES = .129); the remaining effect sizes were between .121 and .060. We found within the physics analysis that the item 'my teacher thinks that I should continue with physics post-16' had the strongest effect size in explaining gender differences at year 10 (ES = .337), although the effect size was almost three times as strong as that found for mathematics (ES = .117).

On average, boys responded more positively than girls about their mathematics teachers. Boys felt to a greater extent than girls that their mathematics teachers: encouraged them to continue with maths post-16 ($t=2.998$, $p<.001$); had high expectations of what students can learn ($t=4.132$, $p<.01$); wanted students to really understand maths ($t=3.239$, $p<.01$); were good at explaining maths ($t=4.559$, $p<.001$) and believed all students could learn maths ($t=2.434$, $p<.05$). The only item for which girls were more positive than boys was 'my maths teacher doesn't only care about students who get good marks' ($t=3.043$, $p<.01$).

When we looked at personal relationships with mathematics teachers, girls were more likely to report that they liked their maths teacher ($t=2.441$, $p<.05$) and that their teacher seemed to like all students ($t=2.027$, $p<.05$).

Intention to Participate and Perceptions of Teachers

A correlation analysis between the items that explored year 8 students' perceptions of their teachers and their intentions to participate in mathematics post-16 further revealed important findings about items that were originally clustered together within an overall construct (see Table 3). The original construct 'perceptions of teachers' was only weakly correlated with intended participation (.277). The correlations in Table 3 demonstrate that students' perceptions of their teachers personally encouraging them to continue with mathematics post-16 (which, as noted above, was originally analysed as part of the 'advice-pressure to study mathematics' construct) is the most strongly associated item with intended participation—more so than items that measure students' perceptions around encouragement in doing mathematics homework. The four strongest correlations between students' perceptions of their teachers and their intention to continue with mathematics post-16 were 'my teacher thinks I should continue to study maths after the age of sixteen' (.425); 'my maths teacher is good at explaining maths' (.227); 'I like my maths teacher' (.226) and 'my maths teacher is interested in me as a person' (.225). This finding reflects what we found with physics. The correlation between year 10 intended participation and 'my teacher thinks I should continue to study physics after the age of sixteen' was .493 and this correlation was also set apart from the rest of the items that explored perceptions of physics teachers (see Mujtaba & Reiss, 2013b). The findings within the year 8 mathematics survey item-based analysis demonstrate that there are a handful of important issues about teachers which are very important in their associations with intended participation and explaining gender differences in participation, that such findings are not apparent when using an overall construct and that there are similarities in findings with the item-based physics analysis.

Students' Perceptions and Emotional Response to Their Mathematics Lessons

The means for the original constructs 'perceptions of lessons' and 'emotional response to lessons' used in the MLM analysis were positive: 4.11 and 4.00, respectively. The MLM analysis indicated that 'perceptions of lessons' despite initially having a significant association, lost statistical influence in explaining year 8 students' intended participation after controlling for a range of other constructs. In our analysis of individual items (see Table 4) we found that the items in these two constructs collectively explored relevance of mathematical concepts, intrinsic value of mathematics lessons, self-concept in mathematics as impacted by mathematics lessons, and emotional response to lessons.

The overall means in Table 4 demonstrate that collectively the students responded positively to items asking them about their mathematics lessons, but with some areas of concern. Collectively, boys and girls were most positive about 'when I am doing maths, I don't get upset' (5.18); 'when I am doing maths, I am learning new skills' (4.73) and seeing the relevance of maths lessons (4.55). They were least positive about looking forward to maths classes (3.41)—a finding which mirrors that for physics in our work reported elsewhere—and not being bored in maths lessons (3.41).

For the large majority of items in Table 4 there were statistically significant differences in responses between boys and girls, with boys responding more positively to questions around mathematics lessons. The effect sizes for gender differences in student perceptions of mathematics lessons were generally larger than those for perceptions of mathematics teachers (Table 2). Again, similar to the findings in physics, we found that the largest statistically significant gender difference was in response to the item, 'thinking about your maths lessons, how do you feel you compare with the others in your group?' (ES = .321), followed by 'I do well in maths tests' (ES = .318). These were followed by 'when I am doing maths, I always know what I am doing' (ES = .238), 'when I am doing maths, I do not get upset' (ES = .196)

Table 4 Year 8 students' perceptions of their mathematics lessons

Overall perceptions of maths / Item	All students[a]				Boys			Girls			Comparison (boys vs. girls)		
	N	Corr.	M	SD	N	M	SD	N	M	SD	t	df	Effect size Cohen's d
Emotional response to maths/physics (original construct)	5,293	0.333***	4.00	0.98	2,338	4.09	0.99	2,921	3.93	0.96	6.080	4,936.096	*0.169****
Perceptions of maths/physics lessons (original construct)	5,302	0.557***	4.11	0.98	2,347	4.17	1.01	2,921	4.06	0.94	4.042	4,858.115	*0.113****
Thinking about your maths lessons, how do you feel you compare with the others in your group?	5,017	0.322***	3.50	1.09	2,241	3.69	1.11	2,744	3.34	1.05	11.225	4,681.128	*0.321****
I do well in maths tests	5,231	0.410***	4.32	1.20	2,325	4.52	1.17	2,873	4.15	1.19	11.439	5,010.675	*0.318****
When I am doing maths, I always know what I am doing	5,243	0.364***	3.58	1.34	2,314	3.75	1.34	2,897	3.44	1.32	8.520	4,923.437	*0.238****
When I am doing maths, I don't get upset	5,204	0.126***	5.18	1.20	2,296	5.31	1.15	2,876	5.07	1.24	7.075	5,058.054	*0.196****
In my maths lessons, my teacher explains how a maths idea can be applied to a number of different situations	5,067	0.371***	3.79	1.28	2,268	3.91	1.34	2,767	3.70	1.23	5.784	4,652.427	*0.165****
When I am doing maths, I don't daydream	5,218	0.288***	3.66	1.63	2,307	3.81	1.67	2,878	3.55	1.59	5.557	4,823.160	*0.156****
I enjoy my maths lessons	5,218	0.473***	3.75	1.51	2,319	3.84	1.54	2,866	3.67	1.48	3.996	4,870.835	*0.112****
When I am doing maths, I am learning new skills	5,226	0.399***	4.73	1.11	2,307	4.80	1.13	2,887	4.68	1.08	3.833	4,844.078	*0.108****
I look forward to maths classes	5,260	0.479***	3.41	1.48	2,330	3.50	1.51	2,896	3.35	1.45	3.463	4,890.205	*0.097****
When I am doing maths, I am not bored	5,215	0.362***	3.41	1.56	2,306	3.49	1.60	2,877	3.36	1.52	2.803	4,811.937	*0.079***
I can see the relevance of maths lessons	5,117	0.440***	4.55	1.28	2,282	4.60	1.33	2,804	4.51	1.24	2.433	4,717.279	*0.069**
I find it difficult to apply most maths concepts to everyday problems	5,173	0.082***	3.74	1.39	2,297	3.78	1.45	2,843	3.71	1.33	1.991	4,714.571	*0.056**
When I am doing maths, I pay attention	5,219	0.368***	4.31	1.23	2,309	4.33	1.26	2,877	4.30	1.21	1.000	4,859.442	*0.028*
In my maths lessons, I have the opportunity to discuss my ideas about maths	5,215	0.330***	4.26	1.35	2,314	4.26	1.39	2,868	4.25	1.31	0.192	4,806.804	*0.005*

Notes: *N* **number,** *M* **mean,** *SD* **standard deviation; comparisons between girls and boys**

[a]Shaded area is a separate set of analyses (correlations (Corr.)) between items that explore all students' perceptions of their lessons with students' intentions to continue to study maths post-16; unshaded area explores gender issues amongst items; ***significant at .001; **significant at .01; *significant at .05

and 'in my maths lessons, my teacher explains how a maths idea can be applied to a number of different situations' (ES$=$.170).

Amongst the perceptions of teacher items, the item 'my teacher is good at explaining maths' had a larger effect size in explaining gender differences than the majority of other items (ES$=$.129). Taking this finding with the effect size of 'my teacher explains how a maths idea can be applied to a number of different situations' (ES$=$.165) demonstrates how important it is for teachers to explain mathematics in a way that engages girls and aids their learning and understanding of mathematics. In order to emphasise our point we refer to the very similar patterns with the physics analysis. Amongst the perceptions of teacher items, 'my teacher is good at explaining physics' (ES$=$.237) had an effect size in line with an item clustered within lessons: 'my teacher explains how a physics idea can be applied to a number of different situations' (ES$=$.265) (see Mujtaba & Reiss, 2013b).

Boys were more likely to report that their teacher explained how maths ideas can be applied to a number of different situations ($t=5.784$, $p<.001$); they saw the relevance of maths lessons ($t=2.443$, $p<.05$) and they found it easy to apply most maths concepts to everyday problems ($t=1.991$, $p<.05$). These items were a part of the 'perceptions of lessons' construct.

Boys were also more positive about looking forward to their maths classes ($t=3.463$, $p<.001$) and enjoying their maths lessons ($t=3.996$, $p<.001$); and gave more favourable answers about doing well in their maths tests ($t=11.439$, $p<.001$); and doing better in their maths lessons than their peers ($t=11.225$, $p<.001$).

Finally, boys were more positive about 'when I am doing maths, I always know what I am doing' ($t=8.520$, $p<.001$); I am learning new skills ($t=3.883$, $p<.001$); I am not bored ($t=2.803$, $p<.01$); I don't get upset ($t=7.075$, $p<.001$) and I do not daydream ($t=5.557$, $p<.001$).

Intention to Participate and Perceptions of Mathematics Lessons

The actual constructs 'perceptions of lessons' and 'emotional response to lessons' were moderately correlated with intended participation (.557 and .333, respectively, see Table 4). We would have expected the associations to be the other way around given that in the final MLM model 'emotional response to lessons' had a statistically significant independent influence in explaining intended post-16 mathematics participation. These associations alone suggest again that our original lessons constructs possibly needed further refinement. Other than the associations with the original constructs, the three strongest item-level associations between intended post-16 participation and these cluster of mathematics lessons items were 'I look forward to maths classes' ($r=.479$); 'I enjoy my maths lessons' ($r=.473$) and 'I can see the relevance of maths lessons' ($r=.440$). It is interesting to note that these were the three strongest associations found with the year 10 physics analysis (Mujtaba & Reiss, 2013b). The item that had the smallest association with intended participation was 'I don't find it difficult to apply most maths concepts to everyday problems' ($r=.082$)—again mirroring our findings with physics.

Students' Perceptions of Mathematics

Items explored five areas concerning students' perceptions of mathematics: useful-ness of mathematics (a part of the extrinsic material gain and social gain motivation constructs); self-concept in mathematics; liking of mathematics; mathematics and social skills; and doing mathematics. Overall student means (see Table 1) indicate that students' responses about mathematics were generally positive though there were some aspects of mathematics that they were not positive about or did not agree with. Students were most positive about or in agreement with 'I think maths is a useful subject' (mean 5.15) and least positive about 'being good at maths makes you popular' (mean 2.35)—this latter finding again mirrored that for physics.

Table 1 demonstrates that there were statistically significant differences in responses between boys and girls for the great majority of items, with year 8 boys responding more positively to questions about their perceptions of mathematics. The findings lend support to existing research that some (but certainly not all) girls typically feel disengaged from mathematics and this may be related to the way it is taught. This is possibly related to (some) girls not feeling there are a range of ways to learn mathematics. The largest significant difference in responses between boys and girls was for the item 'I am good at maths' (ES = .404), followed by 'I don't need help in maths' (ES = .310); both of these findings mirror those found for phys-ics with their respective effect sizes being .583 and .548 (Mujtaba & Reiss, 2013b).

In addition, these are the largest effect sizes reported even when including items that explored perceptions of mathematics teachers and mathematics lessons (see Tables 3 and 4). The next four strongest effect sizes (ranging from .274 to .212) were still larger than the effect sizes found for any of the perception of teacher items: 'I think maths will help me in the job I want to do in the future' (ES = .227); 'to be good at maths you need to be creative' (ES = .274); 'being good at maths makes you popular' (ES = .215) and 'maths is important in making new discoveries' (ES = .212).

Boys were more positive that maths is a useful subject ($t=3.171$, $p<.001$); is more likely to help them get into jobs they want to do in the future ($t=8.094$, $p<.001$); teaches individuals to think logically ($t=5.034$, $p<.001$); helps individu-als to solve everyday problems ($t=2.416$, $p<.01$); is important in making new dis-coveries ($t=7.253$, $p<.001$) and that people who are good at maths get well-paid jobs ($t=6.321$, $p<.001$).

Boys were more likely to report that they are good at maths ($t=14.487$, $p<.001$) and do not need help with maths ($t=11.006$, $p<.001$). They were more positive about maths being an interesting subject ($t=5.897$, $p<.001$); finding maths interest-ing ($t=5.589$, $p<.001$); everyone needing to know some maths ($t=3.248$, $p<.01$); maths being a useful subject ($t=3.171$, $p<.001$) and that it is interesting to find out about the laws of maths that explain different phenomena ($t=5.071$ $p<.001$).

Boys were more likely to report that maths makes individuals popular ($t=7.132$, $p<.001$) and improves social skills ($t=3.411$, $p<.01$). Finally, boys were more likely to report that 'to be good at maths individuals need to be creative' ($t=9.336$, $p<.001$); 'to be good at maths you need to work hard' ($t=6.553$, $p<.001$) and 'those who are good at maths are those who are clever' ($t=5.840$, $p<.001$).

Correlations Between Perceptions of Mathematics and Intended Participation in Mathematics

A correlation analysis was conducted between the items that explored students' perceptions of mathematics and their intention to participate in it post-16. Table 1 demonstrates that for the sample as a whole the three strongest associations between intended participation and perceptions of mathematics were for the items: 'I think maths will help me in the job I want to do in the future' (a part of the 'extrinsic material gain motivation' construct) (.506)—with the associated effect size for gender difference being .227; 'I think maths is an interesting subject' (.568)—with the associated effect size for gender difference being .166 and 'I think maths is a useful subject' (a part of the 'extrinsic material gain motivation' construct) (.515)—with the associated effect size for gender difference being .089.

In Mujtaba and Reiss (2013c) we found that boys and girls who intended to continue with mathematics post-16 had similar levels of 'extrinsic material gain motivation', though they differed in other perceptions of their mathematics education. The correlations between items measuring extrinsic material gain motivation and intended participation in mathematics are not surprising; despite two of the items from the 'extrinsic material gain construct' being the most strongly associated items with intended participation, the gender differences are not as strong as those found in other areas of students' perceptions of their mathematics education. These findings suggest that the differences between boys and girls are in their experiences of their mathematics education rather than girls not appreciating the value of mathematics as much as boys. Table 1 also shows the correlations and gender differences for the original constructs. As can be seen from some of the self-concept items, some are more strongly associated with intended participation than others (for example 'I am good at maths' versus 'I do not need help with maths'). We will discuss this further in the concluding section. The actual 'self-concept' construct was moderately correlated with intended mathematics participation (.444), along with some of the other original constructs: extrinsic social gain motivation (.363), extrinsic material gain motivation (.572) and intrinsic value of mathematics (.516).

Multi-level Re-analysis to Explore the Importance of Students' Perceptions on Intended Post-16 Mathematics Participation (Using Items from the Survey Rather than Constructs)

Finally, a further set of multi-level models were run in a series of stages which had particular conceptual relevance, this time driven by the analysis reported above which included findings from the qualitative work. This final set of analyses tested for items from the year 8 student survey and used survey data that the same students filled out in year 10 (age 15). Table 5 shows the final, best fit model and highlights a number of key messages:

1. Our original construct 'extrinsic material gain motivation', found to be an important construct associated with intended participation at year 8, continues to be important in explaining intended participation at year 10 (even whilst using an item-level analysis).

Table 5 Item-based analysis: estimates of fixed effects on year 10 England students' intentions to study mathematics post-16

Parameter	Estimate	Std error	df	t	Sig.	Effect size
Intercept	6.346	0.180	736.308	35.271	0.001	
Gender	−0.162	0.082	683.859	−1.965	0.050	−0.154
'I think maths will help me in the job I want to do in the future' (comparison group: Strongly agree)						
(Strongly disagree)	−1.007	0.220	800.304	−4.570	0.001	−0.961
(Disagree)	−1.029	0.165	806.611	−6.220	0.001	−0.983
(Slightly disagree)	−0.650	0.178	806.738	−3.653	0.001	−0.621
(Slightly agree)	−0.655	0.125	806.730	−5.252	0.001	−0.625
(Agree)	−0.355	0.097	806.950	−3.641	0.001	−0.338
'My teacher thought that I should continue with maths after my GCSEs' (comparison group: Strongly agree)						
(Strongly disagree)	−0.874	0.261	806.536	−3.353	0.001	−0.835
(Disagree)	−0.340	0.216	806.647	−1.580	0.115	−0.325
(Slightly disagree)	−0.483	0.205	803.362	−2.353	0.019	−0.461
(Slightly agree)	−0.305	0.136	806.885	−2.250	0.025	−0.292
(Agree)	−0.109	0.108	805.839	−1.010	0.313	−0.104
'My friends thought that I should continue with maths after my GCSEs' (comparison group: Strongly agree)						
(Strongly disagree)	−0.714	0.212	805.560	−3.373	0.001	−0.682
(Disagree)	−0.910	0.195	806.288	−4.664	0.001	−0.869
(Slightly disagree)	−0.467	0.178	803.045	−2.630	0.009	−0.446
(Slightly agree)	−0.322	0.140	804.960	−2.300	0.022	−0.308
(Agree)	0.166	0.123	803.267	1.356	0.175	0.159
'I was advised by my family that maths would be a good subject to study after my GCSEs' (comparison group: Strongly agree)						
(Strongly disagree)	−1.337	0.228	800.843	−5.869	0.001	−1.276
(Disagree)	−1.126	0.213	806.288	−5.294	0.001	−1.075
(Slightly disagree)	−0.804	0.198	806.394	−4.066	0.001	−0.767
(Slightly agree)	−0.478	0.139	806.826	−3.435	0.001	−0.457
(Agree)	−0.287	0.100	805.704	−2.864	0.004	−0.274
'I look/looked forward to maths classes' (comparison group: Strongly agree)						
(Strongly disagree)	−0.093	0.185	802.169	−0.504	0.614	−0.089
(Disagree)	−0.404	0.173	806.842	−2.337	0.020	−0.386
(Slightly disagree)	−0.117	0.162	806.502	−0.721	0.471	−0.112
(Slightly agree)	−0.001	0.148	806.688	−0.008	0.993	−0.001
(Agree)	−0.103	0.146	804.504	−0.707	0.480	−0.098

(continued)

Table 5 (continued)

Parameter	Estimate	Std error	df	t	Sig.	Effect size
'When I am/was doing maths, I got upset' (comparison group: Strongly disagree)						
(Strongly agree)	−0.200	0.184	805.048	−1.089	0.276	−0.191
(Agree)	−0.450	0.203	806.505	−2.212	0.027	−0.429
(Slightly agree)	−0.135	0.162	804.089	−0.834	0.405	−0.129
(Slightly disagree)	0.052	0.154	801.152	0.336	0.737	0.050
(Disagree)	−0.051	0.092	806.745	−0.549	0.583	−0.048
'I am good at maths' (comparison group: Strongly agree)						
(Strongly disagree)	−0.001	0.262	805.774	−0.005	0.996	−0.001
(Disagree)	−0.634	0.242	802.824	−2.623	0.009	−0.605
(Slightly disagree)	−0.220	0.218	806.667	−1.013	0.311	−0.210
(Slightly agree)	−0.307	0.140	806.812	−2.186	0.029	−0.293
(Agree)	−0.256	0.116	804.896	−2.218	0.027	−0.245
'I need/needed help with maths' (comparison group: Strongly disagree)						
(Strongly agree)	−0.532	0.189	806.975	−2.812	0.005	−0.507
(Agree)	−0.431	0.160	806.852	−2.695	0.007	−0.411
(Slightly agree)	−0.346	0.146	803.227	−2.377	0.018	−0.330
(Slightly disagree)	−0.154	0.153	804.823	−1.002	0.317	−0.147
(Disagree)	−0.144	0.135	806.657	−1.070	0.285	−0.138
Random-effects parameters						
Variance (Level 2)	0.028	0.018				
Variance (Level 1)	1.097	0.056				
Deviance (−2 × log restricted-likelihood)	2,560.182					

2. Students' views of their lessons and teachers are also important in explaining intended participation. This was missed by our construct-level analysis.
3. Gender becomes an important predictor for intended participation in year 10, whilst at year 8 for the same students the differences between boys and girls were not statistically significant.
4. Students' perceptions and experiences in year 10 are more important in explaining intended participation than in year 8.

The items that formed the original constructs which explored perceptions of mathematics (e.g. extrinsic material gain motivation and self-concept) were added towards the end of the model steps, primarily because it was predicted (given earlier multi-level findings and the associations reported in Table 1) that items from such constructs would wipe away the significant influence of teachers and lessons. We wanted to see what, if any, items were associated with year 10 students' mathematics aspirations in both the preliminary and final model.

Our original construct-based multi-level analysis indicated that underlying personality traits lost significance once more fine-grained measures of mathematics-specific measures were introduced in the models. Given such findings we did not include these (non-mathematics-specific) measures within this analysis. For the same reason, we omitted any non-mathematics-specific items that measured general attitudes/perceptions of learning, support and encouragement.

In this item-based multi-level analysis we tested students' survey responses as year 8 and as year 10 learners of mathematics as predictors of mathematics aspirations in year 10. We found that the students' year 10 survey responses about their mathematics education and support they received were better predictors of year 10 aspirations than the earlier year 8 responses; therefore, in the final model only the year 10 survey measures remain. The final model in many ways supported, built on and shed further light on what we found earlier with the construct-based multi-level analysis when the students were in year 8. Table 4 shows that as year 8 learners of mathematics, the construct 'advice-pressure to study mathematics' (which was a summed score of a range of influences students received) was a strong predictor of year 8 students' mathematics aspirations; some of the items which formed this construct also appear as important predictors of these students' aspirations when they were in year 10 (Table 5).

With respect to the items which formed the 'perception of teachers' construct, prior to the inclusion of items from 'self-concept' or 'extrinsic material gain motivation', we found that the 'my maths teacher is good at explaining maths' and 'my maths teacher is interested in me as a person' both had significant independent influences. However, neither of these items were significant predictors in the final model once we controlled for the items that measured 'advice-pressure to study mathematics' and 'extrinsic material gain motivation'. The item 'my teacher thought that I should continue with maths after my GCSEs' (which was originally a part of the 'advice-pressure to study mathematics' construct) had a significant independent influence in explaining year 10 students' mathematics aspirations, which concurs with the findings from the qualitative work. More generally, it is now clear that the influence of teachers is very important (also taking into account findings from Table 3). Furthermore, our original construct of 'perceptions of teachers' was subsequently found to be composed of a number of distinct sub-constructs. For example, the associations between both students' mathematics teacher being 'interested in them as a person' (.237) and students 'liking their mathematics teacher' (.238) with mathematics aspirations were much stronger than when compared to the items that tapped into homework (.057–.167).

Encouragement (most importantly by teachers and families) appears to be associated with raised mathematics aspirations, as evidenced by both our construct-based and item-based analyses. These findings have implications for policy and practice. In order to increase mathematics aspirations, teachers (given that schools generally have little influence on families) need not only to encourage students but to place an emphasis on the 'extrinsic material gain' of having a post-16 mathematics qualification. In addition, the bivariate item-level analysis and the qualitative work revealed that personal relationships with teachers are important in encouraging students' future mathematics aspirations. Teachers could enhance students' aspirations by actively creating more meaningful relationships with their students within their teaching (cf. Rodd, Reiss, & Mujtaba, 2014).

It was worth separating out and exploring the individual influence of each item that created the overall 'advice-pressure to study mathematics' construct. This was a construct developed and piloted (by ourselves) that proved to be of great value to the research. It was clear that the construct showed a large effect size in explaining year 8 students' intended participation (Table 2). We hope that this construct and the various items within it will prove useful for future studies, both qualitative and quantitative, in exploring mathematics aspirations and in enabling teachers and family members to boost post-compulsory mathematics participation. In the final model of the item-based analysis there was an item which indicated that family influence to continue with mathematics post-16 was quite important, which was in line with the findings within our qualitative work (e.g. Elira). Again, this effect was masked in the original analysis when all of the items formed one overall construct— 'advice-pressure to study mathematics'.

Two of the items that were a part of the original mathematics self-concept construct were found to have a strong independent influence on mathematics aspirations: 'I am good at maths' and 'I don't need help with maths'. In fact, 'I am good at maths' had as strong a correlation with intended participation (.460) as the mathematics 'self-concept' construct (.455). Again, these findings support the construct-based analysis which indicated the importance of self-concept. We find it interesting that these two particular items were also uncovered as being important in a similar item-based multi-level modelling analysis when exploring factors that influence year 10 students' physics aspirations (Mujtaba & Reiss, 2013b).

Methodological Conclusions

Methodologically, this chapter reaches three principal conclusions. First, mathematics-specific measures are better predictors of intended participation in mathematics than more general measures. While hardly surprising, the use of mathematics-specific measures proved vital in helping this research discover more about the factors that shape future aspirations in mathematics. In particular, the mathematics-specific measure of extrinsic material gain motivation was more tightly related to future mathematics aspirations than any of the other measures used within our models that measure motivation.

Second, our work clearly demonstrates that research questions ought to guide and help conceptualise a measure whilst taking into account how students may respond differently to the various items within a construct. We conclude that, valuable as construct-based analyses are, researchers ought, at the very least, to complement such analyses by selected analysis at the level of items.

Third, while it is hardly unusual to combine quantitative and qualitative work within a single study, our work shows the benefit of the two approaches when they truly interdigitate. In the analyses reported above we began with quantitative analyses, then turned to qualitative work and then returned to a new set of quantitative analyses, drawing both on our first sets of quantitative analyses and on our qualitative work. The resulting conclusions are, we believe, more robust than had we relied on only quantitative or qualitative work—a conclusion reinforced by our observation that many of our final mathematics-specific findings are similar to those of our physics-specific investigation (Mujtaba & Reiss, 2013b).

References

Black, L., Mendick, H., & Solomon, Y. (Eds.). (2009). *Mathematical relationships in education: Identities and participation*. New York: Routledge.

Boaler, J. (2009). *The elephant in the classroom: Helping children learn and love maths*. London: Souvenir Press.

Brown, M., Brown, P., & Bibby, T. (2008). 'I would rather die': Reasons given by 16-year-olds for not continuing their study of mathematics education. *Research in Mathematics Education, 10*(1), 3–18.

Connell, R. W. (1995). *Masculinities*. Cambridge, England: Polity Press.

Department for Business, Innovation and Skills. (2009). *Higher ambitions: The future of universities in a knowledge economy*. London: Department for Business, Innovation and Skills.

Eccles, J. (1994). Understanding women's educational and occupational choices: Applying the Eccles et al. model of achievement-related choices. *Psychology of Women Quarterly, 18*, 585–609.

Eccles, J. (2009). Who am I and what am I going to do with my life? Personal and collective identities as motivators of action. *Educational Psychologist, 44*(2), 78–89.

Mendick, H. (2006). *Masculinities in mathematics*. Maidenhead, England: Open University Press.

Middleton, J. A., & Jansen, A. (2011). *Motivation matters and interest counts: Fostering engagement in mathematics*. Reston, VA: National Council of Teachers of Mathematics.

Mujtaba, T., & Reiss, M. J. (under revision). *Girls in the UK have similar reasons to boys for intending to study mathematics post-16*.

Mujtaba, T., & Reiss, M. J. (2013a). A survey of psychological, motivational, family and perceptions of physics education factors that explain 15 year-old students' aspirations to study post-compulsory physics in English schools. *International Journal of Science and Mathematics Education, 12*, 371–393.

Mujtaba, T., & Reiss, M. J. (2013b). Inequality in experiences of physics education: secondary school girls' and boys' perceptions of their physics education and intentions to continue with physics after the age of 16. *International Journal of Science Education, 35*, 1824–1845.

Mujtaba, T., & Reiss, M. J. (2013c). What sort of girl wants to study physics after the age of 16? Findings from a large-scale UK survey. *International Journal of Science Education, 35*, 2979–2998.

Nardi, E., & Steward, S. (2003). Is mathematics T.I.R.E.D? A profile of quiet disaffection in the secondary mathematics classroom. *British Educational Research Journal, 29*(3), 345–367.

Noyes, A. (2003). Mathematics learning trajectories: Class, capital and conflict. *Research in Mathematics Education, 5*(1), 139–153.

Noyes, A. (2009). Exploring patterns of participation in university-entrance level mathematics in England. *Research in Mathematics Education, 11*(2), 167–183.

Palmer, A. (2009). 'I'm not a "maths-person"!' Reconstituting mathematical subjectivities in aesthetic teaching practices. *Gender and Education, 21*(4), 387–404.

Reiss, M. J., Hoyles, C., Mujtaba, T., Riazi-Farzad, B., Rodd, M., Simon, S., et al. (2011). Understanding participation rates in post-16 mathematics and physics: Conceptualising and operationalising the UPMAP Project. *International Journal of Science and Mathematics Education, 9*(2), 273–302.

Rodd, M., Reiss, M., & Mujtaba, T. (2013). Undergraduates talk about their choice to study physics at university: What was key to their participation? *Research in Science & Technological Education, 31*(2), 153–167.

Rodd, M., Reiss, M., & Mujtaba, T. (2014). Qualified, but not choosing STEM at university: Unconscious influences on choice of study. *Canadian Journal of Science, Mathematics, and Technology Education, 14*, 330–345.

Royal Society. (2011). *A 'state of the nation' report on preparing the transfer from school and college science and mathematics education to UK STEM higher education*. London: The Royal Society.

Schunk, D. H., Pintrich, P. R., & Meece, J. L. (2010). *Motivation in education: Theory, research, and applications* (3rd ed.). Upper Saddle River, NJ: Pearson Education.

Addressing Measurement Issues in Two Large-Scale Mathematics Classroom Observation Protocols

Jeffrey C. Shih, Marsha Ing, and James E. Tarr

The challenges as well as the need to engage in research that reflects the reality of classroom teaching and learning are well documented (Confrey et al., 2008). Documenting actual teaching practices requires reliable observational data of teachers in their classrooms but such data are not easy to obtain for many reasons (Chval, Reys, Reys, Tarr, & Chávez, 2006; Hiebert & Grouws, 2007; Hill, Charalambous, & Kraft, 2012). First, external observers represent an intrusion into classrooms and can disrupt the regular classroom routine. Second, the measurement challenges of classroom data are also a concern (e.g., Ing & Webb, 2012). For example, identifying and defining the features of classroom observations that are worth attending to and then properly training observers to focus on these particular behaviors are but a few of the challenges. Furthermore, observing teachers and coding information related to mathematics instruction are both time consuming and costly, thereby limiting the type of measurement that can be done on a large scale. Trade-offs to capturing what mathematics teachers actually do within their classrooms need to be considered (National Research Council, 2004).

Ideally, large-scale classroom observations generate data that accurately characterizes the teaching and learning practices, but the process of collecting such data does not intrude upon teachers or require more observational time and resources than are necessary. Without accurate characterizations, inferences about mathematics education are unfounded. Thus, large-scale observational measures that identify variation in mathematical instructional quality across classrooms

J.C. Shih (✉)
University of Nevada, Las Vegas, Las Vegas, NV, USA
e-mail: jshih@unlv.nevada.edu

M. Ing
University of California, Riverside, Riverside, CA, USA

J.E. Tarr
Department of Learning, Teaching and Curriculum, University of Missouri,
303 Townsend Hall, Columbia, MO 65211-2400, USA

© Springer International Publishing Switzerland 2015
J.A. Middleton et al. (eds.), *Large-Scale Studies in Mathematics Education*,
Research in Mathematics Education, DOI 10.1007/978-3-319-07716-1_16

have tremendous implications for large-scale studies of mathematics education. Generalizability theory is an approach to address these measurement issues in large-scale observational measures. This chapter first describes two large-scale mathematics observational protocols, and presents findings that address measurement issues specific to each observational measure using generalizability theory. We conclude by raising additional technical and conceptual issues around large-scale measures of mathematics classroom instruction.

Methods

Observational Protocols

Classroom Learning Environment (CLE). The purpose of this protocol is to measure particular features of the CLE that are considered important across different types of textbook curricula. The protocol was initially designed as part of a larger study that examined student mathematical learning associated with secondary mathematics curriculum programs of two types: a subject-specific approach and an integrated content approach (Tarr et al., 2008). In the CLE measure, there are ten items that collectively represent the classroom environment. These ten items cluster around three themes: Reasoning about Mathematics, Students' Thinking in Instruction, and Focus on Sense-making. Using a rubric for each of the ten items, observers rendered ratings from 1 to 5, with a 1 indicating the absence of a feature, and a 5 indicating a strong presence of the feature during the observed lesson. The ten items load on a single factor, CLE. The average across the ten elements was created as a common measure across different curriculum types.

Project team members prepared for scoring by viewing videotapes of four mathematics lessons. To simulate an actual classroom visit, each video ran uninterrupted so that project team members could code in real time. When independent coding was completed, discussion commenced and focused on the consistency of coding. Discussion led to modest changes in the user's guides to further enhance the reliability of coding. There was a 70 % exact agreement in coding. When inconsistencies were observed, rubrics in the user's guides were read aloud and discussed to negotiate the optimal code for the given classroom element. Researchers' initial codes differed by no more than plus/minus one from the negotiated code (on a 5-point scale) in 94 % of all cases. After these training sessions, double coding of 15 selected lessons during classroom visits during the data collection phase of the study were conducted. These lessons were selected based on the feasibility of the rater's schedules. Exact agreement for this sample of lessons was 62 %, with more than 90 % of the codes differing by no more than 1 point on a 5-point scale. Raters discussed scoring with each other throughout the data collection phases.

The project team identified one source of variation in the CLE observational protocol: lessons. The content of each lesson varies, which might influence what is observed and how the CLE is characterized. The concern with this particular observational measure, then, is how many lessons need to be observed to obtain an acceptability level of reliability?

Mathematical Quality of Instruction (MQI). The purpose of this protocol is to provide information on teachers' enactment of mathematics instruction. The MQI instrument is based on a theory of instruction that focuses on resources and their use (Cohen, Raudenbush, & Ball, 2003), existing literature on effective instruction in mathematics (e.g., Borko et al., 1992; Ma, 1999; Stigler & Hiebert, 1999), and on an analysis of nearly 250 videotapes of diverse teachers and teaching. The MQI is currently intended for use with videotaped lessons of classroom mathematics instruction (Hill, Kapitula, & Umland, 2011) and provides fine-grained information about instructional practice. It includes four major dimensions, of which three are discussed in this chapter. The first dimension, richness of the mathematics (Richness), captures the depth of the mathematics offered to students, as reflected by the links drawn between different representations, the explanations offered, the discussion of multiple solution approaches, the inductive generation of generalizations, and the richness in the mathematics language used to present the content. The second dimension, teacher errors and imprecision (Errors and Imprecision), captures teacher mathematical errors/oversights and linguistic/notational imprecision when presenting the content; it also pertains to the lack of clarity in teacher's launching of tasks and presentation of the content. The third dimension, student participation in mathematical meaning-making and reasoning (SPMMR), as its name suggests, captures the extent to which students participate in and contribute to meaning-making and reasoning during instruction. This could be evident in provision of explanations, student posing of mathematically motivated questions, offering of mathematical claims and counterclaims, and engagement in cognitively demanding activities.

There are two sources of variation considered in the MQI observational protocol, raters and lessons. Raters refer to the people who are conducting the observations. It is assumed that raters have gone through training on the observational protocol and have achieved a particular level of proficiency with scoring the observational protocol. Ten mathematics education graduate students and former teachers were recruited via emails to colleagues in mathematics education departments. Raters attended a 2-day intensive training on the instrument. At the end of training, raters took a certification exam, in which they were asked to code 16 segments from videotaped lessons taught by four different teachers. Based on these results, one rater whose scores did not meet the certification threshold was excluded from the analysis presented below.

The second source of variation, lessons, refers to the number of lessons teachers are observed. Each time they are observed, they are teaching a different lesson. Each rater watched and scored 24 lessons (three lessons per each of eight teachers). Following the coding protocol, raters skimmed each lesson once. During the second watch each rater assigned scores for each MQI item for every 7.5-min segment of the lesson. The raters did so by using a 3-point scale (low, medium, high). The 24 lessons were selected from videotapes of lessons from 24 middle-school mathematics teachers in one district. From these teachers, eight teachers with different levels of mathematical knowledge for teaching were sampled (see Hill, Kapitula, & Umland, 2011, for more details on the larger study). From the six available videotaped lessons for each teacher, three lessons per teacher that were approximately equal in length were sampled (i.e., each of the sampled lessons contained between six and eight 7.5-min segments). Because of the small sample employed, the results presented in this study are considered exploratory.

Analysis

Generalizability theory (Brennan, 2010; Shavelson & Webb, 1991) was used to investigate the dependability of the observational measures. Generalizability theory estimates the magnitude of multiple sources of error and provides a reliability (generalizability) coefficient for the proposed use of the observational measure. The advantage of using generalizability theory is the ability to interpret variance components such as the effect due to differences from one occasion to another.

We used the information from each generalizability study to conduct a decision study to provide information about what would happen if different levels within each source of variation were modified. For example, would the same conclusions about instructional practice be drawn if the teacher was observed on a single occasion versus 20 occasions? Would the same conclusions about instructional practice be drawn if there were a single rater observing instruction versus 20 raters? Decision studies provide information about these different scenarios to guide future use of the observational protocol (Marcoulides, 1993, 1997; Marcoulides & Goldstein, 1990).

Results

Lessons (CLE)

To examine lessons as a source of error, three lessons taught by 68 teachers were coded by one project team member. Table 1 provides a summary of the generalizability analyses. Most of the variation is between teachers (64 %). Lessons are not

Table 1 Variance decomposition of CLE

Sources of variation	Estimated variance components	Percentage of total variance
Teacher (T)	0.63	64
Lessons (L)	0.00	0
Residual	0.35	36

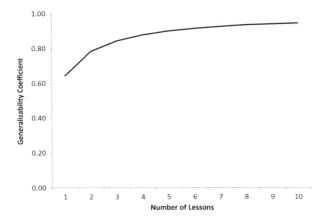

Fig. 1 Absolute generalizability coefficient for different numbers of lessons for CLE

a considerable source of variation, which suggests that the number of different lessons during which teachers are observed is not a measurement issue for this particular observational protocol. However, there is a large proportion of variance due to unexplained systematic and unsystematic sources (36 %) which suggests that there are other sources of error that should be investigated in future administrations.

Due to the lack of variation due to lessons, increasing the number of lessons observed per teacher does not dramatically increase the generalizability coefficient (Fig. 1). With three lessons, there is an acceptable level of reliability (0.84). This protocol focuses on absolute decisions rather than relative decisions. In other words, this protocol is not concerned with how much higher one classroom is compared to another but is more concerned with whether classrooms meet a particular level.

Lessons and Raters (MQI)

To examine lessons and raters as sources of measurement error, we used 24 lessons taught by eight middle-grade teachers (three lessons each) that were coded by nine raters. The raters received an intensive 2-day training on MQI and passed a certification test (for more on the coding and the rater certification process, see Hill et al., 2012). The percent of variation for each source of error for each MQI dimension is presented in Table 2.

A decision study was conducted to determine the number of raters needed per lesson and the number of lessons required to achieve acceptable reliability estimates (Fig. 2). The generalizability coefficient for relative decisions was used, rather than the generalizability coefficient for absolute decisions on the grounds that districts often make relative rather than absolute decisions about teachers, such as rewarding the top 5 % of teachers with merit pay.

Table 2 Percent of total
variation for each source
of variation of teachers'
performance in the three
MQI dimensions

Source of variation	Richness	Errors and imprecision	SPMMR
Teachers (T)	42.52	31.88	32.78
Lessons:teachers (L:T)	10.52	8.81	7.22
Raters (R)	6.17	13.04	28.58
Teachers*raters (T*R)	7.83	6.45	0.00
Residual	32.97	39.82	31.43

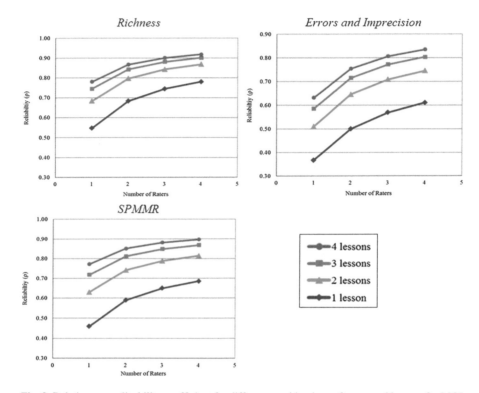

Fig. 2 Relative generalizability coefficient for different combinations of raters and lessons for MQI

As Fig. 2 shows, for all three MQI dimensions, observing more than three lessons yields diminishing returns in terms of the reliability coefficient; instead, adding a second rater to each lesson increases the reliability coefficient markedly.
Figure 2 also shows that the three-lesson/two-rater combination produces relative
reliability coefficients higher than .70 for all three dimensions (Richness = .77,
Errors and Imprecision = .71, SPMMR = .81).

There are differences when comparing lessons as a source of error for the MQI
and CLE protocols. For example, lessons are a larger source of error for the MQI than
for the CLE. This is due in part to the different purpose or intension of each protocol.

For the MQI, the assumption is that differences between lessons can be measured. The CLE, on the other hand, does not assume this. Instead, the CLE is intended to measure more stable or global characteristics of instruction that do not change from lesson to lesson. The lack of variation due to lessons is something that is expected for the CLE but not for the MQI.

Discussion

The purpose of this chapter is to advocate for a closer examination of the measurement qualities of the many available classroom observation tools. Although this chapter presents empirical results for only two mathematics observational protocols, it raises questions about trade-offs between technical quality and practical considerations and about the different purposes and intents of different observational measures. For example, Correnti and Martinez (2012) describe conceptual, methodological, and policy issues in large-scale measures of instruction. The authors identify six purposes of measures of instruction including "characterizing the nature and main feature of instructional practice for groups of teachers or schools" and "understanding and comparing instructional practices and classroom processes across localities, states and countries" (p. 52) that influence the particular conceptual, methodological, and policy issues for each measure of instruction.

The two observational protocols described in this chapter differ in terms of their assumptions about instruction. To measure instruction, the CLE focuses on classroom learning environments that are considered important across different types of curriculum. The type of instruction that is being measured with the CLE is assumed to be relatively stable and not to vary from day to day. In contrast, the MQI does not focus on any particular type of curriculum but instead purports to measure teachers' enactment of mathematics instruction. The MQI measures features of instruction that are not necessarily relatively stable from day to day and require attention to more dimensions of instruction. Thus, the assumptions about instruction differ for these measures, with the CLE measuring stable aspects and MQI measuring less stable aspects.

These protocols also differ in their purpose, which has implications for what sources of variation each protocol identifies. The CLE, for example, does not include raters as a source of error because the raters involved in that study were intensively involved with the development of the observational measure and there was less concern about including a wider circle of raters to conduct the observations. The CLE was not intended for use across all classrooms in the USA but instead was used solely by the research team to measure curriculum implementation for schools participating in a larger study. The MQI, on the other hand, was designed for use in classrooms across the USA. Thus, there is a need to prepare raters with vastly different experiences to use the MQI. The MQI therefore includes raters as a source of variation because the raters are drawn from a much larger sample of raters and is not limited only to raters closely associated with the research team.

With these two observational protocols serving different purposes and designed with different assumptions about teaching and learning, it follows that the development and implementation of each observational tool also varies. This chapter is not meant to prescribe how all generalizability studies in mathematics education should be conducted. This chapter is also not meant to identify the best mathematics observational protocol. Rather this chapter is meant to raise issues in the mathematics education community that require attention when developing or implementing any sort of observational measure. We do not expect to see the same sources of variation being identified and measured with each observational tool. Each observational tool represents choices and assumptions about which sources of variation are most important to capture. Given the increased attention to large-scale measures of mathematics instruction, this chapter provides a significant starting point for a much-needed discussion among mathematics education researchers.

References

Borko, H., Eisenhart, M., Underhill, R., Brown, C., Jones, D., & Agard, P. (1992). Learning to teach hard mathematics: Do novice teachers and their instructors give up too easily? *Journal for Research in Mathematics Education, 23*(3), 194–222.

Brennan, R. L. (2010). *Generalizability theory*. New York: Springer.

Chval, K. B., Reys, R. E., Reys, B. J., Tarr, J. E., & Chávez, O. (2006). Pressures to improve student performance: A context that both urges and impedes school-based research. *Journal for Research in Mathematics Education, 37*(3), 158–166.

Cohen, D., Raudenbush, S., & Ball, D. (2003). Resources, instruction, and research. *Educational Evaluation and Policy Analysis, 25*(2), 1–24.

Confrey, J., Battista, M. T., Boerst, T. A., King, K. D., Reed, J., Smith, M. S., et al. (2008). Situating research on curricular change. *Journal for Research in Mathematics Education, 39*(2), 102–112.

Correnti, R., & Martinez, J. F. (2012). Conceptual, methodological, and policy issues in the study of teaching: Implications for improving instructional practice at scale. *Educational Assessment, 17*, 51–61.

Hiebert, J. S., & Grouws, D. A. (2007). The effects of classroom mathematics teaching on students' learning. In F. K. Lester Jr. (Ed.), *Second handbook on mathematics learning and teaching* (pp. 371–404). Charlotte, NC: Information Age Publishing.

Hill, H. C., Charalambos, C. Y., Blazar, D., McGinn, D., Kraft, M. A., Beisiegel, M., Litke, E., Lynch, K (2012). Validating arguments for observational instruments: Attending to multiple sources of variation. *Educational Assessment, 17*, 88–106.

Hill, H. C., Charalambous, C. Y., & Kraft, M. A. (2012). When rater reliability is not enough: Teacher observation systems and a case for the generalizability study. *Educational Researcher, 41*(2), 58–64.

Hill, H. C., Kapitula, L. R., & Umland, K. L. (2011). A validity argument approach to evaluating value-added scores. *American Educational Research Journal, 48*(3), 794–831.

Ing, M., & Webb, N. M. (2012). Characterizing mathematics classroom practice: Impact of observation and coding choices. *Educational Measurement: Issues and Practice, 31*(1), 14–26.

Ma, L. (1999). *Knowing and teaching elementary mathematics: Teachers' understanding of fundamental mathematics in China and the United States*. Mahwah, NJ: Lawrence Erlbaum.

Marcoulides, G. A. (1993). Maximizing power in generalizability studies under budget constraints. *Journal of Educational and Behavioral Statistics, 18*(2), 197–206.

Marcoulides, G. A. (1997). Optimizing measurement designs with budget constraints: The variable cost case. *Educational and Psychological Measurement, 57*(5), 808–812.

Marcoulides, G. A., & Goldstein, Z. (1990). The optimization of generalizability studies with resource constraints. *Educational and Psychological Measurement, 50*(4), 761–768.

National Research Council. (2004). *On evaluating curricular effectiveness: Judging the quality of K-12 evaluations.* Washington, DC: National Academy Press.

Shavelson, R. J., & Webb, N. M. (1991). *Generalizability theory: A primer.* Newbury Park, CA: Sage.

Stigler, J. W., & Hiebert, J. (1999). *The teaching gap: Best ideas from the world's teachers for improving education in the classroom.* New York: Free Press.

Tarr, J., Reys, R., Reys, B., Chávez, O., Shih, J., & Osterlind, S. (2008). The impact of middle grades mathematics curricula on student achievement and the classroom learning environment. *Journal for Research in Mathematics Education, 39*(3), 247–280.

Engineering [for] Effectiveness in Mathematics Education: Intervention at the Instructional Core in an Era of Common Core Standards

Jere Confrey and Alan Maloney

The Process of "Engineering [for] Effectiveness"

Improving schools has often been cast as a challenge of identifying effective programs, as captured by the general call for "What Works?" (www.whatworks. ed.gov). Many researchers, skeptical of this call, argue that the real question should not be "whether an intervention works," but instead, "what works, when, for whom, and under what circumstances" (Bryk, Gomez, & Grunow, 2011, p. 151). A shift to focus on specific outcomes that accrue under precise conditions and with specified resources rests on the assumption that educational outcomes result from (and often require) adaptations to circumstances; therefore to seek simple broad scientific principles or rules that apply across the board to a curriculum is of limited value. For example, Bryk, Gomez, and Grunow noted, "Treatises on modern causal inference place primacy on the word 'cause' while largely ignoring concerns about the applicability of findings to varied people, places and circumstances. In contrast, *improvement research* must take this on as a central concern if its goal is useable knowledge to inform broad scale change" (Bryk et al., 2011, p. 150; italics added).

Shifting the question to "what works, when, for whom, and under what conditions?" has profound implications for the meaning of effectiveness as a dependent variable. In establishing causal models, one determines, within the restrictions of a particular study's conditions, if an effect, controlling for other factors, can be

---

Based on a paper originally presented to the National Academies Board on Science Education and Board on Testing and Assessment for the conference, "Highly Successful STEM Schools or Programs for K-12 STEM Education: A Workshop"

J. Confrey (✉) • A. Maloney
North Carolina State University, Raleigh, NC, USA
e-mail: jconfre@ncsu.edu

© Springer International Publishing Switzerland 2015
J.A. Middleton et al. (eds.), *Large-Scale Studies in Mathematics Education*,
Research in Mathematics Education, DOI 10.1007/978-3-319-07716-1_17

373

rigorously linked to an antecedent condition. Instead of the causal structure of the phenomenon of interest, this focuses the study on its internal validity—hence "cause" and "effect." While studies typically can and do produce small but statistically significant effects, they often have nested within them more interesting conjectures about interactions and relationships among causes, effects, and co-relational phenomena. Those who demand causal design are often silent on the necessity of replication, which, strictly speaking, is required to realize the benefits of randomization; one study alone does not ensure generalizability.[1] Furthermore, in pursuit of causal models, researchers often rely on average effects, but doing so strips away more robust and potentially relevant differences that may apply to subsets of the whole.

Attempt to identify simple causal chains, and focus on strict control of study conditions, can lead those who attempt to implement research results astray. Too many policy makers and practitioners assume that an established treatment, as "cause," can be simply or directly applied to a practice and guarantee an effect. Perhaps some lack awareness that a study's internal validity does not assure its external validity. Consequently, most studies leave the practitioners themselves responsible to evaluate whether that study generalizes to their own settings. How they are supposed to do this responsibly is seldom addressed.

Regarding randomized field trials as the sole source—or the trump card—of assertions of a program's "effectiveness" poses a major dilemma. They are typically very costly, difficult, and time-consuming to conduct, leaving the public continually awaiting a sufficient set of scientifically "proven" empirical results. Randomized field trials seldom provide timely information in a quickly evolving context (especially for technology-enhanced programs)—by the time the results are available, the program typically is either outdated or has been significantly revised.

In contrast, in this chapter we argue that by developing and deploying explicit means of *engineering [for] effectiveness*, communities of practitioners and researchers can conduct ongoing local experiments at scale, which incorporate adequate design, as well as technologically enabled tools for real-time data collection and continuous analysis of patterns and trends.

Approaches similar to engineering [for] effectiveness have emerged under a variety of names. The study of complex and dynamic systems (Maroulis et al., 2010) has been addressed variously through continuous improvement models (Deming, 2000; Juran, 1962), implementation research (Confrey, Castro-Filho, & Wilhelm, 2000; Confrey & Makar, 2005), improvement research (Bryk et al., 2011), a science of improvement (Berwick, 2008), and Design-Educational Engineering and Development (DEED) (Bryk, 2009; Bryk & Gomez, 2008). When examined through the lenses of these various models, it becomes evident that the improvement of

---

[1] One can, of course, throw five heads in a row in a toss of five coins; only by replicating this experiment multiple times can one be certain that a generalized result of 50–50 emerges. Hence one experiment can never establish any form of cause and effect, a fact too frequently overlooked in discussions of the benefits of randomized field trials.

educational outcomes requires reexamination of approaches to just what is meant by "effectiveness." The following four ideas can be used to frame that reexamination:

1. *Education must be viewed as a complex system, with interlocking parts.* Study of a complex system requires one to locate a focus of attention without losing sight of the broader context. One must also attend to a variety of scales of events and time (Lemke, 2000). For instance, while summative and periodic results (large scale, longer time frames) may be useful as broad but crude policy levers that help in identifying trends and sources of inequities, formative results (smaller grain size, shorter time frames) are crucial to drive classroom processes forward. Measurement issues will vary according to these varying levels and orders of magnitude of phenomena (Lemke, 2000; Maroulis et al., 2010).

2. *Bands and pockets of variability should be expected, examined for causes and correlates, and used as sources of insight, rather than adjusted for, suppressed, or controlled.* Discerning how to characterize variability and its significance is key to knowing how to characterize a particular case or instance. "Most field trials formally assume that there is some fixed treatment effect (aka a standardized effect size) to be estimated. If pressed, investigators acknowledge that the estimate is actually an average effect over some typically nonrandomly selected sample of participants and contexts. Given the well-documented experiences that most educational interventions can be shown to work in some places but not in others, we would argue that *a more realistic starting assumption is that interventions will have variable effects and these variable effects may have predictable causes*" (Bryk et al., 2011, p. 24). Stephen J. Gould (1996) made a similar argument in *Full House*, discussing the diagnosis of his mesothelioma. He pointed out that, as a patient, broad survival rates were of less use to him than the smaller bands of variability that more specifically characterized his situation and provided more insight into his chances of survival. Similarly, analytic frames must therefore take into account patterns of antecedent and coincident conditions that mark potential variation in outcomes. "Effectiveness" is not unifaceted, but only understandable in the context of these causal networks.

3. *Causal or covarying cycles with feedback and interaction are critical elements of educational systems, in which learning is a fundamental process.* Furthermore, feedback loops mediate social cues and their behavioral outcomes, so one expects emergent phenomena (Maroulis et al., 2010). There is a contrast between constructions of simple cause-and-effect on the one hand, and causal cycles on the other. In the case of simple cause-and-effect, one assumes that a curriculum is implemented, and produces knowledge growth among students. In the case of causal cycles, the implementer is already aware of the types of outcomes measured, based on prior feedback, and implements and adapts the curriculum simultaneously, thereby raising the question "to what extent did the curriculum cause the effects, and to what extent did the outcome measures (through anticipation or feedback) cause the curriculum adaptation, and thence the effects (a causal cycle)?"

4. *Education should be treated as an organizational system that seeks, and is expected, to improve continuously.* As such, it is comprised of actors who must coordinate their expertise, set ambitious goals, formulate tractable problems

(Rittell & Webber, 1984), negotiate shared targets and measures of success (Bryk et al., 2011), make design decisions within constraints (Conklin, 2005; Penuel, Confrey, Maloney, & Rupp, 2014; Tatar, 2007), and develop and carry out protocols for inquiry. In such a "networked improvement community" (Bryk et al., 2011), one can position the causal cycles under investigation as "frames of action." Continuous improvement depends on iterations of collecting relevant, valid, and timely data, using them to make inferences and draw conclusions, and take deliberate actions, which, in turn, provide a refined set of data upon which to approximate some meaningful set of outcomes.

In analyzing the following examples of studies of curricular effectiveness, we will refer to these components as (1) complex systems with interlocking parts, (2) expected bands of variability, (3) focus on feedback, causal cycles, interactions, and emergence, and (4) continuous organizational improvement. We seek to show how these four components can inform us in designing and engineering [for] effectiveness and scale.

In this article, we focus our discussion of the redefinition of effectiveness research in the context of curriculum design, implementation, and improvement. We point out complementarities with the call for a change in "protocols for inquiry," in which Bryk et al. (2011) locate a "science of improvement" between models of traditional translational research and action research:

> In its idealized form, translational research envisions a university-based actor drawing on some set of disciplinary theory (e.g., learning theory) to design an intervention. This activity is sometimes described as "pushing research into practice" (see for example Coburn & Stein, 2010, p. 10). After an initial pilot, the intervention is then typically field-tested in a small number of sites in an efficacy trial. If this proves promising, the intervention is then subject to a rigorous randomized control trial to estimate an overall effect size. Along the way, the intervention becomes more specified and detailed. Practitioner advice may be sought during this process, but the ultimate goal is a standard product to be implemented by practitioners as designed. It is assumed that positive effects will accrue generally, regardless of local context, provided the intervention is implemented with fidelity.
>
> In contrast, action research places the individual practitioner (or some small group of practitioners) at the center. The specification of the research problem is highly contextualized and the aim is localized learning for improvement. While both theory and evidence play a role, the structures guiding inquiry are less formalized. Common constructs, measures, inquiry protocols and methods for accumulating evidence typically receive even less emphasis. The strength of such inquiry is the salience of its results to those directly engaged. How this practitioner knowledge might be further tested, refined and generalized into a professional knowledge, however remains largely unaddressed (Hiebert, Gallimore, & Stigler, 2002).
>
> A *science of improvement* offers a productive synthesis across this research-practice divide. It aims to meld the conceptual strength and methodological norms associated with translational research to the contextual specificity, deep clinical insight and practical orientation characteristic of action research. To the point, the ideas … are consistent with the basic principles of scientific inquiry as set out by the National Research Council (Shavelson & Towne, 2002, p. 22).
>
> Entire quote from Bryk et al. (2011), pp. 148–149 (italics added).

By defining a perspective of "engineering [for] effectiveness" we suggest that communities of practice, at a school, district, or state level, can build on what has been learned from studies of curricular effectiveness. We review several studies associated with effectiveness research from mathematics education, and reinterpret their results and implications. Our focus will be the challenge of improving the *instructional core* (derived from Elmore, 2002; Cohen, Raudenbush, & Ball, 2003), by which we refer to *the daily classroom activities of implementing a curriculum, carrying out instruction, and applying formative assessment practices.*

Intervening at the Instructional Core

A model of the instructional core is shown below, in which the instructional core is situated between the Common Core State Standards (CCSS) and High Stakes tests. In general, standards (at the state level) and high stakes summative assessments are the "bookends" that constitute the accountability system. Policy levers of *No Child Left Behind* are designed to drive accountability through external pressure (sanctions and incentives) and to shed light on discrepant subgroup performances or lack of annual yearly progress. However, the bookends neglected and/or avoided the instructional core in relation to professional development, pedagogy, and classroom assessment. The absence of common standards fragmented the attention to curriculum (Reys, Reys, Lapan, Holliday, & Wasman, 2003). By squeezing the educational system by way of the bookends, the accountability system during the past decade and more produced some performance gains from the system. However, it failed to strengthen the instructional core with respect to capacity, unintentionally promoted a narrowing of the content taught, and, while calling for the use of "best practices" it failed to identify means to establish the credibility of practices identified as "best."

We chose the instructional core as a focus for this chapter because it can be readily recognized as a complex system, and should be analyzed as such. Its identifiable interlocking parts act at different levels of the educational system, from the standards and the summative tests to classroom practices and formative feedback. While one could view the instructional core as a temporal sequence of (a) curricular selection, (b) some level of professional development, (c) followed by implementation and assessments (both formative and summative), each of these components also interacts with and can generate (organized and explicit, or de facto and inadvertent) feedback to the other components. For instance, frequent formative results provide regular feedback to classroom practices, while data from high-stakes tests provide intermittent or periodic feedback and a much cruder level of nonspecific but severe institutional pressure. Resulting practices can be customized for groups according to curricular requirements and feedback from measures of learning. We have left the structure of improvement communities intentionally vague. The generality of the model allows for diverse institutional structure, as well as informal relationships among actors. Networked improvement communities are not explicitly identified in Fig. 1, but could be configured such that communities of practice could include practitioners, researchers, and administrators, who can plan together, share experiences, analyze data patterns, and discuss how to revise and adapt instructional approaches, curriculum, and schedules.

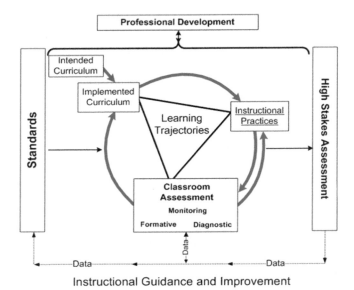

Fig. 1 Model of Classroom Educational System, illustrating position of the instructional core between the accountability "bookends" (Confrey & Maloney, 2012)

The adoption of the CCSS, by most of the states, positions educational communities, writ large, to create policy approaches and to reconsider the importance of focusing on improving the instructional core without overly constraining innovation, over-regulating curricular choice, or de-skilling teaching. By examining exemplars of research on the effectiveness of curricular programs, classroom instructional pedagogies, and formative assessment practices, and defining how these results can inform efforts to engineer [for] effectiveness, researchers could potentially jumpstart a movement towards school improvement in STEM disciplines.

Curricular Effectiveness Studies

"Curriculum matters" (Schmidt et al., 2001). It is the means by which students gain access to the knowledge and skills in a field and also the primary way they are attracted to pursue and persist. Since the publication of the NRC report that one of us (Confrey) chaired, *On Evaluating Curricular Effectiveness* (NRC, 2004), many mathematics educators have worked diligently to strengthen and improve research on and evaluation of curricular effects. That NRC report's framework called for designing evaluations to examine three components of curriculum: the program theory (through content analyses and comparison to standards), the program implementation (through a study of the program's implementation including professional development and on-site staging, resources, and support), and the program outcomes (for alignment to standards and achievement of intended results). The report argued for the use of multiple methods in judging effectiveness, including content analyses, comparative studies, and case studies. It also called for the use of multiple and more sensitive outcome measures, and made a case for increased independence of evaluators, precise identification of comparison programs, and better measures of implementation. We have selected three studies that have taken these recommendations seriously and have moved research to a higher and more nuanced level. We report on their approaches, their principal findings, and identified limitations, and discuss how these can be interpreted to provide a solid foundation to next generation efforts to "engineer [for] effectiveness," that is, to iteratively design, monitor, analyze, and adjust components of the instructional core for more effective teaching and learning.

Case One: Single-Subject vs. Integrated Mathematics (COSMIC Study)

New studies of curricular implementation have advanced our understanding of curricular effectiveness. One such study is "Comparing Options in Secondary Mathematics: Investigating Curriculum," (COSMIC) (Grouws et al., 2010, 2013; Tarr, Grouws, Chávez, & Soria, 2013). The COSMIC project compared the effects of two curricula, one subject-specific and one integrated, on student learning in high school mathematics. Among the important contributions of this large quasi-experimental study was the development of multiple measures of curricular implementation and new types of curricular-appropriate tests to study the effects of curricular content organization on student learning in the first 2 years of high school mathematics.

A goal of the COSMIC study was to improve understanding about the relationships among curricular organization, curricular implementation factors, and gains in student learning. The study's research questions were the following for year 1 (Algebra 1 compared with Integrated Course 2) (Grouws et al., 2013). The research questions for year 2 student learning (Geometry compared with Integrated Course 2) were similar (Tarr et al., 2013):

1. Is there a differential mathematics learning effect when secondary school students study from an integrated textbook (Course 1) and when students study from a subject-specific textbook (Algebra 1)?
2. What are the relationships among curriculum type, curriculum implementation, and student learning? In particular,

   (a) What curriculum implementation factors are associated with high school students' learning in first-year mathematics courses?
   (b) What teacher characteristics and practices are associated with high school students' learning in first-year mathematics courses?

Participating schools all offered both a traditional high school curriculum (algebra 1, geometry, algebra 2) and an integrated curriculum (CORE-Plus), between which students chose freely (i.e., were not tracked by ability level).[2] In all, 11 schools in six districts across five regions of the country participated. The schools' student population demographics varied widely (e.g., the proportion of students eligible for free and reduced lunch (FRL) ranged from 17 to 53 % across the schools in the 2 year-levels of the study). Three distinct measures of student achievement (dependent variables) were used. Student results on those measures were compared to an index of prior achievement based on state-mandated eighth grade tests, normed against NAEP to provide comparability of student preparation across classes and states. Dependent measure data were analyzed using hierarchical linear modeling (HLM).

"Fair test" as essential measure for comparing curricula. The study generated a number of significant advances in research on curricular effectiveness. Researchers incorporated expertise in mathematics content and in learning effectively to design and select the study's outcome measures. They used multiple outcome measures: for each year level, two tests were designed specifically for the project (one of common content and another of reasoning and problem solving). The third test was a nationally normed standardized multiple-choice test, the Iowa Test of Educational Development [ITED]: Mathematics: Concepts and Problem Solving Form B, levels 15 (year 1) and 16 (year 2).

Drawing heavily on the NRC report's recommendations, the project began with content analyses of the printed curricula used in the schools. The project team then designed a "fair test" (NRC, 2004), "developed with the deliberate goal of not being biased towards either of the two curriculum programs studied" (Chávez, Papick, Ross, & Grouws, 2010, p. 4). To create the fair test, items were developed collaboratively by a research mathematician and mathematics educator to include content common to both curricula (i.e., that all students could be expected to have had the opportunity to learn (OTL) in both curriculum types) (Chávez et al., 2010). Items were constructed response instead of multiple choice, and often used realistic situations. Iteratively developed, the items were designed to permit adequate space and time for students to reveal potentially subtle differences in their understanding of underlying constructs, were piloted to ensure high face validity of the items, and were scored using a rubric construction method that assured careful internal and external review, and inter-rater reliability. An overall intent of the fair test was to allow inferences to be made about "student knowledge on constructs underlying the

---

[2] In the COSMIC year 1 study, the textbooks used were Core-Plus Mathematics Course 1 (Coxford et al., 2003) [20 classes], the integrated curriculum, and 5 different single subject curricula, Glencoe Algebra 1 (Holliday et al., 2005), [10]; McDougal Littell Algebra 1 (Larson, Boswell, Kanold, & Stiff, 2001) [6]; Holt Rinehart & Winston Algebra 1 Interactions (Kennedy McGowan, Schultz, Hollowell, & Jovell, 2001) [4]; and Prentice Hall Algebra 1 (Bellman, Bragg, Charles, Handlin, & Kennedy, 2004) [2 classes]. In the year 2 study, the textbooks were Core-Plus Course 2 (Coxford et al., 2003), and one of the following SS curricula Glencoe-McGraw Hill (Boyd, Cummins, Malloy, Carter, & Flores, 2005), Prentice Hall (Bass, Charles, Jonson, & Kennedy, 2004), Holt (Burger et al., 2007), and McDougal Littell (Larson, Boswell, & Stiff, 2001).

content of the tasks on the test, rather than merely…about student ability only on the tasks themselves" (Chávez et al., 2010, p. 8).

Treatment integrity (multiple measures of implementation fidelity). COSMIC researchers also intensified the degree to which they addressed *treatment integrity* (NRC, 2004) using multiple data sources to gauge teachers' implementation of curricular materials. These included Table of Contents Records, Textbook-Use Diaries, an Initial Teacher Survey, a Mid-course Teacher Survey and observations using a Classroom Visit Protocol (McNaught, Tarr, & Sears, 2010, p. 5). The research team was able to examine critical factors such as professional development, familiarity with standards, and teachers' distribution of classroom time among lesson development, non-instruction, practice, and closure. In a sub-study across two consecutive school years, the authors defined, studied, and compared three related indices of curricular implementation: OTL Index, "the percentage of textbook lessons taught without considering teachers' use of supplemental or alternative curricular materials" (the topics or lessons that students thus had an OTL); Extent of Textbook Implementation (ETI) Index, to provide a sense of how closely the textbook was related to the implemented curriculum (a weighted index to indicate the extent to which lessons were taught directly from textbook or with varying degrees of supplementation, including lessons that were not taught at all); and Textbook Content Taught (TCT), representing the extent to which teachers, *when teaching textbook content,* followed their textbook, supplemented their textbook lessons with additional materials, or used altogether alternative curricular materials (McNaught et al., 2010; Tarr et al., 2013). Differences in all these indices could then be folded into the analysis of factors contributing to student learning outcomes.

For example, for the entire study (3 years), for OTL 60.81 % (19.98 SD) of the content of the integrated textbooks was taught while 76.63 % (17.02 SD) of the content of the subject-specific textbooks was taught. The ETI index showed that across all teachers, "(35 %) of the textbook content was taught primarily from the textbook, …(21 %) of the content was taught with some supplementation, a small portion (12 %) was taught from alternative resources, and 32 % of the content was not taught at all." (Overall ETI values were 50.37 (20.20) for integrated and 57.15 (18.94) for single subject (SS)). The TCT index showed that when integrated content was taught, it was more frequently directly from textbook (59 %) as compared to when subject-specific content was taught (46 %). Furthermore, 28 % of integrated lessons were taught with some supplementation, while 33 % of subject-specific lessons were so taught (overall, 81.96 (14.50) for integrated, 74.93 (18.29) for (SS)) (McNaught et al., 2010, pp. 12–13). However, there was considerable variation in curriculum implementation between year-levels 1 and 2. Year 1 teachers' implementation index values were much closer, and higher than the summary values for all teachers in the study, whereas year 2 teachers had wide variation in OTL and ETI, with values for teachers of year 2 integrated much lower that those for teachers of SS. This study provided a major opportunity to interpret student learning outcomes in relation to variation in implementation fidelity, and led to the conclusion that unless information on textbook use is considered, interpreting findings on student learning outcomes related to a curricular treatment can easily lead to unfounded conclusions.

Teacher, classroom, and student data: explaining variation in student outcomes. The COSMIC project design required the accumulation of a wide variety of student- and teacher/classroom-level factors as potential moderators of curricular effects (eventually analyzed using HLM). The project gathered extensive teacher-level data (nearly 30 variables) from an initial and mid-year teacher survey, teachers' self-reports on curriculum implementation (the three indices developed from Table of Contents records), and classroom observations. The teacher data were subjected to principal components analysis and eventually were reduced to seven key teacher-level factors that explained approximately 70 % of the variance in the original data set. The factors clustered around two themes: curriculum implementation (the classroom learning environment, implementation fidelity, use of technology, and OTL) and teacher characteristics (their adherence to and practice of NCTM Standards-based instruction, their teaching and curriculum experience, and professional development) (Grouws et al., 2013; Tarr et al., 2013). Student achievement on the dependent measures was subsequently examined for their relationship to the student- and teacher (classroom)-level factors.

Overall, the extent and richness of student, teacher implementation, and classroom observation data gathered through the curriculum evaluation model, COSMIC was able to develop a more textured understanding of curricular effectiveness than had been accomplished to date.

COSMIC reported on student outcomes by adjusting the scores for students' prior achievement and then aggregating them by teacher (Tarr et al., 2010). The outcomes were reported as residualized gain scores by *teacher*, in recognition that the unit of analysis should not be the individual student (NRC, 2004).

For year 1 course comparisons, the following represent some of the noteworthy results: over all three measures of learning, (1) while several student-level variables were statistically significant predictors of students performance, consistent with previous studies (prior achievement, gender, ethnicities, and special needs); (2) the organization of the curriculum was the single most important factor in the modeling of performance on the tests, with large effect sizes for the test of common objectives and the problem solving test, and somewhat smaller for the Iowa Test of Educational Development. Numerous other factors were statistically significant predictors of performance on some, but not all the measures, and there were statistically significant interactions of factors for performance on one or another of the tests.

For the study of the year 2 courses (Geometry and Core-Plus 2), similar results were seen. However, while many of the individual student level variables were statistically significant predictors of performance on one or more of the measures, for year 2 course students, the CPA index was by far the strongest predictor, with effect sizes greater than 0.5 on all three measures. And perhaps most notably, the curriculum type had little effect on the outcomes on either the test of common objectives or the problem solving test for this year level, but the integrated curriculum had a significant favorable effect on performance on the Iowa Test.

An examination of partial correlations found that when controlling for %FRL, the magnitude of the correlation between Curriculum Type and student outcomes became significantly significant in favor of the integrated curricula, for all three tests.

OTL independent of curriculum was also significantly and positively correlated with higher performance on all three outcome measures.

The importance of OTL is substantially reduced with the partialing out of Class-level %FRL, suggesting that %FRL and OTL may be closely related. While it is possible that the relationship between OTL and %FRL may be attributable to a differential (slower) pace of content coverage in classes with higher percentages of FRL students, the result—less opportunity to have learned the material—suggests there is a need for active intervention to address this resulting inequity of opportunity (note: the study did not address school effects). Since teachers of integrated curricula covered significantly less textbook content than teachers of subject-specific curricula, a difference in coverage (as a percent of the curriculum topics that were taught) may have moderated the effect of Curriculum Type. Further, this study indicates that by controlling for OTL and %FRL, one can more carefully measure the impact of curriculum on student learning.

The year 1 study showed students studying from the integrated curriculum outperforming students studying from single subject curricula on all three measures, but the year 2 results were less clear-cut—while there was a significant effect of the integrated curriculum on the standardized test, there was no significant effect of curriculum on the two project-developed tests. However, prior achievement was a very strong predictor of student learning on all three tests, for both year-level studies. The COSMIC study produced many other results, showing more subtle correlations of student- and teacher-level factors with the student outcomes, as well as more interesting pairwise interactions, than can be discussed here.

No simple answers. Policy makers, administrators, and even practitioners ask whether an integrated program generates (causes) better, worse, or the same learning (outcomes) as a single-subject curriculum. Overall, the COSMIC study illustrates that it is unwise to expect curricular studies to yield such simple answers about curricular effectiveness. The authors note further that the study generalizes only to schools that offer both curricular options, and only if student choice (rather than tracking decisions) determines which students enroll in the two curricula. Unless these conditions are met, the study offers no definitive answer.

However, the COSMIC study yields far more contributions and insight than its statistical "curricular effects." These insights reflect the nature of complex systems. Consider what one could learn from this study that pertains to "engineering [for] effectiveness." COSMIC researchers have provided a protocol for creating and using appropriate multiple outcome measures to compare two curricula, first determining the extent to which they cover the same material, and, second, by selecting common topics by which to create a "fair test." If a district instead wants to know how curricula affect performance on a measure that assesses common standards, such as the Common Core State Standards, the study describes how to recognize and select such a reliable and valid test. It also illustrates how the choice of outcome measure interacts with the curriculum's effects. In systems with causal cycles, measures can also drive the system towards improvement, so such insights into analyzing outcome measures can facilitate important discussions of high-priority goals.

The COSMIC study also illustrates the value of disaggregated data for revealing and identifying relevant bands of variability that may warrant closer inspection. The study reinforces many other findings that the higher the percentage of students eligible for FRL, the lower the OTL. However, OTL was typically a significant moderating effect on student performance on one or more of the tests, while FRL did not have a statistically significant effect. The study further suggests that the effects of the curriculum in favor of integrated math become more evident when FRL measured at the classroom level is partialed out. Arguably, these findings suggest that using integrated mathematics curriculum could be a considerable educational benefit to students with low SES, but may nonetheless require teachers to receive substantial assistance to increase students' "opportunity to learn." At the class-level, experience (in teaching, and in teaching the specific curriculum) was a significant moderating factor, with students taught by experienced teachers (3 or more years of experience) achieving more than students of inexperienced teachers.

Practitioners and policy makers ask whether an integrated program generates better, worse, or the same outcomes as a single-subject approach. The COSMIC study design reflected the complex nature of curriculum organization and implementation, illustrating that it is unwise to expect curricular studies to yield simple general answers. It provides further insight into the inherent weakness of any simple statement that a curriculum is more or less "effective" than another.

The COSMIC study informs readers about the complexity of curricular implementation, as comprising the classroom learning environment (focus on sense-making, reasoning about mathematics, students' thinking in instruction, and presentation fidelity), implementation fidelity (ETI, TCT, textbook satisfaction), technology and collaborative learning, and OTL. These results suggest that in addition to focusing on OTL, school leaders need to help teachers to understand the standards, focus on student reasoning and sense-making, and learn to achieve closure during instruction. In relation to Fig. 1, this suggests that the factors involved in implementation rest within the circle and that their connections to the two bookends in the drawing provide guidance and feedback.

Overall, the COSMIC study results so far suggest that the use of integrated mathematics in year 1, at least, and possibly year 2, may offer considerable learning opportunities for students across the spectrum. Implementation of the integrated curriculum is not a simple matter. In a North Carolina study, based on an analysis of reports from content specialists' monthly observations of teachers' practice, we found that teachers using an integrated mathematics curriculum with low SES students often lost a great deal of time in transitioning to problems in integrated math, tended to be reluctant to turn over authority to students, and missed opportunities to establish closure (Krupa & Confrey, 2010). In a case study of one teacher, instructional coaches engaged in specific and targeted activities with the classroom teacher, and the teacher was able to transform her instructional practices and in fact became a role model for new teachers at the school (Krupa & Confrey, 2012). In studying multiple cases of teachers in these schools, Thomas (2010) showed that providing adequate support to teachers *can* transform practice, but that this is very difficult to accomplish, due to weakness in teacher knowledge and to those teachers' views of instruction. Disentangling these complex relationships may be easier to accomplish in studies seeking improvement over time in the context of smaller studies. Our studies, funded as a Mathematics-Science Partnership through a state department of education, permitted us to form a networked community for improvement, among University researchers, faculty from the state School of Science and Mathematics, a semi-autonomous school organization committed to improving rural education, and—critically—in-service teachers and principals. Our efforts could have greatly benefitted from richer and more continuous data sources informed by research tools such as those developed for COSMIC.

Case Two: Comparing Effects of Four Curricula on First- and Second-Grade Math Learning

A second major study on curricular effectiveness provides another example of the potential contributions of nuanced study that goes beyond simple claims of cause and effect. The study "Achievement Effects of Four Early Elementary School Math Curricula: Findings for First and Second Graders" (Agodini et al., 2009, 2010), examined whether some early elementary school math curricula are more effective than others at improving student math achievement in disadvantaged schools (57 % of schools included in the study were school-wide title 1 eligible, compared to 44 % nationwide). The authors (R. Agodini, B. Harris, M. Thomas, R. Murphy, L. Gallagher, and A. Pendleton) studied the implementation of four contrasting curricula: *Investigations in Number, Data, and Space* ("*Investigations*"), featuring a student-centered approach encouraging metacognitive reasoning and drawing on constructivist learning theory (Wittenberg et al., 2008), *Math Expressions*, blending student-centered and teacher-directed approaches to mathematics (Fuson, 2009a, 2009b), *Saxon Math* (*Saxon*), a scripted curriculum relying heavily on direct instruction in procedures and strategies with guided and distributed practice (Larson, 2008), and *Scott Foresman-Addison Wesley Mathematics* (*SFAW*), a basal curriculum that combines teacher-directed instruction with a variety of differentiated materials and instructional strategies (Charles et al., 2005a, 2005b). *Math Expressions* and *Investigations* are both "reform" curricula whose development had been either initially funded by the National Science Foundation or based on research with considerable NSF funding. A total of 473 districts were invited, but only 12 agreed to participate in the study—a recruitment rate of 2.5 % (Agodini et al., 2010, p. 10).[3] In all, 109 first-grade classes and 70 second-grade classes were randomly assigned to a curriculum within districts.

---

[3] The authors acknowledge that this low rate leaves an "open issue, which cannot be examined with the study's data, is whether the potential differences between participating and nonparticipating sites are related to the study's findings" (p. 14). The conditions of the study, in particular the need for a district to assign different curricula to schools at random, could be viewed by many districts as unacceptably burdensome or arbitrary, and conflict with their own judgment about the most useful curriculum, or simply be at odds with district policy and/or fiscal constraints.

The study addressed three broad questions (Agodini et al., 2010, pp. 4–5):

1. What are the relative effects of the study's four mathematics curricula on first- and second-graders' mathematics achievement in disadvantaged schools?
2. Are the relative curriculum effects influenced by school and classroom characteristics, including teacher knowledge of math content and pedagogy?
3. [Based on subsequent statistical analysis—] What accounts for curriculum differentials that are statistically significant?

Student mathematics achievement outcomes were based on fall and spring administrations (pre- and post-administrations) of the ECLS-K assessment (developed for the National Center for Education Statistics' Early Childhood Longitudinal Study-Kindergarten Class of 1998–1999), a nationally normed adaptive test.[4] Other data were drawn from student demographic and school data, teacher surveys, study-administered assessments of math content and pedagogical content, and scales of instructional practices and approaches derived from classroom observations.

The study results were reported as pairwise comparisons of the curricula, for student outcomes (six pairwise comparisons) for each grade. After 1 year of schools' participation, average first-grade math achievement scores of *Math Expressions* and *Saxon Math* students were similar and higher than those of both *Investigations* and *SFAW* students. In first-grade classrooms, average math achievement scores of *Math Expressions* students were 0.11 standard deviations higher than those of *Investigations* and *SFAW* students. These results were interpreted to mean that, for a first grader at the 50th percentile in math achievement, the student's percentile rank would be 4 points higher if the school had used *Math Expressions* instead of *Investigations* or *SFAW*. In second-grade classrooms, average math achievement scores of *Math Expressions* and *Saxon Math* students were 0.12 and 0.17 standard deviations higher than those of *SFAW* students, respectively. For a second grader at the 50th percentile in math achievement, these results mean that the student's percentile rank would be 5 or 7 points higher if the school used *Math Expressions* or *Saxon Math*, respectively.[5]

---

[4]The test is adaptive in that students are initially administered a short, first-stage routing test that broadly measures each student's achievement level. Based on the first-stage scores, students are then assigned one of three second-stage tests: (1) easy, (2) middle-difficulty, or (3) difficult. Scale calibration among the second-stage is accomplished through overlap of items on the second stage tests and item response theory (IRT) techniques, by which scores from different tests are placed on a single scale.

[5]Another way the authors interpreted these differences was to consider the average score gain by grade in the lowest quintile of SES on ECLS (16 points in first grade) and to convert the .1 effect size into points using the reported standard deviation of 10.9, getting a difference of 1.09 scale points. Comparing 1.09 to an average gain of 16 scale points, they describe an effect size of .10 as having an effect of 7 % of the gain over first grade. Thus the differences in student results reported between curricula account for between 7 and 14 % of the content as measured by the ECLS assessment.

This study, in some ways similar to the COSMIC study, examined curricular implementation, and reported on such factors as the use of the curriculum, the amount, frequency, and stated reasons for supplementation, the availability of support, amount of professional development, distribution of uses of instructional time, and focus on particular content areas. Teachers reported varying coverage of math content areas across the curricula. They determined that variation in coverage (number of lessons on a topic) of 19 out of 20 content areas was significantly different across all four curricula. However, in pairwise comparisons of the curricula, "there was no clear pattern [regarding] which curriculum [coverage] differences are significant." (p. 57): some pairwise differences in coverage were statistically significant and others were not.

For Table 1 below, we selected some implementation differences that could have affected student-learning outcomes. For instance, teachers received twice as much initial (voluntary attendance) professional development for *Expressions* than for other curricula (with >90 % of first-grade teachers reporting attendance at initial training sessions for all the curricula, but 80–97 % of second-grade teachers attending, with Math Expressions having the highest attendance rate). Teachers of *Saxon Math* taught math an additional 20 % of the time each week, teachers of *Math Expressions* used more supplementation materials while *Investigations* teachers used less, and 16.2 % of *Saxon Math* teachers and 21.1 % of *SFAW* teachers had taught with those curricula previously, compared to less than 6 % for each of the other two curricula. It should be noted that *Math Expressions* and *Investigations* are based more intensively on student-centered instructional approaches and represent pedagogical approaches that require extensive teacher preparation. Not surprisingly therefore, implementation reports show that higher percentages of first- and second-grade *Investigations* and *Expressions* teachers report feeling only "somewhat" or "not at all" prepared to teach their curriculum, compared to teachers of *Saxon Math* or *SFAW*.

The study's authors also conducted an analysis of the extent to which teachers adhered to their assigned curriculum. "Adherence" referred to the extent to which a teacher taught the curriculum using practices consistent with the curriculum developers' model. (In the NRC report, the philosophy of a curriculum's designers ("program theory") was distinguished from the application of the curriculum during implementation ("implementation fidelity").) The study measured adherence via a teacher survey and a classroom observation instrument, as the extent to which essential features of the assigned curriculum were implemented. The results shown in Table 2 suggest that teachers were more likely to adhere to designers' intentions in the *Saxon Math* program than in the *Expressions* program.

In an exploratory look at what might account for the relative curricular effects, the researchers examined the instructional practices that occurred across different curricular types (in contrast to adherence) based on the observational data. They conducted a factor analysis, yielding four factors: (1) student-centered instruction, (2) teacher-directed instruction, (3) peer collaboration, and (4) classroom environment. The analysis across the curricular pairs indicated that student-centered instruction and peer collaboration were significantly higher in *Investigations* classrooms

Table 1 Selected differences in implementation variables for different curricular implementations

	Respondents	Investigations	Expressions	Saxon Math	SFAW
Initial PD offered (prior to first day of school)	All teachers	1 day	2 days	1 day	1 day
Responded "Somewhat or not at all" adequately prepared after training (%)	First-grade teachers	23.3	33.7	16.0	10.0
	Second-grade teachers	23.2	56.1	16.4	9.1
Additional training offered	Reported by publishers	3–4 h every 4–6 weeks (group)	Twice a year (one-on-one meetings)	Once in fall (one-on-one meetings)	3–4 h every 4–6 weeks (group)
Follow-up training % participated/days	First-grade teachers	95.5/2.6	90.5/0.6	74.3/0.4	99/2.1
	Second-grade teachers	97.4/2.3	82.4/0.5	66.7/0.4	91.9/2.0
Supplemented curriculum (%)	First-grade teachers	14.8	32.1	24.8	27.5
	Second-grade teachers	11.7	55.6	30.5	24.6
Hours taught per week	First-grade teachers	5.1	5.0	6.1	5.3
	Second-grade teachers	5.4	5.5	6.9	5.5
Used assigned curriculum the previous year (%)	First-grade teachers	5.5	3.6	16.2	21.1

Table 2 Adherence to a curricular program's essential features, as percentage of features implemented

		Investigations	Expressions	Saxon Math	SFAW
Survey (self-report)	First-grade teachers	66 (3)	60 (4)	76 (1)	70 (2)
	Second-grade teachers	67 (3)	54 (4)	76 (1)	68 (2)
Observation of daily essential features	First-grade teachers	56 (2)	48 (4)	63 (1)	54 (3)
	Second-grade teachers	53 (2–3)	47 (4)	65 (1)	53 (2–3)
	Average	60.5 (3)	52.25 (4)	70 (1)	60.75 (2)

Numbers in parentheses indicate the relative ranks of the curricula for each row (pp. 65–67)

than in classrooms using the other three curricula. Teacher-directed instruction was significantly higher in *Saxon Math* classrooms than in classrooms using the other three curricula. The classroom environment did not differ across curricula.

Additional analysis indicated that some of these implementation factors act as mediators of achievement outcomes. The study's design, however, permitted examination of only one mediator at a time. This constraint meant that while differences in professional development for *Expressions* mediated the curricular effect, the authors could not relate this to the mediational effects of less prior experience with, and teachers' reports of less preparedness to teach, the curriculum. Likewise, *Saxon Math* teachers were reported to have had 20 % more instructional time, which mediated the *Saxon Math-SFAW* difference in curricular effect. The study design, however, does not permit assessment of *combined* effects of interactions between instructional time and likelihood of having taught a curriculum before. The authors interjected that a more rigorously designed study of mediation could disentangle the relationships among the mediators (p. 102). In any case, the examinations of implementation variables as mediators of curricular effects make it clear that one must always interrogate the results to understand the nuances in a causal study's assumptions and claims.

Among the many accomplishments of the Agodini et al. (2010) study was the identification of means to measure a considerable number of factors that comprise classroom practice. The study reports on a variety of factors that are worth examining, even if they were not demonstrated to be statistically significant contributors to differentiated curricular effects. For instance, the study reports low levels of mathematical knowledge on the part of elementary teachers, and while this was not differentially related to curricular effectiveness in the study, this is a persistent issue in elementary teaching that needs to be addressed. The study also makes a useful distinction between implementation factors that apply to *any* curriculum, and *adherence*, which pertains to the specific intentions of each curriculum's design; the latter is a curriculum-specific measure of teachers' fidelity of implementation of specific features/activities.

The Agodini study also exhibits limitations and threats to its validity: reliance on only a single student outcome measure (the ECLS-1 and -2), and the absence of a method to check the "fairness" of that outcome measure across the curricula. These are in contrast to the call in *On Curricular Effectiveness* for multiple measures and for outcome measures that demonstrate "curricular validity of measures" (also called "curricular sensitivity") and "curricular alignment with systemic factors" (NRC, 2004, p. 165). Such a notable weakness with regard to the outcome measures unfortunately leads to major problems with the interpretation of the study's conclusions. The size of the curricular effect, 7–14 gain points on the scaled score, could be the result of a few key assessment items.

The study benefits—as an experimental study—from randomized assignment of curricula to teachers (classrooms) within the district, but this feature of the study came at a high cost to its external validity. Few districts were willing to randomly assign curriculum to teachers, calling into question the generalizability of the study's results. Secondly, conducting a study of curricular effectiveness during the first year of a curriculum's implementation, and providing only 1–2 days of professional development for primary teachers, must weaken confidence in the validity of comparisons of curricular effectiveness. For instance, reports of high levels of supplementation by *Expressions* teachers could be due to the teachers' use of prior, more familiar materials. If this were the case, should one draw the conclusion that *Expressions* itself was "effective" under these conditions?

Furthermore, the authors also described teachers' reports, for each curriculum, of the frequency of teaching particular content topics (whole numbers, place value, etc.). If an analysis of the test had been performed, and included in the study, one might have been able to discern patterns in the relationship between students' OTL the different topics and the outcome measure scores.

The Agodini et al. study offers far more insight into curricular effectiveness than is captured by its conclusions of "cause and effect." As with the COSMIC study, it makes progress on establishing implementation factors. Both studies identify similar factors, such as adherence vs. implementation fidelity, the use of student collaboration, and the use of general instructional approaches (student-centered and teacher-directed vs. standards-based instruction). Both examine content variations, one by conducting content analyses and then measuring OTL as teachers implemented, and the other by relying on teacher reports of number of lessons by content area and adherence to essential features of each curriculum. By designing different means of capturing the variations in these factors, these studies help us to progress in our understanding of the complexity of curricular use.

Case Three: The Relationship Among Teacher's Capacity, Quality of Implementation, and the Ways of Using Curricula

A third study, "Selecting and Supporting the Use of Mathematics Curricula at Scale," is a study of curricular impacts on implementation *quality* with respect to teachers' capacity and ways of using the materials, rather than a study of *effectiveness* (as based on student outcomes) (Stein & Kaufman, 2010). The study involved two districts using reform curricula, one using *Everyday Math* (*EM*) and the other using *Investigations*, in order to begin to answer the question of "What curricular materials work best under which kinds of conditions?" (p. 665)

The authors initially analyzed the two curricula with respect to the frequency of two kinds of high cognitive-demand tasks: "procedures with connections to concepts, meaning and understanding" (PWC) tasks and "doing mathematics" (DM) tasks (Stein, Grover, & Henningsen, 1996). They characterized PWC tasks as "... tend[ing] to be more constrained and to point toward a preferred—and conceptual—pathway to follow toward a solution," and identified 79 % of the tasks in *Everyday Math* as PWC tasks. They characterized DM tasks, in contrast, as "...less structured and [not containing] an immediately obvious pathway toward a solution" (Stein & Kaufman, 2010, p. 665), and identified 84 % of the tasks in *Investigations* as DM tasks. Based on these differences, they conjectured that it would be less difficult for teachers to learn to teach with *EM* than with *Investigations*. DM tasks are more difficult to implement faithfully, because they support open-ended discourse, which is often difficult to manage and require more of the teacher's own learning (Henningsen & Stein, 1997). In contrast, PWC tasks are more bounded and predictable, but are susceptible to "losing the connection to meaning" (Stein & Kaufman, 2010). Stein and Kaufman also documented that there is less professional development support embedded in the *EM* materials than in the *Investigations* materials, mirroring the conventional wisdom that teaching with the *EM* is less challenging than with *Investigations* curricula.

From these analyses, the study characterized *EM* as a low-demand, low-support curriculum, and *Investigations* as a high-demand, high-support curriculum. They then investigated how the implementation of these two contrasting reform curricula might differ, particularly with respect to the quality of implementation and its relationship to teacher characteristics.

Using classroom observations, interviews, and surveys, the researchers compared implementation of the two reform curricula in two districts that were similar in terms of the (high) percentage of students eligible for FRL (86 and 88 %). They studied implementation of the curricula by six teachers (one per grade level) in each of four elementary schools in each district over a period of 2 years. Observations (with examples) were conducted on three consecutive lessons in each of fall and spring, and coded for the extent to which teachers were able to (1) sustain high cognitive demand through the enactment of a lesson, (2) elicit and use student thinking, and (3) vest the "intellectual authority in mathematical reasoning," rather than in the text or the teacher. Together, high values on these three dimensions characterized high quality implementation.

Using surveys, observations, and interviews, they examined two teacher characteristics: teachers' *capacity* (defined as comprising years of experience, mathematical knowledge for teaching (MKT), participation in professional development, and educational levels) and their *use of curriculum* (teachers' views of the curriculum's usefulness, percentage of time teachers actually used the curriculum in lessons, and what teachers talked about with others in preparing for lessons—including non-mathematical details, materials needed for the lesson and articulation, and discussion of big ideas.)

In answering their first question, "How does teachers' quality of implementation differ in comparisons between the two mathematics curricula (*Everyday Mathematics* and *Investigations*)?" (Stein & Kaufman, 2010, p. 667), they found that teachers from the district using *Investigations* were more likely to teach high quality lessons than teachers from the district using *Everyday Math* (it must be noted again, however, that this study did not investigate the relationship of instructional performances to student outcome performance, but rather the "less-studied link between curricula and instruction," p. 668). Teachers implementing *Investigations* were more likely to maintain the cognitive demand (6.7 > 4.9, on a scale of 2–8), to utilize student thinking more (1.1 > .5, on a scale of 0–3), and to establish norms for the authority of mathematical reasoning (1.2 > .4, on a scale of 0–2).

Their second question across the two districts and curricula was, "To what extent are teachers' capacity and their use of curricula correlated with the quality of their implementation, and do these correlations vary in comparisons between the two mathematics curricula?" (p. 667). The study found that most of the teacher capacity variables were not consistently and significantly related to the quality of implementation. In the district using *EM*, higher performance on MKT surveys was *negatively* correlated with the use of student thinking and with establishing the authority of mathematical reasoning in the classroom. In the district using *Investigations*, correlations of implementation quality with teacher capacity were positive but not significant. And while no clear relationship was found between either hours or type of professional development to implementation quality in the district using *EM*, in the district using *Investigations*, the amount of professional development was (positively) significantly correlated with all three components of implementation quality.

The study shows that implementation quality cannot be inferred from content *topic* analysis alone but depends also on the kinds of tasks (how the tasks are structured) that are used to promote student learning of those topics. It also suggests that implementation quality appears to relate more strongly to the extent of professional development support both facilitated by the district and afforded within the materials, than to other traditional capacity variables such as teachers' education, experience, and their MKT.

Across the two districts and curricula, the discussion of big ideas during lesson planning was the only teacher's-use-of-curriculum variable that was significantly and positively correlated to implementation quality components (and then to only two of those: attention to student thinking and authority of mathematical reasoning). Further, the authors reported that this tendency was more evident in the district using *Investigations*. In explaining this difference, they reported that teachers using *Everyday Math* indicated that frequent shifts in topics in the spiral curriculum tended to make identification of big ideas more difficult, while in *Investigations*, the "doing math" tasks led teachers to focus more on big ideas. These findings were somewhat counterintuitive because it had been thought that *Investigations* was more difficult to implement because it has a much higher percentage of DM tasks than does *EM*. The consideration of big ideas during instructional planning was strongly linked to high quality implementation of both curricula, and was also more engaged in by teachers implementing the curriculum that focused more extensively on DM tasks (*Investigations*).

Stein and Kaufman (2010) note that this work "provides evidence that one cannot draw a direct relationship between curriculum and student learning" (p. 688).

They asked, "...what elements of teacher capacity interact with particular curriculum features to influence what teachers do with curriculum. Thus, our focus is on *which program leads to better instruction under what conditions*" (p. 668, italics added). They suggest reorienting the concept of teacher capacity to incorporate the interaction of curriculum as tool with how teachers use the curriculum, and that study of how curriculum use over time interacts and promotes improved instruction would be a very fruitful path of research. In essence, by suggesting that "...curricula could be viewed not only as programs to be implemented, but as tools to change practice" (p. 688), they are suggesting that curricular effectiveness might eventually be considered not a static value or a product's claim, but instead a *process of improvement* of instruction through interaction between curriculum and how it is used.

Overall Conclusions from the Three Cases

Juxtaposing the three cases reviewed here provides an opportunity to synthesize advice for the conduct of future effectiveness studies. There has been a strong temptation in the calls for, and the interpretation of, effectiveness studies, to try to identify *some*thing that works—that is, to identify one or more curricula (or in fact, a single most effective curriculum for grade level or range) that can be adopted with the expectation of subsequent, direct major improvements in student learning outcomes. Calls for randomized field trials of curricular effectiveness have carried with them the assumption that such experimental designs will provide the best evidence that a curriculum is "effective." We have asserted, and taken together, the studies discussed in this chapter have shown this approach to be poorly conceptualized, underestimating the collective and cumulative impacts of coverage/OTL, implementation fidelity, and quality of instruction, not to mention differences in curricular structure, pedagogy, and content rigor.

We initially examined the three studies from a perspective of causality, to understand whether and how they might inform us about the results of implementing and comparing two or more curricula. Reviewing these cases, however, demonstrated how tentative causal conclusions are, and reminded us that all studies have flaws and limitations. The quest for the perfect curricular effectiveness study—and a quest for a single most-effective curriculum—is highly unlikely to yield results that are robust or extensive enough to guide practice. Each study provides insight into some *specific conditions* under which certain factors played roles and certain outcomes occurred, and that these depend on how constructs surrounding the implementation of the curricula were defined and measured.

The COSMIC study provides evidence of relative effectiveness of an integrated curriculum compared to subject-specific curriculum when students are provided a choice between those options. However, had multiple curricular alternatives been available, or had ability tracking been used to assign students to the two curricular options, the authors note, we do not know what the results of the study would have been. It could also be the case that if teachers of integrated curricula were able to cover the same percentage of their text during a year as did teachers of a subject-specific curriculum, student performance in integrated math would be even stronger relative to that in the subject-specific curriculum. Practitioners choosing to apply this study to their own curriculum selection decisions must weigh these considerations, and must contextualize the results to develop expectations relevant to their own settings.

Similarly, the Agodini et al. study reported that students taught using *Expressions* outperformed students taught using the other curricula in both first and second grades, with the exception of students using *Saxon Math* in second grade. It is possible however, that this effect may have resulted from the extra day of professional development time or additional supplementation reported to be used by teachers for *Expressions*, or from increased instructional time, in the case of *Saxon Math*. Alternatively, it is feasible that all outcomes of this study could be attributable in large part to the degree of fit of the curricula with the single ECLS outcome measure used; if the study had used a different end-of-year assessment (or multiple measures as in the COSMIC study) the results might have been quite different. Another possible interpretation is that by examining the effectiveness of curricula for only the first year of implementation, the study's results were necessarily skewed in favor of *Saxon Math* and *SFAW*, which had higher levels of prior use and scripting, and that the student outcomes would evolve considerably over a longer study period (allowing more teacher experience with the assigned curricula), potentially re-ordering the student learning results.

All studies are open to multiple interpretations; most are subject to various predictable (or emergent) limits to generalizability. In the Stein and Kaufman study, for example, the stronger implementation quality of *Investigations* could have been attributed to its design of curricular tasks, affordances for focus on big ideas, and/or support for professional development. But perhaps the district that offered *Investigations* simply supported its implementation with higher quality, more extensive professional development.

These studies demonstrate further the complexity of curriculum's relationship to student learning. But some may ask whether the fact that these studies have some conflicting interpretations or that they do not provide generalizable recommendations, means that such investigations are not useful, or even a waste of time and money. Are such studies of limited importance because we cannot know whether a study's results will accrue in a setting that differs from the original—and may require a level of adaptation from the conditions for the study?

If the goal of curricular effectiveness studies were to decide unequivocally whether a single product—a curricular program and its related materials—can be simply dropped into classrooms and be expected to yield predictable learning gains, then these studies fail to establish curricular effectiveness. More to the point, however, these studies instead bolster the recognition that this assumption about curricular effectiveness and its generalizability is mistaken, a false apprehension. Their design is to provide more insight into the factors affecting effectiveness (and possibly leading to redefining the use of the term); their design and their execution make them highly valuable to that end.

We argue that these studies, especially when taken together, demonstrate why simple causality is an insufficient model for judging effectiveness of a curriculum. The message to be taken from them is that the instructional core is a complex system, that many things matter to the implementation of a curriculum and to the learning that students can accomplish with different curricula, and that what matters appears to depend in large degree on multiple factors, and different factors in different situations. Context matters—the extent to which one serves disadvantaged students, requires more resources, or requires teachers with stronger capacity or settings in which professional development is supportive and sustained. Resources matter. The quality of instruction, and the quality of curricular implementation, matter.

Most importantly, these studies contribute substantially to an understanding of the instructional core. By the very fact that the experts who conducted the studies have gained purchase on modeling the instructional core, they provide us insights into the complexity of instructional systems. They identify interlocking factors, loci of possible interventions, and a set of measures and tools that can help in the process of becoming smarter and wiser about *how curricular use in particular settings can improve instructional quality and student outcomes.*

These studies, we believe, provide the following lessons:

1. Outcome measures matter—and with the availability of Common Core State Standards, we have the opportunity and the responsibility to create a variety of measures in a cost effective way across districts and states (this is one of the premises of the Common Core assessment initiatives). The COSMIC study in particular reinforces the notion that implementation or effectiveness studies require multiple outcome measures which should (a) include measures that act as "fair" tests (Chávez et al., 2010) to ensure non-biased comparison of student performance on topics common to all curricula being examined, (b) include project-designed measures of reasoning and problem-solving, (c) be normed against relevant populations (e.g., college-intending students, ELS students) and used to make systemic decisions (such as statewide end-of-course exams or new assessments of Common Core State Standards), (d) assess the development of big ideas over time; learning progressions are one way to conceptualize coherent curricular experiences and their development over time, and (e) assess other dimensions of mathematics learning, such as the mathematical practices in the CCSS, student attitudes, or student intentions to pursue further study or certain STEM careers. The studies showed that the categories by which outcomes were disaggregated were critical, and were sensitive to interactions, such as by ethnicity and socioeconomic factors. At the least, therefore, relevant data gathered in relation to performance measures should include ethnic and racial diversity, gender, ELL, and FRL status, to support the investigation of relevant bands of variability in effects and outcomes.

2. Monitoring what was actually taught, and *why* it was taught, is crucial to making appropriate attributions in examining effectiveness. Monitoring should include measures of curricular coverage (such as OTL and adherence), and of the type and degree of supplementation (and the reasons for choices regarding these variables). Different methods of monitoring curricular coverage and supplementation included table-of-contents reports, surveys of relative emphasis, and textbook use diaries.

3. A better understanding of the factors involved in the implementation of curricula will add a wealth of insight to explanatory frameworks of curricular effectiveness. Some factors should directly reflect the extent to which implementation captures a designer's specific intent, while others should address qualities that apply across all curricula. These studies undertook many innovative methods of data collection: surveys, intermittent and extended classroom observations with various coding schemes, reports of instructional time usage, and interviews. In one case, these were coded in predetermined, theoretically relevant categories— maintaining cognitive demand of tasks, eliciting student thinking, and vesting authority in mathematical reasoning. In the COSMIC and Agodini et al. studies, high numbers of variables were identified a priori, and embedded in other instruments (teacher surveys, for instance). Modeling the factors that can explain the majority of observed variation for different levels of analysis (student, class/ teacher, school level, for instance) requires statistical techniques (factor analysis, principal component analysis) to reduce the dimensionality of the vast amounts of resulting data, and to identify and sort critical variables into appropriate clusters (classroom learning environment, implementation fidelity, peer collaboration, technology use, student-centered instruction, and teacher-directed instruction). Selection of appropriate units of analysis, and hierarchical (multilevel) linear modeling were essential (COSMIC, Agodini et al.) for modeling the relationship and interactions of student- and teacher-level factors and their contributions to the dependent measures of student learning. Research on identifying, defining, and studying implementation factors (perhaps as latent variables) promises to continue to grow and add to our understanding of curricular effects.

4. Issues of teacher capacity and professional development are critical in judging curricular effectiveness, but not necessarily in a predictably simple or straightforward way; their influence varies depending in part on whether they are viewed as a resource within a curriculum and its implementation, or as a factor that interacts with implementation. Teacher capacity, a term that subsumes teacher MKT, experience, education, and professional development, did emerge as influential in two studies (COSMIC,[6] Agodini et al.). In the third study (Stein & Kaufman) however, its influence was mixed: while most teacher capacity factors did not correlate in a significant positive way with implementation quality in one district/curriculum, but some of the component factors correlated *negatively* and significantly in the other. In that study, the amount of professional development time, teachers' access to assistance and support, and the ways in which teachers used materials in planning (i.e., the degree of their focus on big ideas) and communicated with each other about curricular use emerged as the factors most closely associated with implementation quality. On the other hand, professional development was not significantly associated with student outcomes in the

---

[6] Though early results suggested that teacher experience was not significantly correlated with student outcomes (Tarr et al., 2010), completed HLM analyses of year 1 data revealed that teaching experience was a significant predictor of student outcomes on all three measures (Grouws et al., 2013; Tarr et al., 2013).

COSMIC study; teachers reported that they perceived little to no impact of their professional development activities on their teaching practices, in part, because they perceived the activities merely confirmed what they were already doing (Tarr et al., 2013), and the study measured professional development in terms of quantity not quality. The studies incorporate three perspectives on professional development and teacher capacity—one in which these factors are viewed as a resource for curricular implementation, one in which they could be viewed as a factor that interacts with implementation, and one in which curricular implementation is seen as a tool for changing capacity and as a source of professional development. To clarify how professional development and teacher capacity can relate to curricular implementation and effectiveness will require additional investigation.

5. How a study is situated in relation to educational structures and organizations may eventually be important at a meta-level of understanding curricular effects and the conclusions drawn. The location of each of the studies described here was driven by issues of experimental design—for instance, the availability of two curricular options without tracking (COSMIC), the dependence of a study on districts' willingness to randomly assign teachers to treatments (Agodini et al.), to support extended observations over 2 years, and to provide researchers with access to extensive teacher data. These issues were reported as features of the studies' designs, but over time such they may themselves emerge as organizational factors that are as important to curricular implementation as traditional organizational characteristics as governance, decision-making, funding, and data use.

Engineering [for] Effectiveness: Summary and Recommendations

These studies remind us how remarkably complicated are the interplay of curricula, instruction, classroom assessment practices, and professional development. They demonstrate that the instructional core is a complex system, exhibiting the first-order traits of complex systems including interlocking parts, bands of variability, feedback, causal cycles, interactions and emergent phenomena, and the need for focus on continuous improvement. It is incumbent on policy makers, system leaders, teachers, professional development and curriculum designers, and researchers, to treat the entire instructional core accordingly: as a complex system. We suggest therefore that rather than seek any grand causal effect from these or similar studies, one should use them to learn more about possible ways to model and improve the instructional core at the classroom, school, and district level throughout the USA.

We have come to believe that while curricular effectiveness has seemed an important focus for study, we suggest that with the instructional core as the complex system of which curriculum is one part, the focus for improvement should be the functioning of the system itself. The proposal that follows from this is to focus on how to engineer the *instructional core* for improved teaching and learning effectiveness—that is, to iteratively design and improve our way to a greater understanding of the operation and strengthening of the instructional core. The studies recounted here have provided some critical elements of such an endeavor, including identification of a number of critical constructs, and creating measures to gauge and monitor them. Other researchers have argued for the importance of multiple methodologies (NRC, 2004) including such approaches as design studies, which are useful in identifying mechanisms to describe and explain interactions at the classroom level.

Many of the instruments outlined in the studies can be applied using networked technological systems to gather data in real time. For instance, teachers could easily record measures of curricular monitoring and adherence on an on-going basis. Rather than impose lockstep pacing guides, based on external and untested models of sequencing and timing (and instead of focusing on punitive responses if a teacher or class falls off the pace), districts could require teachers to report and interpret how they implement a curriculum, and learn from it. Records of when and why teachers supplement curricular materials, become delayed, or experience difficulty with one or more topics would generate more informative district-wide data about curricular use, and become a means to use ongoing practice to inform future implementation, especially from combining monitoring and supplementation data with disaggregated student and school data. In the near future, along with electronically delivered curriculum, the bulk of such monitoring could even be done automatically.

The studies asked teachers to complete a number of surveys regarding their knowledge of standards, their beliefs about instructions, and their approaches to certain kinds of practices, as well as core information about teacher capacity and about their participation in professional development. Data from such surveys, gathered periodically within technologically networked practitioner communities, could be factored into models of curricular implementation, professional development planning, and overall teacher community organization, with the goal of instructional improvement at the individual teacher or classroom level, and at higher levels of organization such as departments, schools, and districts.

Perhaps the most difficult data gathering tasks will be the collection of the kind of real-time observational data required for analysis of many of the implementation factors. While surveys and teachers' own monitoring reports can shed light on these issues, the collection of observational data, and its analysis via established, reliable rubrics, will continue to be an essential, and costly, element. It will be challenging to gather and use observational data to help define curricular, or, more broadly, *instructional* effectiveness (even with some of new technologies for classroom video recording becoming available). The use of video from such observations to guide professional development may turn out to be a major driver in our efforts to engineer for effectiveness going forward.

In this chapter, we concentrated on measures to permit comparison of curricular implementation and effectiveness, and emphasized the importance of ensuring curricular sensitivity and the alignment of outcome measures to systemic factors. But one can imagine that technological means of data gathering can enhance or transform the kinds of outcomes recorded, measured, and reported.

Treating the instructional core as a complex system will support efficient design and implementation of such new innovations in curricular implementation and prototype systems for gathering and analyzing relevant data. As these are created, with the aim of engineering the instructional core for improved effectiveness, it will be essential to consider the use scenarios of innovations—to ensure that the data gathering fits into the work flow of engaging classroom activities (i.e., does not become onerous for teachers' workloads, *and* in fact reinforces their instructional efforts), that the data neither artificially reduce nor diminish the complexity of the instructional core, and that the statistical analytic approaches are robust and appropriate.

The ongoing improvement of the complex instructional core requires a "capacity to inform improvement" (Bryk, 2009) that establishes regular flow of information, feedback, and consultation within and among different levels of the educational organization. This argues for the establishment, in schools and districts, of networked improvement communities that include practitioners, researchers, technologists, and leaders who all participate throughout the work of achieving common goals, the design, testing, and implementation of the innovations, recognizing patterns and identifying sources of variability (Bryk et al., 2011).

All major complex systems (websites, health systems, communications, consumer marketing, climate analysis, disaster relief) are moving to the use of data-intensive systems with related analytics. What is most compelling in the studies described here is that it is possible to infer from them how we should be developing and deploying technologically enabled systems of data collection that will permit us to (a) gather more complete types and quantities of data about what is happening in classrooms, (b) become aware when a system exhibits patterns or trends toward improvement, stagnation, or deterioration over time, and (c) learn how to drive those systems towards improvement. Learning to undertake this level of analysis would constitute second-order traits of these complex systems.

Several principles continually surface in considering the goal of improving the instructional core: (a) curriculum matters; (b) instructional materials matter, because these best express the enacted curriculum, and their importance grows as the scale of implementation increases to the district level; (c) coherence matters, because it is critical in any complex system that all the moving parts align and mutually support each other; (d) multiple processes combine to result in observed outcomes (Bryk et al., 2011); (e) focusing solely on outcome data is not sufficient to support instructional improvement; and (f) managing and monitoring the implementation of tools/programs/curricula is a key function of school and district leadership.

This review leads us to the conclusion that it should be a high priority to design and implement technologically enabled systems that extend the capability of district and state data systems to gather data that can inform *improvement of the instructional core*, focused on curricular selection, use and implementation.[7]

---

[7] The components outlined here would not be a complete set to drive improvement in the instructional core. In an earlier version of this paper, we sought to discuss formative assessment and tie it to the construct of learning trajectories, diagnostic assessments and instructional practices, but it was too ambitious for a single paper. This second analysis will lead to an additional set of factors and data elements to this system, and we hope to complete that paper as a companion to this one in the near future.

Based on this review, we believe that districts could make significant progress on such an agenda in the areas of outcome measures, curricular monitoring, curricular implementation factors, and professional development and capacity issues. To this end, we outline a set of proposed actions.

Steps in a Strategic Plan to Strengthen the Instructional Core in Relation to Curricular Use, Implementation, and Outcomes

1. Form "networked improvement communities," (Bryk et al., 2011) to define tractable problems on which to focus, establish common targets and develop precise, measurable goals for the instructional core, across multiple levels of the system (teachers and classrooms, researchers, schools, districts).
2. Construct databases of assessment items linked directly to Common Core State Standards (using a set of relevant tags that distinguish among the features and measures), a variety of outcome measures to yield fair tests, and tests aligned to the CCSS. Focus on creating automated means of scoring that support the use of varieties of item types (multiple choice, as well as constructed and extended response) and concentrate on how to get meaningful data to teachers and students.
3. Develop and implement a means of analyzing, documenting, and notating the alignment of a curriculum to the CCSS, and of creating a standardized means of analyzing and representing content analysis of a curricular program.
4. Build a data system to gather and monitor data on curricular use, supplementation, and reasons for supplementation, gathered in real time.
5. Collect data on implementation factors such as those identified in the above studies.
6. Link the data system and various data categories and outcome measures to student, classroom, school, and district demographic data.
7. Link the data system to teacher demographic and survey data.
8. Find/develop/implement ways to conduct valid classroom observations (by teachers, supervisors, principals, specialists) for professional development purposes, and to triangulate these observations with teacher self-reports.

Finally, we argued that the value of the work rested in building models of the complex system known as the instructional core, and in engineering that instructional core for effectiveness by designing and implementing data systems using the constructs and measures developed by the studies. We suggest that treating the instructional core as a complex system, and taking a stance of engineering the instructional core for greater effectiveness of mathematical teaching, learning, and reasoning—studying what is happening in the classrooms in terms of patterns, trends, emergent behaviors, with deliberate sensitivity to variations in contexts—is a means to accelerate improvement in instruction and student learning. Ironically, by doing so, one could create a next generation of "best practices," this time with a focus on a continuously improving community in which research and practice draw more directly and iteratively from each other.

References

Agodini, R., Harris, B., Atkins-Burnett, S., Heaviside, S., Novak, T., Murphy, R., et al. (2009). *Achievement effects of four early elementary school math curricula: Findings from first graders in 39 Schools*. Washington, DC: IES National Center for Education Evaluation and Regional Assistance.

Agodini, R., Harris, B., Thomas, M., Murphy, R., Gallagher, L., & Pendleton, A. (2010). *Achievement effects of four early elementary school math curricula*. Washington, DC: IES National Center for Education Evaluation and Regional Assistance.

Bass, L., Charles, R. I., Jonson, A., & Kennedy, D. (2004). *Geometry*. Upper Saddle River, NJ: Prentice Hall.

Bellman, A. E., Bragg, S. C., Charles, R. I., Handlin, W. G., & Kennedy, D. (2004). *Algebra 1*. Upper Saddle River, NJ: Prentice Hall.

Berwick, D. M. (2008). The science of improvement. *The Journal of the American Medical Association, 299*(10), 1182–1184.

Boyd, C. J., Cummins, J., Malloy, C., Carter, J., & Flores, A. (2005). *Geometry*. Columbus, OH: Glencoe-McGraw Hill.

Bryk, A. S. (2009). Support a science of performance improvement. *Phi Delta Kappan, 90*(8), 597–600.

Bryk, A. S., & Gomez, L. M. (2008). Ruminations on reinventing an R&D capacity for educational improvement. In F. M. Hess (Ed.), *The future of educational entrepreneurship: Possibilities of school reform* (pp. 127–162). Cambridge, MA: Harvard University Press.

Bryk, A. S., Gomez, L. M., & Grunow, A. (2011). Getting ideas into action: Building networked improvement communities in education. In M. Hallinan (Ed.), *Frontiers in sociology of education*. New York: Springer.

Burger, E. B., Chard, D. J., Hall, E. J., Kennedy, P. A., Leinwand, S. J., Renfro, F. L., et al. (2007). *Geometry*. Austin, TX: Holt.

Charles, R., Crown, W., Fennell, F., Caldwell, J. H., Cavanagh, M., Chancellor, D., et al. (2005a). *Scott Foresman-Addison Wesley mathematics: Grade 1*. Glenview, IL: Pearson Scott Foresman.

Charles, R., Crown, W., Fennell, F., Caldwell, J. H., Cavanagh, M., Chancellor, D., et al. (2005b). *Scott Foresman-Addison Wesley mathematics: Grade 2*. Glenview, IL: Pearson Scott Foresman.

Chávez, O., Papick, I., Ross, D. J., & Grouws, D. A. (2010). *The essential role of curricular analyses in comparative studies of mathematics achievement: Developing "fair" tests*. Paper presented at the Annual Meeting of the American Educational Researcher Association, Denver, CO.

Coburn, C. E., & Stein, M. K. (Eds.). (2010). *Research and practice in education: Building alliances, bridging the divide*. Lanham, MD: Rowman & Littlefield.

Cohen, D. K., Raudenbush, S. W., & Ball, D. L. (2003). Resources, instruction, and research. *Educational Evaluation and Policy Analysis, 25*(2), 119–142.

Confrey, J., Castro-Filho, J., & Wilhelm, J. (2000). Implementation research as a means to link systemic reform and applied psychology in mathematics education. *Educational Psychologist, 35*(3), 179–191.

Confrey, J., & Makar, K. (2005). Critiquing and improving the use of data from high stakes tests with the aid of dynamic statistics software. In C. Dede, J. P. Honan, & L. C. Peteres (Eds.), *Scaling up success: Lessons learned from technology-based educational improvement* (pp. 198–226). San Francisco: Jossey-Bass.

Confrey, J., & Maloney, A. P. (2012). Next generation digital classroom assessment based on learning trajectories in mathematics. In C. Dede & J. Richards (Eds.), *Digital teaching platforms: Customizing classroom learning for each student* (pp. 134–152). New York: Teachers College Press.

Conklin, E. J. (2005). *Dialogue mapping: Building shared understanding of wicked problems*. New York: Wiley.

Coxford, A. F., Fey, J. T., Hirsch, C. R., Schoen, H. L., Burrill, G., Hart, E. W., et al. (2003a). *Contemporary mathematics in context: A unified approach (Course 1)*. Columbus, OH: Glencoe.

Coxford, A. F., Fey, J. T., Hirsch, C. R., Schoen, H. L., Burrill, G., Hart, E. W., et al. (2003b). *Contemporary mathematics in context: A unified approach (course 2)*. Chicago: Everyday Learning.

Deming, W. E. (2000). *Out of the crisis*. Cambridge, MA: MIT Press.

Elmore, R. F. (2002). *Bridging the gap between standards and achievement (report)*. Washington, DC: Albert Shanker Institute.

Fuson, K. C. (2009a). *Math expressions: Grade 1*. Orlando, FL: Houghton Mifflin Harcourt Publishing.

Fuson, K. C. (2009b). *Math expressions: Grade 2*. Orlando, FL: Houghton Mifflin Harcourt Publishing.

Gould, S. J. (1996). *Full house: The spread of excellence from Plato to Darwin*. New York: Three Rivers Press.

Grouws, D. H., Reys, R., Papick, I., Tarr, J., Chavez, O., Sears, R., et al. (2010). *COSMIC: Comparing options in secondary mathematics: Investigating curriculum*. Retrieved 2010, from http://cosmic.missouri.edu/

Grouws, D. H., Tarr, J. E., Chávez, Ó., Sears, R., Soria, V. M., & Taylan, R. D. (2013). Curriculum and implementation effects on high school students' mathematics learning from curricula representing subject-specific and integrated content organizations. *Journal for Research in Mathematics Education, 44*(2), 416–463.

Henningsen, M., & Stein, M. K. (1997). Mathematical tasks and student cognition: Classroom-based factors that support and inhibit high-level mathematical thinking and reasoning. *American Education Research Journal, 28*(5), 524–549.

Hiebert, J., Gallimore, R., & Stigler, J. W. (2002). A knowledge base for the teaching profession: What would it look like and how can we get one? *Educational Researcher, 31*(5), 3–15.

Holliday, B., Cuevas, G. J., Moore-Harris, B., Carter, J. A., Marks, D., Casey, R. M., et al. (2005). *Algebra 1*. New York: Glencoe.

Juran, J. M. (1962). *Quality control handbook*. New York: McGraw-Hill.

Kennedy, P. A., McGowan, D., Schultz, J. E., Hollowell, K., & Jovell, I. (2001). *Algebra one interactions: Course 1*. Austin, TX: Holt.

Krupa, E. E., & Confrey, J. (2010). *Teacher change facilitated by instructional coaches: A customized approach to professional development*. Paper presented at the Annual Conference of North American Chapter of the International Group for the Psychology of Mathematics Education.

Krupa, E. E., & Confrey, J. (2012). Using instructional coaching to customize professional development in an integrated high school mathematics program. In J. Bay-Williams & W. R. Speer (Eds.), *Professional collaborations in mathematics teaching and learning 2012: Seeking success for all, the 74th NCTM Yearbook*. Reston, VA: National Council of Teachers of Mathematics.

Larson, N. (2008). *Saxon math*. Orlando, FL: Harcourt Achieve, Inc.

Larson, R., Boswell, L., Kanold, T. D., & Stiff, L. (2001). *Algebra 1*. Evanston, IL: McDougal Littell.

Larson, R., Boswell, L., & Stiff, L. (2001). *Geometry*. Evanston, IL: McDougal Littell.

Lemke, J. L. (2000). *Multiple timescales and semiotics in complex ecosocial systems*. Paper presented at the 3rd International Conference on Complex Systems, Nashua, NH.

Maroulis, S., Guimera, R., Petry, H., Stringer, M. J., Gomez, L. M., Amaral, L. A. N., et al. (2010). Complex systems view of educational policy research. *Science, 330*, 38–39.

McNaught, M., Tarr, J. E., & Sears, R. (2010). *Conceptualizing and measuring fidelity of implementation of secondary mathematics textbooks: Results of a three-year study*. Paper presented at the Annual Meeting of the American Educational Research Association, Denver, CO.

NRC. (2004). *On evaluating curricular effectiveness: Judging the quality of K-12 mathematics evaluations*. Washington, DC: The National Academies Press.

Penuel, W. R., Confrey, J., Maloney, A. P., & Rupp, A. A. (2014). Design decisions in developing assessments of learning trajectories: A case study. *Journal of the Learning Sciences, 23*(1), 47–95.

Reys, R. E., Reys, B. J., Lapan, R., Holliday, G., & Wasman, D. (2003). Assessing the impact of standards-based mathematics curriculum materials on student achievement. *Journal for Research in Mathematics Education, 34*(1), 74–95.

Rittell, H. W. J., & Webber, M. M. (1984). Planning problems are wicked problems. In N. Cross (Ed.), *Developments in design methodology* (pp. 135–144). New York: Wiley.

Schmidt, W. H., McKnight, C. C., Houang, R. T., Wang, H., Wiley, D., Cogan, L. S., et al. (2001). *Why schools matter: A cross-national comparison of curriculum and learning.* San Francisco: Jossey-Bass.

Shavelson, R. J., & Towne, L. (Eds.). (2002). *Scientific research in education.* Washington, DC: National Academy Press.

Stein, M. K., Grover, B. W., & Henningsen, M. (1996). Building student capacity for mathematical thinking and reasoning: An analysis of mathematical tasks used in reform classrooms. *American Education Research Journal, 33*(2), 455–488.

Stein, M. K., & Kaufman, J. H. (2010). Selecting and supporting the use of mathematics curricula at scale. *American Education Research Journal, 47*(3), 663–693.

Tarr, J. E., Grouws, D. H., Chávez, Ó., & Soria, V. M. (2013). The effects of content organization and curriculum implementation on students' mathematics learning in second-year high school courses. *Journal for Research in Mathematics Education, 44*(4), 683–729.

Tarr, J. E., Ross, D. J., McNaught, M. D., Chávez, O., Grouws, D. A., Reys, R. E. et al. (2010). *Identification of student- and teacher-level variables in modeling variation of mathematics achievement data.* Paper presented at the Annual Meeting of the American Educational Research Association, Denver, CO.

Tatar, D. (2007). The design tensions framework. *Human-Computer Interaction, 22*(4), 413–451.

Thomas, S. M. (2010). *A study of the impact of professional development on integrated mathematics on teachers' knowledge and instructional practices in high poverty schools.* Unpublished doctoral dissertation, North Carolina State University, Raleigh, NC.

Wittenberg, L., Economopoulos, K., Bastable, V., Bloomfield, K. H., Cochran, K., Earnest, D., et al. (2008). *Investigations in number, data, and space* (2nd ed.). Glenview, IL: Pearson Scott Foresman.

The Role of Large-Scale Studies in Mathematics Education

Jinfa Cai, Stephen Hwang, and James A. Middleton

The goal for this monograph has been to present the state of the art in large-scale studies in mathematics education. Although the chapters collected here are not intended to be a comprehensive compendium of such research, this selection of work does serve both to represent large-scale studies in the field of mathematics education and to raise awareness of the issues that arise in conducting such studies. As can be seen from the studies included in this volume, the term "large-scale study" can cover a wide variety of research endeavors. Indeed, as we indicated in the introductory chapter, the IES and NSF Common Guidelines (U.S. Department of Education & National Science Foundation, 2013) recognize many different types of research, and large-scale studies can fall into more than one of these categories. The different types of research have correspondingly different purposes which require different sample sizes and methods of analysis. Moreover, the findings from different types of research have different degrees of generalizability. Despite these many differences, there are common features that large-scale studies share. For example, large-scale studies require large sample sizes, although exactly how large "large" is depends on the design of the study. In contrast to small-scale studies, large-scale studies usually employ complex statistical analysis and have the potential to produce more readily generalizable results than small-scale studies.

J. Cai (✉) • S. Hwang
University of Delaware, Ewing Hall 523, Newark, DE 19716, USA
e-mail: jcai@udel.edu; hwangste@udel.edu

J.A. Middleton
Arizona State University, Tempe, AZ, USA
e-mail: jimbo@asu.edu

© Springer International Publishing Switzerland 2015   405
J.A. Middleton et al. (eds.), *Large-Scale Studies in Mathematics Education*,
Research in Mathematics Education, DOI 10.1007/978-3-319-07716-1_18

In this concluding chapter, we draw on some of the common features of large-scale studies in mathematics education to highlight the contributions and considerations associated with this type of research. We focus our discussion in three areas, drawing on the work of the many researchers and research groups that have contributed to this volume. First, we consider some of the key benefits that large-scale studies present to mathematics education. Second, we examine potential pitfalls that face those who conduct large-scale studies in mathematics education. Finally, we look to the future and consider what role large-scale studies might play.

Benefits of Large-Scale Studies in Mathematics Education

Below, we specifically discuss three types of benefits of large-scale studies: their utility for understanding the status of situations and trends, the capacity they provide for testing hypotheses, and the sophisticated analytic methods often employed in such studies. However, we first want to point out that when large-scale studies are designed and conducted with care, these benefits can mean greater power for findings to be generalized with respect to populations. Looking once more to the NSF and IES Common Guidelines (U.S. Department of Education & National Science Foundation, 2013), we see that the last three types of research they discuss are impact studies: efficacy, effectiveness, and scale-up. Although these three types of impact studies differ with regard to the populations to which their findings generalize, the large-scale study design can allow for greater attention to identifying variation in impacts by subgroup, setting, level of implementation, and other mediators.

Understanding the Status of Situations and Trends

To guide research and policy, it is often necessary to better understand the "big picture" of teaching and student learning across a large population. Large-scale studies allow researchers and policy makers to map the current status of teaching and student learning or to understand broad trends in teaching and learning. There is a long history of generating large data sets on teaching and learning through data acquisition studies such as NAEP (e.g., Braswell et al., 2001; National Center for Education Statistics, 2013) and international comparative work such as the IEA studies (e.g., Hiebert et al., 2003; Husén, 1967; Mullis et al., 1997; Robitaille & Garden, 1989). Such studies provide key opportunities to look broadly across time and geographic region to examine features of mathematics education and also how those features have or have not evolved. For example, with respect to student learning, in this volume Kloosterman et al. (2015) made use of NAEP data from between 2003 and 2011 to analyze changes in student understanding of algebraic reasoning. Others have used large-scale data to examine student learning from a comparative lens, whether studying trends in gender differences across the IEA studies (Hanna, 2000), considering gaps in the mathematics achievement on NAEP of students from different ethnic and socio-economic backgrounds (Lubienski & Crockett, 2007) or looking at differential curricular effects on various ethnic groups (Hwang et al., 2015).

In addition to student learning, researchers and policy makers have used data from large-scale studies to look at other features of the educational system. Over the past four decades, the National Science Foundation has supported a series of large-scale surveys of mathematics and science education that have addressed significant questions about instruction, curriculum, and resources in mathematics and science classrooms around the USA (Banilower et al., 2013). These studies have attempted to provide comprehensive and nationally representative data on mathematics and science education by surveying carefully chosen samples of thousands of teachers across the country. This type of large-scale data set has made it possible to, for example, assess the degree to which teaching in the USA aligns with recognized "best practices" (Weiss, Smith, & O'Kelley, 2009). Similarly, the TIMSS 1999 Video Study (Hiebert et al., 2003) documented differences in mathematics teaching across several countries, and the data from that study has informed further research and discussion on the nature of effective mathematics teaching (e.g., Givvin, Jacobs, Hollingsworth, & Hiebert, 2009). In this volume, Zhu (2015) has integrated data from several studies in the TIMSS series to better understand trends in the role, nature, and use of mathematics homework in various countries. In general, the wealth of data from large-scale studies makes space for insightful analyses of the status of and trends in mathematics education.

Testing Hypotheses

In addition to shedding light on the current status and trends in mathematics education, large-scale studies often serve the role of testing hypotheses about theories or interventions for broad samples of students. Of course, smaller-scale studies also investigate and test hypotheses. However, large-scale studies are well positioned to build on earlier research findings and the work of exploratory studies. They can test hypotheses generated from such studies or develop and assess formal interventions that are designed based on the findings of smaller studies. For example, Vig, Star, Dupuis, Lein, and Jitendra (2015) used a large-scale approach to revisit concerns raised in previous research about student use of cross multiplication to solve proportion problems. Although they expected to confirm the findings of existing studies, many of which were conducted in the 1980s and 1990s, that indicated that students often relied on cross multiplication, Vig et al. found that this tendency no longer seemed to hold for the students in their sample.

Large-scale studies can also be effective tools for examining interventions and programs to investigate questions of validity and effectiveness at scale. For example, Lewis and Perry (2015) conducted a randomized, controlled trial of a lesson study intervention designed to improve teachers' and students' mathematical knowledge. Based on earlier work that analyzed the features of lesson study and characterized the cyclic nature of teacher learning through lesson study, Lewis and Perry designed a larger test of scaling-up lesson study with support from centrally designed mathematical resource kits. They hypothesized that these resource kits would act as a substitute for the teacher's manual that is a key resource for lesson study in Japan and that lesson study with these resource kits would develop beliefs and dispositions among the teachers that maximize the effectiveness of lesson study.

With a large pool of participants, it is generally more convenient to test hypotheses about the effectiveness of programs across diverse populations. Several examples of this type of effectiveness research are included in this volume. For example, the LieCal project investigated the effectiveness of the reform-oriented Connected Mathematics Project (CMP) middle school mathematics curriculum as compared with more traditional middle school mathematics curricula (Cai 2014; Cai et al., 2011; Hwang et al., 2015; Moyer et al., 2011). CMP was developed based on a problem-based pedagogy grounded in research and has been field tested. The LieCal project used a quasi-experimental design with statistical controls to examine longitudinally the relationship between students' learning and their curricular experiences. The project was first conducted in 14 middle schools in an urban school district serving a diverse student population in the USA and later followed those students into high school. By taking a large-scale approach, the LieCal Project has been able to investigate the effect of the CMP curriculum as it has been formally implemented in schools, as well as to examine the factors that might contribute to improved student learning.

Working from a base of research showing the significance of success in algebra for future mathematics course-taking and success, Heppen, Clements, and Walters (2015) tested the hypothesis that online Algebra I courses could be an effective way to increase access to algebra for eighth graders in schools that do not offer algebra. This randomized experimental trial involved 68 largely rural schools across Maine and Vermont that did not ordinarily offer an Algebra I course to students in eighth grade. The findings support the conclusion that online courses can be an effective way to increase access to Algebra I.

Employing Sophisticated Analytic Methods

Underlying the strength of large-scale studies to test hypotheses is their potential for analytic and statistical power. Large-scale studies can, and often must, take advantage of advanced statistical techniques to look for causes and correlates of students' learning of mathematics. For example, large-scale research in education frequently involves participants who are nested in hierarchically organized levels such as students grouped into classes. In order to deal with such situations, multilevel modeling techniques are called upon to address the shortcomings of traditional linear regression models. Several of the studies reported in this volume make use of hierarchical linear modeling (Raudenbush & Bryk, 2002) to conduct analyses of various types of nested data, including students nested in classrooms (Hwang et al., 2015), students nested in schools (Heppen et al., 2015), and teachers nested in lesson study groups (Lewis & Perry, 2015).

The statistical power and sophistication of analysis afforded by large-scale studies can be difficult to match in smaller-scale research. Analyses can include attention to intermediate outcomes such as teacher beliefs (e.g., Lewis & Perry, 2015) and, when looking at the effectiveness of a program, can control for factors such as implementation or quality of instruction (e.g., Cai, Ni, & Hwang, 2015; Tarr & Soria, 2015). Thus, large-scale studies can help to reveal the complexities in casual links by attending to the mediating factors and issues that can influence the effects of a treatment in different naturalistic contexts. For example, in their retrospective study of an NSF-funded Local Systemic Change (LSC) initiative in the greater Philadelphia area, Kramer, Cai, and Merlino (2015) synthesized a school-level extension of the ideas of implementation fidelity and buy-in. They combined a host of measures to generate a "will-to-reform" factor that they investigated as a potential intermediate factor in studying the effects of the LSC model on student achievement. This allowed them to determine that the level of school-level buy-in to the LSC model was a critical differentiator between schools that had improved mathematics achievement and schools where mathematics achievement declined.

Not unexpectedly, large-scale studies in mathematics education have traditionally relied on quantitative methods of analysis. However, some researchers have begun to weave together both quantitative and qualitative approaches to provide even more sophisticated analyses. Mujtaba, Reiss, Rodd, and Simon (2015) describe a deft interleaving of quantitative techniques such as multilevel modeling with qualitative analyses of semi-structured interviews of individual students. In studying the factors which contribute to students' decisions whether to participate in post-compulsory mathematics courses, these researchers used an iterative approach. After developing an initial set of findings through multilevel modeling, Mujtaba et al. found that their qualitative interview data suggested that their initial constructs may not have captured the true influence of mathematics teachers and lessons. Guided by their qualitative findings, they deconstructed their original constructs and conducted a second quantitative analysis using item-level rather than construct-level data. This back-and-forth between analytic methods allowed Mujtaba et al. to refine their results and ultimately paint a more complete picture of the factors which influence students' decisions to continue with mathematics.

Finally, the large, rich sets of data that come from large-scale studies allow researchers a great deal of analytic flexibility. For example, analyses carried out on these data can, as Kloosterman et al. (2015) have demonstrated, join assessment items that were originally conceived as disparate and falling into distinct categories into coherent clusters that measure specific mathematical skills. In their work with NAEP items, Kloosterman et al. have examined the algebra-related items used between 2003 and 2011 to produce clusters of items that appear to measure algebra skills that are more specific than the general NAEP algebra scales but also more general than what could be obtained from individual item analyses.

Pitfalls of Large-Scale Studies in Mathematics Education

Although the potential benefits of large-scale studies are compelling, there are a number of difficulties, both structural and operational, that these complex endeavors can fall prey to. We are therefore compelled to ask where and how large-scale studies are prone to failure. Many of these difficulties are directly related to the need to coordinate and manage large numbers of people and large quantities of data across multiple political and geographic settings and over potentially long periods of time. Any project that requires the cooperation of large numbers of people or their governments will be subject to pitfalls simply because they will come to the project with different expectations and priorities. Orletsky, Middleton, and Sloane (2015) aptly remind us of the parallel drawn by Albert Beaton while he was director of TIMSS between sausage making, the legislative process, and the design of international test items (Bracey, 1997). The need to balance the priorities of multiple stakeholders in a complex project can result in critical compromises to the design of the research. However, other challenges to conducting large-scale research are rooted in limitations of resources, time, and methodology.

Resources and Time

If a study is large scale, whether in terms of sample size, geographic coverage, or quantity of data, it goes without saying that it will require great effort that consumes a large amount of time, money, or both. For example, the observational data that Confrey and Maloney (2015) argue is necessary to collect and analyze for judgments of instructional effectiveness is fundamentally resource intensive, and there are methodological trade-offs that must be considered when balancing the quality of the data collected and the required resources (Shih, Ing, & Tarr, 2015). The same is true for multiple, interlocking phases of analysis, such as the iterations of quantitative and qualitative analysis carried out by Mujtaba et al. (2015).

Even when the resources and time are available to carry out a large-scale study, the time lag between the beginning of the study and the availability of the results can be considerable. Indeed, by the time the results are available, the object of study may have changed significantly. This can be particularly challenging for longitudinal studies of curricular effectiveness. For example, the LieCal project described by Cai et al. (2015) and Hwang et al. (2015) compared the effects of reform and traditional middle school mathematics curricula. The project followed students using the Connected Mathematics Program (CMP) and several more traditional curricula through middle school and subsequently through high school. Since the collection of data for that longitudinal study, CMP has undergone two rounds of revisions. Thus, the implications of the LieCal project findings must be interpreted in light of the fact that today's CMP is not entirely the same as the original. This evolution of educational phenomena can be even more pronounced when studying educational interventions that use technology. The technological state of the art changes rapidly and sometimes in evolutionary leaps (witness, for example, the rapidity of the adoption of tablet computers, both in daily life and in classrooms) that confound the ability of researchers to carry out well-planned, large-scale studies of technological tools in education whose results remain relevant after the research is complete.

Another issue related to time is that for large data sets that are generated over time, such as standardized assessments like NAEP, the items, coverage, and sampling can change over the years. This leads to potential complications in interpretation of the data that researchers must plan for. Zhu (2015) has provided a perspective in this volume on some of the challenges that are posed by assessments that change over time. Zhu has noted in detail many changes in the way TIMSS surveys have investigated homework practices over the years. In her analysis of longitudinal trends in the TIMSS data on homework, she compensated for variations in measurement scales over time. Moreover, she was able to use those changes in measurement design to provide a perspective on the shifting role and significance of homework in the TIMSS countries over time.

Complexity of Authentic Research Settings

The complexity of conducting large-scale studies is also related to the "messiness" of the authentic research settings—usually classrooms. Tarr and Soria (2015) had to contend with the fact that, in authentic settings (as opposed to laboratory settings), students move from school to school, making missing data a real issue. Middleton et al. (2015) ended up with only about one-third of their original sample due to high attrition of students in their study in urban schools. Even within classes, tremendous variation exists across students, their prior knowledge and understanding, and the kinds of teaching methods that optimize to different settings (see, for example, Heppen et al., 2015). Scale compounds such issues, but like Tarr and Soria, researchers can be clever in their approaches to handling such variation, first through appropriate sampling and handling of covariates, and second, through use of appropriate statistical modeling.

Methodology

Finally, lest we too-readily ascribe power to the complex statistical methods that can be used in large-scale studies, it is important to remember that there are also methodological pitfalls that can interfere with the effectiveness of any statistical analysis. Orletsky et al. (2015) provide a thorough review of the various threats to validity, both internal and external, that must be guarded against. In addition, both Kloosterman et al. (2015) and Orletsky et al. comment on issues related to sampling in large-scale studies. Both chapters warn readers that, although large-scale studies can offer great statistical power, one must be careful when dealing with subgroups of the study sample (e.g., ethnic groups). To generalize findings to subpopulations, it is necessary to consider the sampling procedure used in the larger study and factor in the appropriate sampling weights. In the context of NAEP data, Kloosterman et al. warn that:

> When analysis is restricted to Hispanic and Black students within a given state, the standard errors for the populations are much larger than they are for national results for all students. When looking at differences in performance on individual items, statistical power decreases further because only those students who complete the item in question can be included in the analysis. This is not to say that analyses for subgroups are inappropriate, but larger differences are needed to make claims involving statistical significance. (p. ???)

In addition, the instruments that researchers use to gather data must bear careful scrutiny. For example, Ebby and Sirinides (2015) describe a large-scale approach used to develop, analyze the characteristics of, and validate TASK, an instrument for measuring teachers' capacity for learning trajectory-based formative assessment. However flawed tools, particularly when used without attending to their constraints and assumptions, can destroy the usefulness of the data gathered with them. At the very least, it is necessary to identify the assumptions underlying a data-gathering instrument and the purposes for which it is intended. Shih et al. (2015) make this point in their analysis of two classroom observation protocols that have been used in large-scale projects. Although both protocols ostensibly measure features of classroom instruction, an analysis showed that they rely on quite different assumptions about that instruction (one assuming relatively stable instruction from day to day and the other attending to more dimensions of instruction that may be less stable). Given the different purposes for which the protocols were designed, Shih et al. found that these two tools captured different sources of variation (although appropriately so for their respective studies). Thus, they warn that, as large-scale measures of mathematics instruction are used more frequently in large-scale studies in mathematics education, careful attention must be paid to their methodological characteristics.

Looking to the Future

This monograph is the first volume explicitly devoted to large-scale studies in mathematics education research. While we certainly value smaller-scale studies in mathematics education and believe in the unique roles that small-scale studies play in educational research, we believe that large-scale studies also have a distinctive role to play. In their survey of over 700 mathematics education articles published between 1995 and 2005, Hart et al. (2008) found that 50 % of the studies used qualitative methods only, 21 % used quantitative methods only, and 29 % mixed qualitative and quantitative methods in various ways. Studies that use quantitative methods are not necessarily large-scale studies; nevertheless, large-scale studies usually do employ quantitative methods. Therefore, we suggest that the findings of Hart and her colleagues show that there remains much room in mathematics education research for more large-scale studies.

Given the benefits and pitfalls that we have considered above, what can we take away from the work presented in this monograph? First, we should ask when it makes sense to design and conduct large-scale studies. The answer to this question hearkens back to the basic training of any researcher—it depends on the kinds of research questions to be answered. The kind of research question we want to answer will dictate whether we choose a large-scale approach. The IES and NSF Common Guidelines (U.S. Department of Education & National Science Foundation, 2013) describe six different types of research, ranging from foundational research and early-stage or exploratory research, to design and development research, to efficacy, effectiveness and scale-up studies. Not every one of these types of research requires a large-scale study, although the distinctions are not precise. Certainly, research questions about the effectiveness of programs and interventions in a wide range of contexts and circumstances would seem to benefit from a large-scale approach. We hope that, the findings from the chapters in this volume and the choices made by their authors will provide guidance for readers when they think about the use of large-scale studies.

We are fully aware of the cost and resources required for designing and conducting large-scale studies. They may go far beyond the reach of dissertation studies. Indeed, it is quite challenging to carry out a large-scale study without a team. However, there is a wealth of existing large-scale data sets available and many researchers have taken advantage of them with fruitful results (e.g., Kloosterman, 2010). As noted above, both Kloosterman et al. (2015) and Zhu (2015) based their work in this volume on analyses of data from existing large data sets (NAEP and TIMSS, respectively).

Moreover, with support from the NSF, the American Educational Research Association (AERA) offers small grants and training programs for research involving large-scale data sets collected or supported by the National Center for Education Statistics (NCES), NSF, or other federal agencies. These include dissertation grants that support advanced doctoral students in undertaking doctoral dissertations using data from these large-scale data sets. AERA also provides research grants for faculty members, postdoctoral researchers, and other doctoral-level scholars to undertake quantitative research using such large-scale data.

The benefits and pitfalls of large-scale studies exhibited in the research included in this monograph may help future researchers when they design and conduct large-scale studies. It is our hope that this monograph will increase awareness in the mathematics education research community of large-scale studies and encourage the community to engage in more such studies. This, in turn, may help the field to move along the "pipeline" of different types of research specified in the NSF and IES guidelines to generate solid evidence that supports the development and implementation of programs that will improve student learning.

References

Banilower, E. R., Smith, P. S., Weiss, I. R., Malzahn, K. A., Campbell, K. M., & Weis, A. M. (2013). *Report of the 2012 national survey of science and mathematics education*. Chapel Hill, NC: Horizon Research, Inc.

Bracey, G. W. (1997). More on TIMSS. *Phi Delta Kappan,* 78(8), 656–657.

Braswell, J. S., Lutkus, A. D., Grigg, W. S., Santapau, S. L., Tay-Lim, B. S.-H., & Johnson, M. S. (2001). *The nation's report card: Mathematics 2000*. Washington, DC: U.S. Department of Education, National Center for Education Statistics.

Cai, J., Wang, N., Moyer, J. C., Wang, C., & Nie, B. (2011). Longitudinal investigation of the curricular effect: An analysis of student learning outcomes from the LieCal project in the United States. *International Journal of Educational Research, 50*, 117–136.

Cai, J. (2014). Searching for evidence of curricular effect on the teaching and learning of mathematics: Some insights from the LieCal project. *Mathematics Education Research Journal, 26*, 811–831.

Cai, J., Ni, Y., & Hwang, S. (2015). Measuring change in mathematics learning with longitudinal studies: Conceptualization and methodological issues. In J. A. Middleton, J. Cai, & S. Hwang (Eds.), *Large-scale studies in mathematics education*. New York: Springer.

Confrey, J., & Maloney, A. (2015). Engineering [for] effectiveness in mathematics education: Intervention at the instructional core in an era of the common core standards. In J. A. Middleton, J. Cai, & S. Hwang (Eds.), *Large-scale studies in mathematics education*. New York: Springer.

Ebby, C. B., & Sirinides, P. M. (2015). Conceptualizing teachers' capacity for learning trajectory-oriented formative assessment in mathematics. In J. A. Middleton, J. Cai, & S. Hwang (Eds.), *Large-scale studies in mathematics education*. New York: Springer.

Givvin, K. B., Jacobs, J., Hollingsworth, H., & Hiebert, J. (2009). What is effective mathematics teaching? International educators' judgments of mathematics lessons from the TIMSS 1999 video study. In J. Cai, G. Kaiser, B. Perry, & N.-Y. Wong (Eds.), *Effective mathematics teaching from teachers' perspectives: National and cross-national studies* (pp. 37–69). Rotterdam, The Netherlands: Sense Publishers.

Hanna, G. (2000). Declining gender differences from FIMS to TIMSS. *Zentralblatt fur Didaktik der Mathematik, 32*, 11–17.

Hart, L. C., Smith, S. Z., Swars, S. L., & Smith, M. E. (2008). An examination of research methods in mathematics education (1995–2005). *Journal of Mixed Methods Research, 3*, 26–41.

Heppen, J. B., Clements, M., & Walters, K. (2015). Turning to online courses to expand access: A rigorous study of the impact of online algebra I for eighth graders. In J. A. Middleton, J. Cai, & S. Hwang (Eds.), *Large-scale studies in mathematics education*. New York: Springer.

Hiebert, J., Gallimore, R., Garnier, H., Givvin, K. B., Hollingsworth, H., Jacobs, J., et al. (2003). *Teaching mathematics in seven countries: Results from the TIMSS 1999 video study. NCES (2003-013)*. Washington, DC: National Center for Education Statistics, U.S. Department of Education.

Husén, T. (1967). *International study of achievement in mathematics: A comparison of twelve countries*. New York: Wiley.

Hwang, S., Cai, J., Shih, J., Moyer, J. C., Wang, N., & Nie, B. (2015). Longitudinally investigating the impact of curricula and classroom emphases on the algebra learning of students of different ethnicities. In J. A. Middleton, J. Cai, & S. Hwang (Eds.), *Large-scale studies in mathematics education*. New York: Springer.

Kloosterman, P. (2010). Mathematics skills of 17-year-olds in the United States: 1978 to 2004. *Journal for Research in Mathematics Education, 41*, 20–51.

Kloosterman, P., Walcott, C., Brown, N. J. S., Mohr, D., Perez, A., Dai, S., et al. (2015). Using NAEP to analyze 8th-grade students' ability to reason algebraically. In J. A. Middleton, J. Cai, & S. Hwang (Eds.), *Large-scale studies in mathematics education*. New York: Springer.

Kramer, S., Cai, J., & Merlino, J. (2015). A lesson for the common core standards era from the NCTM standards era: The importance of considering school-level buy-in when implementing and evaluating standards-based instructional materials. In J. A. Middleton, J. Cai, & S. Hwang (Eds.), *Large-scale studies in mathematics education*. New York: Springer.

Lewis, C. C., & Perry, R. R. (2015). A randomized trial of lesson study with mathematical resource kits: Analysis of impact on teachers' beliefs and learning community. In J. A. Middleton, J. Cai, & S. Hwang (Eds.), *Large-scale studies in mathematics education*. New York: Springer.

Lubienski, S. T., & Crockett, M. (2007). NAEP mathematics achievement and race/ethnicity. In P. Kloosterman & F. Lester (Eds.), *Results from the ninth mathematics assessment of NAEP* (pp. 227–260). Reston, VA: National Council of Teachers of Mathematics.

Middleton, J., Helding, B., Megowan-Romanowicz, C., Yang, Y., Yanik, B., Kim, A., et al. (2015). A longitudinal study of the development of rational number concepts and strategies in the middle grades. In J. A. Middleton, J. Cai, & S. Hwang (Eds.), *Large-scale studies in mathematics education*. New York: Springer.

Moyer, J. C., Cai, J., Wang, N., & Nie, B. (2011). Impact of curriculum reform: Evidence of change in classroom practice in the United States. *International Journal of Educational Research, 50*, 87–99.

Mujtaba, T., Reiss, M. J., Rodd, M., & Simon, S. (2015). Methodological issues in mathematics education research when exploring issues around participation and engagement. In J. A. Middleton, J. Cai, & S. Hwang (Eds.), *Large-scale studies in mathematics education*. New York: Springer.

Mullis, I. V. S., Martin, M. O., Beaton, A. E., Gonzalez, E. J., Kelly, D. L., & Smith, T. A. (1997). *Mathematics achievement in the primary school years: IEA's third international mathematics and science study (TIMSS)*. Chestnut Hill, MA: Center for the Study of Testing, Evaluation, and Educational Policy, Boston College.

National Center for Education Statistics. (2013). *The nation's report card: Trends in academic progress 2012 (NCES 2013 456)*. Washington, DC: IES, U.S. Department of Education.

Orletsky, D., Middleton, J. A., & Sloane, F. (2015). A review of three large-scale datasets critiquing item design, data collection, and the usefulness of claims. In J. A. Middleton, J. Cai, & S. Hwang (Eds.), *Large-scale studies in mathematics education*. New York: Springer.

Raudenbush, S. W., & Bryk, A. S. (2002). *Hierarchical linear models* (2nd ed.). Thousand Oaks, CA: Sage.

Robitaille, D. F., & Garden, R. A. (Eds.). (1989). *The IEA study of mathematics II: Contexts and outcomes of school mathematics*. Oxford, England: Pergamon Press.

Shih, J., Ing, M., & Tarr, J. (2015). Addressing measurement issues in two large-scale mathematics classroom observation protocols. In J. A. Middleton, J. Cai, & S. Hwang (Eds.), *Large-scale studies in mathematics education*. New York: Springer.

Tarr, J. E., & Soria, V. (2015). Challenges in conducting large-scale studies of curricular effectiveness: Data collection and analyses in the COSMIC project. In J. A. Middleton, J. Cai, & S. Hwang (Eds.), *Large-scale studies in mathematics education*. New York: Springer.

U.S. Department of Education, & National Science Foundation. (2013). *Common guidelines for education research and development*. Retrieved from National Science Foundation website www.nsf.gov/publications/pub_summ.jsp?ods_key=nsf13126

Vig, R., Star, J. R., Dupuis, D. N., Lein, A. E., & Jitendra, A. K. (2015). Exploring the impact of knowledge of multiple strategies on students' learning about proportions. In J. A. Middleton, J. Cai, & S. Hwang (Eds.), *Large-scale studies in mathematics education*. New York: Springer.

Weiss, I. R., Smith, P. S., & O'Kelley, S. K. (2009). The presidential award for excellence in mathematics teaching: Setting the standard. In J. Cai, G. Kaiser, B. Perry, & N.-Y. Wong (Eds.), *Effective mathematics teaching from teachers' perspectives: National and cross-national studies* (pp. 281–304). Rotterdam, The Netherlands: Sense Publishers.

Zhu, Y. (2015). Homework and mathematics learning: What can we learn from the TIMSS series studies in the last two decades? In J. A. Middleton, J. Cai, & S. Hwang (Eds.), *Large-scale studies in mathematics education*. New York: Springer.

Index

A
Academic interest
  ANOVA test, 248–250
  PBL approach, 254
  subtest, 241
  survey, 247
Agodini, R., 21, 40, 389
Algebra I courses, 408
Algebraic reasoning
  content strands, 190
  covariation, 183
  eighth-grade students, 192
  equations and inequalities, 194–196
  letter-symbolism, 183
  LTT, 191
  mathematics items, 184
  middle school students, 194
  patterns, relations and functions,
    184–186, 190
  properties and operations, 184
  race/ethnicity, 183
  representations, 184, 187–189
  researchers, 184
  scale scores, 190, 191
  students performance, 183, 406
  variables and expressions, 192, 193
  written explanation, 191
Algebraic thinking, 47, 164
Algebra I for eighth graders, online course
  advanced course sequences, 99, 104
  algebra and general math posttests,
    113–114
  Bernoulli sampling distribution, 106
  Bonferroni correction, 106

CCSSM, 95–96 (*see also* Common Core
    State Standards for Mathematics
    (CCSSM))
Class.com's Algebra I course, 129
course activity, 101–102, 118–120
coursetaking in mathematics, 104
ECLS-K, 96
eligible students
  advanced math course sequence in high
    school, 125
  general math achievement, 125–128
  high-school coursetaking, 124–125
  outcomes, benefits, 128
  in treatment and control schools, 97,
    104, 124
end-of-eighth grade algebra
  achievement, 128
face-to-face version, 130
goals, 96–97, 102
high-school mathematics and science
    courses, 98, 114–115
implementation
  classroom materials, 113
  online course activity data, 112
  proctor logs, 112
  site visits, 112
  teacher survey, 113
in middle school, 99
National Educational Longitudinal Study, 98
non-eligible students
  outcomes, 125–128
  in treatment and control schools,
    104–105
on-site proctors, 118, 120

© Springer International Publishing Switzerland 2015
J.A. Middleton et al. (eds.), *Large-Scale Studies in Mathematics Education*,
Research in Mathematics Education, DOI 10.1007/978-3-319-07716-1

Algebra I for eighth graders, online
        course (*cont.*)
    participating schools, 108–111
    primary and secondary analyses, 129
    randomized controlled trial, 97
    research questions, 103, 104
    school recruitment, 106–108
    in small/rural schools, 96
    students, interaction, 122
    teachers, 117–118
Algebra learning of students
    algebra II course, 47
    algebra readiness/thinking, 47
    ethnicity, effect on CMP and non
        CMP, 54, 56
    HLM (*see* Hierarchical linear models
        (HLM))
    LieCal Project, 46
    method
        assessing, 49–50
        conceptual and procedural
            emphases, 51
        Item Response Theory (IRT) model, 50
        multiple-choice and open-ended
            assessment tests, 49
        quantitative data analysis, 51–52
        QUASAR Project, 50
        sample, 49
    Standards-based curriculam, CMP, 47–48
    Standards-based learning environments, 48
    Standards-documents, NCTM, 45
    state standardized tests, 53
Algina, J., 19
American Educational Research Association
        (AERA), 412, 413
Analysis of student thinking (AST)
    IDM, 173
    lesson study, 132
    misconceptions, 162
    TASK, 173
    teacher responses, 168
Anderson, L.W., 3, 6
ANOVA test
    academic interest, 249–250
    decimal computation, 249
    pretreatment group equivalency, 247
    self-efficacy, 249
    total computation, 247–248
Austin, J.D., 210

B
Balanced Assessment of Mathematics, 50
Ball, D. L., 161

Bandura, A., 236
Bauer, K.W., 4
Baxter, J., 238
Beaton, A.E., 332
Bernoulli sampling distribution, 106
Betts, J., 99
Black, L., 335–336
Blazer, C., 211, 212
Bohrnstedt, G.W., 326
Bonferroni correction, 106
Bozick, R., 323
Bracey, G.W., 315, 328, 331
Brown, M.W., 20
Brown, N.J.S., 179–205
Bryk, A.S., 376
Buy-in concept
    GPSMP Treatment, 35–36
    HLM, 34–35
    Math achievement growth, 35–37
    teacher, 24–25
    Will-to-Reform scale, 34–36

C
Cai, J., 1–12, 17–41, 45–58, 293–306,
        409, 410
Campbell, D., 316
Carpenter, T.P., 285
CCSSM. *See* Common Core State Standards
        for Mathematics (CCSSM)
Chávez, O., 77
Chazan, D., 182–183
Chinese elementary students
    academic interest, 249–250, 253–254
    computation errors, 250–252
    metacognition, effects, 254
    qualitative measurement, 243, 250
    quantitative measurement
        decimal computation, 249
        total computation, 247
    self-efficacy, 249, 253–254
Class.com's Algebra I course,
        101, 116–119, 129
Classroom Educational System, 377–378
Classroom learning environment (CLE),
        77, 88
    generalizability coefficient, 367
    project team members, 364
    purpose, 364
    variance decomposition, 366
    variation source, 364–365
Clements, M., 8, 95–130, 408
CMP. *see* Connected Mathematics Program
        (CMP)

Cogan, L.S., 322–323, 325, 331
Collegial learning effectiveness
  fractions knowledge, 142
  HLM, 151
  mathematics teaching, 138
  student achievement, 142
Common Core State Standards (CCSS)
  achievement measures, 23
  buy-in effects, 34–37, 39
  Comparison schools, 27–29
  Connected Mathematics/Mathematics in
    Context, 39, 41
  construct databases, 400
  expressions and equations, 197
  fidelity of implementation (*see* Curriculum
    implementation)
  instructional core, 377
  "lethal mutations", 20
  mathematics and literacy, 17
  mutual adaptation/co-construction
    perspective, 20
  NAEP assessment, 203
  Next Generation Science Standards, 41
  NSF-funded curricula, 18, 39
  overall treatment effects, 33–34
  productive adaptations, GWP, 20
  statistical model, 30–33
Common Core State Standards for
    Mathematics (CCSSM)
  adoption and implementation, 95–96, 129
  K-7 students, 99–100
  online courses, 129
Comparing Options in Secondary
    Mathematics: Investigating
    Curriculum (COSMIC) Project
  classroom learning environment, 86
  Classroom Visit Protocol, 381
  Common Core State Standards, 383
  Core-Plus Mathematics Program, 75–76
  cross-sectional analyses of student
    learning, 76–77
  curricular effects, 383
  curricular implementation, complexity,
    384, 388
  disaggregated data, 384
  ETI Index, 381
  factor analysis, 387
  fair test, 380
  free and reduced lunch (FRL), 380
  goal, 379
  hierarchical linear models (HLM), 76
  integrated curriculum, 75, 383
  integrated mathematics, 384
  IRT, 76

ITED scores, 76–78
  Mathematics-Science Partnership, 385
  multiple outcome measures, 380
  NAEP scores, 79–81
  National Science Foundation, 90
  opportunity to learn (OTL), 76, 380, 383
  participating schools, 380
  practitioners and policy makers, 384
  research questions, 379
  subject-specific curricula, 75
  teacher, classroom and student data, 382
  teacher knowledge, 384
  Textbook Content Taught (TCT), 381
  treatment integrity, 381
  variation in coverage, 387
Conceptual change approach
  computation errors, 257
  in decimals, 253–256
  misconceptions, 236–237
  PBL approach, 255
  qualitative measurement, 241–242, 250
  quantitative measurement
    decimal computation, 249
    total computation, 247
Confrey, J., 6, 23, 410
Connected Mathematics Program (CMP)
  definition, 46
  HLM, 298
  LieCal Project, 295
  solution strategies, 300
  student performance, 299
Connell, R.W., 335
Constructs
  extrinsic material gain motivation, 344
  final model, 343
  mathematics-specific, 337
  motivation and support for learning, 342
  multi-level analysis, 340–342
  perception of mathematics lessons,
    344, 355
  perception of mathematics teachers, 344
  perceptions of teachers, 348, 351–353
  personality traits, 342
  'self-concept', 344
  social support in mathematics learning, 343
Construct validity
  confounding, 325–327
  experimenter expectancies, 328–329
  inadequate explication, 325
  and internal validity, 326
  large-scale tests, 327
  levels of constructs, 327
  mathematics and science teaching, 326
  NAEP Validity Studies (NVS) panel, 325

Construct validity (*cont.*)
    reactive self-report changes, 328
    resentful demoralization, 329
    teacher effectiveness, 326
    TIMSS, 326–327
Cook, T., 316
Cooper, H., 210, 212
Correnti, R., 369
COSMIC Prior Achievement (CPA) Score,
        79–81
COSMIC Project. *See* Comparing Options in
        Secondary Mathematics:
        Investigating Curriculum
        (COSMIC) Project
Council of Chief State School Officers and
        National Governor's Association
        (CCSSO/NGA), 17
Cross-multiplication (CM) strategy
    definition, 62–63
    Rational Number Project, 71–72
    solving proportion problems, 63–64, 70
    strategy coding, 67–68
Cross-sectional analyses of student learning
    Algebra 1/Integrated I, 76–77
    Algebra 2/Integrated III, 77
    Geometry/Integrated II, 77
Curricular effectiveness
    COSMIC Project, 75–78 (*see also*
        Comparing Options in Secondary
        Mathematics: Investigating
        Curriculum (COSMIC))
    curriculum matters, 378
    educational structures and
        organizations, 397
    factors in implementation, 396
    first-and second-grade math learning,
        385–390
    goal, 394
    impacts on implementation quality,
        390–393
    monitoring, 395–396
    outcome measures matter, 395
    program
        implementation, 378–379
        outcomes, 379
        theory, 378–379
    sensitive outcome measures, 379
    student data, collection and analysis
        FRL-status, 83
        migrated across curricular paths, 83–84
        participated fewer than 3 years, 85–86
        prior achievement, lack of measure,
            78–81
        prior achievement measure, 81–82

student outcomes, modeling
    curriculum type and implementation,
        87–88
    ITED-16 scores, 88–90
    Principal Components Analysis
        (PCA), 88
    "reduced" models, 88
teacher capacity and professional
    development, 396–397
teacher data, collection and analysis
    classroom learning environment, 86
    didactic approaches, 87
    Opportunity to Learn (OTL) Index, 87
    "reform-oriented practices", 86
    self-efficacy, 87
    Textbook Content Taught (TCT)
        Index, 87
Curricular effects
    algebra learning, 294
    classroom practice, 293
    LieCal Project, 294
    reform-orientation, 293
Curricular sensitivity, 390, 398
"Curriculum alignment with systemic
        factors", 23
Curriculum implementation
    fidelity of
        buy-in concept, 20–22
        definition, 19
        GPSMP effects, 22
        mutual adaptation/co-construction
            perspective, 20
        teacher/curriculum interactions, 23
        "Will to Reform", 22–23
    LSC theory of action, 18, 22
    NCTM standards era, 18
    No Child Left Behind Act in 2001, 18
    NSF-funded math curriculum, 18, 27
    "productive adaptations", GWP, 20
    quasi-experimental intent-to-treat
        analysis, 18
    school-level standard deviations, 28, 35–36
    teacher and curriculum materials, 40
    Treatment classrooms
        fidelity to process, 19
        fidelity to structure, 19
    Will-to-Reform (*see* Will-to-Reform scale)

D
Dai, S., 179–205
D'Ambrosio, B.S., 183
Data collection
    attrition, internal validity, 323

classroom instruction, 139
collaborative process, 314–315
COSMIC, 7
curriculum (*see* Curricular effectiveness)
demographic information, 139, 140
ELS:2002, 312–313
NAEP, 200, 313–314
post and pre-assessments, 139
time allocation, 140
TIMSS, 314–315
treatment fidelity, 247
video records, 140
Decimals numbers, Chinese elementary
  students. *See also* Chinese
  elementary students
  academic interest, 249–250
  assessment conditions, 245
  coding and scoring
    decimal computation test, 242–244
    inter-rater agreement, 243
    procedural understanding, qualitative
      measure, 243
    self-efficacy survey, 243
  conceptual change (*see* Conceptual change
    approach)
  control group, 246
  dependent measurement, 241–242
  design, 240
  experimental group, 245–246
  interventions for teaching, 237–238
  participants and setting, 240–241
  PBL (*see* Problem-based
    learning (PBL))
  pretreatment group equivalency, 247
  sample problems from curriculum,
    259–260
  teacher training, 245
  teaching scripts
    new decimal division, 257–258
    reviewing previous contents, 258
  treatment fidelity, 247, 259
  and whole number, 252–253
Design-Educational Engineering and
  Development (DEED), 374
Design study
  contamination, 135
  demographic triads, 135
  educational reform, 134
  experimental treatment, 135
  fractions resource kit, 135, 136
  linear measurement, 137
  materials, resource kit, 135
  mathematics resources, 134, 135
  professional development, 134, 135
  SES, 134–135

Ding, Y., 235–260
Dupuis, D.N., 2, 61–72, 407

E
Early Childhood Longitudinal Study (ECLS),
  96, 386, 389, 394
Ebby, C.B., 10, 159–174, 412
Edelson, D.C., 20
Education Longitudinal Study of 2002
  (ELS:2002)
  data collection, 312
  high school transcripts, 312
  item response theory (IRT) modeling,
    312–313
  mathematics achievement, 312
  mathematics test, 312
Education systems, 217, 218, 222, 230
Ellis, A.B., 183
Empson, S.B., 285
The End of Homework, 211
Engagement
  decision-making processes, 345
  encourage and reinforce, 347–348
  encouragement, 346–347
  long-term support, 346–347
  subject choice, 345
  supportive relationship, 345
  teacher and class environment, 348–351
Engineering for effectiveness
  Common Core State Standards, 400
  components, 376
  curricular monitoring and adherence, 398
  DEED, 374
  definition, 376–377
  feedback and interaction, 375
  instructional core, 397–400
  networked improvement communities, 400
  organizational system, 375–376
  programs identification, 373
  real-time observational data, 398
  surveys, 398
  variability, bands and pockets, 375
Equivalent fractions (EF) strategy
  definition, 63
  and proportion problems, 70–72
  Rational Number Project, 71–72
  strategy coding, 67–68
Extent of Textbook Implementation (ETI)
  Index, 381
External validity
  causal relationship with outcomes,
    interaction, 329
  generalizability, 329
  TIMSS results, 330

F
Face-to-face learning, 101–102, 116, 130
Fidelity of Implementation
     (FIDELITY), 88
Formative assessment
   instructional response, 160
   interpretive process, 159
   mathematics educational research, 160
   MKT, 161
   PCK, 160–161
   student learning, 161
   TASK, 161
Fractions knowledge
   linear measurement, 135, 137
   mathematical resource kit, 135
   rational numbers, 137
   resources, 134
   student assessments, 139
Free and reduced lunch (FRL), 78, 380
   curriculum, 88
   OTL, 383
   partial correlations, 382
   race/ethnicity, 83
   SES, 83

G
Global Warming Project (GWP), 20
Goertz, M., 173
Gould, S.J., 375
Greater Philadelphia Secondary Mathematics
        Project (GPSMP)
   CMP/MiC, 30
   Comparison schools, 27
   LSC theory, 22
   mathematics mentor(s), 24
   NSF-funded middle school, 17–18
   Principal Investigator, 24, 26
   Will-to-Reform point, 35, 36
Grouws, D.A., 76, 77

H
Hallam, S., 212
Hall, L.D., 179–205
Hart, L.C., 412
Harwell, M.R., 77
Heck, R.H., 4
Helding, B., 265–287
Heppen, J., 8
Heppen, J.B., 8, 11, 95–130, 408
Herrig, R.W., 209
Hiebert, J., 237

Hierarchical linear models (HLM),
        2, 76
   experimental treatment, 142
   fractions knowledge, 141
   mathematical resource kits, 142, 146
   student-level and classroom-level, 55–56
   student-level and curriculum cross-
        sectional, 53–55
   students knowledge, 142, 149–150
   teachers knowledge, 142, 147–148
High School Transcript Study (HSTS), 180
Hill, H.C., 161
Hmelo-Silver, C.E., 239
Holliday, B.W., 321, 326
Holliday, W.G., 321, 326
Homework
   education movement, 209
   education systems, 212
   The End of Homework, 211
   grade-level, 210
   guidance and feedback, 229
   in-class activities, 223–226
   levels
        attained curriculum, 213, 232
        implemented curriculum, 229
        intended curriculum, 229, 232
   mathematics classes frequency,
        215–217
   methodological limitations, 212, 229
   parents involvement
        after-school time, 227
        monitoring activities, 228
   PISA, 211
   system-level policy, 214–215
   task types (*see* task (homework) types)
   time, estimated
        Chinese Taipei, 218
        least time-consuming, 219
        5-point Likert scale, 218
   TIMSS investigations, 212–213
Hong, H.K., 322, 325, 331
Horn, L., 99
Huang, H.-C., 179–205
Huang, T.-H., 238
Hwang, S., 1–12, 45–58, 293–306, 410

I
In-class activities
   cross-system, 225
   higher grade levels, 224
   Kruskal–Wallis tests, 225
   monitoring, 224, 226

3-point Likert scale, 223–224
    self-correction, 224
Individualized Educational Program (IEP), 77
Ing, M., 8
Instructional core, intervening, 377–378
Instructional decision making (IDM)
    intervention and responses, 162
    large-scale fields, 168
    LTO, 170
Instructional improvement
    curricular effectiveness, 378–379
    engineering for effectiveness, 373–376
    instructional core, intervening, 377–378
Instructional response
    IDM, 162
    respondents, 165
    student thinking, 159
    TASK, 161
Instrument development, 173
Integrated curriculum
    Algebra 1/Integrated I, 76–77, 84–85
    Algebra 2/Integrated III, 77, 84, 86
    Core-Plus Mathematics Program, 75–76
    COSMIC Project, 75–76
    Geometry/Integrated II, 77, 84
    ITED-15 scores, 77–78
    participating high schools, 83
Intermediate outcomes, 132, 409
Internal validity
    attrition, 323
    conduct of study, 320–321
    instrumentation, 324
    randomization into experimental
        units, 321
    regression, 323
    selection, 321–322
    students, multi-year age differences, 322
    systematic design, 320–321
    TIMSS, 322–323
International Association for the Evaluation of
        Educational Achievement, 271
International Database Analyzer (IDB)
        software, 316
Interviews, teachers via qualitative work,
        345–351
Iowa Tests of Educational Development
        Problem Solving and Concepts,
        Level 15 (ITED-15), 76–77
Item level analysis, 202, 345, 351, 358, 360
Item Response Theory (IRT), 50, 76, 312
    COSMIC Project, 76
    mathematics items, 212
    NAEP, 182

psychometric issues, 204
student assessment data, 50

J
Jackknife repeated replication (JRR)
        technique, 316
Jitendra, A.E., 407
Jitendra, A.K., 2, 61–72
Joint maximum likelihood (JML), 204

K
Kanchanawasi, S., 212
Kastberg, S.E., 183
Kaufman, J.H., 392–393
Kennedy, M.M., 23
Kieran, C., 183, 198
Kim, A., 265–287
Kirk, P.J. Jr., 224
Klangphahol, K., 212
Kloosterman, P., 4, 9, 179–205, 329, 406, 409,
        411, 413
Knowledge of Standards (STANDARDS), 88
Koellner-Clark, K., 131
Kohn, A., 211
Köller, O., 212
K-12 public school students, 100–101, 108
Kramer, S., 5, 9, 12, 17–41, 409
Krutetskii, V.A., 294
K-7 students, 96, 100

L
Lai, E.R., 5
Lakin, J.M., 5
Lakoff, G., 324
Lambdin, D.V., 183
Lane, S., 304
Lange, J.D., 325
Large-scale study
    AERA, 412
    benefits
        situations and trends, 406–407
        sophisticated analytic methods
            employment, 408–409
        testing hypotheses, 407–408
    IES and NSF Common Guidelines, 405, 412
    pitfalls
        authentic research settings,
            complexity, 411
        methodology, 411–412
        resources and time, 410–411

Lauff, E., 323
Learning trajectory orientation (LTO)
  AST, 168–169, 171
  children's thinking, 10
  correlation matrix, 172, 173
  IDM, 169–170
  instructional decisions, 159
  instrument assessments, 159
  instrument properties, 170
  large-scale fields, 168–170
  mathematics instructions, 174
  qualitative studies, 173
  SEM, 172
  sophistication, 169
  statistical modeling, 172
  student achievement, 159
  student ranking, 169
  TASK, 161–169
  teachers interpretation, 173
Lein, A.E., 2, 61–72, 407
Lennex, L., 326
Lesson study
  administrations, 152
  classroom practice, 131
  collaboration, colleagues, 134
  collegial learning effectiveness, 142, 151
  conditional model, 141
  end-of-cycle reflections, 151
  experimental conditions, 151
  fractions knowledge, 138–139
  HLM, 141
  instructional improvement, 153
  learning frequency, 152
  linear measure models, 152
  live practice observation, 132
  mathematical resources, 131
  outcome measures, 141, 151
  ownership, professional learning, 153
  practice-based learning, 132
  professional development program, 131
  student achievement, 151
  teacher inquiry, 134
  validity evidence, 152
Levi, L., 285
Lewis, C.C., 4, 6, 8, 12, 131–155, 407
Linn, R.L., 326
Liu, M., 238
Liu, R.-D., 235–260
Lobato, J., 183
Longitudinal analysis
  COSMIC Project, 76–77
  integrated/subject-specific mathematics
    program, 85
  ITED scores, 77

Longitudinal Investigation of the Effect of
    Curriculum on Algebra Learning
    (LieCal) Project, 46, 408
  aspects, 46
  CMP and non-CMP curricula, 46, 295
  conceptual model, 303
  designs and data analyses, 294
  HLM, 298
  mathematical thinking, 295
Longitudinal study. *See also* Education
    Longitudinal Study of 2002
    (ELS:2002)
  arithmetic operations, 285
  classroom observations, 271
  coded across, 270–271
  conceptual knowledge, 266
  construct validity, 324–329
  cross-sectional studies at different
    ages, 266
  curricular effect, 293–294
  Elias' flexible unitization, 278
  experimental studies, 306
  external validity, 329–330
  fraction problems, 285
  individual students; development,
    267–268
  Inez's equals sign, "colon" ratio symbol,
    280, 281
  Inez's use of unit ratio, 279, 280
  intensive interview and observational
    methods, 286–287
  internal validity, 320–324
  interviews, 269–271
  mathematics achievement, 293, 296
  mathematics achievement over time,
    275–277
  mathematics of fractions and whole
    numbers, 266
  middle-school children, 266
  National/International Sample
    NAEP and TIMSS items, 273
    Part/Whole fraction instruction, 274
    test administration broken out by
      gender, 273, 274
  National Science Foundation, 267
  Part-Whole and unitization, 278–279
  Part-Whole conceptualizations, 284
  vs. performance at different grade
    levels, 274
  powerful iterative methods, 285
  rational number performance (*see* Rational
    numbers)
  ratio subconstruct, 279–281
  reporting changes, 296–302

research, 265
researchers, 293
robust-but-inefficient conceptual
    strategies, 283
setting and participants, 268–269
socioeconomic factors, 305
statistical validity, 318–320
strategy type leading
    non-sensible answers across, 282, 283
    sensible answers across, 282
    technically correct answers across, 282
    technically incorrect answers
        across, 283
student learning, 293–296
students' strategies, 281
subconstructs, 277
teaching, 286
trial-and-error and means-end solution
    methods, 282
in the US children, 284
Long-Term Trend (LTT)
    mathematics programs, 180
    NAEP, 180
    subject-matter, 313
Louisiana Algebra I Online Project, 101–102
Loveless, T., 322
Lubienski, S.T., 48

M
MacGregor, M., 183
Maine Learning Technology Initiative, 107
Maloney, A., 6, 410
Martinez, J.F., 369
Martin, W.G., 182
Mathematical knowledge for teaching (MKT),
    391–392
    content knowledge, 161
    formative assessment, 174
    instrument development, 10
    statistical associations, 170
    TASK, 170
Mathematical Quality of Instruction (MQI)
    classroom mathematics instruction, 365
    and CLE protocols, 368–369
    dimensions, 368
    instrument, 365
    lessons, 365–366
    raters, 365
    SPMMR, 365
    teacher errors and imprecision, 365
Mathematical thinking
    conceptual model
        CMP, 303

cognitive learning outcomes, 304
covariance matrix, 302
curriculum study, 302
HLM, 302–303
instructional tasks, 303
LieCal Project, 302
curricular studies, 302
student equivalence, 302–303
Mathematical validity (MV)
    and Content Knowledge, 170
    IDM, 173
    learning trajectory-oriented formative
        assessment, 162
    student work, 171, 173
Mathematics
    advanced mathematics course sequence,
        99, 102, 104
    Chinese Taipei, 216
    Class.com, 117–118
    eligible students' outcomes, 128
    formal Algebra I course, 129
    grade 9, course sequence, 127
    grade 8 scores, online Algebra I, 126, 127
    Likert scale, 216
    non-eligible students, 128
    posttests, 113–114
    ranks test, 217
    in treatment and control schools, 112–113
Mathematics coursetaking
    advanced course sequence, 115, 124
    Bernoulli sampling distribution, 124
    course titles and grades, 114
    creation of indicators, 114
    eligible students, 115, 124
    non-eligible students, 114
    outcome measures, 113–115
    "ready for algebra", 115
    study design, 103–105
    transcript coding protocols, 114
Mathematics education
    Algebra, 1
    characteristics
        lack of clarity, 9
        mechanical factors, 9
    choices and achievement, 336
    complexity of data analysis
        error variation, 8
        observational methods, 8
    complexity of the measure, 10–11
    constructs, 337
    curriculum, 1
    demographic groups, 4
    departments, 365
    'emotional response to lessons', 353–355

Mathematics education (*cont.*)
  equilibrate, 2
  error of measurement
    algebra-specific content, 9
    proportional concepts, 10
  ethnicity, 4
  gender, 335
  generalizability and transportability,
      standard errors, 7
  instructional core, 6
  large-scale observational measures,
      363–364
  large scale, taxonomy, 3
  level of data analysis, 11–12
  Likert scale, 337
  mathematics lessons, participation and
      perception, 355
  MLM (*see* Multi-level modelling (MLM))
  overall mean response, 337
  perceptions and intention to participate, 357
  'perceptions of lessons', 353–355
  quasi-experiment, 5
  scale-up, 3
  SimCalc, 2
  small-scale studies, 5
  socioeconomic, 1
  student questionnaires, 336, 337
  students' perceptions, 337–339,
      351–352, 356
  target subpopulation, 4
  teacher logs, 2
  teachers, participation and perception,
      352–353
Mathematics in Context (MiC), 22
Mathematics learning
  algebraic reasoning, 197
  COSMIC, 76
  fractions knowledge, 138
  HLM, 151
  homework practice (*see* Homework)
  large-scale studies, 332
  longitudinal (*see* Longitudinal study)
  professional learning, 134
  psychology, 198
  sense of efficacy, 137
  social support, 343
Math instruction, 118, 120, 122, 127, 168
Mayer, R.E., 49, 295
McLaughlin, T., 19
Measurement, classroom observation tools
  analysis, 366
  documenting, 363
  external observers, 363
  generalizability theory, 364, 366

lessons (CLE), 366–367
lessons and raters (MQI), 367–369
protocols, 364–366
teachers and coding information, 363
Measuring changes in student learning
  algorithmic knowledge, 295
  CMP, 295
  cognition and affection, 294
  LieCal Project, 295
  longitudinal analyses, 296
  mathematical thinking, 294
  procedural knowledge, 295
  propensity, 294
  reform-orientation, 296
Megowan-Romanowicz, C., 265–287
Mendick, H., 335–336
Merlino, F.J., 5, 9, 17–41
Merlino, J., 409
Middle schools
  advanced coursetaking sequences, 100
  Algebra I in eighth grade, 95–98
  eligible students, 122
  mathematics courses, 99
  non-eligible students, 126
  rural, 97
  students
    adoption of CCSSM, 96
    advanced coursetaking sequence, 99
    algebra-ready students, 102–103
    higher math skills, 99
Middleton, J., 411
Middleton, J.A., 1–12, 265–287, 410
Mittag, K.C., 326
Mji, A., 325, 329, 330
Mohr, D., 179–205
Moyer, J.C., 4, 45–58
MQI. *See* Mathematical Quality of Instruction
      (MQI)
Mujtaba, T., 9, 10, 335–361, 409, 410
Multi-level modelling (MLM)
  aspirations towards learning, 344
  extrinsic material gain motivation, 344
  final model, 343
  general motivations, 344
  inclusion of constructs, 342–343
  initial construct-based, 340
  learning, motivation and support, 342
  mathematics-specific measures, 343
  personality traits, 342
  procedures, 340
  qualitative and quantitative work, 345
  student background characteristics,
      340–342
  teacher-student relationship, 344

Multiple strategies on students' learning
    intervention, 65
    measures, 65–67
    participants, 64–65
    at pretest, 68–70
    proportional reasoning, 62–64, 71
    strategy coding, 67–68
Munoz, R., 183

N
NAEP Data Explorer, 182
National Assessment Governing Board
    (NAGB), 181, 184, 203, 313
National Assessment of Educational Progress
    (NAEP), 61, 79–81
    algebraic reasoning, 182
    CCSS, 203
    computer administration, 327
    confidence intervals, 180
    constructed response items, 181
    construct validity, 325
    COSMIC Project, 79–81
    curriculum developers, 203
    data collection, 199, 314
    data explorer, 182
    demographic variables, 202
    distractors, 199
    economics, 180
    equal difficulty, 315
    features, 313
    forms, 313
    governing board, 203
    IRT, 182
    item response codes, 202
    items, mathematical achievement
        questions, 313–314
    JML, 204
    LTT, 180
    measurement error, 9
    multiple-choice items, 181
    NAGB, 181
    National/International Sample, 273–274
    problem solving and reasoning, 314
    psychometric analysis, 204–205
    psychometric community, 202
    rational number, 273
    researchers, 179
    scoring, 181–182
    secure data access
        LTT, 200
        NCES, 200
        NCLB, 200
        software license, 200

    state-level reports, 313
    statistical software, 201
    student performance, 182
    students representative samples, 313
    subject-matter, 313
    trend analyses, 181
    the US government program, 179
    validity, 317
    verbatim report, 199
National Center for Education Statistics
    (NCES), 100–101, 413
    data collection, 200
    ECLS, 386
    FRL, 83
    measurement errors, 9
    policy and research, 317
    scale scores, 182
National Council of Teachers of Mathematics
    (NCTM), 17, 45, 235
National Educational Longitudinal
    Study, 98
National School Lunch program, 110
National Science Foundation (NSF), 17
NCES. *See* National Center for Education
    Statistics (NCES)
Ndlovu, M., 325, 329, 330
The New Jersey Grade Eight Proficiency
    Assessment (GEPA), 23
Nie, B., 45–58
Ni, Y., 293–306
No Child Left Behind (NCLB) Act, 18, 180,
    200, 377
Noyes, A., 336
NSF-funded Local Systemic Change
    (LSC), 409
Nuñez, A.M., 99

O
O'Donnell, C.L., 19
O'Dwyer, L.M., 102
Oksuz, C., 265–287
Oláh, L., 173
Online learning. *See also* Algebra I for eighth
    graders, online course
    course activity
        student log-ins, 119
        teacher activities, 119
        teacher-student communications,
            119–120
    course completion rates, 121, 122
    course content
        Class.com course, 116
        in control schools, 121–122

Online learning. *See also* Algebra I for eighth
        graders, online course (*cont.*)
    course material and instruction
        activities, 116–117
    course effectiveness
        Louisiana Algebra I Online Project,
            101–102
        students' access, 101
    courses to expand offerings, 100–101
Opportunity to learn (OTL), 76, 88, 380, 383
    SES, 384
    students learning, 71
    teacher factor scores, 77
Orletsky, D., 4, 11, 410, 411

P
Palmer, A., 336
Participation, intended
    construct-based multi-level analysis,
        359, 361
    encouragement, 360
    extrinsic material gain motivation, 358
    gender differences, 352, 353
    item-based analysis, 358–359
    mathematics-specific measures, 361
    MLM procedures (*see* Multi-level
        modelling (MLM))
    `perception of teachers' construct, 360
    and perceptions of mathematics lessons, 355
    personality traits, 342
    quantitative and qualitative work, 361
    'self-concept' construct, 361
    student questionnaires, 337
PBL. *See* Problem-based learning (PBL)
Pedagogical content knowledge (PCK), 160,
        161, 170, 174
Pennsylvania System of State Assessment
        (PSSA) test, 23, 30, 36
Pérez, A., 179–205
Perry, R.R., 4, 6, 12, 131–155, 407
Pierce, R.U., 238
PISA. *See* Program for international student
        assessment (PISA)
Postlethwaite, T.N., 3, 6
Principal buy-in, 21, 37, 38, 41
Prior research
    learners and teachers, 138
    online courses, 101–102
    professional development, 138
Problem-based learning (PBL), 258, 301
    computation errors, 240, 256–257
    and computation skills, 253
    conceptual change, 255

control group, 243–244
experimental group, 245–246
and self-efficacy, 235, 239–240
students' metacognition, development, 254
teacher training, 245
Professional development (PD)
    COSMIC, 75
    instructional core, 377
    prior research, 138
    TIMSS, 332
    transportability, 7–8
Program for international student assessment
        (PISA), 211
Project team members, 364
Proportional reasoning, 266, 267, 270, 279,
        281. *See also* Multiple strategies on
        students' learning
    advanced mathematics, 61
    CM strategy (*see* Cross-multiplication
        (CM) strategy)
    definition, 62
    EF strategy (*see* Equivalent fractions (EF)
        strategy)
    multiple representations, 62
    Rational Number Project, 62
    STEM curricula, 62
    UR strategy (*see* Unit rate (UR) strategy)

Q
Qin, X., 241
Qualitative research, 24, 153, 351
QUASAR Project, 50

R
Rasch, G., 204
Raters, 365, 367–369
Rational numbers
    assessment, 271
    collecting performance data, 271
    International Association for the
        Evaluation of Educational
        Achievement, 271
    National Assessment Educational
        Progress, 271
    and proportional reasoning, 267
    students' original responses, 273
    test item information, 271–273
    Trends in International Mathematics and
        Science Study, 271
Reiss, M.J., 9, 335–361, 409
Reporting changes
    classroom discourse, 297

CMP, 297
grade band
  CMP, 301
  covariance, 302
  LieCal Project, 301
  mathematics curriculum, 301
  PBL, 301
  problem-solving performance, 302
growth rates, mathematics, 296
HLM, 297
LieCal, 297
open-ended problems, 298
outcome measures, 298
quality
  abstract strategies, 300
  algebraic representation, 300
  CMP, 299–300
  concrete strategy, 300
  generalized strategy, 301
  open-ended assessment, 299, 300
  student performance, 299
reform and non-reform, 298
SES, 297
students achievement, 296
study design, 296
traditional regression analysis, 298
Resources and time, large-scale study,
    410–411
Riggan, M., 173
Roach, M., 179–205
Robinson, R., 238
Rodd, M., 9, 335–361, 409
Roschelle, J., 4
Rose, H., 99
Rubin, D., 28
Rural schools
  broadening access to Algebra I in
      grade 8, 97
  curriculum offerings, 96
  ECLS-K data, 96
  participating schools, 108–109
  randomized controlled trial, 97
  recruitment
    access expansion, 107
    Maine Learning Technology Initiative,
        107
    rural schools in Vermont, 107
    students participation, 108–109
    study requirements, 107
  students in participating schools
    eligible (algebra-ready) students, 109
    Maine and Vermont state
        assessment, 109

non-eligible (non-algebra-ready)
    students, 109
student characteristics, 110–111

S
Sampling issues
  cluster sampling, 315
  data units, 315
  ELS:2002, 315–316
  IDB software, 316
  TIMSS, 315–316
Schilling, S.G., 161
Schmidt, W.H., 322–323, 325, 331
Schneider, B., 114, 115
Schneider, M., 285
Schoenfeld, A.H., 296
School support
  Baseline Score, 30–31
  demographic characteristics, 28, 29
  error terms, 31
  GPSMP school districts, 27
  Hierarchical Linear Models (HLMs), 30
  Math Test Score, 30
  NSF-funded curriculum, main effects, 30
  parameters, definitions, 32–33
  "Priority One" matches, 28
  "Satterthwaite" formula, 30
  Treatment and Comparison groups, 27–29
  Will-to-Reform score, 28, 31–32
Schorr, R.Y., 131
Secondary mathematics. See also Comparing
        Options in Secondary Mathematics:
        Investigating Curriculum
        (COSMIC) Project
  curricular effectiveness, 90
  measures of student achievement, 76
Self-efficacy
  and academic interest, 241, 249, 253–254
  in mathematics computation, 236
  PBL, 239–240
  questionnaire, 247
  scores, 239
  social, 241, 249, 253–254
  students, 249
  survey, 243
Seltiz, C., 9
SEM. See Structural equation modeling (SEM)
Shadish, W., 316, 324
Shih, J., 4, 45–58, 412
Shih, J.C., 8
Shiu, C.-Y., 238
Siegler, R.S., 285

Simon, S., 9, 335–361, 409
Single-subject *vs.* integrated mathematics,
     379–385
Sirinides, P.M., 10, 159–174, 412
Slavin, R.E., 39, 40
Sloan Consortium surveys, 101
Sloane, F., 4, 410
Smith, J.B., 99
Snyder, P., 19
Socioeconomic status (SES)
     FRL, 82
     research design, 83
     subpopulations, 4
Solomon, Y., 335–336
Solution strategies
     non-sensible answers across, 282, 283
     sensible answers across, 282
     students' strategies, 281
     technically correct answers across, 282
     technically incorrect answers
          across, 283
Sophisticated analytic methods
     advanced statistical techniques, 408
     algebra-related items, 409
     NAEP algebra scales, 409
     NSF-funded Local Systemic Change
          (LSC), 409
     quantitative methods of analysis, 409
     statistical power, 409
Soria, V., 7, 75–90, 411
Sowder, J.T., 182
Spielhagen, F.R., 99
SPMMR. *See* Student participation in
          mathematical meaning-making and
          reasoning (SPMMR)
Stacey, K., 183, 238
Standards-based curriculum
     curriculum materials, 17
     NCTM Standards, 17, 18
Star, J.R., 2, 61–72, 407
Statistical validity
     extraneous variance in experimental
          settings, 320
     fishing and error rate problem, 319
     heterogeneity of units, 320
     restriction of range, 319
     statistical tests, violated assumptions, 318
     treatment implementation unreliability,
          319–320
     unreliability of measures, 319
Stedman, L.C., 330
Steinle, V., 238
Stein, M.K., 304, 376, 392–393
Stevenson, D.L., 98, 114

Strategy use
     CM strategy, 68
     EF strategy, 68–69
     and posttest mean scores, 69, 70
     at pretest, 66, 68–69
     proportion, students' learning, 62
     UR strategy, 69
Structural equation modeling (SEM), 10, 172
Strutchens, M., 182
Student learning
     curricular effects, 293–294
     measuring changes, 294
Student participation in mathematical
          meaning-making and reasoning
          (SPMMR), 365, 368
Student performance
     achievements, 18–19, 22–23
     curriculum materials, 23
     educational innovations, 18
     implementation fidelity, 40
     NAEP
          algebraic reasoning, 197
          CCSS, 197
          error patterns, mathematics, 198
          informal settings, 197
          learning theories, 197
          multiple-choice item, 197
          sense of mathematics, 198
          strengths and weaknesses, 197, 198
          test developers, 198
     school organization, 28
     treatment schools, 34
Student thinking
     evidence interpretation, 161
     formative assessment, 159
     instructional improvement, 137
     instructional practice, 161
     lesson study, 134
Swan, M.B., 238
System-level homework policy
     Chi-square test, 214
     feedback, 215
     in Hong Kong SAR and Singapore, 214–215
     Korean policy, 215
     primary responsible parties, 214
     TIMSS 1995/1999, 214
     TIMSS 2007/2011, 215
Sztajn, P., 161

T
Tarr, J.E., 7, 8, 48, 75–90, 326, 411
Task (homework) types
     communication skills, 221

east asian systems, 221
Friedman tests, 223
Mann–Whitney *U* tests, 223
3-point Likert scale, 221
western systems, 221
within-system analyses, 221
Teacher beliefs
   COSMIC, 86
   learning community, 137–138
   reform-oriented practices, 86
Teacher buy-in
   curriculum, 5
   evaluation model, implementation
      fidelity, 21
   GPSMP mentors, 24
   high buy-in, 25
   low buy-in, 24
   medium buy-in, 24
   Principal Support, 37, 38
   SAS proc mixed solution, 38
Teacher community, 141, 398
Teacher knowledge
   formative assessment, 174
   LTO, 174
   pre-post design, 2
   proportional reasoning, 62
   TASK, 161
Teacher learning
   beliefs and dispositions, 137
   collegial learning effectiveness, 155
   HLM, 142, 146
   instructional improvement, 137
   intervention scores, 142
   prior research, 137–138
   professional community, 155
   research relevance, 154–155
   resource kits, 138
   sense of efficacy, 137
   student achievement, 137, 142, 153
   time and conditions, 142, 143
Teacher professional development, 7, 12,
      22, 155
Teachers assessment of student knowledge
      (TASK)
   algebraic reasoning, 162–163
   children's learning, 164
   concept and content knowledge, 162
   formative assessment, 161
   fractional quantities, 164
   instructional response, 161
   MKT, 161
   multiplicative reasoning, 162
   online instruments, 165
   performance assessment, 162

   pilot administration, 167
   ranking, 165
   recruitment, 167
   respondents, 165
   scoring rubrics, 166–167
   student responses, 163
Technology and Collaboration (TECH and
      COLLAB), 88
Textbook Content Taught (TCT), 381
Thomas, S.L., 4
Thomas, S.M., 384
Thompson, C.A., 285
Traiwichitkhun, D., 212
Trautwein, U., 212
Trends in International Mathematics and
      Science Study (TIMSS)
   cognitive abilities and content knowledge,
      314–315
   construct validity, 326–327
   distractors, 315
   encyclopedia, 231
   homework, 212–213
   internal validity, 322–323
   item design, 315
   language, 315
   mathematics and science achievement,
      314–315
   multiple choice answers, 315
   National/International Sample, 273–274
   Professional development, 332
   surveys, 213
Turnbull, B., 20

U
Uekawa, K., 326
Unit rate (UR) strategy
   definition, 63
   solving proportion problems, 66–67, 70
   strategy coding, 67–68

V
Validity
   construct, 324–329
   critiques types, 318
   external, 329–330
   IEA, 318
   independent variable, 318
   internal, 320–324
   NAEP mission, 317
   policy and research issues, 317
   statistical, 318–320
   student progress over time, 318

VanDerHeyden, A., 19,
    21, 40
Vatterott, C., 212
Vig, R., 2, 61–72, 407
Vosniadou, S., 254

W
Walcott, C., 179–205
Walters, K., 8, 95–130, 408
Wang, J., 325, 327, 332
Wang, N., 4, 45–58, 298
Ward, M.E., 224
Wearne, D., 182, 237
Widjaja, W., 238
Will-to-Reform scale
    district coherence, 25–26
    GPSMP mentors, 24
    LSC model, 22–23
    principal support:, 25

subcomponents
    correlations, 38, 39
    principal buy-in effects, 37, 38
    teacher buy-in effects, 37, 38
subscale scores, 26–27
superintendent support, 26
teacher buy-in, 24–25, 39
Woodward, J., 238

Y
Yang, Y., 265–287
Yanik, B., 265–287

Z
Zhang, A., 241
Zhang, D., 235–260
Zhu, Y., 5, 11, 209–232, 407, 411, 413
Zong, M., 235–260

CPSIA information can be obtained at www.ICGtesting.com
Printed in the USA
LVOW01*1023120515

438165LV00001B/1/P